CHAPMAN & HALL/CRC
APPLIED ENVIRONMENTAL STATISTICS

SAMPLING STRATEGIES FOR NATURAL RESOURCES AND THE ENVIRONMENT

CHAPMAN & HALL/CRC
APPLIED ENVIRONMENTAL STATISTICS

Series Editor
Richard Smith, Ph.D.

Published Titles

Timothy G. Gregoire and Harry T. Valentine, Sampling Strategies for Natural Resources and the Environment

Steven P. Millard and Nagaraj K. Neerchal, Environmental Statistics with S Plus

Michael E. Ginevan and Douglas E. Splitstone, Statistical Tools for Environmental Quality

Forthcoming Titles

Daniel Mandallaz, Sampling Techniques for Forest Inventory

Thomas C. Edwards and Richard R. Cutler, Analysis of Ecological Data Using R

Bryan F. J. Manly, Statistics for Environmental Science and Management, 2nd Edition

Song S. Qian, Environmental and Ecological Statistics with R

CHAPMAN & HALL/CRC
APPLIED ENVIRONMENTAL STATISTICS

SAMPLING STRATEGIES FOR NATURAL RESOURCES AND THE ENVIRONMENT

TIMOTHY G. GREGOIRE
HARRY T. VALENTINE

Chapman & Hall/CRC
Taylor & Francis Group

Boca Raton   London   New York

Chapman & Hall/CRC is an imprint of the
Taylor & Francis Group, an **informa** business

Chapman & Hall/CRC
Taylor & Francis Group
6000 Broken Sound Parkway NW, Suite 300
Boca Raton, FL 33487-2742

© 2008 by Taylor & Francis Group, LLC
Chapman & Hall/CRC is an imprint of Taylor & Francis Group, an Informa business

International Standard Book Number-13: 978-1-58488-370-8 (Hardcover)

Library of Congress Cataloging-in-Publication Data

Gregoire, T. G.
   Sampling techniques for natural and environmental resources / Timothy G.
Gregoire and Harry T. Valentine.
      p. cm. -- (Applied environmental statistics ; 1)
   Includes bibliographical references and indexes.
   ISBN 978-1-58488-370-8 (alk. paper)
   1. Environmental sampling. 2. Environmental sciences--Statistical methods.
I. Valentine, Harry T. II. Title. III. Series.

GE45.S75G74 2008
577.072'3--dc22   2007017628

Visit the Taylor & Francis Web site at
http://www.taylorandfrancis.com

and the CRC Press Web site at
http://www.crcpress.com

For George M. Furnival,
our mentor, colleague, and friend.

Contents

Preface

This book is aimed at students and researchers of ecology, natural resources, forestry, or environmental sciences who need to know how to sample natural populations and continuums in a credible manner. The presumed purpose is to estimate, with quantifiable precision, aggregate characteristics of such resources that cannot be measured completely. We present and discuss methods to estimate aggregate characteristics on per unit area basis, say per hectare, as well as on an elemental basis. While some of the sampling designs, e.g., simple random sampling and list sampling, that we present in this book may be found in virtually any text on sampling, others, e.g., randomized branch sampling and 3P sampling, are rather specialized and have evolved specifically for sampling vegetation.

A distinguishing feature of this book is our emphasis on areal sampling designs, including plot sampling, Bitterlich sampling, and line intersect sampling. We also present several recently conceived lesser known areal designs, and we provide solutions to the problem of edge effect, which accompanies the application of many, if not most, areal designs. Our approach to explanation of areal sampling is decidedly geometric, relying heavily upon diagrams and the concept of the inclusion zone.

A second distinguishing feature of this book is our inclusion of sampling designs for continuums. Here we make use of methods of Monte Carlo integration, including crude Monte Carlo, importance sampling, and control variates. These methods originally were used in computer simulation, yet they adapt nicely for the purpose of estimating attributes of continuums such as tree stems, landscapes, lakes, and spans of time.

Because this book is primarily for students of natural resources and the environment, we have chosen to emphasize design and estimation, rather than sampling theory. While we cannot avoid the use of mathematical formulae and expressions, we strive first to provide a conceptual understanding of each sampling design or estimation procedure to which the mathematics apply. For those that are more technically adventurous, we have provided a modicum of derivations and proofs in appendices of chapters. Skipping the technical section will not disadvantage those who lack interest in these seemingly arcane details, but we trust that their mastery will benefit the adventurous by providing a deeper understanding of the underpinnings of sampling theory, estimation, and inference.

As in any sampling text, we impose precise meanings to many common words, and we use some specialized terms that either can be distracting or enlightening, depending on one's disposition. Realizing that the use of this sampling vernacular is an unavoidable necessity in order to establish a common and succinct dialogue, we have attempted to provide a gentle familiarity with essential terms. For example,

Chapter 1 includes a short discussion of what we mean by a 'population' and who gets to decide what the population comprises in a particular setting. Words and phrases we feel to be essential to discourse are indicated by *slant type, like this.* Terms to remember also appear together in a box at the end of each chapter.

Short worked examples appear throughout the book, in the hope that they will provide clarity in cases where the symbolic formulas appear mysterious. In virtually all cases, a simple spreadsheet computer program is sufficient to work through each example on one's own. Graphical displays of data appear throughout the text and we encourage their use as a fundamental step in any analysis of data.

Rarely is there a sampling strategy that is singularly better than any other for a given situation, especially when implementing surveys designed to gather information that will be used to estimate multiple characteristics of the population of interest. Indeed Godambe (1955) established that there is no estimator of the population mean that is best, in a well defined sense. The proof of this result is quite beyond the scope of this book. As we progress through our discussion of sampling and estimation we strive to make comparisons to previously presented methods.

The reader of this text should have some prior knowledge of statistics. We expect familiarity with the normal (Gaussian) and the t distributions, the central limit theorem, elementary notions of probability, the mean and variance of a random variable, and confidence intervals. An introductory course in calculus, or at least a vague memory of one, should provide sufficient foundation for understanding the sections of the text that deal with continuums and Monte Carlo integration.

Material Covered in this Book

Chapter 1 introduces fundamental concepts, and by so doing also establishes some of the vernacular and tenets of probability sampling and design-based inference. In Chapter 2 we develop the notion of a sampling distribution, estimation, and properties of estimators based on the randomization distribution of possible estimates.

In Chapter 3, we introduce five equal probability sampling designs applicable to discretely distributed populations, coupled with various estimators of the population mean value or total value of some attribute of the population. Included in this mix is a discussion of systematic sampling. These topics are then extended to unequal probability sampling designs. Examples from empirical populations are presented to illustrate the applications of these strategies. Chapter 4 parallels Chapter 3 in its presentation of sampling designs and appropriate estimators for continuums.

Stratified sampling is presented in detail in Chapter 5. Related topics which deal with allocation of sampling effort, double sampling to estimate the strata weights, and poststratification are included in this chapter. Generalized ratio and regression estimation is the topic of Chapter 6, which deals more generally with the use of auxiliary information to improve the precision of estimation.

From Chapter 7 onwards we present areal designs that have developed or have been applied in the fields of ecology and natural resource management. Sampling with fixed size plots and quadrats is the subject of Chapter 7. Bitterlich sampling— a.k.a., horizontal point sampling or variable radius plot sampling—is explored in

Chapter 8. Well known and widely used in forestry, there has been considerably less exposure and understanding of the mechanics and utility of the Bitterlich design in other fields. A fixed-population approach to line intersect sampling is presented in Chapter 9. We consider estimators that condition on the orientation of the line transects, as well as unconditional estimators. In all three chapters, special attention is given to edge effects and the possible bias that may ensue.

Chapter 10 draws on the authors' work of the past two decades on the application of Monte Carlo integration to the sampling of continuums. In this chapter we explain how plot sampling, Bitterlich sampling, line intersect sampling, and other areal designs can be formulated as applications of importance sampling. We also deal with edge effect in this context.

Chapter 11 follows with brief presentations of point relascope sampling, variants of horizontal line sampling, transect relascope sampling, ranked set sampling, adaptive cluster sampling, and 3P sampling. Because comprehensive treatments of ranked set and adaptive cluster sampling are available elsewhere, our treatment of these topics is adumbrative rather than comprehensive.

Two-stage sampling designs and cluster designs are considered in Chapter 12. Our treatment of this topic is strategic rather than comprehensive. Two-phase sampling, or double sampling, is also discussed in the book, but rather than devote a separate chapter to this topic, we have chosen to weave it into the fabric of other chapters, where appropriate. Randomized branch sampling, a multistage design, is discussed in Chapter 13. This design conceivably is applicable to any entity that assumes a branched structure, though the design originally was formulated to estimate attributes of individual plants and trees.

Chapter 14, our final chapter, introduces sampling with partial replacement, which is a very economical design for sampling resources on two or more occasions.

Data

Data used in this book can be downloaded from the website at www.crcpress.com. Enter the section for Electronic Products, and then access the Downloads and Updates subsection.

Acknowledgments

"I'm getting tired of citing this book as in preparation!"   Tom Lynch, 2006 NEMO meeting

We have fabricated a list of excuses, some of them valid, for missing our initial and several subsequent target dates for publication, but we decided not to print the list, because the book is already long enough. We do, however, thank our editor, Bob Stern, and the staff at Taylor & Francis Group for their forbearance and good humor in the matter of the moving target, and for their assistance in all other matters.

During the process of proposing and writing of this book, we have benefited from reviews, technical comments, material, and support from Dave Affleck, Lucio Barabesi, Ken Brewer, Tom Burk, Ian Cameron, Ann Camp, Mark Ducey, Marian Eriksson, Zander Evans, Steve Fairweather, Lorenzo Fattorini, Jeff Gove, Ed Green,

Linda Heath, Kim Iles, Adrian Lanz, Tom Lynch, Ross Nelson, Anna Ringvall, Andrew Robinson, Frank Roesch, Christian Salas, Oliver Schabenberger, Chip Scott, Goran Ståhl and his students at the Swedish University of Agricultural Sciences, Al Stage, Jim Thrower, Paul Van Deusen, and Mike Williams.

Data for examples and exercises were graciously provided by Jim Barrett, Paul Doruska, Mark Ducey, Teresa Fonesca, Jeff Gove, Dave Hollinger, Mike Lavigne, Doug Maguire, The Natural Resources Management Corporation, Rich Oderwald, Bernard Parresol, Rick Schroeder, Steve Stehman, Harry Wiant, and Daniel Zarin.

Heather Rose, and Joan and Ben Valentine provided several editorial improvements—some would say translations! Linda Heath proofread the entire book and also provided crucial editorial improvements. Kristin Valentine designed our favorite part of the book, the front cover.

We appreciate the supporting network of colleagues, friends, and family who helped us in countless other ways during the preparation of this book.

TGG
HTV

CHAPTER 1

Introduction

1.1 The need for sampling strategies

A sampling strategy combines procedures for selecting a sample from some larger population with procedures for estimating one or more attributes of that entire population from measurements taken on the sample. As its title implies, the focus of this book is on sampling strategies applicable to environmental and natural populations, including continuums. It is this pointedly biological, ecological, and environmental orientation that distinguishes this book from other sampling texts. Without exception, however, the principles of sampling and estimation described here apply equally well to any disciplinary application.

In almost all cases, sampling is conducted because it is impractical, perhaps even impossible, to completely census the entire population of interest without exhausting available resources. Consider, for example, the impossibility of an attempted census of the below-ground biomass of vegetation in the Amazon basin of South America. In this case not only would the enormity of the task deplete national treasuries, but some of the vegetation itself likely would be destroyed in the process.

Moreover, a population census may be fraught with difficulties such as measurement error, undercounting, and erroneous tabulation. Indeed the time it takes to conduct a 100% inventory of a population may be so long that the population itself may change during the process. In this regard, imagine the implausibility of a complete inventory of the number of rhododendron blossoms during spring in the southeastern United States.

A well designed and executed sampling strategy provides a more efficient alternative to a census, where efficiency relates the amount of information obtained to the resources expended. In view of the potential inaccuracies of a population census, a more accurate accounting of the population may in fact be achievable through sampling.

National political campaigns for public office are always attended by pre-election polls of the electorate, with the results of these polls reported publicly. These polls and their results provide commonplace examples of sampling strategies and their outcomes with which everyone is familiar. Less visible than political polls are the numerous demographic, biological, and official surveys conducted on an ongoing basis by governmental, scientific, and economic bureaus, to name a few sources. Like an electoral poll, a sample, for whatever purpose, is a survey of some portion of a larger group from which one tries to infer meaningful quantitative information descriptive of the larger group, which we designate as the *population* of interest.

We have already insinuated that a *population* is the entire collection of subjects or objects about which we wish to know something. The sampler must decide which collection of objects or subjects constitutes the population of interest. In one case or study it might be a particular species of plant or all plant species in a geographic region; in another case, it might be plants of a particular species and age combination; and in yet another case, attention might be limited to just the vigorous plants, where vigor is defined in some unambiguous and identifiable way. For notational convenience we shall use the symbol $\mathcal{P}$ throughout this book to refer to the population of interest.

Need the population be discrete, or can it be continuous? The seed cones on a Norway spruce (*Picea abies* L.) tree is an example of a discrete population, whereas the forest floor, upon which the cones eventually fall, is an example of a continuous population or continuum. The members of discrete populations are customarily called *elements* or *units*. A continuous population, by contrast, comprises infinitely many *points*, for example, the points comprising the surface of a forest floor or the surface of a lake. More sampling theory and practice has been directed towards discrete populations than continuous populations. Sampling theory for the latter nevertheless has advanced in the disciplines of Monte Carlo integration (Rubinstein 1981), plane sampling (Quenouille 1949; Dalenius et al. 1961), remote sensing (Koop 1990), and geology (de Gruijter & ter Braak 1990). With recent emphasis on environmental monitoring, sampling strategies for continuous populations has garnered increased attention (Cordy 1993; Overton & Stehman 1993; Stevens 1997). Within forestry, Gregoire et al. (1986) defined a continuous population that comprised the continuum of cross sections along the central axis of the bole of a tree.

The population of interest nearly always is determined by the research or informational objectives of the survey, and ideally this is the population that is targeted for sampling. The important point is that the estimate derived from a sample of the targeted population pertains only to that population. To the extent that the intended population differs from that targeted by the sampling design, extrapolation of the sample results to the intended population is inherently risky and may be subject to legitimate criticism. An example may help to clarify the distinction between the intended and targeted population. In the course of sampling to estimate foliar area per hectare of all Amazonian vegetation, suppose we sample plants in such a way that only vegetation with woody stems can be selected and measured. In this scenario, only the woody vegetation is targeted for sampling, and the herbaceous component of the population of interest is given zero chance of being included in the sample. We cannot extend this estimate to include foliar area of herbaceous vegetation without a leap of faith or without a demonstration that the resulting bias is acceptably small. In this text we leave faith to theologians and concern ourselves only with sampling strategies that are statistically defensible, objective, and if not design-unbiased, then estimation with bias that is negligible as the size of the sample increases.

1.2 A medley of sampling scenarios

In the previous section we mentioned the sampling of vegetation in the Amazon for the purpose of estimating foliar area. This section presents a suite of scenarios to illustrate the broad spectrum of uses of probability sampling applied to ecological and natural resources.

1.2.1 Sampling for biomass estimation

Biomass of living vegetation in a region is important for a host of reasons relevant to ecological inquiry or resource management. It might, for example, serve to indicate the region's nutritive capacity and its capability to support vegetation, or it may reflect a site's response following management action such as fertilization, burning, irrigation, or harvesting. Measurement of aboveground biomass of a plant is usually a chore, often involving drying of the plant material to reduce its moisture content to an acceptably small level. For very large plants, say mature trees, the collection, drying, and measurement of biomass of a single individual is an arduous task owing to the sheer amount of living tissue aboveground. While the amount of living tissue below ground may be less, the task of collecting this material for purpose of measurement is challenging because of its difficult access. Moreover, the actual measurement of biomass kills the plant. A well planned sampling of aboveground biomass decreases the amount of effort required, without extirpating the population of interest.

If the population of interest comprises lesser vegetation—shrubs and herbaceous plants—then the size of each individual presents far less of a challenge to measurement of biomass, but this is supplanted by a likely increase in the number of individuals, perhaps a manyfold increase. Here again it makes sense to consider measuring but a portion of the population, and using that sample to estimate the biomass of the entire population. There are innumerable ways in which such a sample can be selected. Regrettably, there is no singular way to collect a sample that is optimal for all populations, circumstances, and information needs. This book presents a few alternative strategies—with comparative advantages and disadvantages which can vary greatly from one setting to another.

1.2.2 Sampling to estimate composition

Providing that you can recognize and distinguish among different plant species in your backyard, it likely would be a straightforward task to tabulate the number of plants of each species and then compute the proportional representation of each species on your property. Doing similarly for the White Mountain National Forest in northern New Hampshire would require either a huge crew of trained labor, or else a reliance on some method of sampling coupled with some way to transform the observations acquired in the sample into estimates applicable to the population. Indeed, even in your backyard, if grasses and forbs were included in your definition of the population, a sampling strategy likely would be required there, also.

In this book, discussions of sampling design involves issues such as the manner in which elements or subgroups are selected for the sample, as well as how large the sample should be. Discussions of estimation involve alternative uses of the quantitative and qualitative information in a sample to arrive at an estimate of an attribute of the population of interest, e.g., the proportion of the population which is coniferous. Discussions of inference embrace issues relating to the reliability of estimation.

1.2.3 Sampling for cover information

How much of the desert floor is covered by pieces of petrified wood in the Petrified Wood National Park in Arizona? Officials of the National Park Service were interested in this question because the amount of petrified wood in its natural state seemed to be diminishing in high traffic areas of the park: some visitors, in defiance of posted prohibitions of the practice, pilfer pieces of petrified wood to take home as souvenirs. In this situation, it seemed to make sense to adopt one sampling strategy for the subpopulation of petrified wood wherein pieces were the size of one's fingernail or smaller, a different sampling method for that subpopulation consisting of log-size pieces, and yet another sampling method for intermediate-size pieces. This technique of stratifying the population into different subpopulations, and then tailoring the sampling design within each subpopulation is commonly employed and can produce tremendous efficiencies of effort. We treat stratified sampling at some length in this book.

Estimation of plant canopy cover has long been of interest to ecologists (cf., Canfield 1941). Curiously, the line intercept technique of sampling canopy cover can be applied, as well, to the estimation of area of forest gaps. We provide details in Chapter 9.

1.2.4 How many trees in Sweden?

Dalenius (1957) mentions an early effort to obtain a probability sampling of the timber resource in Sweden in 1910–1912: it involved counting all trees within strips of land located systematically along parallel lines extending from Sweden's western border to its eastern border. The Varmland survey, as it was called, has historical significance for a number of reasons: from a statistical viewpoint, "... [t]his appears to be the first Swedish survey where the problem of measuring the degree of representativity was approached in terms of probability calculus.... The pioneer character of the Varmland survey was apparent to the forestry statisticians of the time [and it] played a decisive role for the development of sample survey practice in forestry statistics and related fields" (Dalenius 1957, p. 45, 49). Indeed, it served as a pilot survey for subsequent national forest inventories in Sweden initiated in 1923*.

By the start of the 21st century, technology has reduced the requisite field labor and time of large regional and national forest inventories, although the scope of such inventories has expanded enormously. Typically, remotely sensed data from satellite images or aerial photography will be used to stratify the region into noncontiguous

---

* In 2000, the Swedish National Forest Inventory estimated that Sweden contains 65,735,000,000 trees with heights greater than 1.3 m

areas of homogeneous cover, such as water, forest land, non-forest land. The intensity of sampling within each stratum will vary according to stipulated, or perhaps implicitly stated, needs of accuracy, and a subsample of ground locations will be selected for visitation by a field crew. At each ground location, a few measurements may be taken, such as a count of the number of trees in the vicinity of the ground location, the depth of the A and B soil horizons, or the degree of slope. Or, in consideration of the expense of placing a crew in the field, there may be a few dozen, possibly hundreds, of measurements of vegetative, wildlife, topographic, hydrologic, and geologic condition. Moreover, each ground location may serve as a point anchor for a lattice of additional locations around that point. Within the lattice, the measurement protocol may vary for some features, and may be constant for others. We mention this scenario to highlight the fact that many sample surveys of a population embrace multiple attributes, have multiple layers and phases which serve to refine the sample selection process, the estimation process, or both (Nusser & Goebel 1997; Nusser et al. 1998; Rennolls 1989; Rudis 1991; Smith & Aird 1975; FAO 1990; FHM 1998). This is true, also, of many large demographic surveys.

1.3 Probability sample

Our initial focus, in this book, is on sampling designs that select probability samples from discrete populations. A probability sample is underpinned by two probabilities: the selection probability of the sample and the inclusion probabilities of the units in the sample. We discuss the former first.

1.3.1 Selection probability of a sample

Each and every probability sample has a selection probability that is deducible from the strictures of the sampling design or protocol. We adopt the shorthand notation $p(s)$ to denote this selection probability of a sample, s. Although both simple random

ASIDE: When sampling for the purpose of estimating one or more quantitative attributes of a population, one fundamentally aims to select a sample that is representative of the population. What makes a sample representative? The answer to this deceptively simple question depends, of course, on what one means by representative. As Kruskal & Mosteller (1979a,b,c) explore in a very readable and informative series of articles, the term 'representative sampling' admits to a wide range of use not only across the scientific and non-scientific literature, but also within the considerably narrower statistical literature. Indeed in the latter, Kruskal & Mosteller (1979c) elaborate on nine distinctive meanings that have appeared for the term representative sampling. We do not intend to add yet another usage, but will instead follow the sage advice of W.E. Deming, as quoted by (Jones 1958), by phrasing our discourse in terms of probability sampling.

Table 1.1 *Distinct samples of size n = 2 from a population with N = 6 elements.*

Sample	Elements	Sample	Elements	Sample	Elements
s_1	$\mathcal{U}_1, \mathcal{U}_2$	s_6	$\mathcal{U}_2, \mathcal{U}_3$	s_{11}	$\mathcal{U}_3, \mathcal{U}_5$
s_2	$\mathcal{U}_1, \mathcal{U}_3$	s_7	$\mathcal{U}_2, \mathcal{U}_4$	s_{12}	$\mathcal{U}_3, \mathcal{U}_6$
s_3	$\mathcal{U}_1, \mathcal{U}_4$	s_8	$\mathcal{U}_2, \mathcal{U}_5$	s_{13}	$\mathcal{U}_4, \mathcal{U}_5$
s_4	$\mathcal{U}_1, \mathcal{U}_5$	s_9	$\mathcal{U}_2, \mathcal{U}_6$	s_{14}	$\mathcal{U}_4, \mathcal{U}_6$
s_5	$\mathcal{U}_1, \mathcal{U}_6$	s_{10}	$\mathcal{U}_3, \mathcal{U}_4$	s_{15}	$\mathcal{U}_5, \mathcal{U}_6$

sampling and systematic sampling designs (discussed in Chapter 3) ensure that $p(s)$ is identical for each of the possible samples, there is no restriction that the design of a probability sample must ensure that all possible samples be equally likely.

Under most sampling designs, the number of possible samples will far outnumber the elements in the population, because many sampling designs permit a partial overlap in composition, i.e., two distinct samples may have some population elements in common. In practice, it never is necessary to enumerate all possible samples and the composition of each, but we do so here to illustrate this last point. In a well-known primer, *Basic Ideas of Scientific Sampling*, Stuart (1962) examined a six element population with members $\mathcal{U}_1, \mathcal{U}_2, \mathcal{U}_3, \mathcal{U}_4, \mathcal{U}_5, \mathcal{U}_6$. Consider now the various samples each with two distinct elements that may be drawn from this population; these are shown in Table 1.1. Using the symbol Ω to denote the number of possible samples, it is evident from the tabulated display that for this example $\Omega = 15$. While we have shown the first two samples, s_1 and s_2, as comprising the elements $\{\mathcal{U}_1, \mathcal{U}_2\}$ and $\{\mathcal{U}_1, \mathcal{U}_3\}$, respectively, the ordering of the samples in this enumeration is arbitrary. Indeed, the sample labels, s_1, s_2, s_3, and so on, are introduced merely as a notational convenience to help us distinguish one possible sample from another in our discussion. Another person may choose to enumerate the list of possible samples in a different order, one that would lead, for example, to the designation of sample s_3 as containing elements $\{\mathcal{U}_3, \mathcal{U}_4\}$.

In this example from Stuart, we have not stipulated the probability of obtaining, say, sample s_7, i.e., we have not asserted anything about the value of $p(s_7)$ except that it is deducible for any particular probability sampling design. Nonetheless, it is possible to reason that $p(s_7) > 0$, because otherwise $p(s_7)$ must be identically zero, a result that is impossible if, indeed, s_7 is one of the possible samples. Likewise, $p(s_7) < 1$, because otherwise $p(s_7) = 1$, a result which implies that no other samples are possible. We elaborate on this point in order to establish that, under a probability sampling design, $0 < p(s) < 1$ not only for s_7 but for each possible sample. Furthermore, remembering that Ω represents the number of possible samples obtainable under a given design, then $1 = \sum_{k=1}^{\Omega} p(s_k)$.

For discrete populations the magnitude of Ω will depend both on the number of population elements selected for the sample, n, and the number of elements in the population, N. An exception to this generality happens when the sampling design is one that permits the size of the sample, n, to be random; i.e., n cannot be determined in advance of sampling, in which case Ω will depend on N but not n. For this reason, among others, we will be careful to distinguish whether the sample size, n, for each sampling design presented in this book is random or not; the latter will be mentioned as a fixed–n design, whereas the former will be called a random–n design. For infinitely large populations ($N = \infty$) and for continuously distributed populations, Ω likewise will be infinite.

If a probability sample of size n is to be chosen from a population of size N, then the protocol of the design, or more simply stated, the design itself, will determine the probability of obtaining each of the Ω possible samples. There are a few common sampling designs that ensure that $p(s)$ is identical for each of the Ω possible samples, in which case it is evident that $p(s) = 1/\Omega$ for all s permissible under the design. There are many more sampling designs which cause $p(s)$ to vary among samples composed of different elements. Moreover, $p(s)$ for any particular sample, say s_4 from Table 1.1, will depend on the design: the value of $p(s_4)$ under one sampling design will differ from its value under another sampling design, in general.

1.3.2 Inclusion probability

Each element of the population must have a nonzero probability of being included in a sample, which implies that each element must appear in at least one of the possible samples permissible under the design. Otherwise, the sampling is not probability sampling.

We distinguish between the sample probability, $p(s)$, of selecting a particular sample—when sampling according to a specific sampling design—from the probability of including a particular population element into a sample, s, under that design. This distinction perhaps can be made clearer by means of a short example. For the $N = 6$ population whose $\Omega = 15$ possible samples, each of $n = 2$ distinct elements, that are enumerated in Table 1.1, element $\mathcal{U}_5$ appears in samples s_4, s_8, s_{11}, s_{13}, and s_{15}. Without being specific for the moment as to the sampling design, suppose that the corresponding sample probabilities are denoted as $p(s_4)$, $p(s_8)$, $p(s_{11})$, $p(s_{13})$, and $p(s_{15})$. Let π_5 represent the overall probability of including $\mathcal{U}_5$ in a randomly chosen sample under this design. Upon reflection it can be deduced that $\pi_5 = p(s_4) + p(s_8) + p(s_{11}) + p(s_{13}) + p(s_{15})$. In the current parlance, this overall probability of including an element of the population in a randomly selected sample is its *inclusion probability*. Because the sample probabilities, $p(s)$, under one design will differ from those determined for another design, so too will the inclusion probability of $\mathcal{U}_5$. Using a more general notation, if the inclusion probability of the kth population element, $\mathcal{U}_k$, is denoted as π_k, then it must be true that

$$\pi_k = \sum_{s \ni \mathcal{U}_k} p(s), \qquad (1.1)$$

where the notation $s \ni \mathcal{U}_k$ indicates that the summation extends over all samples of which $\mathcal{U}_k$ is a member. We emphasize, again, that because π_k is determined by the sampling design, its value under one design will not be identical, in general, to its value under another design.

ASIDE: Some sampling designs permit the same population element to be selected two or more times for the sample. In essence, once the first sample element is selected from the population, that element is replaced in the population, so that it has the same probability of being selected on the second, and succeeding, selections. Evidently such 'with-replacement' sampling designs imply a sequential selection process, and it is possible to discern yet a different probability than either of the two mentioned above. The *selection probability* of a population element, say $\mathcal{U}_k$, is the probability that this unit will be selected in each of the sequenced selections under a with-replacement sampling design. We will discuss these selection probabilities of population elements in greater detail when we first present a with-replacement sampling design in Chapter 3. We also will show the relationship between a unit's selection probability and its overall inclusion probability at that point in the text.

Another defining characteristic of a probability sample is that the inclusion probability of any unit of the population be deducible. When sampling from a continuously distributed population, the corresponding notion is that the probability density be deducible (we discuss probability densities at length in Chapter 4). Some probability sampling designs ensure that the inclusion probability is constant for all elements of the population, but most do not. For many designs presented in this book, the inclusion probability of a population element may be computed without having to enumerate all possible samples and computing the $p(s)$ of each.

A particular probability sample may not be a miniaturized version of the population, nor will it necessarily comprise elements that are, in some vague sense 'typical' elements of the population. While there are devices, such as stratification and ordering, that can be employed to more nearly match the sample composition to that of the population, the merit of probability sampling lies chiefly with its guarantee of desirable behavior on average, not with desirable behavior of each and every sample that can be selected.

1.3.3 Sampling frame

A sample is selected with the aid of a *sampling frame*, which we regard as any mechanism that allows one to identify and select elements of the population. It might be as simple as a list, somehow numbered or labeled, of all population elements or groups of elements. The frame might be based on area, such that all elements within a selected area are observed; it might constitute a labeled list of elements along with one or more values of auxiliary characteristics corresponding to each. The units constituting the sampling frame may be identical to the population units, or it might contain groupings of units. The collection of population elements included within the sampling frame effectively constitutes the target population, and thus it is a matter of some consequence that the target population coincides with the population about which one wishes to draw an inference.

1.4 Inference

Objective selection is a hallmark of probability sampling. Estimates that derive from probability samples are free from the distorting influences that can be insinuated by subjective selection. Nonetheless, it is natural to question how reliable an estimate from a probability sample can be. How different might another estimate provided from a second sample be from the estimate provided by the first sample? Would one's estimate coincide with the population value in the hypothetical situation where the size of the sample matches the size of the population? Providing answers to questions like these leads one into the area of *statistical inference*.

In common with other established texts on sampling methods, such as the superb works by Cochran (1977) and Särndal et al. (1992), we stay entirely within the realm of *design-based inference*. Succinctly stated, properties of an estimator are deduced from the distribution of all estimates possible under the stipulated sampling design. In the design-based framework the population of interest is regarded as a fixed, not a random, quantity. If, for example, the attribute of interest is aboveground biomass of the N trees in a forest, then the fixed-population concept implies that the biomass of each tree in the population of N trees is likewise a fixed quantity, perhaps unknown until that tree is selected into the sample and measured. For purposes of statistical inference in the design-based framework, nothing is assumed about the manner in which the population elements, the trees, are distributed nor how the characteristics are distributed. In other words, there is nothing random about the population being sampled. Randomness enters only through the sample selection procedure, i.e., *sample design*.

To illustrate the notion of a fixed population, consider the population consisting of all the Douglas-fir trees in British Columbia taller than 1 m. Our aim is to estimate the foliage biomass of this population. Douglas-fir trees can be found along the coast and in the mountains; they grow and compete for nutrients and water among other Douglas-fir trees and among trees of other species. One can imagine grouping together all Douglas-fir trees that are 10 years old, 11 years old, and so on, and then examining how the numbers of trees are distributed across the spectrum of age classes. All of these factors, and numerous others that are deducible from the biology of Douglas-fir growth and that may be related to the amount of foliage supported on each tree, are essentially irrelevant in the system of design-based inference we expound in this text. That is not to say that auxiliary information of this sort cannot be used to great advantage in the design of the sampling procedure, or weaved into the estimator of total foliage, or both. The point of emphasis here is that the process by which we assess the accuracy and reliability of an estimator in no way depends on the aptness of any of these presumed relationships.

By contrast, model-based inference rests on an assumed structure for the population, i.e., a population model. Cassel et al. (1977) and Chaudhuri & Stenger (1992) provide extensive, and quite mathematically rigorous, coverage of model-based inference. A useful way to distinguish these two modes of inference is to remember that in the design-based approach the sample is a realization of a random process, whereas in the model-based approach the population is regarded as a realization of

a random process. Contrasts of the two modes of inference have been penned by
Särndal (1978), de Gruijter & ter Braak (1990), and Gregoire (1998).

Adherents of the model-based mode of inference cite its congruence with much
else that is practiced in statistical analysis: a model is postulated and, conditionally on
the veracity of that model, one draws conclusions about a population, a relationship,
or a stochastic process. Adherents of the design-based mode of inference cite
its freedom from a priori assumptions, i.e., inference is valid irrespective of the
concordance of the presumed model with reality. Both inferential approaches have
merit, and it is important that the basis or mode of inference at all times be
unequivocally stated. Therefore, when one characterizes properties of an estimator
of a population attribute it is important to be quite clear whether these properties are
assessed with regard to the reference distribution of all possible estimates permitted
by the sampling design, or by the stochastic relationship presumed for the population.
An estimator that is unbiased under one mode of inference may well be biased under
the other.

1.5 Population descriptive parameters

1.5.1 Discrete populations

Let y_k denote the value of an attribute of interest, which is associated with the kth
unit, $\mathcal{U}_k$, of a discrete population. The population total obtains by summing the
attribute across all N units, i.e.,

$$\tau_y = \sum_{k=1}^{N} y_k.$$

The population total, τ_y, is a *population descriptive parameter* and often a parameter
of interest in resource management or scientific inquiry. For example, in the context
of the previous section, y_k might be the foliar biomass of the kth tree in the sampling
frame, in which case, τ_y is the total biomass contained in the population of N trees.

The population total often is the *target parameter*, i.e., the population descriptive
parameter to be estimated. An alternative target parameter is the population mean,
μ_y, i.e.,

$$\mu_y = \frac{\tau_y}{N} = \frac{1}{N} \sum_{k=1}^{N} y_k$$

Generally, μ_y is the average amount of attribute per unit. In the biomass context, μ_y
obviously is the average biomass per tree.

There may be occasions where a sample is selected for the purpose of estimating
not only τ_y or μ_y, but also for estimating population parameters of other attributes.
For example, one might be interested in estimating the average biomass per tree, the
average tree height, and the average tree diameter. As needed, we will use x, z, and
other letter symbols to denote additional attributes of interest. In similar fashion to
y_k, x_k will denote the value of the x attribute on the kth unit, and so on.

Another population descriptive parameter of interest sometimes is the ratio of two

> ASIDE: Many will find the notation we use esoteric at first, yet it is a notation that is necessary for precise discourse. The quantitative formula of a population attribute, such as $\tau_y = \sum_{k=1}^{N} y_k$, unambiguously defines the attribute and makes possible a symbolic expression with which the value of this population descriptive parameter may be ascertained, in principle. In practice, we rely on sampling to estimate its value. A population descriptive parameter is one that can be expressed quantitatively in some fashion: for example, the proportion of foliage in a well-defined color class is estimable, whereas the overall color of foliage is not estimable by the sampling methods we expound.

population parameters, e.g.,

$$R_{y|x} = \frac{\tau_y}{\tau_x} = \frac{\mu_y}{\mu_x}.$$

The difference between the minimum and maximum value of y in a population is the *range*, and it is an obvious measure of the spread of values in the population. The range is yet another population descriptive parameter, however, it is one that rarely is of interest. An alternative quantitative measure of how dispersed the values of y are in the population is the *variance* of y, symbolized by σ_y^2, and expressed by

$$\sigma_y^2 = \frac{1}{N-1} \sum_{k=1}^{N} (y_k - \mu_y)^2.$$

If one regards $y_k - \mu_y$ as the 'distance' of y_k from the population mean, then a useful interpretation of σ_y^2 is as the average 'squared distance,' approximately, between individual observed values and their average value. The variance of y also can be interpreted literally: it is a measure of how much the y_ks vary around their mean. The Chapter 1 Appendix provides two alternative, but algebraically equivalent, expressions for σ_y^2.

Because σ_y^2 is a function of squared values of y, its unit of measure is also the square of the unit of measure of y. For example, if y is a measure of weight in kg,

> ASIDE: Various authors define the population variance differently as
>
> $$\sigma_y^2 = \frac{1}{N} \sum_{k=1}^{N} (y_k - \mu_y)^2$$
>
> which differs by a factor of $\frac{N-1}{N}$ from the definition given above, e.g., Cochran (1977). Moreover, some authors will use a symbol other than the Greek letter σ when defining the population variance in the same fashion as we have done above, e.g., Särndal et al. (1992, p. 39) symbolize it as S_{yU}^2.

then σ_y^2 is expressed in kg^2. In contrast, the *standard deviation* of y, namely σ_y, has the same units of measure as y.

At times it is convenient to express variability relative to some benchmark value. The most commonly used benchmark value is the mean, μ_y, and it is used in the population descriptive parameter known as the *coefficient of variation*, γ_y, which is expressed as

$$\gamma_y = \frac{\sigma_y}{\mu_y}$$

or in percentage terms as

$$\gamma_y = \frac{\sigma_y}{\mu_y} 100\%.$$

The coefficient of variation is a meaningful descriptive parameter only when $y > 0$.

Example 1.1

Consider the following three-unit population:

Unit	y	x
	(kg)	(cm^2)
$\mathcal{U}_1$	2	13
$\mathcal{U}_2$	9	25
$\mathcal{U}_3$	17	18

The following results are easily verified:

$$\tau_y = 28 \text{ kg} \qquad\qquad \tau_x = 56 \text{ cm}^2$$
$$\mu_y = 9.3 \text{ kg} \qquad\qquad \mu_x = 18.6 \text{ cm}^2$$
$$\sigma_y^2 = 56.3 \text{ kg}^2 \qquad\qquad \sigma_x^2 = 36.3 \text{ cm}^4$$
$$\sigma_y = 7.5 \text{ kg} \qquad\qquad \sigma_x = 6.0 \text{ cm}^2$$
$$\gamma_y = 80.4\% \qquad\qquad \gamma_x = 32.3\%$$

Verify that the same value for $R_{y|x}$ is obtained regardless of whether it is computed as τ_y/τ_x or as μ_y/μ_x.

1.5.2 Continuous populations

Continuous populations do not naturally divide into discrete units, nor do they lend themselves to simplistic description. For our purposes, a continuous population is a domain of integration, $\mathcal{D}$, comprising infinitely many points. In Chapter 4, we discuss sampling in a one-dimensional domain comprising an interval on a line; a two-dimensional domain comprising a bounded area on a plane; and a three-dimensional domain comprising the volume of a container. Each point in a domain can serve as a 'sample point.' An attribute of interest, which may extend continuously along, across, or throughout the domain, depending on the number of dimensions, has density $\rho(\mathbf{x})$ at point $\mathbf{x}$. In one dimension, the attribute density typically is an amount of attribute per unit length or per unit time. In two and three dimensions, an attribute density typically is an amount attribute per unit area and per unit volume, respectively. Consider, for example, the volume of litter on a forest floor. The attribute density is litter volume per unit land area (m^3 m^{-2}), which reduces to litter depth (m) at a location point, $\mathbf{x}$.

The total amount of attribute extant over the domain of integration is a population descriptive parameter. By definition, this total is equivalent to integral of the attribute density

$$\tau_\rho = \int_{\mathcal{D}} \rho(\mathbf{x})\, d\mathbf{x}.$$

The subscript ρ indicates that the total, τ_ρ, obtains from the integration of an attribute density over a continuous domain. It also serves to distinguish the continuous total from a discrete total, τ_y, which obtains from a summation of N attribute values.

Let D be the size—the length, area, or volume—of domain $\mathcal{D}$, i.e.,

$$D = \int_{\mathcal{D}} d\mathbf{x}.$$

The mean attribute density across the domain is

$$\mu_\rho = \frac{\tau_\rho}{D}$$

The variance of the attribute density across the domain is

$$\sigma_\rho^2 = \frac{1}{D} \int_{\mathcal{D}} \left[\rho(\mathbf{x}) - \mu_\rho \right]^2 d\mathbf{x}.$$

Ratios and coefficients of variation are defined as in discrete populations.

1.6 Historical note

In the opening decades of the twentieth century there was considerable debate as to just how sampling should be conducted. Much of this debate compared the merits of a purposive selection procedure to a probabilistic one, and indeed served to highlight the ambiguity surrounding the term representativeness. While there has been recent renewed interest in purposive selection, unquestionably the latter half of the century has witnessed the near dominance of probability-based sampling in scientific and public forums. In large part this is due to an exceedingly influential paper delivered to the Royal Statistical Society by Neyman (1934). In the words of Bellhouse (1988) the major reason why the paper

"...provides a paradigm in the history of sampling is that a theory of point and interval estimation is provided under randomization that breaks out of an old train of thought and opens up new areas of research."

In addition to the Bellhouse (1988), Hansen et al. (1985) have written a very readable overview of the development of sampling ideas in the twentieth century.

Finally, a dated but still relevant monograph, *The Principles of Sampling*, by Cochran et al. (1954) is worthwhile reading for any who seek to master the techniques presented in this book.

1.7 Terms to remember

Coefficient of variation	Probability sampling
Design-based inference	Sampling design
Elements, units	Sampling frame
Inclusion probability	Sampling strategy
Model-based inference	Standand deviation
Population	Variance

1.8 Appendix

1.8.1 Alternative expressions for σ_y^2

By expanding the squared term, $(y_k - \mu_y)^2$, within the summation of the expression for σ_y^2, and then collecting terms, one gets

$$\sigma_y^2 = \frac{1}{N-1}\left[\left(\sum_{k=1}^{N} y_k^2\right) - N\mu_y^2\right].$$

Taking this last expression and substituting τ_y/N for μ_y yields

$$\sigma_y^2 = \frac{1}{N-1}\left[\left(\sum_{k=1}^{N} y_k^2\right) - \frac{\tau_y^2}{N}\right].$$

CHAPTER 2

Sampling Distribution of an Estimator

2.1 Distribution of values

The concept of a distribution is central to statistical reasoning, so it is worthwhile to understand not only what is meant by the term, but also how to describe and characterize a distribution in quantitative terms through the use of summary statistics, and tabular and graphical displays.

Consider any set of two or more quantitative measurements or, as is common in statistical parlance, observations. If all the observations in this set were identical in value, then the distribution of observations would be concentrated entirely on that one value. In practice, multiple observations resulting from the measurement of a particular feature of biological, ecological, or environmental interest will differ in magnitude. The minimum and maximum observed values can be identified, and the remaining values will be distributed between these two extremes.

To exemplify the notion of a distribution of values, consider the data in Table 2.1. These data are measurements of the diameter of the bole and the aboveground biomass of 29 sugar maple trees (*Acer saccharum* Marsh). Upon scanning the column of diameter measurements, it becomes evident that the diameters of these 29 trees vary. Indeed, it is this variation, i.e., the fact that diameters vary in magnitude, that enables one to examine how these values distribute themselves along the real number line. The examination of Table 2.1 makes evident that the above-ground biomass

Table 2.1 *Measurements of 29 sugar maples (Cunia & Briggs (1984), used with permission of the NRC Research Press of Canada).*

Diameter	Biomass	Diameter	Biomass	Diameter	Biomass
(cm)	(kg)	(cm)	(kg)	(cm)	(kg)
9.1	33.42	26.2	423.80	38.6	1132.69
9.1	27.68	26.4	467.30	38.9	1015.60
9.9	35.60	32.5	799.77	39.4	1222.72
10.2	41.65	32.5	758.63	41.2	1083.17
16.7	144.68	32.5	757.95	41.9	1090.91
17.2	131.94	33.0	759.60	42.2	1189.18
17.6	195.16	33.0	812.37	42.7	1781.14
25.1	371.91	37.3	924.23	43.4	1242.57
25.4	352.38	38.1	1041.42	43.9	1403.90
25.9	420.29	38.6	986.36		

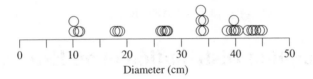

Figure 2.1 *Dot plot display of the distribution of sugar maple diameters. Each open circle represents the diameter of a single tree.*

also varies among these 29 trees. Moreover, the range of biomass—the difference in magnitude between the maximum and minimum values—is very different from the range in diameter values. Diameters are expressed here in cm, whereas biomass is expressed in kg, and this difference in units of measure is one reason why the range of values differs between these two attributes. Inasmuch as the range in values is one aspect of the observed distribution of values, the differing ranges of values between the two attributes implies necessarily that the distribution of diameter values is not identical to the distribution of biomass values. With these data, in particular, it is seen that the tree with the largest diameter is not even the same tree as that with the greatest biomass.

For the purpose of exploring other aspects of distributions of values, we concentrate for the moment on the 29 diameter values listed in Table 2.1, and displayed graphically in the dotplot of Figure 2.1.

Another way to display the distribution of sugar maple diameters graphically is as a histogram, Figure 2.2. Here, the distribution of diameters has been condensed into bins, each of which is 5 cm wide; the height of each bar in Figure 2.2 indicates either the proportional number of trees in each successive 5-cm diameter class, or,

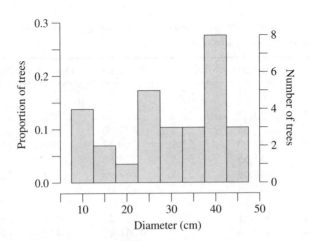

Figure 2.2 *Histogram of the distribution of sugar maple diameters.*

Table 2.2 *Distribution of sugar maple diameters by 5 cm size classes.*

Diameter Class	Frequency	Proportion
(cm)		
7.6 – 12.5	4	0.1379
12.6 – 17.5	2	0.0690
17.6 – 22.5	1	0.0345
22.6 – 27.5	5	0.1724
27.6 – 32.5	3	0.1034
32.6 – 37.5	3	0.1034
37.6 – 42.5	8	0.2759
42.6 – 47.5	3	0.1034

depending on the scaling of the vertical axis, the actual number of trees in each class. The two histograms are evidently identical but for the scaling of the two vertical axes.

A tabular summary of the frequency of trees in discrete classes provides a different way to summarize this distribution of diameter values. In Table 2.2 we have shown the distribution of sugar maple trees binned into 5 cm diameter classes; the columns of frequencies and proportional frequencies are the tabular equivalent of Figure 2.2. A tabular display is useful when one needs to assess exact frequencies or proportions, but the visual impression of a graphical display often suffices, and a figure more easily conveys the 'shape' of the distribution.

In Figure 2.3 the distribution of sugar maple diameters is arranged cumulatively from smallest to largest. In this distributional display, from the horizontal axis at some chosen value one reads up to the line connecting consecutive diameters and across to the vertical axis. The reading on the vertical axis is the proportion of the

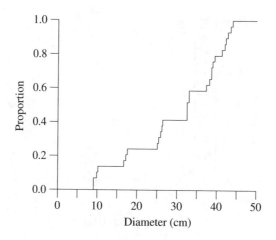

Figure 2.3 *Cumulative distribution of sugar maple diameters.*

ASIDE: Does the variation in diameter among the 29 sugar maple trees in Figures 2.1 and 2.2 reflect, or mimic, or approximate the variation in diameters of all sugar maples in the world? Or in the northeastern region of the U.S.A. where these trees grew? Or in your backyard? Such questions are impossible to answer without knowing how these trees were selected into the sample, i.e., without knowledge of the sampling design.

---

distribution of values smaller or the same size as the one chosen. The cumulative distribution plays an important role in the concept of confidence intervals, which is introduced later in section 2.3.

As a second example to illustrate distributional concepts, we present a histogram of the volume, expressed in m^3, of woody fiber in the bole of 14,443 loblolly pine (*Pinus taeda* L.) trees. The mean bole volume in this population is $\mu_y = 0.622\,\text{m}^3$, the variance is $\sigma_y^2 = 0.561\,\text{m}^6$, and the coefficient of variation, in percentage terms, is 120%. Equivalently, one could assert that this distribution of bole volume has mean $0.622\,\text{m}^3$, variance $0.561\,\text{m}^6$, and coefficient of variation of 120%. The preponderance of small values and the paucity of large values is typical of many biological populations, and it leads to the right (positive) *skewness* evident in the distribution. Fewer than 100 trees have a volume greater than $4\,\text{m}^3$; the maximum value in this distribution is $7.8\,\text{m}^3$.

2.2 Estimation

Much of this book is devoted to the presentation and explication of a variety of sampling designs. With the data obtained by measuring the population elements

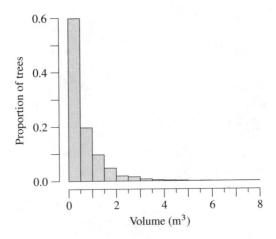

Figure 2.4 *Histogram of the distribution of bole volumes of 14,443 loblolly pines.*

ASIDE: Strictly speaking, an *estimator* is an algebraic expression, or rule, that instructs what to do with sample data, whereas an *estimate* is a number.

that are selected into the sample, one wishes to estimate one or more population descriptive parameters. As presented in Chapter 1, a population descriptive parameter can be viewed as a quantitative or algebraic combination of population values, such as the mean sugar maple diameter or the total bole volume in a population of loblolly pines. In an analogous fashion, an *estimator* is an algebraic expression that one evaluates with the data from the sample in order to provide a quantitative *estimate* of the target parameter. For the moment, let θ represent the population parameter of interest, and let $\hat{\theta}$ represent an estimator of it. We presume that the value of θ is unknown and that it is infeasible to measure all N elements of the population in order to evaluate it. Hence the need to select and measure $n < N$ units in order to estimate θ.

If one were to draw a single sample of n units from the N units in some well-defined population, $\hat{\theta}$ will differ in magnitude from that of θ. Ignorant of the value of θ, it is a futile effort to speculate how close to it that a specific estimate, $\hat{\theta}$, might be. It would be reasonable to expect, however, that if a larger size sample were drawn, i.e., one in which n were closer in size to N, then $\hat{\theta}$ would likely be closer to the targeted value of θ. The property of *consistency* of an estimator relates to the difference between an estimator, $\hat{\theta}$, and the target parameter in the limiting case where $n = N$.

2.2.1 Consistency

There is some variation in the statistical literature as to the meaning of consistency. In this text we adopt the definition that is common in the literature on probability sampling, namely that an estimator is said to be *consistent* if it is identically equal in value to the target parameter whenever the sample includes the entire population. If $\hat{\theta} \neq \theta$ in this situation, then $\hat{\theta}$ is said to be an inconsistent estimator of θ. Initially it may seem surprising that $\hat{\theta}$ could possibly differ from θ when the sample comprises the entire population, but indeed this is the case for some sampling strategies, as noted in the following example.

Example 2.1

Simple random sampling (SRS) will be presented in detail in Chapter 3, but for now it suffices to know that SRS is one design in which each element of the sample is selected with equal probability. Suppose that two characteristics, y and x, are measured on each sample element, and that $\bar{y}$ and $\bar{x}$ are the sample average values of y and x, respectively. With such a design, the estimator $\bar{y}/\bar{x}$ is a consistent estimator of $R_{y|x} = \mu_y/\mu_x$, which was introduced in Chapter 1.

On the other hand, $\bar{y}/\bar{x}$ is an inconsistent estimator of $R_{y|x}$ when sampling according to the Poisson design of §3.3.2.

At first glance, it seems perverse that an estimator can consistently estimate a population parameter in one setting, but not in another, simply due to the way the sample was selected. A purpose of this book is to instill an appreciation and understanding of how the design of the sampling protocol, i.e., how elements are selected into the sample, can affect the statistical properties of an estimator.

Inasmuch as the sample will rarely, if ever, be the same size as the population, it is legitimate to question whether the property of consistency is an important one by which to gauge the goodness of an estimator. We assert that it is, principally because of the disquiet implied by inconsistency: having observed and measured the entire population, if an estimator fails to provide the same value as that of the population parameter, exclusive of measurement error, then its value is especially questionable in a situation where only part of the population is observed, measured, and used to calculate the estimate. Despite the intuitive appeal of consistency of estimation, it would be very much more comforting to know the limiting behavior of $\hat{\theta}$ as n approaches N. For this we appeal to other properties of estimators such as variance and mean square error, both of which are introduced later in this section.

Whenever $n < N$ the estimator $\hat{\theta}$ will be based on a subset of the population. Because the population descriptive parameter, θ, is evaluated with units in the population that are omitted from any particular sample when $n < N$, any estimator $\hat{\theta}$ will, in general, differ from θ because of this omission. Also, the estimate based on the data collected from one sample will differ from that based on data from another sample of the same size from the same population. This variation in estimates among different samples is aptly termed *sampling variation*. This phenomenon, i.e., the fact that different samples generate variation among estimates, gives rise to a distribution of estimates, which is called a *sampling distribution*. To illustrate, we refer to Figure 2.5 in which is shown the sampling distribution of 25,000 estimates of mean bole volume of the loblolly pine population of Figure 2.4. The sampling design was SRS of $n = 20$ trees per sample. From this population of 14,443 elements, there are more than 6×10^{64} possible samples, each with $n = 20$ distinct elements. Obviously the sampling distribution shown in Figure 2.5 is very incomplete. Nonetheless, it is adequate for the purpose of demonstrating that this distribution of estimates bears little resemblance to the distribution of bole volumes

---

ASIDE: In this discussion of properties of estimators, we introduce consistency first because it is a property of an estimator when $n = N$, which implies that there is only one possible sample, i.e., that $\Omega = 1$. In fact, the implication is wrong: when the sample design allows the same population element to be sampled more than once, i.e., a with-replacement design, many different possible samples are possible even when $n = N$.

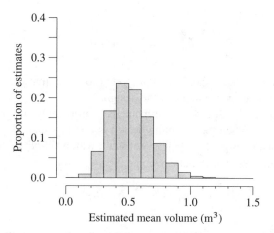

Figure 2.5 *Distribution of estimates of mean bole volume loblolly pine based on 25,000 simple random samples of 20 trees each.*

in the population that was sampled. It is the sampling distribution of an estimator, sometimes called the *randomization distribution*, that is the focus of attention when one is concerned with properties of estimators, and not the distribution of values in the sampled population itself. Nonetheless, the tools used to describe and portray distributions can be used also when examining properties of sampling distributions.

For a stipulated design and target parameter, θ, one often can deduce other properties of an estimator, $\hat{\theta}$, that pertain to how close in value it is to θ, on average. Indeed, estimation of θ by itself is rather easy: one can, for example, routinely (and facetiously) use $\hat{\theta} = 22$ as an estimate of any population parameter in any context! It is quite unlikely that this particular choice of estimator is good in any worthwhile sense, because it is impervious to the sample, i.e., it does not depend on what is observed in the sample.

We generally seek to use estimators of population parameters that have desirable properties, i.e., estimators that behave well in a sense that is made explicit below. Our aim, in the remainder of this section, is to introduce some properties of estimators and associated sampling distributions that generally are considered to be important. As different designs and estimators are introduced throughout this book, we will also comment on these properties of each estimator when used with a particular design.

2.2.2 Expected value

The expected value of an estimator is a weighted average of all possible estimates, where the weight which multiplies the estimate, $\hat{\theta}(s)$, from a particular sample, s, is the probability of selecting that sample, $p(s)$. The expected value of $\hat{\theta}$ can be expressed symbolically as

$$E[\hat{\theta}] = \sum_{s \in \Omega} p(s)\hat{\theta}(s). \tag{2.1}$$

In this expression, $s \in \Omega$ should be interpreted to mean that the summation takes place over all samples possible under the design.

Thus, the expected value of an estimator is a probability-weighted mean of all possible estimates. When the sampling design is one in which all possible samples are equally likely, then $p(s) = 1/\Omega$, a constant, which simplifies the expression of $E[\hat{\theta}]$ to that of a simple arithmetic average:

$$E[\hat{\theta}] = \frac{1}{\Omega} \sum_{s \in \Omega} \hat{\theta}(s).$$

The expected value of an estimator is a parameter, not a statistic. The expression for $E[\hat{\theta}]$ given in (2.1) makes it evident that the expected value of an estimator is a function of the sampling design through its dependence on the $p(s)$. It is also a function of the population being sampled through the sample estimates $\hat{\theta}(s)$. It is not a function of any population descriptive parameter. Semantically, one may speak of the expected value of an estimator as its mean. For equal probability sampling designs, the mean of an estimator is the mean of the sampling or randomization distribution (see Figure 2.5). The mean of the randomization distribution is a parameter, as is the mean of the population itself. However, they are distinctly different distributions, and the mean of one may not be related to, or close in magnitude to, the mean of the other.

The following rudimentary examples further demonstrate the influence of the sampling design on the expected value of an estimator.

Example 2.2

Consider the three unit population of Example 1.1. Only $\Omega = 3$ samples, each with two distinct elements, can possibly be selected from this population. We denote these samples by s_1, s_2, and s_3, where $s_1 = \{\mathcal{U}_1, \mathcal{U}_2\}$, $s_2 = \{\mathcal{U}_1, \mathcal{U}_3\}$, and $s_3 = \{\mathcal{U}_2, \mathcal{U}_3\}$. Imagine selecting one of these samples with an equal-probability sampling design, so that $p(s_1) = p(s_2) = p(s_3) = 1/3$. Finally suppose that one elects to compute the sample average, $\bar{y}$, as an estimate of the population mean, $\mu_y = 9.\bar{3}$. Then

$$E[\bar{y}] = p(s_1)\left(\frac{2+9}{2}\right) + p(s_2)\left(\frac{2+17}{2}\right) + p(s_3)\left(\frac{9+17}{2}\right)$$

$$= \frac{1}{3}\left(\frac{11}{2} + \frac{19}{2} + \frac{26}{2}\right)$$

$$= 9.\bar{3} \text{ kg}$$

$$= \mu_y.$$

Example 2.3

Suppose that the sampling design in the preceding example is one with unequal sample probabilities. In particular, suppose that $p(s_1) = 1/2$, $p(s_2) = 1/3$, and $p(s_3) = 1/6$. Under this design, the expected value of $\bar{y}$ is

$$E[\bar{y}] = p(s_1)\left(\frac{2+9}{2}\right) + p(s_2)\left(\frac{2+17}{2}\right) + p(s_3)\left(\frac{9+17}{2}\right)$$

$$= \frac{1}{2}\left(\frac{2+9}{2}\right) + \frac{1}{3}\left(\frac{2+17}{2}\right) + \frac{1}{6}\left(\frac{9+17}{2}\right)$$

$$= \frac{11}{4} + \frac{19}{6} + \frac{26}{12}$$

$$= 8.08\bar{3} \text{ kg}$$

$$\neq \mu_y.$$

Evidently, the differences in $p(s)$ values from those of the preceding example alter the expected value of the estimator, $\hat{\theta} = \bar{y}$. This will always be the case: whether or not an estimator, $\hat{\theta}$ is 'good' in the sense of providing an estimate that is close in value to that of the target parameter, θ, depends, inter alia, on the sampling design.

The expected value of an estimator that does not depend on the sample data is trivial to compute, as shown in the next example.

Example 2.4

Suppose that the estimator $\hat{\theta} = 22$ is used irrespective of which sample is drawn. Then

$$E[\hat{\theta}] = p(s_1) \times 22 + p(s_2) \times 22 + p(s_3) \times 22$$
$$= 22 \times [p(s_1) + p(s_2) + p(s_3)]$$
$$= 22,$$

where the last result derives from the fact that the sample probabilities must sum to unity. In this example, $\hat{\theta}$ can only take one value, the constant value 22. One's intuition would argue that its 'expected value' can be none other than the only value it is permitted to take, a result which is corroborated by its direct computation, above.

In practice, one can never evaluate $E[\hat{\theta}]$, because to do so implies that one knows the value of all N elements in the population. If that were so, then θ could be evaluated directly, and one would have no need to sample the population in order to estimate its value. Likewise, with non-probability sampling designs for which the sample probabilities, $p(s)$, cannot be deduced, it also is impossible to ascertain the expected value of any sample-based estimator.

The reason we introduce the notion of the expected value of an estimator is that it relates to the bias and variance of an estimator, two properties of estimators that are of fundamental importance. Furthermore, for the designs encountered in this book, $E[\hat{\theta}]$ can be expressed analytically in terms of the population parameter of interest, at least for the estimators of population parameters we consider and present. The reason for this analytical tractability is that we confine ourselves to considering probability sampling, which implies that each possible $p(s)$ is deducible.

2.2.3 Bias

The bias of an estimator is the difference in magnitude between its expected value and the population parameter for which an estimate is desired. Using $B[\hat{\theta}:\theta]$ to symbolize the bias of $\hat{\theta}$ as an estimator of θ, bias is computed as

$$B[\hat{\theta}:\theta] = E[\hat{\theta}] - \theta.$$

When $E[\hat{\theta}] = \theta$, bias is zero, and $\hat{\theta}$ is said to be an *unbiased estimator* of θ.

In contrast to the expected value of an estimator, the bias of an estimator is a function of a particular population parameter. There is a nuance here that is important to remember: one cannot speak of an estimator as being biased or unbiased without identifying not only the sampling design but also the population parameter being estimated. The following examples demonstrate these points.

Example 2.5

For the equal probability design contemplated in Example 2.2, $\bar{y}$ unbiasedly estimates μ_y, whereas for the unequal probability design used in Example 2.3, $\bar{y}$ is a biased estimator of μ_y. Different designs give rise to different properties. Here, unbiasedness of $\bar{y}$ when estimating μ_y under one design does not carry over to another design. For any design, the estimator $\hat{\theta} = 22$ is a biased estimator of not only μ_y but also any other population parameter whose value isn't felicitously 22.

Example 2.6

For the design considered in Example 2.2, $\bar{y}$ is a biased estimator of $\tau_y = 28$ kg.

Bias is not a property of an individual estimate, say $\hat{\theta}(s)$. Specifically, the fact that $\hat{\theta}(s) \neq \theta$ does not indicate that $\hat{\theta}$ is a biased estimator of θ. The quantity $\hat{\theta}(s) - \theta$ is known as *sampling error*. The term 'error' is not meant to imply that a mistake has been made or that the sampling protocol has been erroneously implemented. It is used instead to indicate that the estimate of θ from any particular sample is different, in all likelihood, from the value being estimated, and the source of the error is that θ is being estimated based on measurements of just a fraction of the elements of the population, namely only those elements selected, observed, and measured in the sample. Hence the term sampling error.

With an appreciation of sampling error, one may reasonably wonder how much an estimate, $\hat{\theta}(s)$, from one sample will differ from that calculated from a different sample. In practice one would like to keep this variation in estimated values among different samples small, because in that case there is an assurance that regardless of which single sample you actually choose, the estimated value will be roughly the same. We do not try to prescribe or recommend how small this variation among estimates must be in order to be so assured. Instead we try to provide a quantitative measure of the average deviation of the Ω estimates, $\hat{\theta}(s)$, and their mean value, $E[\hat{\theta}]$.

2.2.4 Variance

In Chapter 1 we introduced σ_y^2 as the variance of the y values in the population and described it as being the average squared distance between individual observed values and their mean. In concert with this meaning, the variance of an estimator is the average squared distance between individual estimates, $\hat{\theta}(s)$, and their mean, $E[\hat{\theta}]$. The variance of an estimator alternately is called the *sampling variance*; it is a parameter of the sampling distribution of the estimator. Using $V[\hat{\theta}]$ to symbolize the variance of $\hat{\theta}$, it is computed as

$$V[\hat{\theta}] = \sum_{s \in \Omega} p(s)\left(\hat{\theta}(s) - E[\hat{\theta}]\right)^2 \tag{2.2}$$

In contrast to bias, the variance of an estimator does not depend on the parameter, θ, being estimated.

Example 2.7

The variance of $\bar{y}$ in Example 2.2 is

$$V[\bar{y}] = p(s_1)\left(\frac{11}{2} - 9.\bar{3}\right)^2 + p(s_2)\left(\frac{19}{2} - 9.\bar{3}\right)^2 + p(s_3)\left(\frac{26}{2} - 9.\bar{3}\right)^2$$

$$= \frac{1}{3}\left(28.1\bar{6}\right)$$

$$= 9.3\bar{8}\,\text{kg}^2.$$

Example 2.8

The variance of $\bar{y}$ in Example 2.3 is

$$V[\bar{y}] = p(s_1)\left(\frac{11}{2} - 8.08\bar{3}\right)^2 + p(s_2)\left(\frac{19}{2} - 8.08\bar{3}\right)^2$$

$$+ p(s_3)\left(\frac{26}{2} - 8.08\bar{3}\right)^2$$

$$= \frac{1}{2}\left(\frac{11}{2} - 8.08\bar{3}\right)^2 + \frac{1}{3}\left(\frac{19}{2} - 8.08\bar{3}\right)^2 + \frac{1}{6}\left(\frac{26}{2} - 8.08\bar{3}\right)^2$$

$$= 3.3368 + 0.6690 + 4.0289$$

$$= 8.035\,\text{kg}^2.$$

Example 2.9

The variance of $\hat{\theta} = 22$ kg in Example 2.4 is identically zero. Being a constant, this estimator simply does not vary from one sample to the next.

The *precision* of an estimator is a qualitative measure of its variability. When comparing the performance of two estimators of θ, the one with the smaller variance is said to be the more precise. Contrast this with bias, which is an absolute property whose magnitude can be determined, at least in principle. A biased estimator is not necessarily an imprecise one, nor is an unbiased estimator necessarily precise. This is borne out in the previous examples: $\bar{y}$ has a variance of $9.3\bar{8}$ kg^2 yet is unbiased in Examples 2.2 and 2.7; its bias is nonzero but its variance, 8.035 kg^2, is smaller in Examples 2.3 and 2.8. Indeed, the estimator $\hat{\theta} = 22$ of Example 2.4 has zero variance, and thus it is infinitely more precise than the unbiased estimator, $\bar{y}$, of Example 2.2.

Recall from Chapter 1 that a sampling strategy is the combination of a sampling design and estimator. A feature of many sampling strategies is that estimation becomes more precise with increasing sample size. This can be seen by comparing the distribution of 25,000 estimates of mean bole volume in Figure 2.6 to that displayed earlier in Figure 2.5. Estimates in the latter were based on samples containing $n = 20$ trees, whereas the former were based on samples twice that size. As a result, the spread, or dispersion, of the sampling distribution in Figure 2.6 is noticeably less than that of Figure 2.5; estimates are much more concentrated in the middle of the distribution.

Like its expected value, the variance of an estimator usually can not be evaluated without knowing the y values for all N units of a discrete population or $\rho(\mathbf{x})$ for the complete domain $\mathcal{D}$ in the case of a continuous population. This obviously is unrealistic, inasmuch as such knowledge would obviate the need to sample from the population in the first place. As a consequence, an estimator, $\hat{\theta}$, of θ is usually accompanied by an estimator, $\hat{v}[\hat{\theta}]$, of its variance, $V[\hat{\theta}]$.

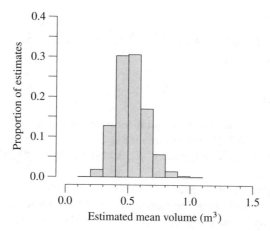

Figure 2.6 *Distribution of estimates of mean bole volume loblolly pine based on 25,000 simple random samples of 40 trees each.*

2.2.5 Standard error

The standard error of an estimator is defined to be the square root of its variance, $\sqrt{V[\hat{\theta}]}$. The standard error of an estimator has the same units of measure as the estimator itself.

2.2.6 Mean square error

Whereas the variance of an estimator is the probability-weighted average squared distance between each estimate, $\hat{\theta}(s)$, and $E[\hat{\theta}]$, the mean square error is the similarly weighted average squared distance separating each $\hat{\theta}(s)$ from θ, the parameter being estimated. Symbolically it is expressed as $\text{MSE}[\hat{\theta}:\theta]$, and it is evaluated as

$$\text{MSE}[\hat{\theta}:\theta] = \sum_{s \in \Omega} p(s) \left(\hat{\theta}(s) - \theta\right)^2.$$

Using the definition of $B[\hat{\theta}:\theta]$ and $V[\hat{\theta}]$, a little algebra (see Chapter 2 Appendix) reveals that $\text{MSE}[\hat{\theta}:\theta]$ can be rewritten as

$$\text{MSE}[\hat{\theta}:\theta] = V[\hat{\theta}] + \left(B[\hat{\theta}:\theta]\right)^2.$$

Thus, mean square error is a measure both of how much $\hat{\theta}$ varies around its mean, and how distant its mean is from the target parameter, θ. Evidently when $\hat{\theta}$ is unbiased, i.e., when $E[\hat{\theta}] = \theta$, $\text{MSE}[\hat{\theta}:\theta]$ and $V[\hat{\theta}]$ are identical. The utility of mean square error is that it enables a more apt assessment of estimator performance because it accounts for both sources of statistical error rather than focusing exclusively on one or the other.

Example 2.10

In Example 2.2, $\mathrm{MSE}\,[\,\bar{y}\!:\!\mu_y\,] \;=\; V[\,\bar{y}\,] \;=\; 9.38\,\mathrm{kg}^2$, because $\bar{y}$ unbiasedly estimates μ_y.

Example 2.11

The bias of $\bar{y}$, when used as an estimator of μ_y in Example 2.3, is $B[\,\bar{y}\!:\!\mu_y\,] = 8.083 - 9.3 = -1.25$ and its variance is $V[\,\bar{y}\,] = 8.035\,\mathrm{kg}^2$. Thus the mean square error of $\bar{y}$ as an estimator of μ_y is $\mathrm{MSE}\,[\,\bar{y}\!:\!\mu_y\,] = 9.597$.

When comparing the performance of two estimators of θ, the one with the smaller mean square error is said to be the more *accurate*. Thus $\bar{y}$ more accurately estimates μ_y with the sampling strategy described in Example 2.10 than it does with the sampling strategy of Example 2.11.

We conclude this section by drawing an important distinction between $\hat{\theta}$ as an estimator of some population parameter, θ, and quantities such as $V[\hat{\theta}]$, which can be used to characterize and describe the sampling distribution of $\hat{\theta}$. Namely, that $\hat{\theta}$ is a random variable, whereas $V[\hat{\theta}]$ is not. Different random samples will produce different estimates, $\hat{\theta}(s)$, whose values cannot be predicted in advance of sampling but instead vary randomly. Hence $\hat{\theta}$ is a random variable. In contrast, $V[\hat{\theta}]$ is a parameter of the sampling distribution of $\hat{\theta}$, which is defined over all possible samples obtainable under the stipulated sampling design. Its value does not change or in any way depend upon the chance selection of any particular sample.

2.3 Interval estimation

The magnitude of the standard error of an estimator provides an indication of how different an estimate is likely to be if one were to select another sample of the same size and according to the same sampling design. For example, if the distribution of an estimator is Gaussian, i.e., the normal distribution introduced in all introductory statistics courses and textbooks, then roughly two-thirds of all possible estimates are within ± 1 standard error of the mean of the distribution. More exactly, if the distribution of $\hat{\theta}$ is normal with mean θ and variance $V[\hat{\theta}]$, then

$$\mathrm{Prob}\left(\theta - \sqrt{V[\hat{\theta}]} \le \hat{\theta} \le \theta + \sqrt{V[\hat{\theta}]} \right) = 0.68. \qquad (2.3)$$

That is, 68% of the distribution of all estimates, $\hat{\theta}$, possible under the stipulated design is within one standard error of the target value θ. Equipped with this assurance, one can reason that an estimate produced from a single sample has a 68% chance of being closer to θ than $\sqrt{V[\hat{\theta}]}$.

If one reaches out two standard errors from θ, again assuming that $\hat{\theta}$ is normally distributed, then

$$\text{Prob}\left(\theta - 2\sqrt{V[\hat{\theta}]} \leq \hat{\theta} \leq \theta + 2\sqrt{V[\hat{\theta}]}\right) = 0.95. \tag{2.4}$$

One can rearrange these probability relationships to craft an 'interval estimate' of the population descriptive parameter based on an estimate of it, $\hat{\theta}$, and an estimate of its standard error. Namely, providing that $\hat{\theta}$ is normally distributed, then one can reason that the random interval

$$\hat{\theta} - \sqrt{\hat{v}[\hat{\theta}]} \leq \theta \leq \hat{\theta} + \sqrt{\hat{v}[\hat{\theta}]} \tag{2.5}$$

includes or 'covers' the value θ with probability 0.68, approximately. Similarly, the interval

$$\hat{\theta} - 2\sqrt{\hat{v}[\hat{\theta}]} \leq \theta \leq \hat{\theta} + 2\sqrt{\hat{v}[\hat{\theta}]} \tag{2.6}$$

ought to include the unknown value θ with approximate probability 0.95. One reason for asserting that these probability levels are approximate is that $V[\hat{\theta}]$ in (2.3) and (2.4) have been replaced by estimates in (2.5) and (2.6), thereby making the probability results of (2.3) and (2.4) somewhat inexact.

The smaller $\sqrt{V[\hat{\theta}]}$ is, the narrower is the distribution of $\hat{\theta}$, i.e., the less variable will be estimates from different samples. Presuming that $\hat{v}[\hat{\theta}]$ is a 'good' estimate of $V[\hat{\theta}]$, then these intervals, (2.5) and (2.6), will be narrower, also, which is the desired result. One would prefer to be in a position of proclaiming θ to be within the interval 50 ± 10, rather than 50 ± 30.

The intervals described above are widely known as *confidence intervals*, or, more specifically, 'normal-based confidence intervals.' The general form for their construction is

$$\hat{\theta} \pm t_{n-1}\sqrt{\hat{v}[\hat{\theta}]} \tag{2.7}$$

where t is the $1 - (\alpha/2)$ percentile of the Student t distribution with $n - 1$ degrees of freedom, which depend on both the sample size and the sample design. Tabulated values of t are widely available. One generated by the authors can be downloaded from the book's website at the URL identified on page xvii. The interpretation of (2.7) is as follows: if $\hat{\theta}$ is normally distributed with mean θ and variance $V[\hat{\theta}]$, then θ is contained within this interval with probability $1 - \alpha$, i.e.,

$$\text{Prob}\left(\hat{\theta} - t_{n-1}\sqrt{\hat{v}[\hat{\theta}]} \leq \theta \leq \hat{\theta} + t_{n-1}\sqrt{\hat{v}[\hat{\theta}]}\right) = 1 - \alpha. \tag{2.8}$$

When $\alpha = 0.10$, one speaks of a $(1 - \alpha)100\% = 90\%$ confidence interval for θ, or, to use another example, when $\alpha = 0.05$, one speaks of a $(1 - \alpha)100\% = 95\%$ confidence interval for θ. Once a sample has been selected and both $\hat{\theta}$ and $\hat{v}[\hat{\theta}]$ have been computed from the sample data, then the population descriptive parameter, θ, is either within the computed interval, or not. The probabilistic assertion in (2.8) pertains to the distribution of the random variable $\hat{\theta}$, not to any single estimate: if one were to repeatedly sample a population and estimate θ from each sample, then

$(1 - \alpha)100\%$ of the intervals of the form shown in (2.7) would include the value of θ (see Example 2.12).

In practice, a confidence interval, or interval estimate, of the form (2.7) is more informative than simply reporting the value of $\hat{\theta}$ based on the data from a particular sample. One may be 'confident' that the range of values in the confidence interval includes or covers the unknown value of the target parameter, θ. For a given level of confidence, the narrower the interval, the better.

Example 2.12

The 14,443 loblolly pine volumes displayed in Figure 2.4 have a distribution that is quite skewed to the right. In contrast, however, the distribution of $\bar{y}$ displayed in Figure 2.5 is so slightly skewed as to be barely noticeable. With each of the 25,000 samples of size $n = 20$, a 90% confidence interval for μ_y of the type (2.7) was computed. Of these intervals, 90.2% included $\mu_y = 0.015\,\text{m}^3$; the average width of these 25,000 intervals was $0.013\,\text{m}^3$.

Example 2.13

The increased precision when sampling with larger samples was noted earlier when comparing the distribution of $\bar{y}$ shown in Figure 2.6 to that of Figure 2.5. Correspondingly, the intervals based on the larger samples will be narrower, on average, than those based on smaller ones. The average width of the 25,000 intervals based on samples of size $n = 40$ was $0.009\,\text{m}^3$. In contrast, the average width of 25,000 intervals based on samples of size $n = 5$ was $0.031\,\text{m}^3$.

Despite the appeal of providing an interval estimate of the population parameter, there are two reasons to view confidence intervals cautiously. First, the distribution of $\hat{\theta}$ is never truly distributed as a normal random variable when sampling from finite populations, even when sampling is conducted with the SRSwR design. The deviation from the normal distribution may be especially severe when the sample size, n, is small, and when the population being sampled is very asymmetric. Thompson (2002, §3.2) provides details of a finite population version of the central limit theorem that asserts that the distribution of the sample mean will approach that of a normal distribution when certain limiting conditions are met. It is unlikely, however, that applied samplers will find these conditions and the results which flow from them to be of much comfort, or even intuitively sensible.

With a particular sample in hand and with the interval of (2.7) evaluated on the basis of the data from the sample, one will never know whether θ is included in that particular interval, or not. The α of a $(1 - \alpha)100\%$ confidence interval is the proportional number of intervals that 'miss' the target. One of ten 90% confidence intervals will fail to cover the intended parameter value. One of twenty 95% confidence intervals miss.

Could you have been so unlucky as to have selected a probability sample that provided an interval estimate that missed the intended target? Yes. The second reason for interpreting confidence intervals cautiously is that rare events do occur: the interval you have constructed may not in fact cover θ, despite your best efforts to design and execute an efficient probabilistic sampling plan.

These caveats notwithstanding, the distribution of estimators of population descriptive parameters of typical interest are approximately normal for reasonably large samples. How large is 'reasonably large' depends, *inter alia*, on how asymmetric the population distribution of y values is—the more skew, the larger n must be to ensure that $\hat{\theta}$ is approximately normal and that (2.7) achieves its nominal $(1 - \alpha)100\%$ coverage. Cochran (1977, §2.15) provided a crude rule which linked the size of the sample needed when sampling an asymmetric population to a quantitative measure of its asymmetry. Raj (1968, §2.11) discusses the effect of bias in $\hat{\theta}$ on confidence interval coverage of θ, and concludes that it is negligible whenever the sampling distribution is approximately normal and the ratio of estimator bias to standard error is less than 0.1.

Although 95% confidence intervals ($\alpha = 0.05$) are commonly reported, there is no cogent reason to prefer this level of confidence over the 99% level, or the 90% level, or any other. For a given sample size, the value of t increases with increasing confidence level. Thus, the price one pays for increased confidence is a wider range of plausible values for θ. Conversely, in exchange for less confidence, one can proclaim a narrower range of likely values for θ.

Example 2.14

Refer back to Example 2.12. The average width of the 25,000 95% confidence intervals based on samples of size $n = 20$ was $0.016\,\mathrm{m}^3$, compared to the average width of $0.013\,\mathrm{m}^3$ for 90% intervals and $0.010\,\mathrm{m}^3$ for 80% intervals.

Finally, if one is concerned that the nominal confidence level overstates the actual coverage, one can rely on Tchebysheff's Theorem (vide: Mendenhall & Schaeffer 1973, §3.11) to put a lower bound on the actual coverage level. This theorem holds for any distribution, and, in particular, its use does not require faith in the restrictive assumption that the distribution of an estimator is normal. This theorem proclaims that

$$\mathrm{Prob}\left(\hat{\theta} - k\sqrt{V[\hat{\theta}]} \le \theta \le \hat{\theta} + k\sqrt{V[\hat{\theta}]}\right) \ge 1 - \frac{1}{k^2}. \tag{2.9}$$

Thus, when $k = 2$, an approximate 75% confidence interval is

$$\hat{\theta} \pm 2\sqrt{\hat{v}[\hat{\theta}]}, \tag{2.10}$$

where the approximation is introduced by the use of $\hat{v}[\hat{\theta}]$ instead of the unknown $V[\hat{\theta}]$. For an example of the use of Tchebysheff's Theorem for interval estimation, see Fowler & Hauke (1979).

2.4 The role of simulated sampling

The ubiquity of affordable, high-speed computing has made it possible to select samples repeatedly from an electronic data file. If one regards the data in the file as measurements from a complete population of interest, then this repeated sampling enables one to construct empirically the sampling distribution of estimates and to examine the distributional properties of the estimator. This is what we have done in generating the distributions displayed in Figures 2.5 and 2.6. This tactic of repeatedly drawing samples from some artificial or contrived population of interest has great pedagogic value, because it enables us to display the shape of the randomization distribution, and its location relative to the value of the target parameter being estimated. We shall use results of simulated sampling throughout this text as we introduce various sampling strategies. We call this 'simulated sampling' because it simulates via the computer what would result if one were to repeatedly sample a population of actual interest: in practice one would draw but a single sample. This pedagogic and research tool is a simulation in another sense, too, because in practice one would not, of course, have the set of population y values in an electronic file, even after having sampled the population.

The value of simulated sampling is to explore the performance of competing and alternative sampling strategies with data that one has conveniently available. Its limitation is that aside from unrealistically small populations such as that used in the examples of this chapter, the simulated sampling distributions are never exactly identical to the complete sampling distribution of all possible estimates. Nonetheless, the simulations we report in this book are sufficiently extensive to ensure that the simulated sampling distribution is very similar in shape and location to the complete sampling distribution.

Schabenberger & Gregoire (1994) and Gregoire & Schabenberger (1999) illustrate the utility of simulated sampling to explore comparative properties of alternative estimators. Kraft et al. (1995) provide a similar type of comparative analysis using a known population of pronghorn (*Antilocapra americana*).

2.5 Other considerations

Measurement errors may have an insidious and degrading effect on the interpretation of sampling results. While realizing that few measurements can be made perfectly—perhaps counts are exceptions—we assume throughout this book that the magnitude of measurement error is negligibly small in relation to the measurement itself. Where this is not the case, we defer to Fuller (1995); or to Särndal et al. (1992, Chap. 16); or to Sukhatme et al. (1984, Chap. 11).

The bootstrap method of resampling survey data is a flexible tool by which an empirical distribution of an estimator can be generated. By so doing, the first two moments, mean and variance, of the distribution can be computed directly, as an alternative to relying on analytical results. We defer to Efron & Tibshirani (1986) or Dixon (2002) for the basic precepts of this method and to Sitter (1992) or Davison & Hinkley (1997) for its application to sampling of finite populations.

2.6 Terms to remember

Accuracy	Expected value	Sampling variation
Bias	Interval estimate	Sampling distribution
Confidence interval	Mean square error	Simulated sampling
Consistency	Observation	Skewness
Distribution	Precision	Standard error
Estimator	Sampling error	Variance
Estimate		

2.7 Exercises

1. Let $\hat{R}_{y|x} = \bar{y}/\bar{x}$. Refer to the three-element population of Example 1.1 and the sampling design described in Example 2.2. Compute MSE $[\,\hat{R}_{y|x}\,]$.

2. Repeat the previous exercise but for the sampling design described in Example 2.3. Under which design is $R_{y|x}$ more accurate for this population?

3. Explain how the property of consistency differs from that of bias.

4. Explain how the variance of an estimator differs from that of mean square error. Can the mean square error ever be less than the variance? Can the mean square error ever be more than the variance?

5. Explain why the variance of an estimator can never be less than zero.

2.8 Appendix

2.8.1 *Derivation of the relationship between mean square error, variance, and squared bias*

$$
\begin{aligned}
\mathrm{MSE}\,[\hat{\theta}:\theta] &= \sum_{s\in\Omega} p(s)\left(\hat{\theta}(s)-\theta\right)^2 \\
&= \sum_{s\in\Omega} p(s)\left(\hat{\theta}(s)-E[\hat{\theta}]+E[\hat{\theta}]-\theta\right)^2 \\
&= \sum_{s\in\Omega} p(s)\left[\left(\hat{\theta}(s)-E[\hat{\theta}]\right)+\left(E[\hat{\theta}]-\theta\right)\right]^2 \\
&= \sum_{s\in\Omega} p(s)\left[\left(\hat{\theta}(s)-E[\hat{\theta}]\right)^2+\left(E[\hat{\theta}]-\theta\right)^2\right. \\
&\qquad\qquad \left.+\,2\left(\hat{\theta}(s)-E[\hat{\theta}]\right)\left(E[\hat{\theta}]-\theta\right)\right] \\
&= \sum_{s\in\Omega} p(s)\left(\hat{\theta}(s)-E[\hat{\theta}]\right)^2+\sum_{s\in\Omega} p(s)\left(E[\hat{\theta}]-\theta\right)^2 \\
&\qquad\qquad +\,2\sum_{s\in\Omega} p(s)\left(\hat{\theta}(s)-E[\hat{\theta}]\right)\left(E[\hat{\theta}]-\theta\right) \\
&= V[\hat{\theta}]+\left(B[\hat{\theta}:\theta]\right)^2 \\
&\qquad\qquad +\,2\left(E[\hat{\theta}]-\theta\right)\sum_{s\in\Omega} p(s)\left(\hat{\theta}(s)-E[\hat{\theta}]\right) \\
&= V[\hat{\theta}]+\left(B[\hat{\theta}:\theta]\right)^2 \\
&\qquad\qquad +\,2\left(E[\hat{\theta}]-\theta\right)\left(\sum_{s\in\Omega} p(s)\hat{\theta}(s)-\sum_{s\in\Omega} p(s)E[\hat{\theta}]\right) \\
&= V[\hat{\theta}]+\left(B[\hat{\theta}:\theta]\right)^2 \\
&\qquad\qquad +\,2\left(E[\hat{\theta}]-\theta\right)\left(\sum_{s\in\Omega} p(s)\hat{\theta}(s)-E[\hat{\theta}]\sum_{s\in\Omega} p(s)\right) \\
&= V[\hat{\theta}]+\left(B[\hat{\theta}:\theta]\right)^2+2\left(E[\hat{\theta}]-\theta\right)\left(E[\hat{\theta}]-E[\hat{\theta}]\right) \\
&= V[\hat{\theta}]+\left(B[\hat{\theta}:\theta]\right)^2+0 \\
&= V[\hat{\theta}]+\left(B[\hat{\theta}:\theta]\right)^2
\end{aligned}
$$

CHAPTER 3

Sampling Designs for Discrete Populations

3.1 Introduction

Designs for selecting a sample from populations comprising discrete elements are presented in this chapter. For each design, one or more estimators of the population total, τ_y, or mean value per element, μ_y, are also presented. Integral to this discussion is the consideration of the probability with which samples are selected and the probability with which individual elements of the population are included into the sample.

3.2 Equal probability designs

Equal-probability designs impose the same inclusion probability on each element of a population. In this section we present three such designs: simple random sampling, systematic sampling, and Bernoulli sampling. Simple random sampling may be applied with or without the replacement of population elements. Both systematic sampling and Bernoulli sampling are applied without the replacement of population elements.

3.2.1 Simple random sampling

The simple random sampling design is one in which all possible samples of fixed size, n, are equally likely. Conversely, any design which ensures that all possible samples of n elements are equally likely is a simple random sampling design.

The selection probability of a sample, $p(s)$, is constant under the simple random sampling and the inclusion probability, π_k, of each and every population element is identical. However, that the inclusion probabilities are constant and equal is not the defining feature of simple random sampling because some other sampling designs also possess this feature.

Simple random sampling with replacement (SRSwR) permits a population element, say $\mathcal{U}_k$, to be included into a sample more than once. Simple random sampling without replacement (SRSwoR) permits each $\mathcal{U}_k$ to be included in a sample no more than once. We discuss the latter first.

Both SRSwoR and SRSwR are fixed-n designs, which means that the size of the sample, n, is decided upon in advance of the selection of the sample. Thus, n is an integral part of the sampling design. We consider only the case where $n < N$. Typically n will be but a tiny fraction of N. With SRSwoR one is assured that the n elements of the selected sample will all be distinct. This is not assured with SRSwR.

> ASIDE: The simple relationship between $p(s)$ and Ω is true of a few designs other than SRSwoR, but it does not hold in general.

Simple random sampling without replacement

The number of possible without-replacement samples, each of size n, that can be selected from a population of N discrete elements is $\Omega = {}_N C_n$, where

$$\begin{aligned} {}_N C_n &= \frac{N!}{n!(N-n)!} \qquad &(3.1) \\ &= \frac{N(N-1)\cdots(N-n+1)}{n(n-1)(n-2)\cdots 3 \times 2 \times 1}. \qquad &(3.2) \end{aligned}$$

The factorial terms, $N!$, and so on, are elucidated in the Chapter 3 Appendix. Since these Ω samples are equally likely under SRSwoR, the probability of each one is

$$p(s) = \frac{1}{\Omega}.$$

Example 3.1

Suppose SRSwoR of a sample of size $n = 1$ from a population of size $N = 6$. The number of possible samples is

$$\Omega = \frac{6!}{1!\,5!} = 6,$$

a result which accords with one's intuition. Moreover, if the six possible samples are equally likely, then each has a probability of $p(s) = 1/6$ of being selected.

Example 3.2

Suppose SRSwoR of a sample of size $n = 4$ from a population of size $N = 6$. There are

$$\Omega = \frac{6!}{4!\,2!} = 15$$

equally likely samples, each with a probability of $p(s) = 1/15$ of being drawn.

In these simple examples it would be possible to enumerate all possible samples and the composition of each one. In practice this is infeasible because Ω becomes very large even with moderately sized populations and modest sized samples.

Example 3.3

With $N = 100$ and the SRSwoR sampling design, there are

$$\Omega = 17,310,309,456,440$$

samples of size $n = 10$, each of which differs by at least one element from any other.

With SRSwoR, the inclusion probability of each unit, $\mathcal{U}_k$, is

$$\pi_k = \frac{n}{N}, \tag{3.3}$$

where n/N is the fraction of the population that is selected into the sample and measured (see Chapter 3 Appendix). Under SRSwoR, this sampling fraction is identical to the probability of including each and every unit of the population into the sample.

An unbiased estimator of a population total, $\tau_y = \sum_{k=1}^{N} y_k$, is

$$\hat{\tau}_{y\pi} = \sum_{\mathcal{U}_k \in s} \frac{y_k}{\pi_k} \tag{3.4}$$

where the notation $\mathcal{U}_k \in s$ indicates that the summation extends over all elements $\mathcal{U}_k$ that have been selected into the sample. A proof of the unbiasedness of $\hat{\tau}_{y\pi}$ when estimating τ_y from a SRSwoR sample is given in the Chapter 3 Appendix, page 80. This estimator is known as the Horvitz–Thompson estimator in honor of the path-breaking contribution of Horvitz & Thompson (1952). This estimator is very general and will be encountered repeatedly, and with a variety of sampling designs, throughout this book. Because $0 < \pi_k < 1$, $\hat{\tau}_{y\pi}$ expands or prorates each y_k measured in the sample and sums these expanded values together to serve as an estimator of τ_y. We will refer variously to this estimator as the HT estimator or as the simple expansion estimator, henceforth.

For the SRSwoR design, the HT estimator can be expressed more transparently by substituting (3.3) into (3.4) to obtain

$$\hat{\tau}_{y\pi} = \frac{N}{n} \sum_{\mathcal{U}_k \in s} y_k$$

$$= N\bar{y}.$$

Because the population total can be expressed as $\tau_y = N\mu_y$, many find $N\bar{y}$ to be an intuitively appealing estimator of τ_y with the SRSwoR design.

From (2.2), the variance of the sampling distribution of $\hat{\tau}_{y\pi}$ under the SRSwoR design is

$$V[\hat{\tau}_{y\pi}] = V[N\bar{y}]$$

$$= N^2 V[\bar{y}]$$

$$= N^2 \sum_{s \in \Omega} \frac{1}{\Omega} (\bar{y}(s) - \mu_y)^2.$$

Equivalently,

$$V[\hat{\tau}_{y\pi}] = V[N\bar{y}]$$

$$= N^2 \left(\frac{1}{n} - \frac{1}{N}\right)\sigma_y^2. \tag{3.5}$$

The derivation of (3.5) appears in the Chapter 3 Appendix. For the SRSwoR sampling design, the variance of the estimator of τ_y is an explicit function of σ_y^2, the variance of the y-values in the population itself, as defined in §1.5. For other designs and estimators, this rarely will be the case.

The parenthesized term in (3.5) can never be less than zero because $n \leq N$. However, the variance of $N\bar{y}$ is inversely related to n, a design parameter under the control of the sampler: the larger the sample one chooses to use, the more precisely one can estimate τ_y with this sampling strategy. Two alternative ways to express the variance of $N\bar{y}$, which are algebraically equivalent to (3.5), are

$$V[N\bar{y}] = N^2 \left(\frac{N-n}{N}\right)\frac{\sigma_y^2}{n} \tag{3.6}$$

and

$$V[N\bar{y}] = N^2 (1-f)\frac{\sigma_y^2}{n}, \tag{3.7}$$

where $f = n/N$ is the fraction of the population included in the sample. Because $N\bar{y}$ unbiasedly estimates τ_y under SRSwoR, the mean square error of this estimator of τ_y is identical to its variance.

In practice, σ_y^2 cannot be evaluated from a sample of $n < N$ observations. As a consequence, none of the equivalent expressions for $V[N\bar{y}]$ can be evaluated after having selected a single sample by SRSwoR. However, with this design the estimator

$$s_y^2 = \frac{1}{n-1} \sum_{\mathcal{U}_k \in s} (y_k - \bar{y})^2 \tag{3.8}$$

unbiasedly estimates σ_y^2 (see Chapter 3 Appendix for proof). Consequently,

$$\hat{v}[N\bar{y}] = N^2 \left(\frac{1}{n} - \frac{1}{N}\right)s_y^2$$

$$= N^2 \left(\frac{N-n}{N}\right)\frac{s_y^2}{n}$$

$$= N^2 (1-f)\frac{s_y^2}{n} \tag{3.9}$$

unbiasedly estimates $V[N\bar{y}]$.

A commonly used estimator of the standard error of $\hat{\tau}_{y\pi}$ under this design is

$\sqrt{\hat{v}[N\bar{y}]}$. While this is a biased estimator of $\sqrt{V[N\bar{y}]}$, the bias is usually ignored when using the estimated standard error to provide interval estimates (vide: §2.3).

Example 3.4

To monitor patterns of weekly water use, a town conducted a SRSwoR of $n = 100$ homes. The sample was chosen with a sampling frame compiled from records at offices of the municipal government. Town records showed that there were $N = 5392$ residential dwellings with water meters in town.

Water consumption was recorded in 100-gallon units. The sample average consumption was $\bar{y} = 12.5$ 100-gallon units per week. The sample variance was $s_y^2 = 1352$ (100-gallon units)^2.

Using the HT estimator, $\hat{\tau}_{y\pi}$, the estimated weekly residential water use in town totals $5392 \times 12.5 = 67,400$ 100-gallon units. The estimated standard error is $5392 \times \sqrt{1352 \times [1 - (100/5392)]/100} = 19,641$ 100-gallon units. Expressed as a percentage of $\hat{\tau}_{y\pi}$, the estimated standard error of estimate is $(19,641/67,400) \times 100\% = 29\%$.

Because $\mu_y = \tau_{y\pi}/N$, and because $\hat{\tau}_{y\pi}$ unbiasedly estimates τ_y with variance given by (3.5), it follows that

$$\bar{y} = \frac{\hat{\tau}_{y\pi}}{N} \qquad (3.10)$$

unbiasedly estimates μ_y, and that its variance is given by

$$V[\bar{y}] = \frac{V[N\bar{y}]}{N^2} \qquad (3.11)$$

The variance of $\bar{y}$ is estimated unbiasedly by

$$\hat{v}[\bar{y}] = \frac{\hat{v}[N\bar{y}]}{N^2}$$

$$= \left(\frac{1}{n} - \frac{1}{N}\right) s_y^2 \qquad (3.12)$$

$$= (1 - f)\frac{s_y^2}{n}.$$

The usual estimator of standard error of $\bar{y}$ is $\sqrt{\hat{v}[\bar{y}]}$.

Example 3.5

Following Example 3.4, the estimated mean water usage per residential dwelling is $\bar{y} = 12.5$ and estimated standard error of 3.6, both in units of 100 gallons.

ASIDE: None of the properties of the estimators given to this point necessarily transfer to other sampling designs. In particular, $\hat{\tau}_{y\pi}$ and $N\bar{y}$ will not necessarily coincide: while $\hat{\tau}_{y\pi} = \sum_{u_k \in s} y_k/\pi_k$ provides an unbiased estimator of τ_y under any sampling design, $N\bar{y}$ does not. By extension, neither does $\bar{y}$ necessarily estimate μ_y unbiasedly for most other sampling designs. However, μ_y can always be estimated unbiasedly by $\hat{\tau}_{y\pi}/N$ for any design. It is for that reason that when we present the Horvitz-Thomspon estimator, we customarily present the estimator for the population total, τ_y, first, and from that derive the unbiased estimator of the population mean, μ_y.

Example 3.6

Radon gas in houses can adversely affect the health of its occupants. The community of Blueridge conducted a simple random sample without replacement of $n = 140$ houses from the $N = 13{,}895$ houses in the community. The cost to test for radon was approximately \$200 per house, so the town spent nearly \$29,000 on the survey. The sampling fraction for this survey was $f = n/N = 0.010076$, in other words about 1% of the houses in Blueridge were sampled. Of the 140 houses in the sample, 79 had basements with concrete walls.

The average level of radon was $\bar{y} = 9.04$ pCi L^{-1}, and the estimated standard error was $\sqrt{\hat{v}[\bar{y}]} = 0.971$ pCi L^{-1}. A 90% confidence interval ($t = 1.66$ with $n - 1 = 139$ degrees of freedom) for μ_y is 9.04 ± 1.61 pCi L^{-1}. Expressing the half-width of the interval as a percentage of $\bar{y}$ leads to an expression of the 90% interval estimate as 9.04 pCi L$^{-1} \pm 18\%$. The corresponding 95% interval ($t = 1.98$) is 9.04 pCi L$^{-1} \pm 22\%$.

In reporting the results of a sample, one normally would not present alternative interval estimates, as doing so likely would be more confusing than informative. We do so here, however, to illustrate that greater confidence comes with a price, namely a greater range of uncertainty as indicated by the increased width of the interval.

Example 3.7

With the SRSwoR of the preceding example, the proportion of houses with basements in the sample unbiasedly estimates the proportion of houses in the population with basements. To understand the basis for this assertion, partition N into the N_0 houses ($\mathcal{U}_k$) without basements and N_1 houses ($\mathcal{U}_k$) with basements. Hence,

$$N = N_0 + N_1.$$

ASIDE: In Example 3.7, suppose one is interested in estimating the average radon concentration in houses in Blueridge with basements, which we denote by the symbol μ_{y1}. An obvious estimator is

$$\bar{y}_1 = \frac{1}{n_1} \sum_{\mathcal{U}_k \in s_1} y_k, \qquad (3.14)$$

where $\mathcal{U}_k \in s_1$ indicates the subset of the sample s consisting of the n_1 houses with basements. However, $\bar{y}_1$ does not unbiasedly estimate μ_{y1}, because n_1 is a random variable—unlike n, which is a fixed constant under SRSwoR. Estimators such as $\bar{y}_1$, which are the ratio of two random variables, constitute an important class of estimators, which we consider in Chapter 6.

Dividing both sides by N yields

$$1 = \frac{N_0}{N} + \frac{N_1}{N}$$

$$= P_0 + P_1, \qquad (3.13)$$

where P_0 is the proportion of Blueridge houses without basements and P_1 the proportion with basements. Let x_k take on one of two values, as follows:

$$x_k = \begin{cases} 1, & \text{if the } k\text{th house has a basement;} \\ 0, & \text{otherwise.} \end{cases}$$

Then

$$\tau_x = \sum_{k=1}^{N} x_k = N_1,$$

because N_1 of the $\mathcal{U}_k$ have $x_k = 1$ and the remaining $x_k = 0$. As always, $\mu_x = \tau_x/N$, which evaluates here to

$$\mu_x = \frac{N_1}{N} = P_1.$$

Therefore, following SRSwoR,

$$\bar{x} = \frac{\hat{\tau}_{x\pi}}{N} = \frac{1}{n} \sum_{\mathcal{U}_k \in s} x_k = \frac{n_1}{n}$$

$$= \hat{p}_1, \text{ say,}$$

where $n = n_0 + n_1$, n_0 is the number of houses in the sample (i.e., $\mathcal{U}_k \in s$) without basements, and n_1 is the number of $\mathcal{U}_k \in s$ with basements.

The preceding example illustrates that μ_x is unbiasedly estimated by $\bar{x}$ following a SRSwoR of size n, even when x is a binary-valued variate, in which case $\mu_x = P_1$

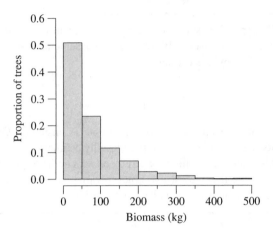

Figure 3.1 *Total aboveground biomass (kg) of* 1058 *balsam fir, black spruce, white birch, and white spruce trees.*

and $\bar{x} = \hat{p}_1$. The variance of $\bar{x}$, namely $V[\bar{x}]$, is identical to that of $V[\bar{y}]$ given in (3.11), with an obvious change in notation from y to x. Likewise, the customary estimator of $V[\bar{x}]$ is analogous to $\hat{v}[\bar{y}]$ given in (3.12). However, when y or x are binary variates, the expressions for their variances and variance estimators can be written in terms of P_1 and $\hat{p}_1$, as detailed in the Chapter 3 Appendix. The Appendix also provides details on a nearly exact confidence interval estimate for a population proportion.

Example 3.8

The total aboveground biomass was measured on a total of $N = 1,058$ balsam fir, black spruce, white birch, and white spruce trees, providing the distribution displayed in Figure 3.1. The average biomass per tree was $\mu_y = 72.2$ kg. Treating this as a 'population' for sake of example, a SRSwoR of $n = 52$ trees were selected. Among the 52 sample trees were 17 balsam fir, 14 black spruce, 10 white birch, and 11 white spruce. On the basis of this sample, the estimated average biomass was $\hat{\mu}_y = \bar{y} = 81.0$ kg, and the estimated standard error of $\hat{\mu}_y$ was 12.9 kg. The 90% confidence interval extended from 59.4 kg to 102.6 kg, which evidently includes the value of μ_y.

Selecting a SRSwoR sample

Many statistical software programs and electronic spreadsheets provide a capability of drawing a simple random sample either with or without replacement. Because the details for usage differ from one software product to another, we do not attempt to explain how to use any of them. Instead we outline methods that may be implemented in any programming language or software product or, with some effort, a handheld

---

ASIDE: Under the SRSwoR design, population elements are not selected independently. The probability that both $\mathcal{U}_k$ and $\mathcal{U}_{k'}$ are included in the same sample is

$$\frac{n(n-1)}{N(N-1)}.$$

---

calculator or printed table of random numbers. Instructions for use of the latter are provided *inter alia* in §2.2 of Cochran (1977).

SRSwoR Method I:

1. Generate a discrete random integer value between 1 and N, inclusive. One way to do this is to generate a uniformly distributed random number between 0 and 1. Let such a uniform random number be denoted as u, and henceforth let the notation $u \sim \mathrm{U}[0, 1]$ indicate that u is generated as a uniform random number between 0 and 1 (many handheld calculators, electronic spreadsheets, and statistical software programs have a function that can produce such random, or pseudo-random numbers). Then multiply u by the desired sample size, n, and add 1. In other words, calculate $u^* = 1 + nu$. Truncate u^*, keeping only its integer portion. Denote this truncated value as $[u^*]_{\mathrm{giv}}$, where the label 'giv' is a mnemonic reminder that the value saved is the 'greatest integer value' of u^*.

2. Select that unit $\mathcal{U}_k$ for which $k = [u^*]_{\mathrm{giv}}$, disregarding any repeated selections of $\mathcal{U}_k$.

3. Repeat steps 1 and 2 until n distinct units, $\mathcal{U}_k$, are selected.

SRSwoR Method II:

1. Corresponding to each of the N population elements in the sampling frame, generate a uniform random number. For sake of notation, we let u_k denote the U[0,1] number generated and associated with unit $\mathcal{U}_k$.

2. Sort the N u_ks from smallest to largest, making sure that all the other information about $\mathcal{U}_k$ is carried along in the sort.

3. Select those population units, $\mathcal{U}_k$, into the sample that correspond to the n smallest, or largest, u_k numbers. The n units so selected constitute a simple random sample from the population of N units.

Example 3.9

An unsorted list frame consisting of $N = 6$ population units is shown in the first column of the following table followed by the U[0,1] number generated for each unit. The third and fourth columns display the frame sorted by increasing u_k. Thus, $\mathcal{U}_2$ and $\mathcal{U}_6$ constitute a valid SRSwoR sample of size $n = 2$, as does $\mathcal{U}_3$ and $\mathcal{U}_4$. Indeed, any sequence of n consecutive units from the sorted list constitutes a SRSwoR of size n.

	Unsorted frame		Sorted frame
Unit $\mathcal{U}_k$	Uniform random number u_k	Unit $\mathcal{U}_k$	Uniform random number u_k
$\mathcal{U}_1$	0.38200018	$\mathcal{U}_2$	0.10068056
$\mathcal{U}_2$	0.10068056	$\mathcal{U}_6$	0.25846431
$\mathcal{U}_3$	0.59648426	$\mathcal{U}_1$	0.38200018
$\mathcal{U}_4$	0.89910580	$\mathcal{U}_5$	0.42910584
$\mathcal{U}_5$	0.42910584	$\mathcal{U}_3$	0.59648426
$\mathcal{U}_6$	0.25846431	$\mathcal{U}_4$	0.89910580

A third selection method, devised by Bebbington (1975), is a sequential method of selection where one considers each element in the sampling frame sequentially until n elements have been selected. The proof that this procedure results in the selection of a SRSwoR sample was given by Chaudhuri & Vos (1988, p. 202).

SRSwoR Method III:

1. Beginning with $\mathcal{U}_1$, generate a U[0,1] random number, u.

2. Set $n^* = n$ and $N^* = N$.

3. If $u \leq n^*/N^*$, select the unit into the sample, and reduce the value of n^* by one.

4. Reduce the value of N^* by one and consider the next population element in the sampling frame.

5. Repeat steps 3 and 4 until n units have been selected.

In the three methods just described for generating a SRSwoR, the size of the population, N, must be known prior to sampling. Bissell (1986) proposed an improvement to the Bebbington technique that is more efficient yet slightly more complicated. McLeod & Bellhouse (1983) and Pinkham (1987) devised a method of sequential selection similar to Bebbington's that does not require prior knowledge of the value of N.

Simple random sampling with replacement

With SRSwR, one permits each population element, $\mathcal{U}_k$, to be selected more than once into the sample, as first mentioned in §1.3. Evidently this implies that the n sample elements will not necessarily be distinctly different population elements. With large populations, the probability that a population unit will be selected more than once into the sample is so small that SRSwR is practically equivalent to SRSwoR. In the sampling of natural and environmental resources, SRSwR is rarely used as a sampling design. Our reason for presenting this design is that it provides a foundation for other sampling designs, which we present later.

With SRSwR and other with-replacement designs considered in this book, we assume, unless stipulated otherwise, that if an element, $\mathcal{U}_k$, is selected more than once, then its value, y_k, is used the same number of times in whatever estimators of population parameters are considered. One might correctly intuit that estimation following SRSwR must be less efficient than that following a similarly sized SRSwoR sample. For reasonably large samples and populations, however, the loss of efficiency is small.

With the SRSwoR, the sequence in which population elements are selected into the sample is irrelevant: for example, in Table 1.1, s_1 consists of the set $\{\mathcal{U}_1, \mathcal{U}_2\}$ regardless of whether $\mathcal{U}_1$ was selected first or second. With SRSwR, all of the possible sequences of n elements are equally likely, and there are $\Omega = N^n$ such sequences. Therefore, $\{\mathcal{U}_1, \mathcal{U}_1\}$, $\{\mathcal{U}_1, \mathcal{U}_2\}$, and $\{\mathcal{U}_2, \mathcal{U}_1\}$, $\{\mathcal{U}_2, \mathcal{U}_2\}$ are considered to be four, not three, distinct SRSwR samples, even though $\{\mathcal{U}_1, \mathcal{U}_2\}$ and $\{\mathcal{U}_2, \mathcal{U}_1\}$ obviously comprise the same set of population elements. With SRSwR, the order of selection matters: not all sets of values are equally likely, but all sequences of n elements have the same probability of occurrence, which is $p(s) = 1/\Omega = 1/N^n$. The number of sample sequences of size $n > 1$ possible under a SRSwR design exceeds the number of SRSwoR samples of the same size.

Example 3.10

For the six-element Stuart population introduced in §1.3, the $\Omega = 15$ possible samples, each with $n = 2$ distinct population elements, are listed in Table 1.1. In contrast, there are $\Omega = 6^2 = 36$ possible with-replacement samples of size $n = 2$.

Example 3.11

For a population of size $N = 100$ considered in Example 3.3, there are

$$\Omega = 100{,}000{,}000{,}000{,}000{,}000{,}000 \tag{3.15}$$

possible SRSwR samples of size $n = 10$.

When sampling without replacement, it is evident that the selection of a particular unit on one draw gives that unit zero probability of being selected on a subsequent draw. This is not true for SRSwR, however, because with this and other with-replacement designs, the n sample selections are independent. Not only can $\mathcal{U}_k$ be selected again after having been selected once, its probability of being selected on a second, third, or greater selection is in no way affected by its earlier selection or lack of selection. As mentioned in §1.3, with-replacement designs give rise to the notion of the *selection probability* of a population element $\mathcal{U}_k$. It is the probability that $\mathcal{U}_k$ will be selected on each of the sequenced selections. We denote the selection probability of $\mathcal{U}_k$ as p_k. With the SRSwR design, each element of the population has a selection probability of $p_k = 1/N$. Moreover, the selection probability, p_k, remains constant for each and every one of the n selections. Evidently, $\sum_{k=1}^{N} p_k = 1$.

The inclusion probability, π_k, of $\mathcal{U}_k$ is, as before, the probability that $\mathcal{U}_k$ will be included in a sample. With the SRSwR design, $\mathcal{U}_k$ can be included by being selected once, or twice, or three times, and onward. It is possible that $\mathcal{U}_k$ could be selected on all of the n selections from the population. This is admittedly an unlikely event, though it is no less likely than any of the remaining $\Omega = N^n$ sequences of sample selections that are possible with SRSwR. The inclusion probability of each unit, $\mathcal{U}_k$, under SRSwR, is

$$\pi_k = 1 - (1 - p_k)^n$$
$$= 1 - \left(1 - \frac{1}{N}\right)^n \tag{3.16}$$

as derived in the Chapter 3 Appendix, page 84.

Although an unbiased estimator of τ is $\hat{\tau}_{y\pi}$ of (3.4), an alternative, and more customarily used estimator is

$$\hat{\tau}_{yp} = \frac{1}{n} \sum_{\mathcal{U}_k \in s} \frac{y_k}{p_k}$$
$$= \frac{N}{n} \sum_{\mathcal{U}_k \in s} y_k \tag{3.17}$$
$$= N\bar{y},$$

which is identical in appearance to the $\hat{\tau}_{y\pi}$ estimator under SRSwoR. We emphasize that under SRSwR, $\hat{\tau}_{y\pi} \neq \hat{\tau}_{yp}$, because $\pi_k = 1 - (1 - p_k)^n$ and $p_k = 1/N$ under this design. The variance of $\hat{\tau}_{yp}$ is

$$V[\hat{\tau}_{yp}] = \frac{1}{n} \sum_{k=1}^{N} p_k \left(\frac{y_k}{p_k} - \tau_y\right)^2, \tag{3.18}$$

which reduces to

$$V[N\bar{y}] = N^2 \left(\frac{N-1}{N}\right) \frac{\sigma_y^2}{n} \tag{3.19}$$

under the SRSwR design (see Chapter 3 Appendix). The variance of $\hat{\tau}_{yp}$ can be estimated unbiasedly by

$$\hat{v}[\hat{\tau}_{yp}] = \frac{1}{n(n-1)} \sum_{\mathcal{U}_k \in s} \left(\frac{y_k}{p_k} - \hat{\tau}_{yp}\right)^2, \tag{3.20}$$

which reduces to

$$\hat{v}[N\bar{y}] = N^2 \left(\frac{s_y^2}{n}\right)$$
$$= \frac{N^2}{n(n-1)} \sum_{\mathcal{U}_k \in s} (y_k - \bar{y})^2 \tag{3.21}$$

when $p_k = 1/N$.

Comparing (3.6) to (3.19) reveals that, under SRSwoR, the variance of $N\bar{y}$ has the

> ASIDE: While under SRSwR, $\hat{\tau}_{y\pi} \neq \hat{\tau}_{yp}$, both unbiasedly estimate τ_y. That is, the mean of both distributions of all possible estimates computed as $\hat{\tau}_{y\pi}$ and as $\hat{\tau}_{yp}$ coincides identically with τ_y. Although the estimates produced by $\hat{\tau}_{y\pi}$ differ from those produced by $\hat{\tau}_{yp}$, their differences are small, because $\pi_k = 1 - (1 - p_k)^n = np_k + \phi$, where ϕ involves terms in p_k^2 and higher powers. Thus $\pi_k \approx np_k$.

term $N - n$ in place of the term $N - 1$ in the variance of $N\bar{y}$ under SRSwR. Hence for the same sample and population size, the latter variance will always exceed the former, as illustrated by the following example.

Example 3.12

Refer back to Example 3.1 wherein 100 homes had been selected as part of a SRSwoR design. Had the design actually been that of SRSwR, the estimated total residential water use would have been $\hat{\tau}_{yp} = N\bar{y} = 67{,}400$ 100-gallon units, i.e., the same as in Example 3.1. However the estimated standard error of this estimate under a SRSwR design would have been 19,826 100-gallon units, which is slightly greater than in Example 3.1.

From $\hat{\tau}_{yp}$ one can derive an unbiased estimator of μ_y, i.e.,

$$\hat{\mu}_{yp} = \frac{\hat{\tau}_{yp}}{N} \tag{3.22}$$

$$= \bar{y}$$

The variance of $\bar{y}$ following SRSwR is

$$V[\bar{y}] = \frac{V[\hat{\tau}_{yp}]}{N^2}$$

$$= \frac{1}{nN^2} \sum_{k=1}^{N} p_k \left(\frac{y_k}{p_k} - \tau_y \right)^2 \tag{3.23}$$

$$= \left(\frac{N-1}{N} \right) \frac{\sigma_y^2}{n},$$

which is estimated unbiasedly by

$$\hat{v}[\bar{y}] = \frac{s_y^2}{n}$$

$$= \frac{1}{n(n-1)} \sum_{u_k \in s} (y_k - \bar{y})^2. \tag{3.24}$$

Implicit in this last result is the fact that, under SRSwR, the sample variance, s_y^2, unbiasedly estimates

$$\left(\frac{N-1}{N}\right)\sigma_y^2 = \frac{1}{N}\sum_{k=1}^{N}\left(y_k - \mu_y\right)^2.$$

As with SRSwoR, the usual estimator of standard error of $\bar{y}$ following SRSwR is $\sqrt{\hat{v}[\bar{y}]}$.

When N is infinitely large, or unknown, then estimation of τ_y is impossible with simple random sampling, either with or without replacement. However, it still is possible to estimate the mean value per unit. Moreover, any terms involving $1/N$ in expressions of variance are effectively zero, so that the estimation of sampling variance is identical to that appropriate for SRSwR, namely (3.24).

Example 3.13

Eight random samples of water were taken from a popular swimming area. In each sample the number of colonies of Coliform bacteria per 100 ml were counted with these results: 513, 82, 414, 887, 241, 97, 200, 382. The estimated mean number of colonies per 100 ml is 352, and an estimated standard error of 94. This example was excerpted from Barrett & Nutt (1979, p. 77).

Selecting a SRSwR sample

SRSwR Method I:

1. Use Method I as outlined for SRSwoR, but in Step 2, do not disregard any repeated selections of $\mathcal{U}_k$.

SRSwR Method II:

1. In a list frame of the population, for each unit also list its selection probability, $p_k = 1/N$.

2. Working from the top to the bottom of the list, record the cumulative probability which is evaluated as

$$c_k = \sum_{j=1}^{k} p_j.$$

Thus $c_1 = p_1, c_2 = p_1 + p_2, c_3 = p_1 + p_2 + p_3$, and so on, until $c_N = 1$.

3. Generate n U[0,1] random deviates, $u_1, u_2, \ldots, u_n$.

4. For each random deviate, say u_j, select the first $\mathcal{U}_k$ into the sample for which $c_{k-1} \leq u_j < c_k$.

This procedure will provide n sample units, $\mathcal{U}_1, \mathcal{U}_1, \ldots, \mathcal{U}_n$, each selected with probability $1/N$. Some of the $\mathcal{U}_k$ may be duplicates. This method also can be used with the unequal probability method known as list sampling in section §3.3.1.

Example 3.14

Consider the following four-unit population:

Unit	Selection probability p_k	Cumulative probability c_k
$\mathcal{U}_1$	0.25	0.25
$\mathcal{U}_2$	0.25	0.50
$\mathcal{U}_3$	0.25	0.75
$\mathcal{U}_4$	0.25	1.00

We generate two random U[0,1] numbers for a sample of size $n = 2$. The first is $u = 0.6489$, which serves to select $\mathcal{U}_3$ into the sample; and second is $u = 0.2330$, which serves to select $\mathcal{U}_1$. Had the second u been in the range $0.50 \leq u < 0.75$, $\mathcal{U}_3$ would have been selected again into the sample.

When sampling with replacement, the number of distinct units, say v, selected into the sample is a random variable. Following SRSwR, the expected number of distinct units is

$$E[v] = N \left(1 - \frac{(N-1)^n}{N^n} \right)$$

(cf., Tillé 1998). Thus, when $N = 100$ and $n = 10$, the expected number of distinct units is $v = 9.6$.

3.2.2 Systematic sampling

The existence of a sampling frame from which to select a SRS implies that the population units, namely $\{\mathcal{U}_1, \mathcal{U}_2, \ldots, \mathcal{U}_N\}$, are arrayed in some orderly fashion. Särndal et al. (1992, p. 9) aptly defines a sampling frame to be "any material or device used to obtain observational access to the finite population of interest." Often this device might be a list, such as a list of names, possibly alphabetized, or the list might be a column of personal identification numbers in an electronic spreadsheet or data base. Or the frame might comprise a physical arrangement of unit identifiers such as folders in the drawers of file cabinets. Also easy to envision and employ in some contexts is an areal frame wherein a geographic region is subdivided into sampling units, and the location of each subdivisional unit serves as the unit identification. A chronological frame may arise in certain contexts, e.g., hours during which to measure water flow over a river dam. In an areal frame the number of population units depends on the mesh of the grid used to subdivide the landscape. Similarly the discretization of a chronological frame directly determines the number of temporal units in the frame. By contrast, a list frame almost always is composed of inherently discrete population units.

In nearly every situation where a sampling frame is available from which to select a simple random sample, that same frame can be used alternatively to systematically select every ath unit from the frame. The sampling interval, a, represents the number of units between successive selections into the sample beginning with the first unit chosen at random from among the first a units, as these units are arrayed in the sampling frame. Sample selection ceases when the frame is exhausted. Commonly this is described as a 1-in-a systematic sample. As the following examples make clear, sometimes the sample planner determines the desired size, n, of the sample, in which case this choice determines the sampling interval, a. Other times a will be determined in advance, which then implicitly determines the size of the sample. A sampling interval of $a = 1$ corresponds to a population census, and so we will dispense with any consideration of this special case.

Example 3.15

A 1-in-100 systematic sample from an electronic list of $N = 29,587$ Wildlife Fund contributors is sought. The list is arranged in order of decreasing monetary donation to the Fund over the past 10 years. A sample of $n = 295$ or 296 elements is selected and contacted by the administrators of the fund. A different 1-in-100 systematic sample would be chosen from an alphabetized list of contributors, or from a list arranged in order of increasing size of donation.

Example 3.16

Remotely sensed satellite data of British Columbia, Canada are interpreted by trained experts. The interpretation consists of delineation of the landscape into distinct polygons of homogeneous land cover. The resulting $N = 5643$ polygons are classified according to many additional vegetative and land use attributes, and then assigned a unique alphameric code, such as A0134c or F2199g. This identifying information is stored in an electronic database such as a geographic information system, which sequences the storage of information according to the identifying labels of the polygon. A systematic sample of size $n = 50$ polygons is chosen from the sequenced file. Since $N/n = 5643/50 \approx 113$, the resulting sample is a 1-in-113 systematic sample.

Example 3.17

In a study of the water quality and mineral content of an aquifer in Finnish Lapland, the level of the aquifer was measured at 3-hour intervals over a period of 90 days. The objectives were to study the diurnal fluctuation in aquifer level, as well as to estimate its average level. The design just described is a 1-in-3 systematic sample from a frame of $N = 90 \times 24 = 2160$ hourly units arranged in natural (chronological) sequence.

When working sequentially through a frame, the size of the population, N, need not be known in advance, but presumably it would become known at the conclusion of sampling if the systematic selection process is carried out over the entire frame. We assume that this is the case.

When planning a 1-in-a systematic sample, the size of the sample is determined implicitly to be the largest integer number less than N/a (symbolically we represent this as $[N/a]_{giv}$), or one larger than this ($[N/a]_{giv} + 1$). Alternatively, when the planner predetermines the sample size, n, the sampling interval is implicitly set as $a = [N/n]_{giv}$ or $a = [N/n]_{giv} + 1$. When $N = na$ exactly, then n is constant for all possible 1-in-a systematic samples. In most realistic situations, the product na will differ from N by some integer quantity, $c < a$. Symbolically, this is expressed as

$$N = na + c.$$

Whenever N is not an integer multiple of a, and hence $c \neq 0$, then there will be some slight variation in size among the set of possible 1-in-a systematic samples. The number of units in some samples will be $n = [N/a]_{giv}$ while the remaining samples will have $n = [N/a]_{giv} + 1$ elements. No sample can ever exceed the size of another by more than one unit. The expected sample size in this circumstance is the non-integer value N/a (see the subsection entitled 'Expected sample size' in the Chapter 3 Appendix).

Example 3.18

Suppose a 1-in-3 systematic sample is to be selected from a population with $N = 10$ units. Suppose further that the population units are arranged in natural order in the sampling frame, namely in sequence $\mathcal{U}_1, \mathcal{U}_2, \ldots, \mathcal{U}_{10}$. One possible systematic sample includes $\mathcal{U}_1, \mathcal{U}_4, \mathcal{U}_7, \mathcal{U}_{10}$; another is $\mathcal{U}_2, \mathcal{U}_5, \mathcal{U}_8$; the last is $\mathcal{U}_3, \mathcal{U}_6, \mathcal{U}_9$. No other 1-in-3 systematic sample is possible with this population and frame. Obviously, the first sample enumerated above has 4 elements, one more element than the other two samples.

This variable sample size of 1-in-a systematic sampling, whenever $na \neq N$, serves to highlight one difference between this design and SRS because n in SRS is constant for all samples permitted by the design.

Whenever $a > 1$, $\mathcal{U}_1$ cannot be selected in the same sample as the adjacent unit in the sampling frame. Once $\mathcal{U}_1$ is chosen, the remainder of the sample comprise $\mathcal{U}_{1+a}$, $\mathcal{U}_{1+2a}$, etc. The sample just enumerated is the only one in which these units can appear. Similarly, $\mathcal{U}_2$ appears in one and only one systematic sample. Indeed with 1-in-a systematic sampling the number of possible samples is $\Omega = a$, and each population element appears in but one sample. This highlights another salient difference between a systematic design and SRS: in SRSwoR, there are $_NC_n$ possible samples and each population units appears in each of $_{N-1}C_{n-1}$ distinct samples.

Often the rationale for favoring a systematic sampling design over a simple random sampling design is the more even distribution of the sample over the sampling frame than is likely to occur with SRS. This perhaps is easiest to visualize with an areal frame.

Example 3.19

Genetically improved Douglas-fir seedlings were planted in nursery beds 10 m wide. The collective length of the beds was 625 m. The beds have a narrow, rectangular metal grate at 5 m intervals to drain excess water. These grates are perpendicular to the sides of the beds, and serve naturally to partition the beds into 125 cells, each of dimension 10×5 m. The managers of the nursery wished to estimate the mortality rate of these seedlings. A systematic sample of every 20th cell was selected, and the proportion of dead seedlings was measured in 31 cells starting with number 7 from one end of the nursery, and proceeding with cells $27, 47, \ldots, 607$ near the other end.

In the above example, cell 7 was chosen randomly from among the first $a = 20$ cells in the bed. This probabilistic selection of the first sample unit in a systematic sample is crucial. Because every other element in a 1-in-a systematic sample is determined once the first element is picked, it is only through the selection of this first element that probability enters into the design. This design is called systematic sampling with a random start. While it is not necessary that the starting unit be chosen from among the first a units in the frame, this is the usual practice, outside of circular systematic sampling which is described later. The actual selection is quite easy: from a table, a handheld calculator, electronic spreadsheet or other computer program, generate a U[0,1] random number, u; multiply u by a, add 1 to the product, and select $\mathcal{U}_k$, the kth unit in the sampling frame, where $k = [ua + 1]_{\text{giv}}$.

Since all of the first a units in the frame have the same probability of being selected as the start, the probability of each of the possible samples is $p(s) = 1/a$, regardless of whether the sample includes $n = [N/a]_{\text{giv}}$ or $n = [N/a]_{\text{giv}} + 1$ elements. In contrast to SRSwoR, the inclusion probability of each population unit is also $1/a$, that is $\pi_k = p(s)$ for all $\mathcal{U}_k$ in a systematic sampling design.

As usual, the population total is unbiasedly estimated by the Horvitz-Thompson estimator, which takes the following form after 1-in-a systematic sampling:

$$\hat{\tau}_{y\pi} = a \sum_{\mathcal{U}_k \in s} y_k. \qquad (3.25)$$

Dividing by N provides an unbiased estimator of μ_y, i.e.,

$$\hat{\mu}_{y,\text{sys}} = \frac{a}{N} \sum_{\mathcal{U}_k \in s} y_k, \qquad (3.26)$$

which will differ from the sample mean, $\bar{y}$, whenever $N \neq na$. Indeed, following 1-in-a systematic sampling, $\bar{y}$ is unbiased for μ_y only when $N = na$, whereas $\hat{\mu}_{y,\text{sys}}$ always estimates μ_y unbiasedly. The bias arises because both $1/n$ and $\sum_{\mathcal{U}_k \in s} y_k$ are random variables, whereas in SRS, n is fixed, not random. In many circumstances the bias of $\bar{y}$, as an estimator of μ_y, or the bias of $N\bar{y}$, as an estimator of τ_y, will be negligibly small following systematic sampling.

For convenience, let the sample total be denoted by t_s, i.e.,

$$t_s = \sum_{\mathcal{U}_k \in s} y_k \tag{3.27}$$

Hence, $\hat{\tau}_{y\pi}$ can be written as

$$\hat{\tau}_{y\pi} = a t_s. \tag{3.28}$$

The variance of $\hat{\tau}_{y\pi}$ is a measure of the spread of the distribution of all a estimates of τ_y generated by the 1-in-a systematic design, and it is expressed as

$$V[\hat{\tau}_{y\pi}] = V[a t_s]$$

$$= \frac{1}{a} \sum_{s=1}^{a} (a t_s - \tau_y)^2$$

$$= a \sum_{s=1}^{a} t_s^2 - \tau_y^2 \tag{3.29}$$

$$= a \sum_{s=1}^{a} (t_s - \bar{t})^2$$

where $\bar{t} = \tau_y/a$. The corresponding variance of $\hat{\mu}_{y,\text{sys}}$ is

$$V\left[\hat{\mu}_{y,\text{sys}}\right] = \frac{V[\hat{\tau}_{y\pi}]}{N^2}$$

$$= \frac{a}{N^2} \sum_{s=1}^{a} (t_s - \bar{t})^2. \tag{3.30}$$

If the arrangement of the $\mathcal{U}_k$s in the frame is such that the y_k values exhibit a random pattern, then the variance of $\hat{\tau}_{y\pi}$ will be similar in magnitude to its variance under SRSwoR. In this case any potential advantage of systematic sample selection over a simple random sampling procedure would lie with the ease of execution and the more even distribution of the sample over the sampling frame. However, the precision of $\hat{\tau}_{y\pi}$ following systematic sampling can be much greater than when following SRSwoR if the y_ks can be arrayed in a linearly increasing or decreasing order. When the y_ks can be linearly trended in the sampling frame, the sample totals, t_s for $s = 1, \ldots, a$ will be fairly uniform in value, and not too far removed from $\bar{t}$. Since the y_ks are unknown while the frame is being constructed, the ordering may be based on auxiliary information x_k, that is strongly correlated with the characteristic of interest.

Example 3.20

Consider the population of $N = 236$ red oak trees from a published report by Beers & Gingrich (1958). The volumes of the tree boles in m^3 versus the corresponding diameters at breast height in cm are plotted in Figure 3.2.

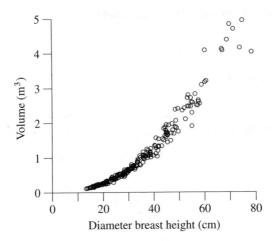

Figure 3.2 *Pennsylvania red oak data of Beers & Gingrich (1958).*

The population total volume is $\tau_y = 230$ m^3. It is plausible to expect that the diameter of each tree has been measured, as this characteristic is much more economical to measure than volume. Moreover, diameter and volume are strongly and positively correlated. For sake of example we suppose that the bole volumes have not been measured, and that the 236 tree population is to be sampled in order to estimate τ_y. The estimator $\hat{\tau}_{y\pi} = N\bar{y}$ has variance $V[N\bar{y}] = 4332$ m^6 following SRSwoR of $n = 12$ trees, although this value would not be known in a practical setting. Here it provides us with a means to compare the advantages of systematic sampling to SRSwoR. A 1-in-20 systematic sample is contemplated as an alternative to SRSwoR, and the estimator $\hat{\tau}_{y\pi}$ is to be used to estimate τ_y. Since $[N/a]_{\text{giv}} = 11$, we know in advance that $c = 16$ of the possible 1-in-20 samples will have 12 trees, and the remaining $a - c = 4$ samples will contain $n = 11$ trees. The sampling frame as originally constructed was in no discernible order with respect to the tree diameters or bole volumes.

The $\Omega = 20$ systematic samples from the originally ordered frame had sample t_s values:

12.0	11.7	12.2	9.4	11.8
9.4	15.6	12.0	8.4	11.0
7.8	10.4	9.0	10.9	13.3
9.5	14.8	12.8	12.1	15.6

The average of the 20 resulting estimates calculated by $\hat{\tau}_{y\pi} = at_s$ coincides exactly with $\tau_y = 230$ m^3, and the variance, computed according to $V[\hat{\tau}_{y\pi}] = \frac{1}{a}\sum_{s=1}^{a}(at_s - \tau_y)^2$ is 1980 m^6. Expressed as a percentage of $\hat{\tau}_{y\pi}$, the standard error of $\hat{\tau}_{y\pi}$ is $(44.5/230)100\% = 19\%$, which is quite an appreciable gain over the 29% realized with $\hat{\tau}_{y\pi}$ and SRSwoR.

The frame was then ordered by increasing diameter values. The $\Omega = 20\,t_s$ values from the ordered frame were:

9.6	10.4	10.0	10.7	10.6
11.2	12.7	11.1	12.8	12.7
12.8	13.1	13.9	13.3	13.7
12.7	9.0	9.4	10.5	10.0

The variance of $\hat{\tau}_{y\pi}$ following 1-in-20 systematic sampling from the ordered frame is 939.9 m^6, and the relative standard error is $(30.7/230)100\%=13\%$. Thus, the simple task of ordering the sampling frame in this situation allows one to decrease the relative standard error of $\hat{\tau}_{y\pi}$ from 19% to 13%.

For this systematic design, the expected sample size was $E[n] = 236/20 = 11.8$ trees, and the bias of the estimator $N\bar{y}$ as an estimator of τ_y is 0.5%.

The above example serves to illustrate both the potential increase in precision offered by systematic sampling compared to SRSwoR, as well as the further increase to be realized by ordering the sampling frame, when possible. Only when the average within-sample variance exceeds σ_y^2 will one realize the former gain. Another way to view the beneficial effects of ordering the frame is that it increases the heterogeneity within each systematic sample, thereby minimizing the variance among the a samples.

Nonetheless, there is no guarantee that the strategy of systematic sampling and the estimator $\hat{\tau}_{y\pi}$ will always be more precise than SRSwoR with $\hat{\tau}_{y\pi}$. One situation where it may do worse occurs when the y_ks in the order implied by the sampling frame exhibit a periodicity that coincides with the sampling interval, a. Cochran (1977) provides further details on this phenomenon and other aspects of systematic sampling.

Inasmuch as the 1-in-a systematic selection device partitions the population into a non-overlapping samples, one can think of it as a data reduction device wherein the population is concentrated into a meta-observations, $t_1, t_2, \ldots, t_a$, and but a single one of these meta-observations is selected as the sample. Therein lies one major drawback to systematic sampling, namely that it is impossible to estimate the variance of $\hat{\tau}_{y\pi}$ unbiasedly on the basis of a single observation. One often resorts to using the variance estimator $\hat{v}[\hat{\tau}_{y\pi}] = N^2(1/n - 1/N)s_y^2$ which unbiasedly estimates $V[\hat{\tau}_{y\pi}]$ following a SRSwoR. The rationale for using it when sampling systematically is that it is likely to overestimate $V[\hat{\tau}_{y\pi}]$. In other words, although $\hat{v}[\hat{\tau}_{y\pi}] = N^2(1/n - 1/N)s_y^2$ is a biased estimator of $V[\hat{\tau}_{y\pi}]$ following systematic sampling, it is unlikely to give an unwarranted impression of greater precision than was actually obtained. In that sense it is said to be a conservative estimator of the variance of $\hat{\tau}_{y\pi}$.

As an alternative to $\hat{v}[\hat{\tau}_{y\pi}]$, Meyer (1958) suggested the successive differences estimator given by

$$\hat{v}_{sd}[\hat{\tau}_{y\pi}] = N^2\left(\frac{1}{n} - \frac{1}{N}\right)\sum_{k=2}^{n}\frac{(\Delta y_k)^2}{2(n-1)}$$

where $\Delta y_k = y_k - y_{k-1}$.

In an extensive comparison of the performance of $\hat{v}[\hat{\tau}_{y\pi}]$, $\hat{v}_{sd}[\hat{\tau}_{y\pi}]$, and six other estimators of $V[\hat{\tau}_{y\pi}]$ following systematic sampling, Wolter (1985, p. 283) concluded that the successive differences estimator was best overall, but noted that it tends to underestimate the variance when a linear trend exists. For the red oak population in Example 3.20, the relative bias of both $\hat{v}[\hat{\tau}_{y\pi}]$ and $\hat{v}_{sd}[\hat{\tau}_{y\pi}]$ was 139% when sampling from the unordered frame. These biases were 396% and -37%, respectively, when sampling from the frame ordered by tree diameter.

When auxiliary information is available to order the sampling frame, as in Example 3.20, a sensible tactic might be to see which of these two variance estimators, or perhaps some other, works best when estimating the known population total of the auxiliary variate, and then to use that estimator for $V[\hat{\tau}_{y\pi}]$.

Yet another alternative course of action is to take multiple independent, but smaller, systematic samples which collectively entail the same overall level of effort as one large sample. The variance among these independent estimates then serves as an unbiased estimate of $V[\hat{\tau}_{y\pi}]$. The chief drawback to this approach, however, is the reduced precision of the estimator $\hat{\tau}_{y\pi}$ owing to the smaller sample size.

Circular systematic sampling

Recall that sample selection ceases when the frame is exhausted in the 1-in-a systematic sampling with a random start described earlier. A variant of this method, known as circular systematic sampling, is to permit the selection mechanism to cycle around to the beginning of the sampling frame, if needed, and to continue selecting units from the population until the desired sample size, say n, is reached. Moreover, to ensure that all elements in the population have positive probability of being included in the sample, the starting unit is chosen randomly from all N units of the population, not just the first a units as listed in the frame. As a consequence, the number of possible samples is $\Omega = N$. The advantage of this method is that the variation in size among the possible systematic samples is removed, and thus $\bar{y}$ is an unbiased estimator of μ_y.

Example 3.21

Consider again the $N = 10$ element population of Example 3.1, with the added stipulation that we wish to obtain a sample with $n = 3$ units systematically selected on a 1-in-3 interval. Under the circular design, the following $\Omega = 10$ are possible:

$$
\begin{array}{ll}
\mathcal{U}_1, \mathcal{U}_4, \mathcal{U}_7 & \mathcal{U}_6, \mathcal{U}_9, \mathcal{U}_2 \\
\mathcal{U}_2, \mathcal{U}_5, \mathcal{U}_8 & \mathcal{U}_7, \mathcal{U}_{10}, \mathcal{U}_3 \\
\mathcal{U}_3, \mathcal{U}_6, \mathcal{U}_9 & \mathcal{U}_8, \mathcal{U}_1, \mathcal{U}_4 \\
\mathcal{U}_4, \mathcal{U}_7, \mathcal{U}_{10} & \mathcal{U}_9, \mathcal{U}_2, \mathcal{U}_5 \\
\mathcal{U}_5, \mathcal{U}_8, \mathcal{U}_1 & \mathcal{U}_{10}, \mathcal{U}_3, \mathcal{U}_7
\end{array}
$$

With circular systematic samples of size n, $\pi_k = n/N$, and

$$V\left[\hat{\tau}_{y\pi}\right] = V\left[\frac{Nt_s}{n}\right]$$

$$= \frac{1}{N}\sum_{s=1}^{N}\left(\frac{Nt_s}{n} - \tau_y\right)^2 \tag{3.31}$$

Särndal et al. (1992) indicate that circular systematic sampling performs similarly to the conventional method of systematic sampling whenever the sampling fraction, n/N, is small. In our experience, it almost always entails some loss of efficiency: the estimator of the population total is less precise following circular systematic sampling than it is under 1-in-a systematic sampling with a random start.

For certain values of N and n, the circular systematic selection rule will choose the same population unit more than once into a single sample unless care is taken. Sudakar (1978) suggests a way to set a to avoid this possibility.

3.2.3 Bernoulli sampling

Introduced as binomial sampling by Goodman (1949), Bernoulli sampling is an equal-probability, without-replacement, sampling design in which population elements are selected independently and with constant inclusion probability, say π. That is, $\pi_k = \pi$ for all $\mathcal{U}_k$. A straightforward way to select a Bernoulli sample with inclusion probability π is to generate a uniform random variate $u_k = U[0, 1]$ for each $\mathcal{U}_k$ in the population. Include $\mathcal{U}_k$ if $u_k < \pi$, otherwise exclude $\mathcal{U}_k$ from the sample.

Evidently, the actual size of the sample that is selected in this fashion will vary from one application to another, i.e., under Bernoulli sampling, the size of the sample, n, is a (binomial) random variable. As shown in (3.74) of the Chapter 3 Appendix, the expected sample size in Bernoulli sampling is

$$E[n] = N\pi.$$

The number of possible samples under Bernoulli sampling is

$$\Omega = {}_NC_0 + {}_NC_1 + {}_NC_2 + \cdots + {}_NC_N$$

$$= \sum_{n=0}^{N} {}_NC_n.$$

For a specific value of n in the range of possible values $0 \le n < N$, the probability of selecting each distinct sample of that size is

$$p(s) = \pi^n(1 - \pi)^{N-n}.$$

Example 3.22

Suppose a Bernoulli sample with $\pi = 0.1$ is contemplated for a population of size $N = 10$. There are ${}_NC_9 = 10$ distinct samples of size $n = 9$, ${}_NC_1 = 10$

distinct samples of size $n = 1$, whereas there are $_N C_4 = {}_N C_6 = 210$ distinct
samples of size $n = 4$ and of size $n = 6$, respectively. In all, there are $\Omega = 1024$
possible samples.

Example 3.23

Suppose a Bernoulli sample design as in Example 3.22. The probability of a null
(empty) sample is $0.1^0 \times 0.9^{10} = 0.349$, whereas the probability of selecting
a sample with $n = 9$ elements is a minuscule $0.1^9 \times 0.9^1 = 9 \times 10^{-10}$. The
probability of selecting a sample with $n = 4$ elements is $0.1^4 \times 0.9^6 = 0.0005$,
whereas that of a sample with $n = 6$ elements is $0.1^6 \times 0.9^4 = 6.561 \times 10^{-7}$.

The HT estimator, (3.4), under Bernoulli sampling, simplifies to

$$
\begin{aligned}
\hat{\tau}_{y\pi} &= \frac{1}{\pi} \sum_{u_k \in s} y_k \\
&= \frac{N}{E[n]} \sum_{u_k \in s} y_k
\end{aligned}
\tag{3.32}
$$

provided that $n \geq 1$. Presumably, if an empty sample results, the sampling effort
would be repeated, and τ_y would be estimated by $\hat{\tau}_{y\pi}/(1 - \text{Prob}(n = 0))$.

An unusual feature about $\hat{\tau}_{y\pi}$ with Bernoulli sampling is that when $n = N$,
$\hat{\tau}_{y\pi} \neq \tau_y$, so that it is an inconsistent estimator despite being design-unbiased. This
is due to the randomness of the size of the sample selected with this design.

The variance of $\hat{\tau}_{y\pi}$ under this design is

$$
V\left[\hat{\tau}_{y\pi}\right] = \frac{1-\pi}{\pi} \sum_{k=1}^{N} y_k^2.
\tag{3.33}
$$

Another unusual feature about $\hat{\tau}_{y\pi}$, under the Bernoulli sampling design, is that its
variance, i.e., the spread of the sampling distribution of all possible estimates, is no
greater for small sample size than for large sample sizes. In contrast to the three
fixed-n designs considered to this point, for which the variance of the HT estimator
decreases with increasing n, the variance of $\hat{\tau}_{y\pi}$ under the Bernoulli sampling design
is impervious to the actual size of sample that is selected.

An unbiased estimator of $V\left[\hat{\tau}_{y\pi}\right]$ is

$$
\hat{v}\left[\hat{\tau}_{y\pi}\right] = \frac{1-\pi}{\pi^2} \sum_{u_k \in s} y_k^2.
\tag{3.34}
$$

Dividing $\hat{\tau}_{y\pi}$ by the population size yields

$$\hat{\mu}_{y\pi} = \frac{\hat{\tau}_{y\pi}}{N}$$

$$= \frac{1}{\pi N} \sum_{\mathcal{U}_k \in s} y_k \qquad (3.35)$$

$$= \frac{1}{E[n]} \sum_{\mathcal{U}_k \in s} y_k$$

as an unbiased but inconsistent and imprecise estimator of μ_y. The variance of $\hat{\mu}_{y\pi}$ under Bernoulli sampling is

$$V[\hat{\mu}_{y\pi}] = \frac{V[\hat{\tau}_{y\pi}]}{N^2}, \qquad (3.36)$$

which is estimated unbiasedly by

$$\hat{v}[\hat{\mu}_{y\pi}] = \frac{\hat{v}[\hat{\tau}_{y\pi}]}{N^2}$$

$$= \frac{1-\pi}{N^2\pi^2} \sum_{\mathcal{U}_k \in s} y_k^2. \qquad (3.37)$$

As an alternative to $\hat{\tau}_{y\pi}$ following Bernoulli sampling, consider incorporating the HT estimator of the population size, N, to adjust $\hat{\tau}_{y\pi}$. The general expression of the former is

$$\hat{N}_\pi = \sum_{\mathcal{U}_k \in s} \frac{1}{\pi_k}, \qquad (3.38)$$

which is simply $\hat{N}_\pi = n/\pi$ under Bernoulli sampling. The adjustment involves the multiplication of $\hat{\tau}_{y\pi}$ by the ratio of the known population size, N, to the estimated size, $\hat{N}_\pi$. This provides

$$\hat{\tau}_{y\pi,\text{rat}} = \left(\frac{N}{\hat{N}_\pi}\right) \hat{\tau}_{y\pi} \qquad (3.39a)$$

$$= \frac{N}{n/\pi} \sum_{\mathcal{U}_k \in s} \frac{y_k}{\pi} \qquad (3.39b)$$

$$= \frac{E[n]}{n} \hat{\tau}_{y\pi} \qquad (3.39c)$$

$$= \frac{N}{n} \sum_{\mathcal{U}_k \in s} y_k \qquad (3.39d)$$

which for this design simplifies to

$$= N\bar{y} \qquad (3.39e)$$

Moreover,

$$\hat{\mu}_{y\pi,\text{rat}} = \frac{\hat{\tau}_{y\pi,\text{rat}}}{N}$$

$$= \frac{E[n]}{n}\hat{\mu}_{y\pi},$$ (3.40)

which also can be expressed as

$$\hat{\mu}_{y\pi,\text{rat}} = \frac{\hat{\tau}_{y\pi}}{\hat{N}_\pi}$$ (3.41a)

$$= \tilde{\mu}_{y\pi,\text{rat}}$$ (3.41b)

$$= \bar{y}$$ (3.41c)

where $\tilde{\mu}_{y\pi,\text{rat}}$ is used here, and henceforth, to represent the estimator $\hat{\tau}_{y\pi}/\hat{N}_\pi$, because $\tilde{\mu}_{y\pi,\text{rat}}$ will not always be identical to $\hat{\tau}_{y\pi,\text{rat}}/N$ following sampling designs which yield a random sample size.

The ratio estimator, $\hat{\tau}_{y\pi,\text{rat}}$, is almost always more precise than $\hat{\tau}_{y\pi}$ following Bernoulli sampling. The greater precision of $\hat{\tau}_{y\pi,\text{rat}}$ over $\hat{\tau}_{y\pi}$ is due solely to its use of the sample size, n, actually realized in the sample, rather than the expected sample size, $E[n]$: smaller than expected size samples usually will have smaller sample sums, $\sum_{u_k \in s} y_k$, than average. When $n < E[n]$, therefore, the ratio $E[n]/n$ in (3.39c) will be greater than unity, with the result that $\hat{\tau}_{y\pi,\text{rat}}$ will be increased towards τ_y. In the opposite case, $E[n]/n < 1$, thereby adjusting $\hat{\tau}_{y\pi}$ down to account for the larger than average size sample and its consequent inflation of the value of $\hat{\tau}_{y\pi}$.

This ratio estimator is biased when used in conjunction with Bernoulli sampling. This and other properties of ratio-type estimators under various designs is explained more fully in Chapter 6.

Särndal et al. (1992, p. 65) present an approximate variance of $\hat{\tau}_{y\pi,\text{rat}}$ following Bernoulli sampling, i.e.,

$$V_a\left[\hat{\tau}_{y\pi,\text{rat}}\right] = \frac{1-\pi}{\pi}\sum_{k=1}^{N}(y_k - \mu_y)^2$$ (3.42a)

$$\approx N^2\left(\frac{1}{E[n]} - \frac{1}{N}\right)\sigma_y^2.$$ (3.42b)

The approximate variance of $\hat{\tau}_{y\pi,\text{rat}}$ in (3.42a) is itself approximated in (3.42b). There is an obvious resemblance of (3.42b) following Bernoulli sampling to the variance of $\hat{\tau}_{y\pi}$ following SRSwoR, (3.5). In like fashion,

$$V_a\left[\hat{\mu}_{y\pi,\text{rat}}\right] \approx \left(\frac{1}{E[n]} - \frac{1}{N}\right)\sigma_y^2$$ (3.43)

following Bernoulli sampling.

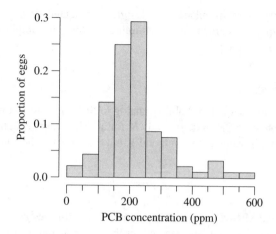

Figure 3.3 *Distribution of PCB concentration (ppm) in 92 brown pelican eggs (Risebrough 1972).*

Example 3.24

Figure 3.3 displays the distribution of PCB (polychlorinated biphenyl, an industrial pollutant) in parts per million measured in 92 eggs of brown pelicans (Risebrough 1972). If we treat these $N = 92$ eggs as the population for sampling under a Bernoulli design with $E[n] = 10$, then the standard error of $\hat{\mu}_{y\pi}$ is 71.4 ppm. By contrast, with SRSwoR for a fixed sample size $n = 10$, $\bar{y} = \hat{\tau}_{y\pi}/N$ with standard error of 29.6 ppm, which is approximately the same as the standard error of $\hat{\mu}_{y\pi,\text{rat}}$ under Bernoulli sampling.

An unbiased estimator of $V_a\left[\hat{\tau}_{y\pi,\text{rat}}\right]$ following Bernoulli sampling is

$$\hat{v}\left[\hat{\tau}_{y\pi,\text{rat}}\right] = \frac{1-\pi}{\pi^2}(n-1)s_y^2 \qquad (3.44)$$

for (3.42a), and for (3.42b),

$$\hat{v}_2\left[\hat{\tau}_{y\pi,\text{rat}}\right] = N^2\left(\frac{1}{E[n]} - \frac{1}{N}\right)s_y^2. \qquad (3.45)$$

3.3 Unequal probability designs

In contrast to the theme of §3.2, we now focus on designs for which the inclusion probabilities of population elements are not all equal. While there are multiple reasons for choosing an unequal probability design, a frequently encountered motivation for this choice is the potential increase in precision of estimation under certain conditions that may be under the control of the sampler or sample designer. Brewer & Hanif (1982) and Chaudhuri & Vos (1988) are useful compendia of unequal probability sampling designs. We present four unequal probability sampling strategies of practical utility. A with-replacement design is presented first, followed by three without-replacement designs: Poisson sampling, systematic sampling, and a design based on random groupings.

3.3.1 List sampling

List sampling is the unequal probability analog to SRSwR, simple random sampling with replacement. With SRSwR, $p_k = 1/N$ for all $\mathcal{U}_k$, whereas in list sampling the p_k can be unequal in value between 0 and 1, but must satisfy the constraint that

$$\sum_{k=1}^{N} p_k = 1.$$

List sampling, like SRSwR, is a with-replacement, fixed-n design.

One reason for favoring an unequal probability design is to purposely give population elements with large attribute values greater chance of being selected into the sample and hence measured. In other words, if $\mathcal{U}_k$ contributes more to the population total, τ_y, than does some other unit, $\mathcal{U}_{k'}$, i.e., $y_k > y_{k'}$, then $\mathcal{U}_k$ ought to be given a greater chance of being included in the sample. The matter of determining desirable selection probabilities, p_ks, then becomes an issue in the sampling design, which we discuss below.

For the moment, assume that selection probabilities have been determined in some fashion. To use probabilities to select an unequal-probability sample with replacement, we may follow the same procedure as Method II for SRSwR (see page 48), except that the p_k will not equal $1/N$.

Example 3.25

Consider the following four-unit population:

Unit	Selection probability p_k	Cumulative probability c_k
$\mathcal{U}_1$	0.36	0.36
$\mathcal{U}_2$	0.25	0.61
$\mathcal{U}_3$	0.14	0.75
$\mathcal{U}_4$	0.25	1.00

Suppose that for a sample of size $n = 2$ we generate first $u = 0.8801$ and then $u = 0.2064$. The first selects $\mathcal{U}_4$ into the sample; the second selects $\mathcal{U}_1$.

There are other similarities of list sampling to the SRSwR design. Namely, the inclusion probability of $\mathcal{U}_k$ is $\pi_k = 1-(1-p_k)^n \approx np_k$ and the number of possible sample sequences is $\Omega = N^n$. However, it is no longer true that $p(s) = 1/\Omega$, because with list sampling $p(s)$ will vary from one possible sample to another, depending on the composition of the sample. In general, $p(s) = \prod_{\mathcal{U}_k \in s} p_k$.

ASIDE: Because units in list sampling, indeed in SRSwR also, are selected independently, each ratio y_k/p_k is an independent estimator of τ_y. What does one do with n independent estimates? Average them together, as in (3.46).

Example 3.26

For the sample selected in Example 3.25, its probability of selection is $p(s) = p_1 \times p_4 = 0.36 \times 0.25 = 0.09$.

As with SRSwR, the customary estimator of τ_y is $\hat{\tau}_{yp}$ of (3.17), repeated here for convenience along with the variance of $\hat{\tau}_{yp}$, previously given in (3.18):

$$\hat{\tau}_{yp} = \frac{1}{n} \sum_{\mathcal{U}_k \in s} \frac{y_k}{p_k} \qquad (3.46)$$

$$V\left[\hat{\tau}_{yp}\right] = \frac{1}{n} \sum_{k=1}^{N} p_k \left(\frac{y_k}{p_k} - \tau_y\right)^2. \qquad (3.47)$$

This estimator is attributed to Hansen & Hurwitz (1943). To estimate μ_y unbiasedly one can use

$$\hat{\mu}_y = \frac{\hat{\tau}_{yp}}{N}. \qquad (3.48)$$

The variance of $\hat{\mu}_y$ is

$$V\left[\hat{\mu}_y\right] = \frac{V\left[\hat{\tau}_{yp}\right]}{N^2} \qquad (3.49)$$

Manipulating selection probabilities to increase precision

Suppose that the selection probabilities have been determined so that for each $\mathcal{U}_k$, $p_k = \kappa y_k$, where κ is some fixed constant. Thus, p_k is proportional to y_k, where κ is the constant of proportionality. To satisfy $\sum_{k=1}^{N} p_k = 1$, κ must equal $1/\tau_y$. Substituting κy_k for p_k in (3.47) yields

$$V\left[\hat{\tau}_{yp}\right] = \frac{1}{n} \sum_{k=1}^{N} \kappa y_k \left(\frac{y_k}{\kappa y_k} - \tau_y\right)^2$$

$$= \frac{1}{n} \sum_{k=1}^{N} \frac{y_k}{\tau_y} \left(\tau_y - \tau_y\right)^2 \qquad (3.50)$$

$$= 0.$$

In other words, if population units, $\mathcal{U}_k$, can be selected into the sample with probability, p_k, proportional to y_k, then $\hat{\tau}_{yp}$ has zero variance. In this situation, $\hat{\tau}_{yp}$ will always evaluate to the same number irrespective of which elements are selected.

Since $\hat{\tau}_{yp}$ unbiasedly estimates τ_y, then the implication of this is that $\hat{\tau}_{yp} = \tau_y$, and hence τ_y would be estimated perfectly, if the p_k can be so determined.

In practice, this result is unattainable because we do not know the values, y_k, in advance of sampling, and hence are unable to determine the value of the proportionality constant, κ, needed to determine the values of $p_k = \kappa y_k$. However, if auxiliary information about each $\mathcal{U}_k$ is available in the form of some other characteristic of $\mathcal{U}_k$, say x_k, such that x_k and y_k are strongly and positively correlated, then we could set $p_k = x_k/\tau_x$ and hope to achieve approximate proportionality between the p_ks and the corresponding y_ks. By extension, we would expect that the variance of $\hat{\tau}_{yp}$, while not being identically zero, would be reduced over what it would be if the selection probabilities were all equal or completely arbitrary.

Consider, for example, sampling the leaves on a sapling or bush for the purpose of estimating the total surface area of leaves. For irregularly shaped leaves, measurement of leaf area is an arduous chore, at least without optical scanning equipment. On the other hand, weighing a leaf is easy and quick. As shown in Figure 3.4, leaf area and weight are positively correlated, meaning that as leaf weight increases, leaf area tends to increase, too.

If the $N = 64$ leaves in Figure 3.4 were all the leaves on the sapling, one could weigh each and use weight as the auxiliary variate, x_k, to conduct a list sample with $p_k = x_k/\tau_x$.

Example 3.27

The HT estimator of total leaf area for the $N = 64$ leaf population (see Figure 3.4) has variance 12,792 cm^4 following SRSwoR with a sample of size $n = 8$ units. Following a list sample of the same size and with $p_k = x_k/\tau_x$, $\hat{\tau}_{yp}$ has a variance of 8884 cm^4. For comparison, under a SRSwR design, $\hat{\tau}_{yp}$ has variance 14,619 cm^4.

The variance of $\hat{\tau}_{yp}$ is given by (3.47) regardless of how the selection probabilities, the p_ks, are determined. When the p_ks are proportional to the x_ks, and the x_ks are, in turn, positively correlated with y_ks, then $\hat{\tau}_{yp}$ under list sampling is more precise than it would be for a similarly sized SRSwR sample. If the x_ks and y_ks are negatively correlated, then $\hat{\tau}_{yp}$ under SRSwR will be more precise. Negative

ASIDE: List sampling is one form of unequal probability sampling with replacement. When the selection probability is exactly proportional to an auxiliary variate, x_k, this sampling design is also called 'probability proportional to size' sampling, abbreviated as just pps sampling. List sampling is not synonymous with pps sampling, not only because the list sampling p_k may not be related to any particular measure of size, but also because there are other forms of pps sampling that do not require a list frame.

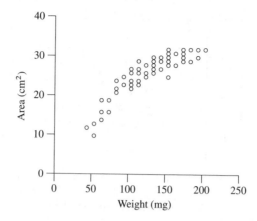

Figure 3.4 *The relation between the surface area and weight of 64 leaves.*

correlation is revealed graphically as a negatively sloping trend in a scatter plot of y_k on x_k. If x_ks and y_ks are uncorrelated, or weakly correlated, then the precision of $\hat{\tau}_{yp}$ will be similar under both designs. Lack of correlation is revealed in a scatter plot by a horizontal scatter of points, or else by an increasing trend for some range of x_k values and a decreasing trend outside that range.

Following list sampling, $V[\hat{\tau}_{yp}]$ is estimated unbiasedly by

$$\hat{v}\left[\hat{\tau}_{yp}\right] = \frac{1}{n(n-1)} \sum_{\mathcal{U}_k \in s} \left(\frac{y_k}{p_k} - \hat{\tau}_{yp}\right)^2, \tag{3.51}$$

given earlier as (3.20).

In the presentation of systematic sampling we introduced the notion of using auxiliary information in the form of a variate, x_k, to induce a linear trend into the sampling frame. When this can be done, the precision of $\hat{\tau}_{y\pi}$ is improved. In list sampling, x_k is used in a different manner to assist in the sample selection. In both cases, x_k must be known for each and every $\mathcal{U}_k$ in the population, which implies that the auxiliary information must be convenient and inexpensive to acquire. Otherwise, the time and effort needed to obtain the N values of x_ks might be better spent in selecting a larger sample of just the y_ks. Either strategy will usually bring about a decrease in variance, i.e., an increase the precision of estimation. Thus, there are two issues that bear on the question of whether, or not, to incorporate auxiliary information into the sampling design. One is the practical issue of the cost of obtaining such information. The other is the statistical issue requiring that x_ks and y_ks be sufficiently well correlated to effect an increase in precision of estimation. With systematic sampling, the trend between the x_ks and y_ks may be either positive or negative; either trend improves the precision of $\hat{\tau}_{y\pi}$. With pps sampling, the trend must be positive to improve the precision of both $\hat{\tau}_{y\pi}$ and $\hat{\tau}_{yp}$. If it is negative, units with small y_k will be favored for selection more than units with large y_k, thereby reducing the precision of estimation.

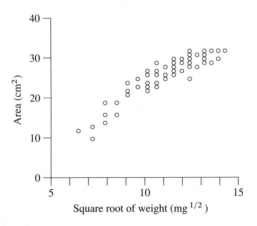

Figure 3.5 *The relation between the surface area and the square root of weight of 64 leaves.*

In Figure 3.4 there is a noticeable curve in the trend of leaf area and weight values. Oftentimes the precision of $\hat{\tau}_{yp}$ can be improved by transforming the auxiliary variate so that the plotted relationship between y_k and the transformed auxiliary variate is straighter. With the leaf data, using the square root of leaf weight as the auxiliary variate makes the relationship with leaf area less curvilinear, as seen in Figure 3.5. Further improvement is obtained by using the transformation $x_k = \ln(\text{leaf weight})$ (see Figure 3.6).

Example 3.28

Using $x_k = \sqrt{\text{leaf weight}}$ as the auxiliary variate, the variance of $\hat{\tau}_{yp}$ is reduced to 2849 cm^4. Using the transformation $x_k = \ln(\text{leaf weight})$ (see Figure 3.6), lowers the variance of $\hat{\tau}_{yp}$ to 1921 cm^4.

From a statistical standpoint, any transformation that increases the linear correlation between the variate of interest and the auxiliary variate will improve the performance of $\hat{\tau}_{yp}$. With these data, the linear correlation coefficient between leaf area and weight is $\hat{\rho} = 0.892$, that between area and $\sqrt{\text{leaf weight}}$ is $\hat{\rho} = 0.920$, and that between area and $\ln(\text{leaf weight})$ is $\hat{\rho} = 0.939$.

With the logarithmic transformation, care is needed to avoid logarithmically transformed values that are less than zero, or else the computed τ_x will be deceptively small.

When the $p_k = x_k/\tau_x$ so that the sample units are selected with probability

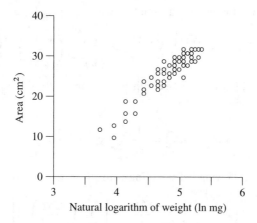

Figure 3.6 *The relation between the surface area and the natural logarithm of leaf weight (+ 4) of 64 leaves.*

proportional to the size of x_k, $\hat{\tau}_{yp}$ of (3.46) can be written in terms of x_k, namely

$$\hat{\tau}_{yp} = \frac{1}{n} \sum_{\mathcal{U}_k \in s} \frac{y_k}{p_k} \qquad (3.52a)$$

$$= \frac{\tau_x}{n} \sum_{\mathcal{U}_k \in s} \frac{y_k}{x_k} \qquad (3.52b)$$

$$= \tau_x \, \bar{r}, \qquad (3.52c)$$

where $\bar{r}$ is the sample average of the ratio of y_k to x_k. The simplification of $\hat{\tau}_{yp}$ in (3.52a) to a ratio adjustment of τ_x in (3.52b) holds only for pps sampling by x_k, and does not generalize to other strategies of unequal probability sampling with replacement.

A special case of $\hat{\tau}_{yp} = \tau_x \bar{r}$ occurs when $x_k = 1$ for all $\mathcal{U}_k$. In this case $\tau_x = N$ and $p_k = 1/N$, so that list sampling is identical to SRSwR and, upon substituting for τ_x and x_k in (3.52b), $\hat{\tau}_{yp}$ of (3.52c) becomes the familiar $N \bar{y}$.

A counterexample in which the equivalence of $\hat{\tau}_{yp}$ to $N \bar{y}$ does not hold occurs when $p_k = x_k / \tau_x$ but $x_k \neq 1$. In this case the unequal probability of selection among units makes $N \bar{y}$ a biased estimator of τ_y; its bias is

$$B\left[N \bar{y} : \tau_y \right] = E\left[N \bar{y} \right] - \tau_y$$

$$= \frac{N}{\tau_x} \sum_{k=1}^{N} y_k x_k - \tau_y \qquad (3.53)$$

$$= \frac{1}{\tau_x} \sum_{k=1}^{N} y_k \left(N x_k - \tau_x \right).$$

The variance of $N\bar{y}$ following list sampling with $p_k = x_k/\tau_x$ and $x_k \neq 1$ is

$$V[N\bar{y}] = \frac{N^2}{n} \left\{ \sum_{k=1}^{N} \frac{y_k^2 x_k}{\tau_x} - \left[\sum_{k=1}^{N} \left(\frac{y_k x_k}{\tau_x} \right)^2 \right] \right\}. \qquad (3.54)$$

Note that while $V[N\bar{y}]$ decreases with increasing n, $B[N\bar{y}:\tau_y]$ does not. In an analogous fashion,

$$B[\bar{y}:\mu_y] = \frac{1}{\tau_x} \sum_{k=1}^{N} y_k (x_k - \mu_x). \qquad (3.55)$$

The variance of $\bar{y}$ following list sampling with $p_k = x_k/\tau_x$ and $x_k \neq 1$ is

$$V[\bar{y}] = \frac{1}{n} \left\{ \sum_{k=1}^{N} \frac{y_k^2 x_k}{\tau_x} - \left[\sum_{k=1}^{N} \left(\frac{y_k x_k}{\tau_x} \right)^2 \right] \right\}. \qquad (3.56)$$

3.3.2 Poisson sampling

Poisson sampling is a generalization of Bernoulli sampling. Like Bernoulli sampling, population units are selected into the sample independently and without replacement. With Poisson sampling, however, the inclusion probability varies from one unit to another, and very often it is proportional to the auxiliary variate x_k. For example, in a variant of Poisson sampling in used in forestry, the probability of including a tree, $\mathcal{U}_k$, in a sample is proportional to an auxiliary variate comprising an ocular estimate of the tree's volume (§11.7, p. 382). For distinct population elements $\mathcal{U}_k$ and $\mathcal{U}_{k'}$, generally $\pi_k \neq \pi_{k'}$. Sample selection is identical to that described for Bernoulli sampling: for each $\mathcal{U}_k$ generate $u_k \sim U[0, 1]$, and include $\mathcal{U}_k$ into the sample if $u_k < \pi_k$.

The size of the sample, n, is a random variable, i.e., it cannot be determined beforehand how many units will be selected into the sample. The expected value and variance of n is $E[n] = \sum_{k=1}^{N} \pi_k$, and

$$V[n] = \sum_{k=1}^{N} \pi_k (1 - \pi_k), \qquad (3.57)$$

respectively. With a Poisson design, $\Omega = \sum_{n=0}^{N} {}_N C_n$, as with Bernoulli sampling, however, the expression for the probability of any particular sample, $p(s)$, is a bit more complicated, namely

$$p(s) = \prod_{\mathcal{U}_k \in s} \pi_k \prod_{\mathcal{U}_k \notin s} (1 - \pi_k).$$

The HT estimator, $\hat{\tau}_{y\pi}$, under Poisson sampling is the same as shown in (3.4), namely

$$\hat{\tau}_{y\pi} = \sum_{\mathcal{U}_k \in s} \frac{y_k}{\pi_k}, \qquad (3.58)$$

ASIDE: In practice, π_k^2 will be much smaller than π_k, so that $V[n] \approx \sum_{k=1}^{N} \pi_k = E[n]$. A Poisson random variable has the feature that its mean and variance are equal. We speculate that the name Poisson sampling derives from this approximate relationship between the mean and variance of n under this sampling protocol.

which is an unbiased but inefficient estimator of τ_y. The variance of $\hat{\tau}_{y\pi}$ under Poisson sampling is

$$V\left[\hat{\tau}_{y\pi}\right] = \sum_{k=1}^{N} y_k^2 \left(\frac{1 - \pi_k}{\pi_k}\right), \tag{3.59}$$

irrespective of the size sample actually selected.

One advantage of Poisson sampling, shared by Bernoulli sampling, is that a list frame of the target population is not needed in advance of sampling, provided that each element of the population can be accessed sequentially.

One would expect that $\hat{\tau}_{y\pi,\text{rat}} = (N/\hat{N}_\pi)\hat{\tau}_{y\pi}$, where $\hat{N}_\pi = \sum_{u_k \in s} 1/\pi_k$, would be less variable than $\hat{\tau}_{y\pi}$ following Poisson sampling. The rationale is the same as explained in the presentation of Bernoulli sampling. However, a further improvement may be possible if there is an auxiliary variate, x_k, that is well and positively correlated with y_k and known for all $\mathcal{U}_k$ in the population. Thus τ_x is known without error. Nonetheless, if y_k is measured and x_k is recorded for each $\mathcal{U}_k$ selected into a Poisson sample, then τ_x can be unbiasedly estimated by $\hat{\tau}_{x\pi} = \sum_{u_k \in s} x_k/\pi_k$, which can be used to adjust $\hat{\tau}_{y\pi}$:

$$\hat{\tau}_{y\pi,\text{rat}} = \left(\frac{\tau_x}{\hat{\tau}_{x\pi}}\right) \hat{\tau}_{y\pi} \tag{3.60}$$

Similar to the use of auxiliary information with list sampling, here too the x_ks are used both to guide the sample selection process and in the estimation of τ_y. Similar to the rationale for adjusting $\hat{\tau}_{y\pi}$ by the factor $N/\hat{N}_\pi$, if the Poisson sample provides a HT estimate that overestimates τ_y, then the corresponding HT estimate of τ_x is likely too big, also. In that case the ratio of $\hat{\tau}_{y\pi}$ to $\hat{\tau}_{x\pi}$ in (3.60) will tend to be rather less variable than either one alone. The gain in precision of $\hat{\tau}_{y\pi,\text{rat}} = (\tau_x/\hat{\tau}_{x\pi}) \hat{\tau}_{y\pi}$ over $\hat{\tau}_{y\pi}$ depends on how strongly y_k and x_k are correlated. If they are perfectly correlated, i.e., $y_k = \kappa x_k$ for some constant of proportionality κ, then the variance of $\hat{\tau}_{y\pi,\text{rat}} = (\tau_x/\hat{\tau}_{x\pi}) \hat{\tau}_{y\pi}$ is identically zero. Brewer & Hanif (1982, p. 7) term this the *ratio estimator property*.

This estimator, (3.60), is known as the generalized ratio estimator, which we discuss in more detail in Chapter 6. Figure 3.7 shows empirical sampling distributions of $\hat{\tau}_{y\pi}$ and $\hat{\tau}_{y\pi,\text{rat}}$. Both are based on the selection of 100,000 Poisson samples of expected size $E[n] = 24$ trees from the population of loblolly pine trees of Figure 2.4. The very much smaller variation of $\hat{\tau}_{y\pi,\text{rat}}$ among these samples is strikingly evident.

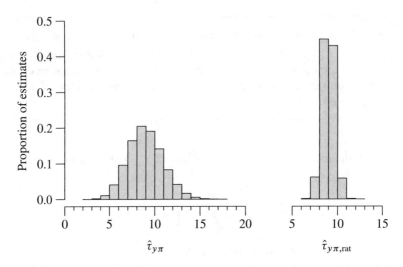

Figure 3.7 *On the left is the empirical sampling distribution of $\hat{\tau}_{y\pi}$ based upon the selection of* 100,000 *Poisson samples from the* $N = 14{,}443$ *population of loblolly pine trees. On the right is the empirical sampling distribution of $\hat{\tau}_{y\pi,\text{rat}}$ based on these same samples. All Poisson samples had an expected sample size of* $E[n] = 24$ *trees. The target parameter,* τ_y, *is aggregate bole volume of the population.*

The approximate variance of $\hat{\tau}_{y\pi,\text{rat}}$ following Poisson sampling is

$$V_a\left[\hat{\tau}_{y\pi,\text{rat}}\right] = \sum_{k=1}^{N} \left(y_k - R_{y|x}x_k\right)^2 \left(\frac{1 - \pi_k}{\pi_k}\right), \tag{3.61}$$

where $R_{y|x} = \tau_y/\tau_x$. Moreover, letting $\hat{\mu}_{y\pi,\text{rat}} = \hat{\tau}_{y\pi,\text{rat}}/N$ as before,

$$V_a\left[\hat{\mu}_{y\pi,\text{rat}}\right] = \frac{1}{N^2} \sum_{k=1}^{N} \left(y_k - R_{y|x}x_k\right)^2 \left(\frac{1 - \pi_k}{\pi_k}\right), \tag{3.62}$$

Example 3.29

Consider the leaf population displayed in Figure 3.4 and used in Example 3.27. Following Poisson sampling with expected sample size of 8 leaves, the HT estimator of total leaf area has a variance of 280,821 cm^4, yielding a relative standard error of 33.5%. Using leaf weight as the auxiliary variate, the approximate variance of $\hat{\tau}_{y\pi,\text{rat}} = (\tau_x/\hat{\tau}_{x\pi})\,\hat{\tau}_{y\pi}$ is 7796 cm^4, which has a relative standard error of 5.6%, slightly lower than the standard error of $\hat{\tau}_{yp}$ following list sampling ($n = 8$) with replacement in Example 3.27.

Example 3.30

Again consider Poisson sampling with expected sample size of 8 leaves. Using $x_k = \ln(\text{leaf weight (mg)})$ as the auxiliary variate, the approximate variance of $\hat{\tau}_{y\pi,\text{rat}}$ is 6254 cm^4, which has a relative standard error of 5.0%.

3.3.3 Unequal probability systematic sampling

With equal probability systematic sampling, ordering the sampling frame by increasing or decreasing value of x can produce large gains in the precision of $\hat{\tau}_{y\pi}$ of (3.25) when the auxiliary variate, x, and the variate of interest, y, are well correlated. An alternative way to incorporate available auxiliary information, which is positively correlated with y, is to sample systematically with probability proportional to x. Hartley & Rao (1962) generally are credited with providing the rigorous mathematical justification for this sampling strategy, which, they asserted, was already widely used.

For this strategy of unequal probability systematic sampling, the frame is ordered randomly. Indeed, Brewer & Gregoire (2000) call this method 'randomly ordered systematic sampling.' One straightforward way to randomly arrange population elements in a list sampling frame involves generating a U[0,1] random number, u_k, for each $\mathcal{U}_k$ in the population. Then order the frame by increasing value of u_k. This is identical to steps 1 and 2 of the Method II for drawing a SRSwoR sample, as presented on page 43. For each $\mathcal{U}_k$ in the randomly ordered frame, compute x_k/τ_x. Let c_k denote the cumulative value $n \sum_{j=1}^{k} x_j/\tau_x$ in the randomly ordered frame, i.e.,

$$c_0 = 0$$

$$c_1 = \frac{n}{\tau_x} x_1$$

$$c_2 = \frac{n}{\tau_x}(x_1 + x_2)$$

$$\vdots$$

$$c_n = \frac{n}{\tau_x}(x_1 + x_2 + \cdots + x_n) = n$$

where n is the size of the unequal-probability systematic sample that is desired. Sample selection proceeds by generating a single random number, $u \sim$ U[0,1]. The first unit is selected into the sample is that unit $\mathcal{U}_k$ in the randomly ordered list frame whose c_k value satisfies

$$c_{k-1} \leq u < c_k.$$

The second sample element is that $\mathcal{U}_k$ whose c_k satisfies

$$c_{k-1} \leq u + 1 < c_k.$$

ASIDE: In populations where there are a few $\mathcal{U}_k$ with large x_k values, care must be taken to check that $x_k/\tau_x < 1/n$ for all $\mathcal{U}_k$. This is the same as requiring that $x_k < \tau_x/n$. For those units that are so large that $x_k > \tau_x/n$, the customary tactic is to separate these large units from the other $\mathcal{U}_k$, and to measure these units' y_k. Suppose that there are N^L such large units and that the aggregate x_k and y_k of these N^L units are τ_x^L and τ_y^L, respectively. Therefore the target parameter can be partitioned as $\tau_y = \tau_y^S + \tau_y^L$, where τ_y^S is the population total y_k for the $N^S = N - N^L$ 'small units' whose $x_k < \tau_x/n$. Sampling then proceeds to select the remaining $n^S = n - N^L$ sample units from the randomly ordered frame of N^S listed units, with each of the remaining N^S $\mathcal{U}_k$ being assigned a value x_k/τ_x^S where $\tau_x^S = \tau_x - \tau_x^L$. If there are some $\mathcal{U}_k$ among the N^S remaining units with $x_k > \tau_x/n$, then these, also, will have to be grouped with the 'large units' that are completely measured.

The third sample element is that which satisfies

$$c_{k-1} \leq u + 2 < c_k,$$

and so on until the nth unit selected into the sample is the $\mathcal{U}_k$ whose c_k satisfies

$$c_{k-1} \leq u + n - 1 < c_k.$$

This procedure ensures that n different $\mathcal{U}_k$ will be selected providing that $x_k/\tau_x < 1/n$ for all $\mathcal{U}_k$. With this design there are

$$\Omega = \frac{N!}{n!(N-n)!}$$

possible distinct samples. Moreover, the randomization of the frame prior to sample selection gives each pair of population elements, say $\mathcal{U}_k$ and $\mathcal{U}_{k'}$, positive probability of being included in the same sample, in contrast to the equal probability systematic sampling design presented earlier.

The inclusion probability of each $\mathcal{U}_k$ is $\pi_k = nx_k/\tau_x$, provided that no $x_k > \tau_x/n$ (see previous ASIDE). Consequently, the unbiased HT estimator of τ_y is

$$\hat{\tau}_{y\pi} = \sum_{\mathcal{U}_k \in s} \frac{y_k}{\pi_k}$$

$$= \frac{\tau_x}{n} \sum_{\mathcal{U}_k \in s} \frac{y_k}{x_k}. \tag{3.63}$$

Hartley & Rao (1962, p. 369) derived an approximation to the variance of $\hat{\tau}_{y\pi}$ under this design, viz.,

$$V\left[\hat{\tau}_{y\pi}\right] \approx V_a\left[\hat{\tau}_{y\pi}\right] = \sum_{k=1}^{N} \pi_k \left(1 - \frac{n-1}{n}\pi_k\right)\left(\frac{y_k}{\pi_k} - \frac{\tau_y}{n}\right)^2. \tag{3.64}$$

Results from simulated sampling confirm that this approximation, $V_a[\hat{\tau}_{y\pi}]$, to $V[\hat{\tau}_{y\pi}]$ is very close. They also proposed the following estimator of $V[\hat{\tau}_{y\pi}]$:

$$\hat{v}[\hat{\tau}_{y\pi}] = \frac{1}{n-1} \sum_{k<k'}^{n} \left[1 - \pi_k - \pi_{k'} + \frac{\phi}{n}\right]\left(\frac{y_k}{\pi_k} - \frac{y_{k'}}{\pi_{k'}}\right)^2, \qquad (3.65)$$

where $\phi = \sum_{k=1}^{N} \pi_k^2$.

Example 3.31

Consider again the data on red oak bole volumes used in Example 3.20. The population consists of $N = 236$ red oak trees whose aggregate bole volume, $\tau_y = 230$ m^3, was the target parameter. In that example, the variance of $\hat{\tau}_{y\pi}$ with a 1-in-20 equal probability systematic sample was 1980 m^6 (relative standard error of 19.3%). When the original list sampling frame was arranged in order of increasing bole diameter, the variance of $\hat{\tau}_{y\pi}$ following 1-in-20 systematic sampling from the ordered frame is 939.9 m^6 (13.3%). Of the 20 possible samples, there were 16 of size $n = 12$ trees and four of size $n = 11$ trees. When using diameter as the auxiliary variate, x, the variance of $\hat{\tau}_{y\pi}$ following an unequal probability systematic sample of the $n = 11$ trees is $V[\hat{\tau}_{y\pi}] = 1267.6$ m^6 (15.5%); for such samples with $n = 12$ tree each, it is $V[\hat{\tau}_{y\pi}] = 1163.1$ m^6 (14.8%). In this instance, a sampling strategy consisting of ordered, equal probability systematic sampling with the HT estimator is more efficacious than 1-in-20 systematic sampling from the ordered frame. In both cases, bole diameter is the source of auxiliary information, which is used to guide sample selection: in Example 3.20 it is used to order the frame, whereas in this example it is used to sample with probability proportional to diameter.

If one were to use the circular cross-sectional, or basal, area of the bole at breast height as the auxiliary variate to order the frame, $V[\hat{\tau}_{y\pi}]$ following 1-in-20 systematic sampling from the ordered frame remains unchanged from what it is when diameter is the auxiliary variate, because the ordering of the frame remains unchanged and because the auxiliary variate is not used in the estimator of τ_y. In contrast, the variance of $\hat{\tau}_{y\pi}$ following unequal probability systematic sample from a randomly ordered frame shrinks remarkably to 146.0 m^6 (5.3%) when using basal area as the auxiliary size variate. As has been noted before, whenever an auxiliary variate can be algebraically transformed to create a more nearly straight line relationship between it and the variate of interest, y, the precision of the HT estimator of τ_y is very likely to be improved. This holds for other estimators of τ_y, too, as discussed in Chapter 6.

Example 3.32

With the help of a computer program to simulate repeated unequal probability systematic sampling, we conducted a simulated sampling experiment to assess

ASIDE: In the presentation of equal-probability systematic sampling on p. 49 we stipulated that the first unit be chosen from among the first a units in the list frame. Had we instead stipulated a procedure analogous to the one above, a fixed-n, equal-probability design results. Specifically, one can choose d such that $0 \leq d < N/a$; then one chooses as the first unit to enter the sample that $\mathcal{U}_k$ whose c_k value satisfies

$$c_{k-1} \leq d < c_k.$$

This would ensure that n units would always be selected in the equal-probability systematic sampling case, thereby dispensing with the slight problematic feature of having a random sample size in the ordinary application of equal probability systematic sampling of discrete populations. However, the variance of $\hat{\tau}_{y\pi}$ is no longer as given in (3.29).

how well $V[\hat{\tau}_{y\pi}]$ as given by (3.64) approximates the actual sampling variance of $\hat{\tau}_{y\pi}$ for this design. We instructed the procedure to select 100,000 samples of size $n = 12$ trees from the $N = 236$ red oak population of Example 3.31. When tree basal area was used as the auxiliary variate, the observed variance of the 100,000 $\hat{\tau}_{y\pi}$ estimates was 145.4 m^6, which is nearly identical to the variance, 146.0 m^6, computed in Example 3.31.

For each of the 100,000 samples we estimated the variance of $\hat{\tau}_{y\pi}$ by computing $\hat{v}[\hat{\tau}_{y\pi}]$ according to (3.65). The average estimate of variance was 145.9 m^6.

Repeated simulations confirmed that the average, or expected, value provided by the variance estimator (3.65) was always within a fraction of a percentage point of its target value given by (3.64), and that the variance observed among the 100,000 estimates was identical to (3.64), barring the variance of the simulation process itself.

3.3.4 Rao, Hartley, Cochran sampling strategy

For designs other than systematic, a fixed $n > 2$, without-replacement design with unequal probabilities is problematic because the probability with which $\mathcal{U}_k$ is included into the sample depends in a complex fashion upon whether it is selected as the first unit to enter the sample, the second, the third, and so on. By contrast, with Poisson sampling the inclusion probabilities do not depend on the order of selection of units into the sample, but the size of the sample selected cannot be determined in advance of sampling. Randomly ordered systematic sampling yields a fixed size sample, but no design unbiased estimator of the variance of $\hat{\tau}_{y\pi}$ has yet been developed. Citing these and other limitations of without-replacement, unequal-probability sampling designs, Rao et al. (1962) proposed a method that circumvents these difficulties and has smaller variance than $\hat{\tau}_{yp}$ with list sampling.

The design proposed by Rao et al. (RHC) consists of two stages. The first stage entails a random partitioning of the sampling frame into n groups, where n, as usual, is the desired sample size. These n groups need not each have the same number of population elements, indeed it will be unusual to have N be an integer multiple of n, which is a necessary condition to be able to form equal-size groups. Let N_i denote the size of the ith group, so that $N = \sum_{i=1}^{n} N_i$. Let G_i symbolize the ith group itself. The total of y and of x in each group are found by summing over only the elements selected into that group. Specifically,

$$\tau_{yi} = \sum_{\mathcal{U}_k \in G_i} y_k$$

is the total of y in the ith randomly formed group. Similarly

$$\tau_{xi} = \sum_{\mathcal{U}_k \in G_i} x_k$$

is the total of x in G_i.

Example 3.33

Consider the following five-unit population divided as indicated into two groups:

Unit	y_k	x_k	Group
$\mathcal{U}_1$	25	0.15	2
$\mathcal{U}_2$	14	0.30	1
$\mathcal{U}_3$	57	0.28	2
$\mathcal{U}_4$	32	0.82	2
$\mathcal{U}_5$	44	0.23	1

Thus, $\tau_{y1} = 58$ and $\tau_{y2} = 114$; and $\tau_{x1} = 0.53$ and $\tau_{x2} = 1.25$. In application of the RHC procedure, the τ_{yi} totals would be unknown, whereas the τ_{xi} could be calculated after the groups had been formed.

The second stage of the RHC sampling design consists of selecting one element from each group, and since there are n groups, this will yield a final sample of exactly n elements. For any $\mathcal{U}_k$ belonging to G_i, symbolically denoted by $\mathcal{U}_k \in G_i$, $\mathcal{U}_k$ has probability of being selected into the sample at the second stage of $p_k = x_k / \tau_{xi}$. A design-unbiased estimator of τ_{yi} is

$$\hat{\tau}_{yi} = \tau_{xi} \left(\frac{y_k}{x_k} \right). \tag{3.66}$$

Therefore, a design-unbiased estimator of the population total, τ_y, is

$$\hat{\tau}_{y,\text{RHC}} = \sum_{i=1}^{n} \hat{\tau}_{yi}. \tag{3.67}$$

The variance of $\hat{\tau}_{y,\text{RHC}}$ is

$$V\left[\hat{\tau}_{y,\text{RHC}}\right] = \left[\frac{\sum_{i=1}^{n} N_i^2 - N}{N(N-1)}\right]\left(\sum_{k=1}^{N} \frac{y_k^2}{p_k} - \tau_y^2\right), \tag{3.68}$$

which depends implicitly on the size of the sample through the summation in the first term.

Both list sampling and the RHC method are fixed n sampling designs, the former a with-replacement and the latter a without-replacement design. Rao et al. (1962) derived the relation between $V[\hat{\tau}_{y,\text{RHC}}]$ and $V[\hat{\tau}_{yp}]$ in (3.18):

$$V\left[\hat{\tau}_{y,\text{RHC}}\right] = \left[\frac{n\left(\sum_{i=1}^{n} N_i^2 - N\right)}{N(N-1)}\right] V[\hat{\tau}_{yp}]. \tag{3.69}$$

Thus, for the same size sample, $V[\hat{\tau}_{y,\text{RHC}}] < V[\hat{\tau}_{yp}]$. Moreover, the variance of $\hat{\tau}_{y,\text{RHC}}$ is minimized when the sizes of the groups are equal, i.e., $N_i = N/n$, or nearly so.

An unbiased estimator of $V[\hat{\tau}_{y,\text{RHC}}]$, derived by Rao et al. (1962), is

$$\hat{v}\left[\hat{\tau}_{y,\text{RHC}}\right] = \left(\frac{\sum_{i=1}^{n} N_i^2 - N}{N^2 - \sum_{i=1}^{n} N_i^2}\right) \sum_{\mathcal{U}_k \in s} \frac{\tau_{xi}}{\tau_x}\left(\frac{y_k}{x_k/\tau_x} - \hat{\tau}_{y,\text{RHC}}\right)^2. \tag{3.70}$$

In a case study, Pontius (1996) showed empirically that $\hat{v}[\hat{\tau}_{y,\text{RHC}}]$ was a more reliable estimator of $V[\hat{\tau}_{y,\text{RHC}}]$ than $\hat{v}[\hat{\tau}_{yp}]$ was of $V[\hat{\tau}_{yp}]$.

Schabenberger & Gregoire (1994) compared the RHC strategy to two other unequal-probability, without-replacement designs proposed by Sunter (1986, 1989). They concluded that the RHC method performed quite favorably, especially because it does not depend on a possibly complicated ordering of the sampling frame. Moreover, RHC can never be less precise than list sampling with the same size of sample.

Example 3.34

Using the red oak data again with tree basal area as the auxiliary variate, the variance of $\hat{\tau}_{y,\text{RHC}}$ for a sample of size $n = 12$ trees is $V[\hat{\tau}_{y,\text{RHC}}] = 149.3$ m^6. Thus, the RHC strategy is nearly as precise as unequal probability systematic sampling when using basal area as the auxiliary variate in Example 3.31.

Example 3.35

In Example 3.27, a list sample of $n = 8$ leaves provided a variance $V[\hat{\tau}_{yp}] = 8884$ cm^4 when estimating total leaf area, τ_y with $\hat{\tau}_{yp}$ and when using leaf weight as the auxiliary variate, x, to enable the list sampling. For the same size sample and choice of auxiliary variate, the RHC design with groups of size $N_i = 8$, coupled with $\hat{\tau}_{y,\text{RHC}}$ yields $V[\hat{\tau}_{y,\text{RHC}}] = 7897$ cm^4. Since

$\tau_y = 1582$ cm^2 for this $N = 64$ leaf population, the relative standard error of $\hat{\tau}_{yp}$ was $100\sqrt{8884 \text{ cm}^4}/(1582 \text{ cm}^2) = 6.0\%$ and for $\hat{\tau}_{y,\text{RHC}}$ it is $100\sqrt{7897 \text{ cm}^4}/(1582 \text{ cm}^2) = 5.6\%$.

3.4 Terms to remember

Bernoulli sampling	Random sample size
Circular systematic sampling	Ratio estimator property
Equal probability sampling	Simple random sampling
Expansion estimator	Systematic sampling
Horvitz–Thompson estimator	Sampling frame
List sampling	Sampling interval
Poisson sampling	Systematic sampling with a random start
Pps sampling	Unequal probability sampling

3.5 Exercises

1. Verify the number of possible SRSwoR samples, Ω, of size $n = 10$ that can be drawn from a population of size $N = 100$.

2. For a population of size $N = 10,000$ units with $\sigma_y^2 = 2$, compute $V[N\bar{y}]$ following SRSwoR with samples of size $n = 10$. Repeat for samples of size $n = 20$. How many times more precise is the latter sampling strategy?

3. Enumerate the 36 possible with-replacement samples of size $n = 2$ that can be selected from the $N = 6$ element Stuart population.

4. Use the radon data of Example 3.7 to construct a 90% confidence interval for $\hat{p}_1$ using the method described in §3.6.11, p. 90.

5. Use the red oak data of Example 3.20 to draw each of the possible 1-in-20 systematic samples. For each one estimate the variance of $\hat{\tau}_{y\pi}$ using $\hat{v}[\hat{\tau}_{y\pi}]$, and then compute the 90% confidence interval for τ_y. How many of the twenty intervals cover the value of τ_y?

6. Repeat the previous exercise using $\hat{v}_{sd}[\hat{\tau}_{y\pi}]$ as the estimator of variance. How many of these intervals cover the value of τ_y?

7. In Example 3.26 the $p(s)$ for one of the $\Omega = N^n = 4^2 = 16$ possible list samples of size $n = 2$ was computed. Compute $p(s)$ for the other 15 possible samples, and verify that $\sum_{s \in \Omega} p(s) = 1$.

8. Use the biomass population of Example 3.8 in order to draw another SRSwoR of size $n = 52$ trees. Compare the distribution of trees you obtain in this sample to that obtained in the Example. Construct a 90% confidence interval for the average total aboveground biomass per tree. Also, construct a 95% confidence interval

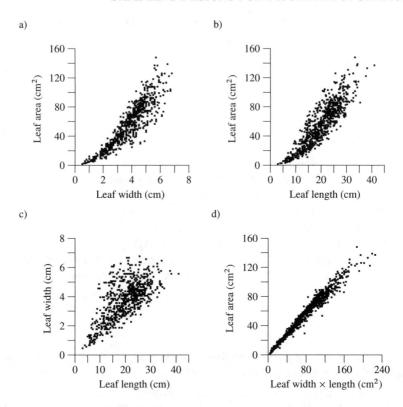

Figure 3.8 *Area, width and length of 742 Eucalyptus leaves.*

based on these same sample data. Which interval is wider? Explain the reason
why one interval is wider than the other.

9. Use the biomass population of Example 3.8 in order to draw an equal probability
   systematic sample of size $n = 52$ trees, and compute a 90% confidence interval
   for the average total aboveground biomass per tree.

10. Consider the set of $N = 742$ Eucalyptus leaves which is displayed in Figure 3.8.
    Treat this as a population which is to be sampled for the purpose of estimating
    the average area per leaf, $\mu_y = 55.1 \mathrm{cm}^2$. Because the product of leaf width and
    length is has more of a straight-line relationship with leaf area, it should be a more
    useful auxiliary variate than either leaf width or leaf length individually.

    a) Select a sample of size $n = 37$ leaves by list sampling using $p_k = x_k/\tau_x$
    with leaf width as the auxiliary variate, and using $\hat{\tau}_{yp}$ as the estimator of τ_y;
    b) Repeat a) using an unequal probability systematic selection using the same
    auxiliary variate, and using $\hat{\tau}_{y\pi}$ as the estimator of τ_y.

11. Repeat the preceding exercise with the rectangular area (length $\times$ width/100) of
    each leaf as the auxiliary variate.

12. Randomly partition the population of Eucalyptus leaves into 37 groups of roughly

equal size. Select a sample of $n = 37$ leaves by the Rao-Hartley-Cochran scheme and proceed to estimate τ_y by $\hat{\tau}_{y,\text{RHC}}$.

13. Compare the difference in precision between $\hat{\tau}_{y\pi}$ and $\hat{\tau}_{y\pi,\text{rat}}$ to estimate total leaf area of the Eucalyptus leaf population when relying on a Poisson sample of expected sample size $E[n] = 37$. That is, set $\pi_k = E[n]x_k/\tau_x$, where the auxiliary variate, x_k, is the leaf's rectangular area (length $\times$ width/100), and then compute and compare $V[\hat{\tau}_{y\pi}]$ given in (3.59) to $V_a[\hat{\tau}_{y\pi,\text{rat}}]$ given in (3.61).

14. Follow up on Example 3.35 by selecting a sample from the $N = 64$ leaf population using the RHC sampling design with groups of size $N_i = 8$ each. Compute a 90% confidence interval for total leaf area using $\hat{\tau}_{y,\text{RHC}}$ as an estimator of τ_y.

15. Repeat the preceding exercise using six groups of size $N_i = 9$ and one group of size 10.

16. Repeat the preceding exercise using eight groups of size $N_i = 7$ and one group of size 8.

3.6 Appendix

3.6.1 Factorial and combinatorial notation

For any positive integer A, the factorial $A!$ is defined as

$$A! = A \times (A - 1) \times (A - 2) \times \cdots \times 3 \times 2 \times 1.$$

Thus,

$$N! = N \times (N - 1) \times (N - 2) \times \cdots \times (N - n + 1)$$
$$\times (N - n) \times (N - n - 1) \times \cdots \times 3 \times 2 \times 1.$$

$$n! = n \times (n - 1) \times (n - 2) \times \cdots \times 3 \times 2 \times 1.$$

$$(N - n)! = (N - n) \times (N - n - 1) \times (N - n - 2) \times \cdots \times 3 \times 2 \times 1.$$

The expression for Ω on page 36 thus resolves to,

$$\frac{N!}{n!(N-n)!} = \frac{N}{n}\left(\frac{N-1}{n-1}\right)\left(\frac{N-2}{n-2}\right)\cdots\left(\frac{N-n+2}{2}\right)\left(\frac{N-n+1}{1}\right),$$

which might be easier to calculate than (3.2).

3.6.2 Derivation of the inclusion probability of an element for SRSwoR.

Since all samples are equally likely, π_k can be computed as the following proportion:

$$\pi_k = \frac{\text{number of samples which include } \mathcal{U}_k}{\text{total number of samples possible}}$$

The denominator is just $\Omega = {}_NC_n$.

For those samples which include $\mathcal{U}_k$ as a member, there are $n - 1$ other elements in the sample chosen from the remaining $N - 1$ elements of the population. There are

$$N-1 C_{n-1} = \frac{(N-1)!}{(n-1)!\,(N-n)!}$$

such samples possible. Hence,

$$\pi_k = \frac{N-1 C_{n-1}}{N\, C_n} = \frac{n}{N} \tag{3.71}$$

3.6.3 Proof of unbiasedness of $\hat{\tau}_{y\pi}$ as an estimator of τ_y

From 2.1 we have

$$E[\hat{\tau}_{y\pi}] = \sum_{s \in \Omega} p(s)\hat{\tau}_{y\pi}(s)$$

$$= \sum_{s \in \Omega} \left[p(s) \sum_{\mathcal{U}_k \in s} \frac{y_k}{\pi_k} \right]$$

$$= \sum_{s \in \Omega} \left[\sum_{s \ni \mathcal{U}_k} p(s)\frac{y_k}{\pi_k} \right]$$

$$= \sum_{k=1}^{N} \frac{y_k}{\pi_k} \left[\sum_{s \ni \mathcal{U}_k} p(s) \right]$$

$$= \sum_{k=1}^{N} \frac{y_k}{\pi_k}\pi_k \qquad \text{(by 1.1)}$$

$$= \sum_{k=1}^{N} y_k = \tau_y.$$

where the notation $s \ni \mathcal{U}_k$ indicates that the summation extends over all samples of which $\mathcal{U}_k$ is a member.

An alternative proof relies on a device that we will use repeatedly henceforth. We define the random variable I_k to indicate whether or not $\mathcal{U}_k$ is included in a sample. Let

$$I_k = \begin{cases} 1, & \text{if } \mathcal{U}_k \in s; \\ 0, & \text{otherwise.} \end{cases} \tag{3.72}$$

Then the HT estimator as written in (3.4) can be written alternately as

$$\hat{\tau}_{y\pi} = \sum_{k=1}^{N} \frac{y_k I_k}{\pi_k}.$$

In this expression, I_k is the only term that is random, as it indicates by its value whether or not $\mathcal{U}_k$ is selected into the random sample. Hence, its value will vary from one sample to another. As mentioned in §1.4, the y_k value associated with $\mathcal{U}_k$ is a regarded as a fixed value, irrespective of whether $\mathcal{U}_k$ is selected into any particular sample. Likewise, π_k is determined by the sampling design, and is independent of the particular sample chosen, too.

The expected value of I_k is

$$E[I_k] = 0 \times \text{Prob}[I_k = 0] + 1 \times \text{Prob}[I_k = 1]$$
$$= 0 + \pi_k = \pi_k.$$

Thus, the expected value of $\hat{\tau}_{y\pi}$ is

$$E\left[\hat{\tau}_{y\pi}\right] = E\left[\sum_{k=1}^{N} \frac{y_k I_k}{\pi_k}\right]$$

$$= \sum_{k=1}^{N} \frac{y_k E[I_k]}{\pi_k}$$

$$= \sum_{k=1}^{N} \frac{y_k \pi_k}{\pi_k}$$

$$= \sum_{k=1}^{N} y_k = \tau_y.$$

The bias of $\hat{\tau}_{y\pi}$ as an estimator of τ_y is $B[\hat{\tau}_{y\pi} : \tau_y] = E[\hat{\tau}_{y\pi}] - \tau_y = 0$, i.e., $\hat{\tau}_{y\pi}$ unbiasedly estimates τ_y.

3.6.4 Derivation of $V[N\bar{y}]$ in (3.6) following SRSwoR

We start with the well known identity for the variance of a random variable., ψ: $V[\psi] = E[\psi^2] - E[\psi]^2$. This identity holds for any random variable, and in particular for $\hat{\tau}_{y\pi}$.

Furthermore, define the random variable, $I_{kk'}$ to indicate whether both $\mathcal{U}_k$ and $\mathcal{U}_{k'}$ are included in the same sample. Let

$$I_{kk'} = I_k I_{k'} = \begin{cases} 1, & \text{if } \mathcal{U}_k \text{ and } \mathcal{U}_{k'} \in s; \\ 0, & \text{otherwise.} \end{cases}$$

There are

$$_{N-2}C_{n-2} = \frac{(N-2)!}{(n-2)!(N-n)!}$$

possible samples of size n that include both $\mathcal{U}_k$ and $\mathcal{U}_{k'}$ as elements. Thus,

$$E[I_{kk'}] = \frac{{}_{N-2}C_{n-2}}{{}_{N}C_{n}}$$

$$= \frac{n(n-1)}{N(N-1)}$$

$$= \pi_{kk'}, \text{ say.}$$

To derive $V[\hat{\tau}_{y\pi}] = E[\hat{\tau}_{y\pi}^2] - E[\hat{\tau}_{y\pi}]^2$ it is necessary to derive both terms on the right side of the identity. The square of $\hat{\tau}_{y\pi}$ is

$$\hat{\tau}_{y\pi}^2 = \left(\sum_{k=1}^{N} \frac{y_k I_k}{\pi_k}\right)^2$$

$$= \sum_{k=1}^{N} \frac{y_k^2 I_k^2}{\pi_k^2} + \sum_{k=1}^{N} \sum_{\substack{k' \neq k \\ k'=1}}^{N} \frac{y_k y_{k'} I_k I_{k'}}{\pi_k \pi_{k'}}$$

$$= \sum_{k=1}^{N} \frac{y_k^2 I_k}{\pi_k^2} + \sum_{k=1}^{N} \sum_{\substack{k' \neq k \\ k'=1}}^{N} \frac{y_k y_{k'} I_{kk'}}{\pi_k \pi_{k'}}.$$

Thus,

$$E[\hat{\tau}_{y\pi}^2] = \sum_{k=1}^{N} \frac{y_k^2 \pi_k}{\pi_k^2} + \sum_{k=1}^{N} \sum_{\substack{k' \neq k \\ k'=1}}^{N} \frac{y_k y_{k'} \pi_{kk'}}{\pi_k \pi_{k'}}$$

$$= \frac{N}{n} \sum_{k=1}^{N} y_k^2 + \frac{\frac{n(n-1)}{N(N-1)}}{\frac{n^2}{N^2}} \sum_{k=1}^{N} \sum_{\substack{k' \neq k \\ k'=1}}^{N} y_k y_{k'}$$

$$= \frac{N}{n} \sum_{k=1}^{N} y_k^2 + \frac{N(n-1)}{n(N-1)} \sum_{k=1}^{N} \sum_{\substack{k' \neq k \\ k'=1}}^{N} y_k y_{k'}.$$

Since $E[\hat{\tau}_{y\pi}] = \tau_y$, and $\tau_y^2 = \sum_{k=1}^{N} y_k^2 + \sum_{k=1}^{N} \sum_{k \neq k'=1}^{N} y_k y_{k'}$, then

$$V[\hat{\tau}_{y\pi}] = E[\hat{\tau}_{y\pi}^2] - E[\hat{\tau}_{y\pi}]^2$$

$$= \sum_{k=1}^{N} y_k^2 \left(\frac{N}{n} - 1\right) + \sum_{k=1}^{N} \sum_{\substack{k' \neq k \\ k'=1}}^{N} y_k y_{k'} \left(\frac{N(n-1)}{n(N-1)} - 1\right)$$

$$= \frac{N-n}{n} \left(\sum_{k=1}^{N} y_k^2 - \frac{1}{N-1} \sum_{k=1}^{N} \sum_{\substack{k' \neq k \\ k'=1}}^{N} y_k y_{k'} \right)$$

$$= \frac{N-n}{n} \left(\frac{N \sum_{k=1}^{N} y_k^2 - \tau_y^2}{N-1} \right)$$

$$= \frac{N-n}{n} \left(N\sigma_y^2 \right)$$

$$= N^2 \left(\frac{1}{n} - \frac{1}{N} \right) \sigma_y^2.$$

3.6.5 Proof of unbiasedness of s_y^2 as estimator of σ_y^2 following SRSwoR

Define

$$s^2 = \frac{1}{n-1} \sum_{\mathcal{U}_k \in s} (y_k - \bar{y})^2.$$

Then,

$$(n-1)E[s_y^2] = \sum_{k=1}^{N} y_k^2 E[I_k] - nE[\bar{y}^2]$$

$$= [(N-1)\sigma_y^2 + N\mu_y^2] \frac{n}{N} - n[V[\bar{y}] + \mu_y^2]$$

$$= \left[\frac{n(N-1)}{N} - n\left(\frac{N-n}{nN}\right) \right] \sigma_y^2 + n\mu_y^2 - n\mu_y^2$$

$$= [nN - n - N + n] \frac{\sigma_y^2}{N}$$

$$= (n-1)\sigma_y^2.$$

Thus,

$$E\left[s_y^2\right] = \sigma_y^2.$$

3.6.6 Derivation of the inclusion probability of an element for SRSwR

Because all possible sequences are equally likely, π_k could be computed as the ratio of the number of samples in which $\mathcal{U}_k$ appears at least once to the number of possible samples, Ω. Such a computation would be an arduous chore, however. Instead we take the following approach. If π_k is the probability that $\mathcal{U}_k \in s$, then $1 - \pi_k$ is the probability that $\mathcal{U}_k \notin s$. The latter probability is $\prod_{i=1}^{n}$ (probability that $\mathcal{U}_k$ is not chosen on the ith draw), since the n draws are independent. Now if p_k is the probability that $\mathcal{U}_k$ is selected on each and every draw, then its complement, $1 - p_k$ is the probability of not being selected on each draw. Thus, the probability of $\mathcal{U}_k \notin s$ is $\prod_{i=1}^{n}(1 - p_k) = (1 - p_k)^n$, which leads to

$$\pi_k = 1 - (1 - p_k)^n.$$

3.6.7 Derivation of $E[n]$ and $V[n]$ following Bernoulli sampling

Consider first the general case where elements from a discrete population are sampled by a design with inclusion probability for $\mathcal{U}_k$ of π_k. Regardless whether the size of the sample is fixed or random, the number, n, of elements included in the sample can be expressed as

$$n = \sum_{k=1}^{N} I_k, \qquad (3.73)$$

where I_k is defined in (3.72). The result $E[I_k] = \pi_k$ leads directly to

$$E[n] = \sum_{k=1}^{N} \pi_k. \qquad (3.74)$$

The identity in (3.74) holds for any probability sampling design applicable to discrete populations.

When $\pi_k = \pi$ for each element of the population, the expected number of elements selected into a Bernoulli sample is $N\pi$, a result deducible directly from (3.74). Furthermore, $E[I_k I_{k'}] = E[I_k]E[I_{k'}] = \pi^2$, owing to the independence of sample inclusions under Bernoulli sampling. Using

$$n^2 = \sum_{k=1}^{N} I_k + \sum_{k=1}^{N} \sum_{\substack{k' \neq k \\ k'=1}}^{N} I_k I_{k'}, \qquad (3.75)$$

we obtain

$$V[n] = E[n^2] - \left(E[n]\right)^2 \tag{3.76a}$$

$$= \left(N\pi + N(N-1)\pi^2\right) - (N\pi)^2 \tag{3.76b}$$

$$= N\pi(1-\pi). \tag{3.76c}$$

3.6.8 Derivation of the expected value and variance of $\hat{\tau}_{yp}$

Define the indicator of the selection of unit $\mathcal{U}_k$ on the jth draw as

$$I_{kj} = \begin{cases} 1, & \text{if } \mathcal{U}_k \text{ selected on } j\text{th draw}; \\ 0, & \text{otherwise.} \end{cases}$$

Then

$$E\left[\hat{\tau}_{yp}\right] = E\left[\sum_{\mathcal{U}_k \in s} \frac{y_k}{p_k}\right]$$

$$= E\left[\frac{1}{n} \sum_{j=1}^{n} \sum_{k=1}^{N} \frac{y_k I_{kj}}{p_k}\right]$$

$$= \frac{1}{n} \sum_{j=1}^{n} \sum_{k=1}^{N} \frac{y_k E\left[I_{kj}\right]}{p_k}$$

$$= \frac{1}{n} \sum_{j=1}^{n} \sum_{k=1}^{N} \frac{y_k p_k}{p_k}$$

$$= \frac{1}{n} \sum_{j=1}^{n} \sum_{k=1}^{N} y_k$$

$$= \frac{1}{n} \sum_{j=1}^{n} \tau_y$$

$$= \left(\frac{n}{n}\right) \tau_y.$$

Thus,

$$E\left[\hat{\tau}_{yp}\right] = \tau_y.$$

Furthermore,

$$
n^2 \hat{\tau}_{yp}^2 = \left(\sum_{j=1}^{n} \sum_{k=1}^{N} \frac{y_k I_{kj}}{p_k} \right)^2
$$

$$
= \left(\sum_{k=1}^{N} \sum_{j=1}^{n} \frac{y_k I_{kj}}{p_k} \right)^2
$$

$$
= \left[\sum_{k=1}^{N} \left(\frac{y_k}{p_k} \right) (I_{k1} + I_{k2} + \cdots + I_{kn}) \right]^2
$$

$$
= \sum_{k=1}^{N} \left(\frac{y_k}{p_k} \right)^2 (I_{k1} + I_{k2} + \cdots + I_{kn})^2
$$

$$
+ \sum_{k=1}^{N} \sum_{\substack{k' \neq k \\ k'=1}}^{N} \left(\frac{y_k y_{k'}}{p_k p_{k'}} \right) (I_{k1} + \cdots + I_{kn}) (I_{k'1} + \cdots + I_{k'n})
$$

$$
= \sum_{k=1}^{N} \left(\frac{y_k}{p_k} \right)^2 \left(\sum_{j=1}^{n} I_{kj} + \sum_{j=1}^{n} \sum_{\substack{j' \neq j \\ j'=1}}^{n} I_{kj} I_{kj'} \right)
$$

$$
+ \sum_{k=1}^{N} \sum_{\substack{k' \neq k \\ k'=1}}^{N} \left(\frac{y_k y_{k'}}{p_k p_{k'}} \right) \left(\sum_{j=1}^{n} I_{kj} I_{k'j} + \sum_{j=1}^{n} \sum_{\substack{j' \neq j \\ j'=1}}^{n} I_{kj} I_{k'j'} \right).
$$

Therefore,

$$
n^2 E \left[\hat{\tau}_{yp}^2 \right] = \sum_{k=1}^{N} \left(\frac{y_k}{p_k} \right)^2 \left(n p_k + n(n-1) p_k^2 \right)
$$

$$
+ \sum_{k=1}^{N} \sum_{\substack{k' \neq k \\ k'=1}}^{N} \left(\frac{y_k y_{k'}}{p_k p_{k'}} \right) (0 + n(n-1) p_k p_{k'})
$$

$$
= n \sum_{k=1}^{N} \frac{y_k^2}{p_k} + n(n-1) \left(\sum_{k=1}^{N} y_k^2 + \sum_{k=1}^{N} \sum_{\substack{k' \neq k \\ k'=1}}^{N} y_k y_{k'} \right)
$$

$$= n \sum_{k=1}^{N} \frac{y_k^2}{p_k} + n(n-1)\tau_y^2.$$

Hence,

$$E\left[\hat{\tau}_{yp}^2\right] = \frac{1}{n}\left[\sum_{k=1}^{N} \frac{y_k^2}{p_k} + (n-1)\tau_y^2\right].$$

Putting these results together leads to (3.47), i.e.,

$$V\left[\hat{\tau}_{yp}\right] = E\left[\hat{\tau}_{yp}^2\right] - \left(E\left[\hat{\tau}_{yp}\right]\right)^2$$

$$= \frac{1}{n}\left[\sum_{k=1}^{N} \frac{y_k^2}{p_k} + (n-1)\tau_y^2\right] - \frac{n}{n}\tau_y^2$$

$$= \frac{1}{n}\left[\sum_{k=1}^{N} \frac{y_k^2}{p_k} + (n-1-n)\tau_y^2\right]$$

$$= \frac{1}{n}\left(\sum_{k=1}^{N} \frac{y_k^2}{p_k} - \tau_y^2\right)$$

$$= \frac{1}{n}\sum_{k=1}^{N} p_k\left(\frac{y_k^2}{p_k} - \tau_y\right)^2.$$

For the case when $x_k = 1$ for all $\mathcal{U}_k$, then $p_k = 1/N$. Substituting this result into the above expression, and recognizing that $(N-1)\sigma_y^2 = \sum_{k=1}^{N} y_k^2 - N\mu_y^2$, one gets (3.19).

An alternative derivation of $V[\hat{\tau}_{yp}]$ relies on the fact that the variance of a sum of independent random variables is the sum of their variances. For the jth selection of either a SRSwR or list sampling,

$$V\left[\sum_{k=1}^{N} \frac{y_k I_{kj}}{p_k}\right] = E\left[\left(\sum_{k=1}^{N} \frac{y_k I_{kj}}{p_k}\right)^2\right] - \tau_y^2$$

$$= E\left[\sum_{k=1}^{N} \frac{y_k^2 I_{kj}^2}{p_k^2} + \sum_{k=1}^{N}\sum_{\substack{k'\neq k \\ k'=1}}^{N} \frac{y_k y_{k'} I_{kj} I_{k'j}}{p_k p_{k'}}\right] - \tau_y^2$$

$$= \sum_{k=1}^{N} \frac{y_k^2 E\left[I_{kj}^2\right]}{p_k^2} + \sum_{k=1}^{N} \sum_{\substack{k' \neq k \\ k'=1}}^{N} \frac{y_k y_{k'} E\left[I_{kj} I_{k'j}\right]}{p_k p_{k'}} - \tau_y^2$$

$$= \sum_{k=1}^{N} \frac{y_k^2 p_k}{p_k^2} - \tau_y^2 \qquad \left(\text{since } E\left[I_{kj} I_{k'j}\right] = 0\right)$$

$$= \sum_{k=1}^{N} \frac{y_k^2}{p_k} - \tau_y^2.$$

Therefore,

$$V\left[\sum_{j=1}^{n} \sum_{k=1}^{N} \frac{y_k I_{kj}}{p_k}\right] = n\left(\sum_{k=1}^{N} \frac{y_k^2}{p_k} - \tau_y^2\right),$$

which leads to

$$V\left[\hat{\tau}_{yp}\right] = V\left[\frac{1}{n} \sum_{j=1}^{n} \sum_{k=1}^{N} \frac{y_k I_{kj}}{p_k}\right]$$

$$= \frac{1}{n^2} V\left[\sum_{j=1}^{n} \sum_{k=1}^{N} \frac{y_k I_{kj}}{p_k}\right]$$

$$= \frac{1}{n^2}\left[n\left(\sum_{k=1}^{N} \frac{y_k^2}{p_k} - \tau_y^2\right)\right]$$

$$= \frac{1}{n}\left(\sum_{k=1}^{N} \frac{y_k^2}{p_k} - \tau_y^2\right).$$

3.6.9 Proof of the unbiasedness of $\hat{v}[\hat{\tau}_{yp}]$ as an estimator of $V[\hat{\tau}_{yp}]$

$$n(n-1)E\left\{\hat{v}\left[\hat{\tau}_{yp}\right]\right\} = E\left[\sum_{u_k \in s}\left(\frac{y_k}{p_k} - \hat{\tau}_{yp}\right)^2\right]$$

$$= E\left[\sum_{u_k \in s}\frac{y_k^2}{p_k^2} - 2\hat{\tau}_{yp}\sum_{u_k \in s}\frac{y_k}{p_k} + n\hat{\tau}_{yp}^2\right]$$

$$= E\left[\sum_{u_k \in s}\frac{y_k^2}{p_k^2}\right] - nE\left[\hat{\tau}_{yp}^2\right]$$

Continuing,

$$n(n-1)E\left\{\hat{v}\left[\hat{\tau}_{yp}\right]\right\} = \sum_{j=1}^{n}\sum_{u_k \in s}\frac{y_k^2 E\left[I_{kj}\right]}{p_k^2} - nE\left[\hat{\tau}_{yp}^2\right]$$

$$= n\sum_{k=1}^{N}\frac{y_k^2}{p_k} - n\left\{V\left[\hat{\tau}_{yp}\right] + \tau_y^2\right\}$$

$$= n\left[\sum_{k=1}^{N}\frac{y_k^2}{p_k} - \tau_y^2\right] - nV\left[\hat{\tau}_{yp}\right]$$

$$= n\left\{nV\left[\hat{\tau}_{yp}\right]\right\} - nV\left[\hat{\tau}_{yp}\right]$$

$$= n(n-1)V\left[\hat{\tau}_{yp}\right].$$

Hence,

$$E\left\{\hat{v}\left[\hat{\tau}_{yp}\right]\right\} = V\left[\hat{\tau}_{yp}\right].$$

3.6.10 Variance of estimated proportions

When all y_k, or x_k, can have a value of 1 or 0, depending on the presence or absence, respectively, of some attribute, then $\mu_y = \frac{1}{N}\sum_{k=1}^{N}y_k = P_1$, the proportional number of elements in the population possessing the attribute. Also, $\sum_{k=1}^{N}y_k = \sum_{k=1}^{N}y_k^2 = \tau_y = NP_1$. Recalling the definition of the population variance from

Chapter 1,

$$\sigma_y^2 = \frac{\sum_{k=1}^{N} (y_k - \mu_y)^2}{N - 1}$$

$$= \frac{\sum_{k=1}^{N} y_k^2 - N\mu_y^2}{N - 1}.$$

Algebraic substitution of NP_1 for the $\sum_{k=1}^{N} y_k^2$ term and P_1 for μ_y yields

$$\sigma_y^2 = \frac{N}{N - 1} P_1 (1 - P_1).$$

Following a SRSwoR of n elements, the sample variance of the y_ks in the sample reduces from

$$s_y^2 = \frac{1}{n - 1} \sum_{u_k \in s} (y_k - \bar{y})^2$$

to

$$s_y^2 = \frac{n}{n - 1} \hat{p}_1 (1 - \hat{p}_1).$$

Thus,

$$V[\bar{y}] = V[\hat{p}_1]$$

$$= \frac{(N - n)}{(N - 1)} \frac{P_1 (1 - P_1)}{n},$$

which is estimated unbiasedly by

$$\hat{v}[\hat{p}_1] = \frac{(N - n)}{N} \frac{\hat{p}_1 (1 - \hat{p}_1)}{(n - 1)}.$$

Following a SRSwR of n elements, the corresponding variance and unbiased estimator are

$$V[\bar{y}] = V[\hat{p}_1]$$

$$= \frac{N}{(N - 1)} \frac{P_1 (1 - P_1)}{n},$$

and

$$\hat{v}[\hat{p}_1] = \frac{\hat{p}_1 (1 - \hat{p}_1)}{n - 1}.$$

3.6.11 Nearly exact confidence intervals for proportions

When N is large, nearly exact $(1 - \alpha)100\%$ confidence intervals can be estimated with the following with endpoints, as described by:

Lower endpoint of interval: $\dfrac{1}{1 + \phi_L}$, where

$$\phi_L = \frac{n - n\hat{p}_1 + 1}{n\hat{p}_1 \, \text{FINV}\,[1 - \alpha/2, \, 2n\hat{p}_1, \, 2(n - n\hat{p}_1 + 1)]}.$$

Upper endpoint of interval: $\dfrac{1}{1 + \phi_U}$, where

$$\phi_U = \frac{n - n\hat{p}_1}{(n\hat{p}_1 + 1) \, \text{FINV} \, [\alpha/2, \, 2\,(n\hat{p}_1 + 1), \, 2\,(n - n\hat{p}_1)]}.$$

FINV is the inverse of the distribution function of an F random variable. Three arguments shown for FINV are:

1. The probability level: for lower limit, use $1 - \alpha/2$; for upper limit, use $\alpha/2$;

2. Numerator degrees of freedom: for lower limit, use $2n\hat{p}_1$; for upper limit, use $2(n\hat{p}_1 + 1)$;

3. Denominator degrees of freedom: for lower limit, use $2(n - n\hat{p}_1 + 1)$; for upper limit, use $2(n - n\hat{p}_1)$.

Refer to Leemis & Trivedi (1996) for further details.

3.6.12 Expected sample size

As introduced earlier, let the random variable I_k indicate whether or not $\mathcal{U}_k$ is included in a sample:

$$I_k = \begin{cases} 1, & \text{if } \mathcal{U}_k \in s; \\ 0, & \text{otherwise.} \end{cases}$$

Evidently, the actual size of the sample that is selected must be

$$n = \sum_{k=1}^{N} I_k.$$

When n is not fixed a priori by the sampling design, as in systematic sampling, Bernoulli sampling, and Poisson sampling, its expected value is

$$E[n] = E\left[\sum_{k=1}^{N} I_k\right]$$

$$= \sum_{k=1}^{N} E[I_k]$$

$$= \sum_{k=1}^{N} \pi_k. \tag{3.77}$$

This identity relating the expected sample size to the sum of the inclusion probabilities holds for any probability sampling design applicable to discrete populations. For those designs in which n is fixed, it is trivially true that $n = E[n]$ for all $s \in \Omega$, and the identity of (3.77) holds in this case, also.

CHAPTER 4

Sampling Designs for Continuous Populations

4.1 Introduction

We now consider the problem of estimating attributes of continuums, objects, or entities that do not naturally divide into smaller discrete units. These entities include, for example, organs of plants and animals, plants and animals themselves, landscapes, lakes, the atmosphere, and spans of time. Our approach is to treat each entity as a continuous population of points and to define the total amount of attribute possessed by the entity as an integral of a continuous attribute density function.

To whet our appetites for this subject matter, we consider a cucumber with length L with the aim of estimating the cucumber's volume, τ_ρ. We recognize a cucumber as a solid object, yet we may define its volume in terms of a continuous population of points along the cucumber's axis of length, the ordinate of each point serving to locate a cross section with measurable area. Let $\rho(x)$ denote the cross-sectional area of the cucumber at any point x, where $0 \leq x \leq L$. Hence $\rho(x)$ is a continuous function whose integral is equivalent to the volume of the cucumber, i.e.,

$$\tau_\rho = \int_0^L \rho(x)\, dx.$$

The continuous function, $\rho(x)$, describes how the cucumber's volume is distributed along its length. We have defined $\rho(x)$ as the cross-sectional area at x, but we may also interpret this quantity as the *attribute density* at x. In this example, the attribute density is a 'volume density,' which is measured in units of volume per unit length (since $\rho(x) = d\tau_\rho(x)/dx$). These units, of course, reduce to units of area because the volume density is equivalent to cross-sectional area. The average volume density of the cucumber is $\mu_\rho = \tau_\rho/L$, which is equivalent to the average cross-sectional area of the cucumber.

As a second example, we consider a frozen lake, $\mathcal{A}$, with horizontal surface area A, which is covered, for the most part, by snow. We allow for some windswept patches on the frozen surface where the snow depth may be zero. Of interest is the total volume of snow, τ_ρ, which we treat as an attribute of the frozen surface. We identify a location point on the frozen surface by its cartesian coordinates, (x, z). If we let $\rho(x, z)$ denote the depth of snow at any point $(x, z) \in \mathcal{A}$, then the volume of snow on the frozen lake can be expressed as an integral, i.e.,

$$\tau_\rho = \iint_{\mathcal{A}} \rho(x, z)\, dz\, dx.$$

The function $\rho(x, z)$ is continuous because it is defined for all points on the surface of the frozen lake (otherwise known as the domain of integration). However, $\rho(x, z) = 0$ at location points in any windswept patches that are clear of snow. The attribute of interest, snow volume, is distributed over the entire surface of the frozen lake with average volume density $\mu_\rho = \tau_\rho / A$. This average attribute density is measured in units of snow volume per unit lake surface area and is equivalent to the average snow depth. Similarly, the volume density at any location point (x, z) is equivalent to the snow depth at that point.

Several strategies, which have proven useful for sampling continuous populations, originally were advanced in the 1950's as techniques of Monte Carlo integration (see e.g., Hammersley & Handscomb 1979; Rubinstein 1981). Sampling strategies for continuous populations have also appeared in the statistics literature (e.g., Bartlett 1986; Cordy 1993; Stevens 1997), though readers without training in mathematical statistics may find these papers somewhat beyond their ken. In this chapter, we introduce three strategies from the Monte Carlo literature: (*i*) crude Monte Carlo, which is also called the sample mean method, (*ii*) importance sampling, and (*iii*) control variate estimation. In the taxonomy of sampling, crude Monte Carlo is a continuous analog of simple random sampling with replacement (§3.2.1) and importance sampling is a continuous analog of sampling with unequal selection probabilities or, more specifically, sampling with probability proportional to size (§3.3.1). Control variate estimation leads naturally to difference estimation.

Because continuous populations comprise infinitely many points, selection probabilities cannot be defined and assigned by the sampling design. In their place is the notion of a probability density, which is inherent to the mechanisms that we use to select 'sample points.' Moreover, since the target parameter is an integral, we utilize estimators that incorporate measurements of attribute densities, rather than measurements of attributes. These differences notwithstanding, strategies for sampling continuous populations are no more complicated than their discrete counterparts, and often are quite intuitive. For example, crude Monte Carlo yields the result that the volume of an internode of a plant is unbiasedly estimated by the product of internodal length and cross-sectional area measured at a point selected at random.

4.2 Crude Monte Carlo

Let τ_ρ denote the definite integral of a continuous function $\rho(x)$, i.e.,

$$\tau_\rho \equiv \tau_\rho(b) - \tau_\rho(a) = \int_a^b \rho(x)\,dx. \tag{4.1}$$

In a graph of $\rho(x)$ versus x (Figure 4.1), τ_ρ is the area under the curve $\rho(x)$ between $x = a$ and $x = b$. Indeed, τ_ρ is measured in units of area (e.g., m^2) if both x and $\rho(x)$ are measured in units of length (m). However, if $\rho(x)$ is measured in units of area, then τ_ρ is measured in units of volume.

Areas and volumes of many entities can be specified as definite integrals. The strategies developed in this chapter will allow us to estimate such areas from measurements of lengths, and such volumes from measurements of areas or lengths at

a)

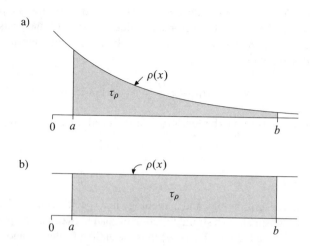

b)

Figure 4.1 a) *The attribute or target parameter, τ_ρ, is represented by the shaded area under the attribute density function, $\rho(x)$; b) a uniform attribute density is described by a 'flat function.'*

sample locations selected at random. We also touch on other problems, for example, the estimation of mass and the estimation of time integrals.

Crude Monte Carlo is one of the simplest strategies for estimating a target parameter that can be represented by an integral. We could easily present a recipe for crude Monte Carlo in a single paragraph, the essential ingredients being equations (4.4), (4.8), and (4.10). However, we develop the strategy gradually, introducing concepts that underlie not only crude Monte Carlo, but also all of the other strategies discussed in the chapter.

4.2.1 Definitions

For our purposes, x, in equation (4.1), is a point on an axis of length. When measured in the units of x, the entity extends from $x = a$ to $x = b$ or, to put it another way, the entity is wholly contained in the interval $[a, b]$. For example, if x is length in m, then the entity has length $(b - a)$ m; if x is time in seconds, then the entity is a span of time of length $(b - a)$ seconds. In either case, the interval $[a, b]$ comprises a continuous population of points.

The symbol τ_ρ denotes an attribute of the entity and $\rho(x) = d\tau_\rho(x)/dx$ is the attribute density function that describes how the attribute is distributed in the interval $[a, b]$. In a sampling context, τ_ρ is the target parameter, so $\rho(x)$ is necessarily measurable everywhere in $[a, b]$, i.e., at any point x where $a \leq x \leq b$; otherwise, we can not unbiasedly estimate τ_ρ. The dimensions of the attribute density are the same as the dimensions of τ_ρ/x. For example, if τ_ρ is a volume and x a length, then $\rho(x)$ has dimensions of volume per unit length, which, of course, reduces to area. Attribute densities usually are positive or zero (i.e., $\rho(x) \geq 0$) in $[a, b]$, though negative densities do sometimes occur as we shall see in Example 4.9.

Figure 4.1a depicts an attribute density function that is always positive, but decidedly variable in $[a, b]$. By the mean-value theorem of integrals, the mean attribute density in $[a, b]$ is $\mu_\rho = \tau_\rho/(b - a)$. If the attribute density is constant, then $\rho(x)$ is described by

$$\rho(x) = \mu_\rho = \frac{\tau_\rho}{b - a} \qquad a \le x \le b,$$

and the attribute, τ_ρ, is said to be 'uniformly distributed' in the interval $[a, b]$ (Figure 4.1b).

Example 4.1

To interpret the terminology and symbology in a specific context, let us consider the volume of asphalt on a straight section of road between given points a and b. The section of road of length $(b - a)$ m is the entity of interest, and its centerline comprises a continuous population of points, $a \le x \le b$. The volume of asphalt, τ_ρ (m^3), is an attribute of the road and the target parameter. The attribute density, $\rho(x) = d\tau_\rho(x)/dx$, is the volume of asphalt per unit length of road (m^3 m^{-1}). These dimensions reduce to a dimension of area (m^2) because, at a specific point on the road, e.g., $x = x_s$ where $a \le x_s \le b$, the attribute density is equivalent to the area of the vertical cross-section of asphalt measured perpendicular to the centerline. If the road is paved in its entirety, we should expect the cross-sectional area of asphalt to be fairly uniform along the length of the road section, so a graph of $\rho(x)$ versus x would resemble the graph in Figure 4.1b. However, only a portion of the section of road may be paved with asphalt. If asphalt occurs at point x_s, then $\rho(x_s) > 0$; otherwise, $\rho(x_s) = 0$.

Example 4.2

Consider the number of growing degree-days over some given span of time. The span of time (days) is the entity of interest, extending from time a to time b, and the attribute, τ_ρ, is the number of growing degree-days. Hence, $\rho(x)$ is the density of growing degree-days per day at time x ($a \le x \le b$). The attribute density reduces to air temperature above some threshold, i.e., $\rho(x) = \max[0, T(x) - T_0]$, where $T(x)$ is air temperature at time x and T_0 is the threshold temperature. Thus, this attribute density equals zero when $T(x) < T_0$.

If the attribute density function is known, then we can use analytical or numerical procedures to integrate $\rho(x)$ and calculate τ_ρ, so we have no need of sampling designs. However, if the mathematical form of $\rho(x)$ is not known, we can use Monte Carlo integration, specifically crude Monte Carlo, to estimate τ_ρ from measurements of the attribute density $\rho(x)$ at sample points selected at random between a and b. This assumes, of course, that $\rho(x)$ can be measured to an acceptable degree of accuracy.

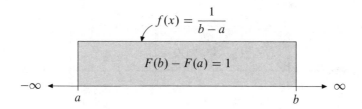

Figure 4.2 *In crude Monte Carlo, the probability density function, $f(x)$, is constant, or uniform, in the interval of interest, $[a, b]$, and zero elsewhere.*

4.2.2 Selection

In order to select a particular point, x, at which to measure $\rho(x)$, we define a probability density function, $f(x)$, over the interval of integration. Analogous to the attribute density function, the probability density function describes how 'a unit of probability' is distributed over the interval $[a, b]$. For unbiased estimation, $f(x)$ must adhere to certain constraints, i.e.,

$$f(x) > 0, \qquad a \le x \le b;$$
$$f(x) = 0, \qquad \text{otherwise.}$$

The target parameter is the integral of $\rho(x)$ from a to b, so the integral of $f(x)$ from a to b must equal 1, i.e.,

$$F(b) - F(a) = \int_a^b f(x)\,dx = 1.$$

The probability density is the amount of probability per unit length, so the integral of the probability density function provides the probability that a point, x, is contained in a particular finite interval. For example, $F(b) - F(a) = 1$ is the probability that a point x, selected at random, occurs in the interval $[a, b]$. Let $a \le x_s \le b$, then $u_s = F(x_s) - F(a)$ is the probability that x occurs in the interval $[a, x_s]$, i.e.,

$$F(x_s) - F(a) = \int_a^{x_s} f(x)\,dx = u_s. \tag{4.2}$$

Equation (4.2) is key to the *inverse-transform method* (e.g., Rubinstein 1981), a mechanism for selecting sample points. In effect, we select $u_s \sim U[0, 1]$ and then solve $F(x_s) - F(a) = u_s$ for x_s. The result is the selection of a sample point $x = x_s$ in the interval $[a, b]$ with probability density $f(x_s)$. In a sampling context, the attribute density at x_s, namely $\rho(x_s)$, then would be measured. For a sample of size n, this process is repeated until n distinct sample points, x_s, $s = 1, \ldots, n$ have been selected. That is, for $u_s \sim U[0, 1]$, $s = 1, \ldots, n$, solve $F(x_s) - F(a) = u_s$ for x_s.

By design, crude Monte Carlo sampling uses a uniform density function over the

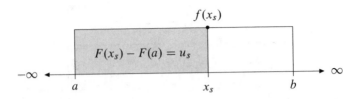

$$f(x_s)$$

Figure 4.3 *The area under the probability density function, $f(x)$, between a and x_s is the probability, u_s, that a value of x, selected at random, falls in the interval $[a, x_s]$. The probability density $f(x)$ is uniform from a to b, so $u_s = (x_s - a)/(b - a)$ and, therefore, $x_s = a + (b - a) u_s$.*

interval of integration (Figure 4.2), i.e.,

$$f(x) = \frac{F(b) - F(a)}{b - a} = \frac{1}{b - a} \qquad a \leq x \leq b. \qquad (4.3)$$

Substituting $f(x) = 1/(b - a)$ into equation (4.2),

$$F(x_s) - F(a) = u_s = \int_a^{x_s} \frac{1}{b - a} \, dx.$$

Integrating,

$$u_s = \frac{x_s - a}{b - a}.$$

Solving for x_s yields the crude Monte Carlo selection formula (see Figure 4.3):

$$x_s = a + (b - a) u_s. \qquad (4.4)$$

This selection formula is a continuous analog of selection method II for simple random sampling with replacement (see page 48).

Example 4.3

Suppose that a uniform variate, $u_s = 0.63602$, is generated in order to select a sample point between $a = 1$ and $b = 5$. Solving (4.4) provides $x_s = 3.54408$.

4.2.3 Estimation

An unbiased estimator of τ_ρ, based on the jth selection, is the quotient of the attribute density and the probability density at $x = x_s$, i.e.,

$$\hat{\tau}_{\rho s} = \frac{\rho(x_s)}{f(x_s)}. \qquad (4.5)$$

See the Appendix for a proof of unbiasedness. For the uniform density used in crude Monte Carlo, $\hat{\tau}_{\rho s}$ simplifies to

$$\hat{\tau}_{\rho s} = (b - a) \rho(x_s), \qquad (4.6)$$

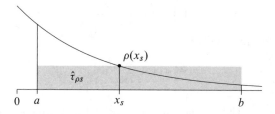

Figure 4.4 *An estimate of τ_ρ provided by the estimator $\hat{\tau}_{\rho s}$, equation (4.6), is represented by the rectangular shaded area.*

which is the area of a rectangle of length $(b - a)$ and height $\rho(x_s)$ (see Figure 4.4).

A combined estimator, $\hat{\tau}_\rho$, uses measurements from $n > 1$ independent selections, i.e.,

$$\hat{\tau}_\rho = \frac{1}{n} \sum_{s=1}^{n} \hat{\tau}_{\rho s}. \tag{4.7}$$

Substituting (4.6) into (4.7) provides a combined estimator specifically for crude Monte Carlo, viz.,

$$\hat{\tau}_\rho = \frac{b - a}{n} \sum_{s=1}^{n} \rho(x_s). \tag{4.8}$$

The variance of $\hat{\tau}_\rho$ is (see Chapter 4 Appendix for derivation)

$$V[\hat{\tau}_\rho] = \frac{1}{n} \left(\int_a^b \frac{\rho^2(x)}{f(x)} \, dx - \tau_\rho^2 \right). \tag{4.9}$$

The variance of $\hat{\tau}_\rho$ is estimated unbiasedly by

$$\hat{v}[\hat{\tau}_\rho] = \frac{\sum_{s=1}^{n} \left(\hat{\tau}_{\rho s} - \hat{\tau}_\rho \right)^2}{n(n-1)} \qquad n > 1. \tag{4.10}$$

The mean attribute density in the interval $[a, b]$ is μ_ρ, which is unbiasedly estimated by

$$\hat{\mu}_{\rho s} = \frac{\hat{\tau}_{\rho s}}{b - a}. \tag{4.11}$$

For crude Monte Carlo, this estimator is simply the measured attribute density at x_s,

$$\hat{\mu}_{\rho s} = \rho(x_s). \tag{4.12}$$

A combined estimator of the mean density uses $n > 1$ independent selections, i.e.,

$$\hat{\mu}_\rho = \frac{1}{n} \sum_{s=1}^{n} \hat{\mu}_{\rho s}. \tag{4.13}$$

The variance of $\hat{\mu}_\rho$ is

$$V\left[\hat{\mu}_\rho\right] = \frac{V[\hat{\tau}_\rho]}{(b-a)^2}, \tag{4.14}$$

which is estimated unbiasedly by

$$\hat{v}\left[\hat{\mu}_\rho\right] = \frac{\hat{v}[\hat{\tau}_\rho]}{(b-a)^2}. \tag{4.15}$$

The estimators, $\hat{\tau}_\rho$ and $\hat{\mu}_\rho$, have zero variance whenever the attribute density, like the probability density, is uniform in the interval $[a, b]$, i.e., if $\rho(x) = \tau_\rho/(b-a) = \mu_\rho$. In this case, $\hat{\tau}_{\rho s} = \tau_\rho$, irrespective of which sample point, $x = x_s$, is selected, because

$$\hat{\tau}_{\rho s} = \frac{\rho(x_s)}{f(x_s)} = \frac{\tau_\rho/(b-a)}{1/(b-a)} = \mu_\rho(b-a) = \tau_\rho.$$

Inasmuch as τ_ρ is a constant, $\hat{\tau}_{\rho s} = \tau_\rho$ necessarily implies that $\hat{\tau}_{\rho s}$ is also constant, i.e., it will not vary from one sample to another. Of course, a uniform or constant attribute density is apt to be very rare in nature. Suffice it to say, the closer the attribute density to uniformity, the more precise the estimates from crude Monte Carlo.

Example 4.4

The curve in Figures 4.1a and 4.4 is $\rho(x) = e^{-x}$. The integral from $a = 0.25$ to $b = 2.75$ is easily calculated, i.e., $\tau_\rho = \int_a^b e^{-x}\,dx = -e^{-2.75} - (-e^{-0.25}) = 0.7149$.

To implement a crude Monte Carlo sample for the purpose of estimating τ_ρ, six random numbers $u_s \sim U[0, 1]$, $s = 1, 2, \ldots, 6$ were generated. With these, sample points were calculated with (4.4) as $x_s = 0.25 + (2.75 - 0.25)u_s$. The attribute density at the sample point x_s is $\rho(x_s) = e^{-x_s}$ and the estimate of τ_ρ, based on this selection is $\hat{\tau}_{\rho s} = (2.75 - 0.25)\rho(x_s)$. The results of these calculations are provided in the following table:

s	u_s	x_s	$\rho(x_s)$	$\hat{\tau}_{\rho s}$
1	0.06573	0.41433	0.66079	1.6520
2	0.83402	2.33505	0.09681	0.2420
3	0.39638	1.24095	0.28911	0.7228
4	0.09605	0.49012	0.61255	1.5314
5	0.16908	0.67270	0.51033	1.2758
6	0.62471	1.81178	0.16336	0.4084

By equation (4.8), the combined estimate of the target integral is

$$\hat{\tau}_\rho = \frac{1.6520 + 0.2420 + \cdots + 0.4084}{6} = 0.9721,$$

and, using (4.9), the variance of $\hat{\tau}_\rho$ with $n = 6$ is calculated to be $V[\hat{\tau}_\rho] = 0.04033$.

The estimated variance of $\hat{\tau}_\rho$, by (4.10), is 0.05934, and, therefore, the estimated standard error of $\hat{\tau}_\rho$ is 0.2436. The 95% confidence interval for τ_ρ is

$$0.3459 \le \tau_\rho \le 1.5983$$

which evidently includes $\tau_\rho = 0.7149$ in this instance.

Example 4.5

In this example we contrast the estimation of the volume of the bole of a tree by (a) simple random sampling with replacement (SRSwR) and (b) crude Monte Carlo. Elements of our notation are specific to this example.

Assume that the length of the tree bole, from butt to tip, equals the height of the tree (H). Assume, also, that the bole consists of a stack of N connected segments, each of length $\Delta h = H/N$. The kth segment, $\mathcal{U}_k$, $k = 1, 2, \ldots, N$, has volume y_k (Figure 4.5). In other words, the tree bole constitutes the population of interest, which has been divided into N discrete non-overlapping units, each of which is one segment of the tree bole. The target parameter is the total volume of the bole, $\tau_y = \sum_k y_k$.

From this discrete population consider selecting a sample segment by generating $u_s \sim U[0, 1]$, so that $x_s = u_s H$ is a random height between 0 and H. The segment that occurs at x_s is selected for the sample. By generating n distinct heights, $x_s, s = 1, \ldots, n$, a sample of n segments is selected for the sample, although these may not all be distinctly different segments.

This method of selection is equivalent to simple random sampling with replacement, in which the selection probability of each segment, $\mathcal{U}_k$, is $p_k = 1/N = \Delta h/H$. The Hansen-Hurwitz estimator of the volume of the bole is

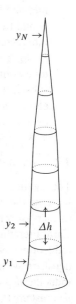

Figure 4.5
Bole segments.

$$\hat{\tau}_{yp} = \frac{1}{n} \sum_{\mathcal{U}_k \in s} \frac{y_k}{p_k}$$

$$= \frac{N}{n} \sum_{\mathcal{U}_k \in s} y_k. \tag{4.16}$$

Estimation of the volume of the bole evidently requires the measurement of the volumes of the discrete segments of the bole that are selected for the sample, a task which is difficult on standing trees. However, the length of each segment, Δh, obviously depends on the choice of N. If we assume that N approaches ∞, then Δh approaches 0 and, at the limit, the population is continuous because the segments are vanishingly thin wafers of wood. Thus, as an alternative to dividing the bole into a sequence of N discrete segments, we view the target parameter as

$$\tau_\rho = \int_0^H \rho(h)\, dh,$$

where the attribute density, $\rho(h) = d\tau_\rho(h)/dh$, is volume per unit length at height h, which is equivalent to cross-sectional area at height h. As a consequence, crude Monte Carlo sampling can be used to estimate τ_ρ.

Each of n sample points, h_s ($s = 1, 2, \ldots, n$), is drawn with uniform probability density, $f(h) = 1/H$, so that $h_s = u_s H$, as in (4.4). By (4.5) and (4.8),

$$\hat{\tau}_\rho = \frac{1}{n} \sum_{j=1}^n \frac{\rho(h_s)}{f(h_s)}$$

$$= \frac{H}{n} \sum_{j=1}^n \rho(h_s). \tag{4.17}$$

It is obvious that the continuous strategy is more practical for estimating bole volume than its discrete analog because the former eliminates the need to measure volumes of discrete bole segments. Instead, cross-sectional areas are measured at heights selected at random. In practice, it is likely that bole diameter would be measured and then used to calculate cross-sectional area for insertion into the estimator.

Note: In this example, both τ_y and τ_ρ denote the same target parameter, the total volume of the bole, i.e., $\tau_y \equiv \tau_\rho$. In keeping with established notation, the subscript y indicates that the total obtains from a summation across N discrete elements and the subscript ρ indicates that the total obtains from an integration over a continuous domain.

4.2.4 Antithetic variates

The method of *antithetic variates* (Hammersley & Morton 1956) can be used to increase the efficiency of an estimation whenever the attribute density, $\rho(x)$, tends to increase or decrease in a monotone fashion in the interval $[a, b]$. In the Monte Carlo literature, the method of antithetic variates falls under the rubric of 'variance-reduction methods.' Heuristically, the method works by averaging large attribute densities with small ones. More technically, the method reduces the variance by inducing negative covariance among pairs of measured attribute densities (see, e.g., Rubinstein 1981).

'Antithetic selection' involves the selection of two sample points from $[a, b]$ with a single random number, $u_s \sim U[0, 1]$, i.e.,

$$x_s = a + (b - a)u_s \tag{4.18}$$

and

$$x_s' = a + (b - a)(1 - u_s). \tag{4.19}$$

The estimator of τ_ρ, based on the sth antithetic selection of the sample points x_s and x'_s, is

$$\hat{\tau}_{\rho s} = (b - a) \left[\frac{\rho(x_s) + \rho(x'_s)}{2} \right]. \tag{4.20}$$

For n pairs of antithetic selections,

$$\hat{\tau}_\rho = \frac{1}{n} \sum_{j=1}^{n} \hat{\tau}_{\rho s}$$

$$= \frac{b - a}{2n} \sum_{j=1}^{n} \left[\rho(x_s) + \rho(x'_s) \right]. \tag{4.21}$$

The variance of $\hat{\tau}_\rho$ is estimated by equation (4.10).

A single antithetic selection with (4.18) and (4.19) would estimate τ_ρ without error if $\rho(x)$ were linearly related to x. In this case, a graph of $\rho(x)$ versus x would be a trapezoid. The target integral (by the trapezoidal rule) is

$$\tau_\rho = (b - a) \left[\frac{\rho(a) + \rho(b)}{2} \right] = (b - a) \rho(\bar{x}),$$

where $\bar{x} = (a + b)/2$. Because $x_s + x'_s = a + b$, the trapezoidal rule also ensures that $\hat{\tau}_{\rho s} = (b - a)\rho(\bar{x}_s)$, where $\bar{x}_s = (x_s + x'_s)/2$. Hence, any antithetic selection would yield the same result with (4.20), i.e., $\hat{\tau}_{\rho s} = \tau_\rho$. Of course, straight lines are rare in nature, and attribute densities tend to change in an irregular or bumpy fashion. However, antithetic selection should be a reasonable approach if the major tendency in $\rho(x)$ is to increase or decrease in value in the interval $[a, b]$.

Example 4.6

In Example 4.4, we used crude Monte Carlo to estimate the integral $\tau_\rho = \int_a^b \rho(x)\,dx$, where $\rho(x) = e^{-x}$, $a = 0.25$, and $b = 2.75$. The attribute density was measured at six sample points, which were selected independently. In this example we select three pairs of sample points antithetically. The results from reuse of the first three random numbers of Example 4.4 are provided in the following table:

j	u_s	$1 - u_s$	x_s	x'_s	$\rho(x_s)$	$\rho(x'_s)$	$\hat{\tau}_{\rho s}$
1	0.06573	0.93427	0.41433	2.58567	0.66079	0.07361	0.9202
2	0.83402	0.16598	2.33505	0.66495	0.09681	0.51430	0.7639
3	0.39638	0.60362	1.24095	1.75905	0.28911	0.17221	0.5766

Three independent estimates of τ_ρ, which derive from the three independent antithetic selections, are listed in the last column of the table. The combined estimate,

$$\hat{\tau}_\rho = \frac{0.9202 + 0.7639 + 0.5766}{3} = 0.7536,$$

is quite close to the target parameter value, $\tau_\rho = 0.7149$. The estimated standard error of $\hat{\tau}_\rho$, based on the three antithetic selections, is 0.09930 and the 95% confidence interval is $0.32632 \leq \tau_\rho \leq 1.18081$. This result is more precise than what was obtained in Example 4.4 with crude Monte Carlo, but without antithetic selection.

4.2.5 Systematic selection

Systematic selection, in conjunction with crude Monte Carlo, involves taking measurements of $\rho(x)$ at a fixed interval from a random start in the interval $[a, b]$. For example, divide the interval into N equal sub-intervals of length Δx. Select $x_s = a + u_s \times \Delta x$ and take measurements $\rho(x_s)$, $\rho(x_s + \Delta x)$, $\rho(x_s + 2\Delta x)$, and so forth. Based on these measurements,

$$\hat{\tau}_{\rho s} = (b - a) \sum_{i=0}^{N-1} \frac{\rho(x_s + i \Delta x)}{N}. \qquad (4.22)$$

Although $\hat{\tau}_{\rho s}$ is calculated from two or more measurements, these measurements do not derive from independent selections. Consequently, the variance of $\hat{\tau}_{\rho s}$ cannot be estimated unbiasedly.

Two independent selections would involve two random starts, in which case we could use (4.7) to calculate a combined estimate, $\hat{\tau}_\rho$, and (4.10) to calculate $\hat{v}[\hat{\tau}_\rho]$.

Example 4.7

Let us return to the integration problem of Examples 4.4 and 4.6, viz., $\tau_\rho = \int_a^b \rho(x)\, dx$, where $\rho(x) = e^{-x}$, $a = 0.25$, and $b = 2.75$. This time we estimate the integral with systematic selection with two random starts. We divide the interval of integration, $[0.25, 2.75]$, into $N = 3$ segments, each of length $\Delta x = 2.5/3 = 0.83333$. Each of the two systematic samples requires three measurements of $\rho(x)$. This is six measurements in total, the same as in Examples 4.4 and 4.6. The results from reuse of the first two random numbers of Example 4.4 are provided in the following table:

s	i	u_s	$x_s + i\Delta x$	$\rho(x_s + i\Delta x)$
1	0	0.06573	0.30477	0.73729
1	1	0.06573	1.13811	0.32043
1	2	0.06573	1.97144	0.13926
2	0	0.83402	0.94502	0.38867
2	1	0.83402	1.77835	0.16892
2	2	0.83402	2.61168	0.07341

The entries in the last column of the table yield two independent estimates of τ_ρ

$$\hat{\tau}_{\rho 1} = 2.5 \times \frac{0.73729 + 0.32043 + 0.13926}{3} = 0.9975$$

$$\hat{\tau}_{\rho 2} = 2.5 \times \frac{0.38867 + 0.16892 + 0.07341}{3} = 0.5258$$

The combined estimate is $\hat{\tau}_\rho = (0.9975 + 0.5258)/2 = 0.7617$ and the estimated standard error of $\hat{\tau}_\rho$ is 0.2359. If we treat the six sample points as independent, the resultant estimate of τ_ρ is, of course, unchanged, but the (biased) estimate of the standard error increases slightly to 0.2474. Comparing the results in Examples 4.4 and 4.6 with this result, it appears that antithetic selection is more efficient than simple random selection or systematic selection for estimating the target parameter $\tau_\rho = \int e^{-x} dx$.

Example 4.8

The following table contains systematic (half-hourly) measurements of the flux of carbon, $\rho(t)$ (μmol C m^{-2} s^{-1}), from vegetation to the atmosphere over a 12-hour period on 20 September 2001 at an Ameriflux site near Howland, Maine, U.S.A.

Time	Flux	Time	Flux	Time	Flux	Time	Flux
1200	−10.5	1500	−9.1	1800	1.9	2100	5.0
1230	−14.5	1530	−4.6	1830	3.4	2130	6.0
1300	−11.6	1600	−4.1	1900	3.6	2200	2.4
1330	−7.4	1630	−2.3	1930	3.4	2230	4.2
1400	−9.4	1700	−0.5	2000	5.6	2300	4.7
1430	−7.8	1730	2.5	2030	4.6	2330	5.1

By meteorological convention, negative numbers denote influx of carbon to the vegetation and positive numbers denote efflux. Of interest is the amount of carbon sequestered by the vegetation over this period of 12 hours or 43200 seconds. Carbon sequestration (τ_ρ, μmol C m^{-2}), by definition, is the time integral of the flux, i.e.,

$$\tau_\rho = \int_{t_0}^{t_0 + 43200} \rho(t)\, dt. \tag{4.23}$$

This quantity is estimated by application of equation (4.22), i.e.,

$$\hat{\tau}_\rho = 43200 \times \frac{(-10.5) + (-14.5) + \cdots + 4.7 + 5.1}{24} = -52920.0$$

Thus, over the 12-hour period, we estimate that the vegetation sequestered 52920.0 μmol C (or 0.635 g C) per m^2 of land area.

For the sake of example, let us assume that measurements taken on the hour are systematic measurements from one random start and measurements taken on the half hour are systematic measurements from a second independent random start. The two 'independent estimates' of τ_ρ are

$$\hat{\tau}_{\rho 1} = 43200 \times \frac{(-10.5) + (-11.6) + \cdots + 2.4 + 4.7}{12} = -79198.6$$

and

$$\hat{\tau}_{\rho 2} = 43200 \times \frac{(-14.5) + (-7.4) + \cdots + 4.2 + 5.1}{12} = -26641.4$$

The combined estimate is

$$\hat{\tau}_\rho = \frac{(-79198.6) + (-26641.4)}{2} = -52920.0$$

Because we have just two 'independent estimates' of τ_ρ, we can calculate the standard error of $\hat{\tau}_\rho$ thus:

$$\sqrt{\hat{v}[\hat{\tau}_\rho]} = \frac{|\hat{\tau}_{\rho 1} - \hat{\tau}_{\rho 2}|}{2} = \frac{|(-79198.6) - (-26641.4)|}{2} = 26278.6$$

Example 4.9

Investigators often operate under the assumption that systematic selections are independent. Applying this assumption to the 24 measurements of the previous example, the 24 estimates of τ_ρ are

$$\hat{\tau}_{\rho 1} = 43200 \times (-10.5) = -453600.0$$
$$\hat{\tau}_{\rho 2} = 43200 \times (-14.5) = -626400.0$$
$$\vdots$$
$$\hat{\tau}_{\rho 24} = 43200 \times 5.1 = 220320.0$$

Calculation of $\hat{v}[\hat{\tau}_\rho]$ by equation (4.10) and then taking the square root yields $\sqrt{\hat{v}[\hat{\tau}_\rho]} = 57569.0$, which is more than twice as large as our previous result.

4.3 Importance sampling

The method of importance sampling seems to date from the 1950s. Rubinstein (1981) cited a symposium paper by Marshall (1956) as a source of the method. Hammersley & Handscomb (1979) described the method without attribution. The domain of application of importance sampling was extended to the estimation of attributes of physical objects—i.e., the branches and boles of botanical trees—by Valentine et al. (1984) and Gregoire et al. (1986). The method has many potential applications in connection with the estimation of natural and environmental resources. As was noted, importance sampling is a continuous analog of list sampling with probability proportional to size (§3.3.1). A comparison of the formulae that pertain to list sampling of discrete populations and importance sampling of continuums is provided in Table 4.1 in the Appendix (§4.9).

In our discussion of importance sampling, we shall use the same notation for the target integral as in crude Monte Carlo (Figure 4.6a), viz.,

$$\tau_\rho \equiv \tau_\rho(b) - \tau_\rho(a) = \int_a^b \rho(x)\,dx.$$

Importance sampling can be used to estimate τ_ρ if $\rho(x)$ exists everywhere in the interval $[a, b]$. First, we must formulate a probability density function, $f(x)$, as in §4.2.2. As was noted, importance sampling is a continuous analog of sampling with probability proportional to size. Ideally, the probability density function, $f(x)$, is proportional to the attribute density function, $\rho(x)$. If $f(x)$ is constant, then the importance sampling reduces to crude Monte Carlo.

4.3.1 Proxy function

In many applications, $f(x)$ can be developed from a model of $\rho(x)$. We call such models *proxy functions*. The shape of a proxy function, $g(x)$, should resemble the shape of $\rho(x)$ in the interval $[a, b]$. Or, to put it another way, the proxy function should provide a good approximation of $\rho(x)$ in the interval $[a, b]$ or it should be proportional to a model that provides a good approximation of $\rho(x)$. Division of the proxy function by its integral yields the needed probability density function:

$$f(x) = \begin{cases} \dfrac{g(x)}{G} & \text{if } a \leq x \leq b, \\ 0 & \text{otherwise,} \end{cases}$$

where

$$G \equiv G(b) - G(a) = \int_a^b g(x)\,dx.$$

Because G is a constant, the resultant probability density function has the same shape as the proxy function in the interval $[a, b]$.

Example 4.10

Suppose that we choose to approximate $\rho(x)$ in Figure 4.6a with a linear model, given measurements (or prior knowledge) of $\rho(a)$ and $\rho(b)$ (Figure 4.6b). Then,

$$g(x) = \alpha' + \beta' x,$$

where α' and β' are constants:

$$\beta' = \frac{\rho(b) - \rho(a)}{b - a}$$
$$\alpha' = g(a) - \beta' a$$
$$= \rho(a) - \beta' a.$$

We can calculate G by the trapezoidal rule, i.e.,

$$G = (b - a)\left[\frac{\rho(a) + \rho(b)}{2}\right].$$

a)

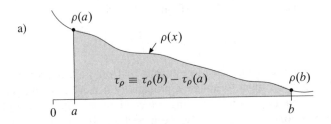

b)

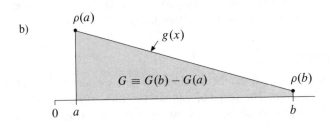

c)

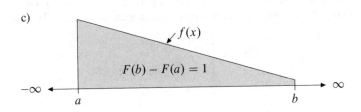

Figure 4.6 a) *The target integral,* τ_ρ*; b) A proxy function,* $g(x)$*, constructed from measurements (or prior knowledge) of* $\rho(a)$ *and* $\rho(b)$*; c) The probability density function,* $f(x) = g(x)/G$.

The probability density function is

$$f(x) = \alpha + \beta x$$

where

$$\alpha = \frac{\alpha'}{G} \qquad \beta = \frac{\beta'}{G}.$$

This function (see Figure 4.6c) has a shape identical to the proxy function, $g(x)$, and similar to the attribute density function, $\rho(x)$.

4.3.2 Selection by the inverse-transform method

Sample units at $x = x_s$ ($s = 1, 2, \ldots, n$) ordinarily are selected by the inverse-transform method or the acceptance-rejection method. We have already used the inverse-transform method in connection with crude Monte Carlo, i.e., the solution of equation (4.2) selects $x = x_s$ by the inverse transform method. This method can be employed with either a probability density function or a proxy function. In the latter case, equation (4.2) converts to

$$G(x_s) - G(a) = \int_a^{x_s} g(x)\,dx = u_s\,G. \qquad (4.24)$$

Thus, x_s is a 'root' of $G(x_s) - G(a) - u_s\,G = 0$ or, identically, a root of $F(x_s) - F(a) - u_s = 0$, where u_s is drawn from U[0, 1].

Example 4.11

From Example 4.10, the probability density function is $f(x) = \alpha + \beta x$, therefore,

$$\int_a^{x_s} (\alpha + \beta x)\,dx = u_s.$$

The integration yields a quadratic equation in x_s:

$$\alpha(x_s - a) + \frac{\beta}{2}\left(x_s^2 - a^2\right) = u_s.$$

Hence,

$$x_s = \frac{-\alpha \pm \sqrt{\alpha^2 + \beta[\beta a^2 + 2\alpha a + 2u_s]}}{\beta}.$$

4.3.3 Selection by the acceptance-rejection method

Von Neumann's acceptance-rejection method (cf., Rubinstein 1981) is a simple general method for selecting sample units at x_s ($s = 1, 2, \ldots, n$) when the probability density function varies in $[a, b]$. The method can be employed with a proxy function or a probability density function.

Let g_{max} denote the greatest value of $g(x)$ in the interval $a \leq x \leq b$, then

1. Draw independent random numbers u_1 and u_2 from U[0, 1].

2. Calculate $x_s = a + (b - a)u_1$.

3. If $u_2 \times g_{max} \leq g(x_s)$, then accept x_s; otherwise, reject x_s and repeat from step 1.

Alternatively, we can use the probability densities f_{max} and $f(x_s)$ instead of g_{max} and $g(x_s)$ in the procedure. The result is exactly the same.

One way to visualize the acceptance-rejection method is to graph $g(x)$ versus x for $a \leq x \leq b$ and draw the bounding box of this graph (Figure 4.7). The acceptance-rejection method provides coordinates $(x_s, u_2 \times g_{max})$, which fall somewhere within the bounding box. If they fall within the area under the proxy function, then x_s is accepted; otherwise, x_s is rejected. If x_s is accepted, then it is done so with probability density $f(x_s) = g(x_s)/G$.

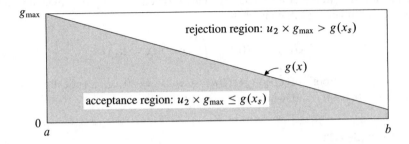

Figure 4.7 *Graphical depiction of the acceptance-rejection method, where* $x_s = a + (b - a)u_1$; $u_1, u_2 \sim U[0, 1]$.

4.3.4 Estimation

The estimator of τ_ρ, based on the sth selection is

$$\hat{\tau}_{\rho s} = \frac{\rho(x_s)}{f(x_s)}$$

$$= G\left[\frac{\rho(x_s)}{g(x_s)}\right]. \tag{4.25}$$

A combined estimate can be calculated from $n \geq 2$ independent selections, i.e.,

$$\hat{\tau}_\rho = \frac{1}{n}\sum_{s=1}^{n}\hat{\tau}_{\rho s}$$

$$= \frac{G}{n}\sum_{s+1}^{n}\frac{\rho(x_s)}{g(x_s)}. \tag{4.26}$$

The variance of $\hat{\tau}_\rho$ is provided by (4.9). Expressed in terms of $g(x)$,

$$V[\hat{\tau}_\rho] = \frac{1}{n}\left(G\int_a^b \frac{\rho^2(x)}{g(x)}\,dx - \tau_\rho^2\right).$$

This variance is estimated with equation (4.10).

As hinted above, a good proxy (or probability density function) is key to precise estimation following importance sampling. Ideally, $g(x)$ should be proportional to $\rho(x)$, in which case $\rho(x)/g(x)$ would be constant in the interval $[a, b]$ and, therefore, the estimate $\hat{\tau}_{\rho s}$ would equal the target integral τ_ρ for all values of $x = x_s$. To see this, assume that $g(x) = c\,\rho(x)$ and, therefore, $G = c\,\tau_\rho$. Substituting into (4.25),

$$\hat{\tau}_{\rho s} = c\,\tau_\rho\left[\frac{\rho(x_s)}{c\,\rho(x_s)}\right] = \tau_\rho.$$

The variance under proportionality is zero. Generally, however, proportionality between $g(x)$ and $\rho(x)$ is an unrealizable goal. Otherwise, there would be no need for importance sampling. Nevertheless, some thought should be given to making $g(x)$ as nearly proportional to $\rho(x)$ as possible to reduce the variance of $\hat{\tau}_\rho$.

Example 4.12

Let us reconsider the estimation of the volume of a tree bole (see Example 4.5). Recall that bole volume can be estimated from measurements of cross-sectional area at heights selected at random. Because tree boles generally decrease in cross-sectional area from butt to tip, the selection of heights of cross sections with probability density proportional to a proxy of cross-sectional area should be more efficient than a selection with a uniform density. Moreover, from a practical standpoint, this selection should yield sample heights that are more concentrated in the lower half of the bole.

Recall that $\rho(h)$ denotes the cross-sectional area of a bole at height h, and H denotes the total height of the bole. Let τ_ρ denote the volume of the bole between heights a and b, where $0 \le a < b \le H$. For example, a might be the usual height of a stump and b the upper height of merchantability. Let us call the bole of interest the 'real bole.' Thus, the volume of the real bole to be estimated is

$$\tau_\rho \equiv \tau_\rho(b) - \tau_\rho(a) = \int_a^b \rho(h)\,dh.$$

Our strategy shall be to mathematically define a 'proxy bole' with a height equal to, and a shape similar to, the real bole. The proxy bole is defined by a proxy function that furnishes cross-sectional area at any height between 0 and H. This function should be integrable between 0 and H for the calculation of volume of the proxy bole to any height. Let $g(h)$ denote the cross-sectional area of a proxy bole at height h and let G denote the volume of the proxy bole between heights a and b, i.e.,

$$G \equiv G(b) - G(a) = \int_a^b g(h)\,dh.$$

Consider the following proxy function (cf., Gregoire et al. 1986):

$$g_1(h) = \left[\frac{\rho(1.37)}{H - 1.37}\right](H - h), \tag{4.27}$$

where $\rho(1.37)$ is the cross-sectional area of the real bole at 1.37 m, a height commonly referred to as 'breast height.' H is the height of the real bole and the defined height of the proxy bole. Note that if $h = 1.37$, then $g_1(h) = \rho(h)$; if $h = H$, then $g_1(h) = \rho(h) = 0$. Let us also consider a simpler proxy function:

$$g_2(h) = H - h. \tag{4.28}$$

With this second proxy function, the cross-sectional area of proxy bole is not scaled to that of the real bole. Yet, we assert that the two proxy functions are equivalent to each other for the purpose of an importance sampling of a tree bole because (a) the sampling is proportional to size and (b) the two proxy functions are proportional to each other, since $\rho(1.37)/(H - 1.37)$ is constant.

The estimator of τ_ρ, equation (4.25), uses a measurement of $\rho(h)$ at height h_s. This height is the root of

$$G(h_s) - G(a) - u_s G = 0. \tag{4.29}$$

Assuming that $g(h) = g_2(h) = H - h$, then,

$$G(h_s) - G(a) = \int_a^{h_s} (H - h) \, dh$$

$$= H(h_s - a) - \frac{(h_s^2 - a^2)}{2}$$

$$= \frac{(H - a)^2 - (H - h_s)^2}{2}.$$

and, similarly,

$$G \equiv G(b) - G(a) = \frac{(H - a)^2 - (H - b)^2}{2}.$$

Substituting into equation (4.29) and solving for h_s, we obtain

$$h_s = H - \sqrt{(1 - u_s)(H - a)^2 + u_s (H - b)^2}. \tag{4.30}$$

We can rewrite the estimator, equation (4.25), as

$$\hat{\tau}_{\rho s} = \frac{\left[(H - a)^2 - (H - b)^2\right] \rho(h_s)}{2(H - h_s)}. \tag{4.31}$$

A practitioner, who may have no knowledge of attribute or probability densities, should be able to use (4.30) and (4.31) to estimate the volume of almost any tree bole. A combined estimate can be calculated from n independent estimates with (4.26).

The dry weight of bole wood is highly correlated with volume and can be estimated with a little additional effort (e.g., Van Deusen & Baldwin 1993). An increment core is extracted at each sample height h_s and the bulk density—dry weight per unit wet volume—of the core, less bark, is measured. Multiplication of the bulk density by cross-sectional area (inside bark) yields the desired attribute density, i.e., dry weight per unit length of wood. If we let $\rho(h_s)$ denote this density at height h_s, then equation (4.17) estimates the dry weight of the bole. The resultant estimate will be biased if the bulk density of the wood is not uniform over the length of a core.

Example 4.13

As an alternative to equation (4.30) in Example 4.12, consider the acceptance-rejection method to determine a measurement height, h_s. Suppose that $a = 0.3$ m, $b = 21$ m, and $H = 30$ m. Because $a \le h \le b$,

$$g_{max} = g(a) = H - a = 29.7 \text{ m}.$$

ASIDE: Alternatives to the simple generic proxy function used in Example 4.12 are readily available in the form of bole-taper models. A bole-taper model provides a three dimensional description of a 'model bole,' or as we have termed it, a proxy bole. This description may be inside or outside bark. Bole-taper models date from the late 18th century (cf., Gray 1943), the earliest models based on frusta of simple geometric solids (e.g., Assmann 1970). In the last half century, forest mensurationists have fitted and published alternative bole-taper models for hundreds of species (see, e.g., Clark et al. 1991).

We draw two random numbers from U[0, 1], say, $u_1 = 0.21717$ and $u_2 = 0.18446$. We calculate

$$h_s = a + u_1 \times (b - a)$$
$$= 0.3 + 0.21727 \times (21 - 0.3)$$
$$= 6.618 \text{ m.}$$

Finally, we test if $u_2 \times g_{max} \le g(h_s)$, where $g(h_s) = H - h_s$. Indeed,

$$0.18446 \times 29.7 \le 30 - 6.618$$

so $h_s = 6.618$ m is accepted.

Example 4.14

Consider the estimation of the total volume of coarse woody fuels (m^3) lying on the ground in a mapped tract of land in a region where forest fires are common. In this case, the tract of land is the entity that is sampled, and the aggregate volume of the discrete pieces of fuel, which are scattered over the land, is the attribute of interest (τ_ρ).

Let us define locations on a tract in terms of an x-axis that runs west to east and a z-axis that runs south to north. Let the most western point of the tract define $x = 0$ and the most eastern point define $x = x_{max}$; and let the most southern point define $z = 0$ and the most northern point define $z = z_{max}$. Thus, we assume that the tract of interest fits within a rectangle with dimensions $x_{max} \times z_{max}$ (Figure 4.8). Let $L(x_s)$ (m) denote the horizontal length of a line from the southern boundary to the northern boundary at $x = x_s$ where $0 \le x_s \le x_{max}$. If the tract contains coves or concavities, then $L(x_s)$ is the sum of lengths of the line segments contained within the tract at $x = x_s$. Thus, we assume that the tract is spanned by infinitely many parallel lines that run south to north. By definition, the area of the tract, A (m^2), is

$$A \equiv A(x_{max}) - A(0) = \int_0^{x_{max}} L(x) \, dx.$$

Let $\rho(x_s)$ denote the cross-sectional area (m^2) of coarse woody fuel intercepted by the line at $x = x_s$. In other words, $\rho(x_s)$ is the vertical cross-sectional area of

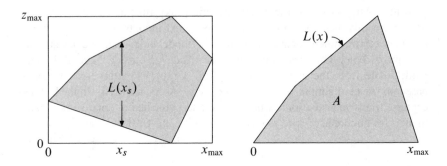

Figure 4.8 *A map (or GIS polygon) of a tract of land (left) can serve to define the shape (right) of a probability density function,* $f(x) = L(x)/A$, *for the selection of a line that spans the tract.*

wood that would be exposed if we were to cut along the line, from the southern border to the northern border, with a chainsaw. Hence, the target parameter, τ_ρ (m^3), can be expressed as the integral

$$\tau_\rho \equiv \tau_\rho(x_{max}) - \tau_\rho(0) = \int_0^{x_{max}} \rho(x)\,dx.$$

It seems reasonable to assume that the longer the line at $x = x_s$, the greater amount of intercepted coarse woody fuel, in which case $L(x_s)$ should be a reasonable proxy for $\rho(x_s)$.

A sample line is selected by the acceptance-rejection method: generate coordinates (x_s, z_s), where $x_s = u_1 \times x_{max}$ and $z_s = u_2 \times z_{max}$. If (x_s, z_s) fall within the tract, then the line at x_s is selected with probability density $f(x_s) = L(x_s)/A$. Otherwise, generate a new set of coordinates with new random numbers.

An estimator of coarse woody fuel on the tract is

$$\hat{\tau}_\rho = \frac{\rho(x_s)}{f(x_s)} = A\left[\frac{\rho(x_s)}{L(x_s)}\right]. \tag{4.32}$$

Note that $\rho(x_s)/L(x_s)$ is an unbiased estimate of the volume of coarse woody fuel per unit land area ($m^3\,m^{-2}$).

4.4 Control variate estimation

As with importance sampling, the efficiency of control variate estimation depends auxiliary information in the form of a good proxy function. However, instead of estimating τ_ρ directly, we estimate the difference between τ_ρ and its known proxy, G. Recall that

$$\tau_\rho \equiv \tau_\rho(b) - \tau_\rho(a) = \int_a^b \rho(x)\,dx$$

and

$$G \equiv G(b) - G(a) = \int_a^b g(x) \, dx.$$

Subtracting the second equation from the first,

$$\tau_\rho - G = \int_a^b \rho(x) \, dx - \int_a^b g(x) \, dx$$

or

$$\tau_\rho = G + \int_a^b [\rho(x) - g(x)] \, dx.$$

In this approach, the proxy function, $g(x)$, is called a *control variate* for $\rho(x)$. The estimator of τ_ρ, based on the sth selection, is

$$\hat{\tau}_{\rho s} = G + \frac{\rho(x_s) - g(x_s)}{f(x_s)}. \qquad (4.33)$$

Crude Monte Carlo ordinarily is used to estimate the integral, so

$$f(x) = \frac{1}{b - a}$$

and $x_s = a + (b - a) u_s$. Hence,

$$\hat{\tau}_{\rho s} = G + (b - a) [\rho(x_s) - g(x_s)]. \qquad (4.34)$$

A combined estimate, $\hat{\tau}_\rho$, is calculated with (4.7) with $n \geq 2$ independent selections, and $\hat{v}[\hat{\tau}_\rho]$ is calculated with (4.10).

Control variate estimation reduces to direct crude Monte Carlo if $g(x)$ is constant. The method is equivalent to importance sampling if $f(x) = g(x)/G$, as equation (4.33) reduces to equation (4.25). As was noted, importance sampling is most efficient when $g(x)$ is proportional to $\rho(x)$. By contrast, the method of sampling with a control variate, with $f(x) = (b - a)^{-1}$, is most efficient when the difference between $g(x)$ and $\rho(x)$ is constant everywhere in the interval $[a, b]$. To see this, let $g(x) = \rho(x) + k$. Then $G = \tau_\rho + (b - a)k$. Letting $x = x_s$ and substituting into (4.34),

$$\hat{\tau}_{\rho s} = \tau_\rho + (b - a) k + (b - a) [\rho(x_s) - (\rho(x_s) + k)] = \tau_\rho. \qquad (4.35)$$

With $g(x)$ as a control variate, the sampling error of $\rho(x) - g(x)$ will be less than the sampling error of $\rho(x)$ if $\rho(x)$ and $g(x)$ are sufficiently correlated. Antithetic selection may improve efficiency.

In repeated sampling, we expect that $G - (b-a)g(x_s)$ will average zero. Therefore, if we let β be an arbitrary constant, then $\beta[G - (b - a)g(x_s)]$ will also average zero. Thus, substituting into (4.34),

$$\hat{\tau}_{\rho s} = \beta G + (b - a) [\rho(x_s) - \beta g(x_s)]$$

unbiasedly estimates τ_ρ. See §13.6.5 for some instances where this form of the estimator is useful.

ASIDE: In two simulation studies (Van Deusen 1990; Valentine et al. 1992), Control variate estimation proved to be more a precise strategy for estimating the volume of a tree bole than importance sampling. Despite this result, importance sampling may be the method of choice for standing trees because of the concentration of the sample heights low on the bole, where locating sample heights and measuring cross-sectional areas are relatively easy.

In a test of 11 different methods, Wolf et al. (1995) found that antithetic selection combined with control variate estimation was the superior method of estimating daily whole-tree photosynthesis with a relative sampling error of 6% for a sample size of two measurements during the day.

Example 4.15

In Example 4.12, we used equation (4.28) for the proxy function in connection with the estimation of the volume of a tree bole by importance sampling. For control variate estimation, we ordinarily strive for a control variate that is parallel to the attribute density function. Accordingly, equation (4.27) would be a better choice than (4.28) if sampling with a control variate is used to estimate bole volume.

4.5 Sampling in two or three dimensions

In Example 4.14, we considered the selection of a line with length $L(x_s)$ that spanned a tract of land with area A. The line was selected at $x = x_s$ with probability density $f(x_s) = L(x_s)/A$. Now we consider the selection of the coordinates, (x_s, z_s), of a location point within a tract. 'Tract' is used in the generic sense to signify something with a closed boundary, for example, a tract of land, a lake, or the top surface of a leaf.

Our purpose for selecting a location point is to estimate an attribute of the tract from a measurement of the attribute density at the point. Let A denote the area of the horizontal projection of the surface of tract $\mathcal{A}$, and let τ_ρ denote an attribute of interest. Moreover, let $\rho(x, z)$ denote the attribute density—the amount of attribute per unit area—at a location point with rectangular coordinates (x, z). Thus, the total amount of attribute within $\mathcal{A}$ is

$$\tau_\rho = \iint_{\mathcal{A}} \rho(x, z) \, dz \, dx.$$

Example 4.16

Suppose that the attribute of interest is the volume of topsoil (m^3) on a tract of land. The attribute density at a location point, i.e., the volume of topsoil per unit land area ($m^3 \, m^{-2}$), equals the depth of the topsoil (m) at that point.

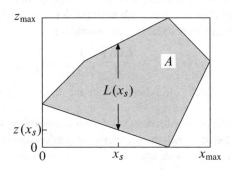

Figure 4.9 *A tract with area A*

Example 4.17

The number of leaves intersected by a vertical line over a location point is the horizontally projected leaf surface area per unit land area ($m^2 \ m^{-2}$) at the location point. This attribute density, which may be measured by counting the number of leaves that touch a taut vertical string, is sometimes called 'projected leaf area index.'

4.5.1 Selection of a sample point

It is instructive to consider the selection of the coordinates of a sample point in two stages: (*i*) the x-ordinate ($x = x_s$) of a line that spans the tract is selected with probability density $f(x_s)$, and (*ii*) the z-ordinate ($z = z_s$) of a point is selected somewhere on that line with probability density $f(z_s|x_s)$. The joint probability density of the coordinates (x_s, z_s) of the resultant sample point is

$$f(x_s, z_s) = f(x_s) \, f(z_s|x_s).$$

The tract depicted in Figure 4.9 has area A and the bounding box of the tract is a rectangle with area $x_{max} \times z_{max}$; $z(x_s)$ denotes the z-ordinate of the southern boundary of the tract at $x = x_s$ and $L(x_s)$ is the distance from the southern to the northern boundary at x_s.

Consider uniform selection of a sample point in the tract in two stages. First, select x_s with uniform density $f(x_s) = 1/x_{max}$ and then select z_s with uniform density $f(z_s|x_s) = 1/L(x_s)$, i.e., $x_s = u_1 x_{max}$ and $z_s = z(x_s) + u_2 L(x_s)$. The joint probability density of (x_s, z_s) is

$$f(x_s, z_s) = \frac{1}{x_{max}} \frac{1}{L(x_s)} = \frac{1}{x_{max} L(x_s)}.$$

With this 'two-stage uniform' approach, the resultant joint probability is uniform only if $L(x)$ is constant. If $L(x)$ varies in $[0, x_{max}]$, as in Figure 4.9, then the joint density varies inversely with $L(x)$.

Alternatively, were we to select x_s with probability density $f(x_s) = L(x_s)/A$, as in Example 4.14, and then select z_s with uniform density $f(z_s|x_s) = 1/L(x_s)$, the resultant joint probability density at (x_s, z_s) would be

$$f(x_s, z_s) = \frac{L(x_s)}{A} \frac{1}{L(x_s)} = \frac{1}{A}.$$

With this second approach, all points within the tract have uniform probability density $1/A$.

In practice, selection of coordinates (x_s, z_s) with uniform probability density $1/A$ is most easily accomplished with the acceptance-rejection method: assuming that the tract fits within a rectangle with area $x_{max} \times z_{max}$, generate $x_s = u_1 \times x_{max}$ and $z_s = u_2 \times z_{max}$. If the coordinates (x_s, z_s) fall within $\mathcal{A}$, accept them; otherwise, reject them and try again with new random numbers.

More generally, we may specify a 'non-uniform' joint probability density function, $f(x, z) = f(x)f(z|x)$, where $f(x, z) > 0$ for all $(x, z) \in \mathcal{A}$, and

$$\iint\limits_{(x,z)\in\mathcal{A}} f(x, z)\, dz\, dx = 1.$$

A sample point, (x_s, z_s), obtains from the inverse transform method, which involves solving $F(x_s) = u_1$ for x_s and $F(z_s|x_s) = u_2$ for z_s.

4.5.2 Estimation

If we independently select the coordinates, (x_s, z_s), of the sth sample point within the tract with probability density $f(x_s, z_s)$, then τ_ρ is unbiasedly estimated by

$$\hat{\tau}_{\rho s} = \frac{\rho(x_s, z_s)}{f(x_s, z_s)}.$$

If we specify $f(x, z) = 1/A$ for all $(x, z) \in \mathcal{A}$, then the estimator simplifies to

$$\hat{\tau}_{\rho s} = A\, \rho(x_s, z_s). \tag{4.36}$$

Equation (4.7) provides a combined estimate, $\hat{\tau}_\rho$, given ≥ 2 independent sample points, and (4.10) provides $\hat{v}[\hat{\tau}_\rho]$.

Example 4.18

An unbiased estimator of the volume of topsoil on a tract of land is provided by the product of the depth of the topsoil at a location point selected uniformly at random and the land area of the tract. Note that the depth of the topsoil (m) at the sample point is an unbiased estimator of the volume of topsoil per unit land area ($m^3\ m^{-2}$).

Example 4.19

Suppose that we are interested in the area of land, τ_ρ, within a given region

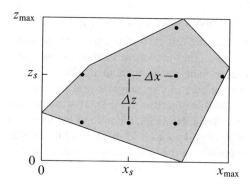

Figure 4.10 *A systematic grid of sample points, which is anchored to one point (x_s, z_s) selected at random.*

with area A that is occupied by a particular type of forest community, e.g., northern hardwood. Numerous (n) points in the region are selected at random, each with uniform probability density $1/A$. The attribute density at a selected location point is $\rho(x_s, z_s) = 1$ (ha northern hardwood forest (ha land)$^{-1}$), if the point falls within a northern hardwood forest; or $\rho(x_s, z_s) = 0$, otherwise. An unbiased estimate of the area occupied by northern hardwood forest is, of course, $\hat{\tau}_\rho = (A/n) \sum_s \rho(x_s, z_s)$. Moreover, $\hat{\tau}_\rho/A = (1/n) \sum_s \rho(x_s, z_s)$ is an unbiased estimate of the fraction of regional land area that is northern hardwood forest.

4.5.3 Systematic selection

In systematic selection, one 'anchor point' is selected at random (or decided upon prior to sampling), and additional points are found by the application of a formula or recipe. For example, a rectangular grid of sample points may be formed with coordinates $(x_s \pm i\Delta x, z_s \pm k\Delta z)$, where $i = 1, 2, \ldots$ and $k = 1, 2, \ldots$ (see Figure 4.10). Suppose that the sth grid contains N points, including the anchor point (x_s, z_s), which is selected by the acceptance-rejection method. The unbiased estimator of the target integral, τ_ρ, is

$$\hat{\tau}_{\rho s} = \frac{A}{N} \sum \rho(x_s \pm i\Delta x, z_s \pm k\Delta z). \tag{4.37}$$

where the summation is over all the grid points. Two grids afford the calculation of an unbiased combined estimate, $\hat{\tau}_\rho$, with (4.7) and an unbiased estimate of the variance with (4.10). The common practice, however, is to go with one grid and assume that the grid points are independent.

Example 4.20

Radtke & Bolstad (2001) used a laser rangefinder, which was mounted on a monopod, to measure the vertical distance from the ground to a leaf or branch in a broad-leaved forest. The instrument emitted an audible warning if the laser beam traveled more than 500 m without intercepting a leaf or branch. Field workers traversed an area in a rectangular grid pattern, stopping every step to take a measurement. The objective of the study was an estimation of the vertical profile of leaf area index. However, the ratio of the number of audible warnings to the total number of grid points is an estimate of the 'gap fraction' of the forest canopy, i.e., the fraction of the land area not covered by the forest canopy.

4.5.4 Three dimensions

In theory, we apply the two-dimensional strategy to any number of dimensions. Practical applications arise in the sampling of three-dimensional containers, for example, vernal ponds, lakes, segments of an ocean, or segments of the atmosphere.

Let $\mathcal{V}$ identify a container of interest and let $|V|$ denote its volume. Let τ_ρ denote an attribute of interest and let $\rho(x, z, h)$ denote the attribute density at location point (x, z, h) within the container. Thus, the total amount of attribute within the container is

$$\tau_\rho = \iiint_{\mathcal{V}} \rho(x, z, h) \, dh \, dz \, dx.$$

Example 4.21

Suppose that the attribute of interest is the quantity of some chemical species (mol) in a body of water. The location points within the body of water comprise the continuous population. The attribute density at a location point is the concentration of the chemical species (mol l^{-1}) at that point.

The acceptance-rejection method furnishes a sample point (x_s, z_s, h_s) with uniform probability density $f(x_s, z_s, h_s) = 1/|V|$. Assuming that the container of interest fits within a box with volume $x_{max} \times z_{max} \times h_{max}$, generate $x_s = u_1 \times x_{max}$, $z_s = u_2 \times z_{max}$, and $z_s = u_3 \times h_{max}$. If the coordinates (x_s, z_s, h_s) fall within the container of interest, accept them; otherwise, reject them and try again with new random numbers. If sample point s is selected with uniform probability density, then τ_ρ is estimated by

$$\hat{\tau}_{\rho s} = |V| \rho(x_s, z_s, h_s). \tag{4.38}$$

As usual, equation (4.7) provides a combined estimate, $\hat{\tau}_\rho$, given ≥ 2 independent sample points, and (4.10) provides $\hat{v}[\hat{\tau}_\rho]$.

4.6 General notation

In the present chapter, we have used the familiar notation of elementary calculus texts to discuss the sampling of continuums that comprise populations of infinitely many points in one, two, or three dimensions. By contrast, in Chapter 1, we introduced a general notation pertaining to the sampling of continuums in any number of dimensions. We shall use the general notation in the appendix (§4.9.3) and in other chapters. The purpose of this section is to reconcile the two notations.

In the general notation, we let $\mathcal{D}$ denote the domain of a target integral, τ_ρ, i.e.,

$$\tau_\rho = \int_{\mathcal{D}} \rho(\mathbf{x})\, d\mathbf{x},$$

where $\mathbf{x}$ is any location point in the domain $\mathcal{D}$. In the context of the present chapter, the domain, $\mathcal{D}$, is equivalent to the interval $[a, b]$ in one dimension; $\mathcal{D}$ is equivalent to the bounded planar region $\mathcal{A}$ in two dimensions and equivalent to the container $\mathcal{V}$ in three dimensions. Letting D denote the size—the length, area, or volume—of $\mathcal{D}$, the mean attribute density in $\mathcal{D}$ is $\mu_\rho = \tau_\rho/D$.

Let $\mathbf{x}_1, \mathbf{x}_2, \ldots, \mathbf{x}_n$ be a set of n sample points selected from $\mathcal{D}$ according to a design with a probability density function $f(\mathbf{x})$, where

$$f(\mathbf{x}) > 0, \text{ for all } \mathbf{x} \in \mathcal{D};$$
$$f(\mathbf{x}) = 0, \text{ otherwise;}$$

and

$$\int_{\mathcal{D}} f(\mathbf{x})\, d\mathbf{x} = 1.$$

In the context of the present chapter, $\mathbf{x}_s \equiv x_s$ in the one-dimensional problem; $\mathbf{x}_s \equiv (x_s, z_s)$ in the two-dimensional problem; and $\mathbf{x}_s \equiv (x_s, z_s, h_s)$ in the three-dimensional problem. Hence, the combined estimator of τ_ρ, in the general notation is

$$\hat{\tau}_\rho = \frac{1}{n} \sum_{s=1}^{n} \frac{\rho(\mathbf{x}_s)}{f(\mathbf{x}_s)}.$$

The other estimators can be rewritten in general notation in an analogous fashion.

4.7 Terms to remember

Acceptance-rejection method	Importance sampling
Antithetic variates	Inverse-transform method
Attribute density	Monte Carlo Integration
Continuous population	Probability density
Control variate	Proxy function
Crude Monte Carlo	

4.8 Exercises and projects

1. Use crude Monte Carlo to estimate the volume of some object for which you can define an axis of length. Choose an object that tapers from thick to thin over its length (for example, your right leg). Estimate volume first with four independent selections of measurement points, and then with two antithetic selections. If need be, calculate areas of cross sections from circumference. Estimate the standard error for the combined estimate of volume for each selection method. Do the results accord with theory?

2. Estimate the volume of the object used in Exercise 1 with importance sampling. Use a linear proxy function, as in Example 4.10, and select two sample points by the inverse transform method. Choose two more sample points with the acceptance-rejection method. Calculate an estimate of the standard error for the combined estimate of volume (from the four measurements). Why is it valid to calculate a combined estimate if two different methods are used to obtain the measurement points?

3. Use the four independent measurement points from Exercise 1 and use the proxy from Exercise 2 as a control variate to estimate the volume of the object. Calculate the standard error for the combined estimate.

4. Review Example 4.14. What is the continuous population that is sampled? Suppose that a line is selected from a tract and it turns out to have a different length than the map indicates. Why *wouldn't* this map error affect the unbiasedness of the estimates? Suppose that the tract turned out to be wider along the x-axis than expected. Why *would* this map error bias the estimates?

5. Explain why the count of leaves touching a taut vertical string is an estimate of projected leaf area per unit land area, as noted in Example 4.17.

4.9 Appendix

Table 4.1 *A comparison of formulae for list sampling of a discrete population and importance sampling of a continuum.*

	List Sampling	Importance Sampling
Population total	$\tau_y = \sum\limits_{k=1}^{N} y_k$	$\tau_\rho = \int_a^b \rho(x)\, dx$
Select	$\mathcal{U}_k$, if $\sum\limits_{i=0}^{k-1} p_i < u_k \le \sum\limits_{i=0}^{k} p_i$	x_s, where $\int_a^{x_s} f(x)\, dx = u_s$
Estimator of total	$\hat{\tau}_{yp} = \dfrac{1}{n} \sum\limits_{\mathcal{U}_k \in s} \dfrac{y_k}{p_k}$	$\hat{\tau}_\rho = \dfrac{1}{n} \sum\limits_{s=1}^{n} \dfrac{\rho(x_s)}{f(x_s)}$
Variance of estimator	$V\left[\hat{\tau}_{yp}\right] = \dfrac{\sum\limits_{k=1}^{N} \dfrac{y_k^2}{p_k} - \tau_y^2}{n}$	$V\left[\hat{\tau}_\rho\right] = \dfrac{\int_a^b \dfrac{\rho^2(x)}{f(x)}\, dx - \tau_\rho^2}{n}$
Estimator of variance	$\hat{v}\left[\hat{\tau}_{yp}\right] = \dfrac{\sum\limits_{\mathcal{U}_k \in s} \left[\dfrac{y_k}{p_k} - \hat{\tau}_{yp}\right]^2}{n(n-1)}$	$\hat{v}\left[\hat{\tau}_\rho\right] = \dfrac{\sum\limits_{s=1}^{n} \left[\dfrac{\rho(x_s)}{f(x_s)} - \hat{\tau}_\rho\right]^2}{n(n-1)}$
Confidence interval	$\hat{\tau}_{yp} \pm t_{n-1} \sqrt{\hat{v}\left[\hat{\tau}_{yp}\right]}$	$\hat{\tau}_\rho \pm t_{n-1} \sqrt{\hat{v}\left[\hat{\tau}_\rho\right]}$
Population mean	$\mu_y = \dfrac{\tau_y}{N}$	$\mu_\rho = \dfrac{\tau_\rho}{b-a}$
Estimator of mean	$\hat{\mu}_{yp} = \dfrac{\hat{\tau}_{yp}}{N}$	$\hat{\mu}_\rho = \dfrac{\hat{\tau}_\rho}{b-a}$
Variance of estimator	$V\left[\hat{\mu}_{yp}\right] = \dfrac{V[\hat{\tau}_{yp}]}{N^2}$	$V\left[\hat{\mu}_\rho\right] = \dfrac{V[\hat{\tau}_\rho]}{(b-a)^2}$
Estimator of variance	$\hat{v}\left[\hat{\mu}_{yp}\right] = \dfrac{\hat{v}[\hat{\tau}_{yp}]}{N^2}$	$\hat{v}\left[\hat{\mu}_\rho\right] = \dfrac{\hat{v}[\hat{\tau}_\rho]}{(b-a)^2}$
Confidence interval	$\hat{\mu}_{yp} \pm t_{n-1} \sqrt{\hat{v}\left[\hat{\mu}_{yp}\right]}$	$\hat{\mu}_\rho \pm t_{n-1} \sqrt{\hat{v}\left[\hat{\mu}_\rho\right]}$

4.9.1 Proof of the unbiasedness of $\hat{\tau}_\rho$ as an estimator of τ_ρ

When $n = 1$ the expected value of

$$\hat{\tau}_{\rho s} = \frac{\rho(x_s)}{f(x_s)}, \qquad a \le x_s \le b,$$

is

$$E\left[\hat{\tau}_{\rho s}\right] = E\left[\frac{\rho(x_s)}{f(x_s)}\right]$$

$$= \int_a^b f(x) \frac{\rho(x)}{f(x)}\, dx$$

$$= \int_a^b \rho(x)\, dx$$

$$= \tau_\rho.$$

With antithetic selection,

$$E\left[\hat{\tau}_{\rho s}\right] = E\left[\frac{1}{2}\left(\frac{\rho(x_s)}{f(x_s)} + \frac{\rho(x_s')}{f(x_s)}\right)\right]$$

$$= \frac{1}{2}\left\{E\left[\frac{\rho(x_s)}{f(x_s)}\right] + E\left[\frac{\rho(x_s')}{f(x_s)}\right]\right\}$$

$$= \frac{1}{2}\left(\tau_\rho + \tau_\rho\right) = \tau_\rho.$$

If a control variate is used, then

$$E\left[\hat{\tau}_{\rho s}\right] = E\left[G + \frac{\rho(x_s)}{f(x_s)} - \frac{g(x_s)}{f(x_s)}\right]$$

where $G = \int_a^b g(x)\, dx$. Hence,

$$E\left[\hat{\tau}_{\rho s}\right] = G + E\left[\frac{\rho(x_s)}{f(x_s)}\right] - E\left[\frac{g(x_s)}{f(x_s)}\right]$$

$$= G + \tau_\rho - \int_a^b f(x) \frac{g(x)}{f(x)}\, dx$$

$$= G + \tau_\rho - G = \tau_\rho.$$

Therefore, when $n > 1$,

$$E\left[\hat{\tau}_\rho\right] = E\left[\frac{1}{n}\sum_{s=1}^{n}\hat{\tau}_{\rho s}\right]$$

$$= \frac{1}{n}\sum_{s=1}^{n}E\left[\hat{\tau}_{\rho s}\right]$$

$$= \frac{1}{n}\sum_{s=1}^{n}\tau_\rho = \tau_\rho.$$

4.9.2 Derivation of $V\left[\hat{\tau}_\rho\right]$

$$V\left[\hat{\tau}_\rho\right] = E\left[\hat{\tau}_\rho^{\,2}\right] - \left(E\left[\hat{\tau}_\rho\right]\right)^2$$

$$= E\left[\hat{\tau}_\rho^{\,2}\right] - \tau_\rho^2$$

$$= E\left[\left(\frac{1}{n}\sum_{s=1}^{n}\frac{\rho(x_s)}{f(x_s)}\right)^2\right] - \tau_\rho^2$$

$$= \frac{1}{n^2}\sum_{s=1}^{n}E\left[\left(\frac{\rho(x_s)}{f(x_s)}\right)^2\right] + \frac{1}{n^2}\sum_{s=1}^{n}\sum_{\substack{s'\neq s \\ s'=1}}^{n}E\left[\frac{\rho(x_s)}{f(x_s)}\frac{\rho(x_{s'})}{f(x_{s'})}\right] - \tau_\rho^2$$

$$= \frac{1}{n^2}\sum_{s=1}^{n}\int_a^b f(x)\left(\frac{\rho(x)}{f(x)}\right)^2 dx$$

$$+ \frac{1}{n^2}\sum_{s=1}^{n}\sum_{\substack{s'\neq s \\ s'=1}}^{n}E\left[\frac{\rho(x_s)}{f(x_s)}\right]E\left[\frac{\rho(x_{s'})}{f(x_{s'})}\right] - \tau_\rho^2$$

$$= \frac{1}{n^2}\sum_{s=1}^{n}\int_a^b \frac{\rho^2(x)}{f(x)}\,dx + \frac{1}{n^2}\sum_{s=1}^{n}\sum_{\substack{s'\neq s \\ s'=1}}^{n}\tau_\rho^2 - \tau_\rho^2$$

$$= \frac{1}{n^2}\left(n\int_a^b\frac{\rho^2(x)}{f(x)}\,dx\right) + \frac{n-1}{n}\tau_\rho^2 - \tau_\rho^2,$$

which reduces to

$$V\left[\hat{\tau}_\rho\right] = \frac{1}{n}\left(\int_a^b \frac{\rho^2(x)}{f(x)}\,dx - \tau_\rho^2\right).$$

4.9.3 The Horvitz-Thompson estimator of τ_ρ and its variance

Cordy (1993) presented both the Horvitz-Thompson estimator of τ_ρ and its variance. Both formulae were stated in terms of an inclusion density, $\pi(\mathbf{x}_s)$, and a joint inclusion density, $\pi(\mathbf{x}_s, \mathbf{x}_{s'})$, where $\mathbf{x}_s$ and $\mathbf{x}_{s'}$ are any two distinct sample points in $\mathcal{D}$, the domain of integration. A replicated sample comprising n points is assumed.

For importance sampling, which includes crude Monte Carlo as a special case,

$$\pi(\mathbf{x}_s) = nf(\mathbf{x}_s) \tag{4.39}$$

and

$$\pi(\mathbf{x}_s, \mathbf{x}_{s'}) = n(n-1)f(\mathbf{x}_s)f(\mathbf{x}_{s'}) \tag{4.40}$$

where $f(\mathbf{x})$ is the probability density function. The Horvitz-Thompson estimator of τ_ρ is

$$\hat{\tau}_{\rho\pi} = \sum_{s=1}^n \frac{\rho(\mathbf{x}_s)}{\pi(\mathbf{x}_s)}$$

Substituting $nf(\mathbf{x}_s)$ for $\pi(\mathbf{x}_s)$,

$$\hat{\tau}_{\rho\pi} = \frac{1}{n}\sum_{s=1}^n \frac{\rho(\mathbf{x}_s)}{f(\mathbf{x}_s)},$$

which is equivalent to $\hat{\tau}_\rho$. Let $\mathbf{x}$ and $\mathbf{x}'$ be any two location points in $\mathcal{D}$. The variance of $\hat{\tau}_{\rho\pi}$ is

$$V[\hat{\tau}_{\rho\pi}] = \int_{\mathcal{D}} \frac{\rho^2(\mathbf{x})}{\pi(\mathbf{x})}\,d\mathbf{x} + \iint_{\mathcal{D}} \left[\frac{\pi(\mathbf{x},\mathbf{x}') - \pi(\mathbf{x})\pi(\mathbf{x}')}{\pi(\mathbf{x})\pi(\mathbf{x}')}\right]\rho(\mathbf{x})\rho(\mathbf{x}')\,d\mathbf{x}\,d\mathbf{x}'.$$

Substituting (4.39) and (4.40) and reducing,

$$V[\hat{\tau}_{\rho\pi}] = \frac{1}{n}\left(\int_{\mathcal{D}} \frac{\rho^2(\mathbf{x})}{f(\mathbf{x})}\,d\mathbf{x} - \tau_\rho^2\right),$$

which is equivalent to $V[\hat{\tau}_\rho]$.

CHAPTER 5

Stratified Sampling Designs

5.1 Introduction

A stratified sampling design purposely partitions the target population, $\mathcal{P}$, into two or more non-overlapping subpopulations, called *strata*, which are sampled separately. In this chapter we present the rationale for stratified sampling and estimators for the population total, τ_y, and related population parameters, as well as corresponding estimators for the individual strata. A design issue which arises with stratified sampling is the allocation of the overall sampling effort to the various strata, a topic which is addressed following the section on estimation. The sections at the end of the chapter discuss matters related to stratified sampling: incorrect strata assignment, double sampling for stratification, and poststratification.

5.2 Rationale for stratified sampling

Stratification is often motivated by a desire or requirement to estimate the total or average value of some attribute, y, for each stratum of interest. In the U.S.A., for example, natural resource and agricultural surveys are mandated and conducted by the federal government. Survey results are reported separately by state, so each state serves as a stratum, and sampling within each state is conducted independently of sampling in any other. Separate reporting is not necessarily contingent upon stratification because results may be calculated separately for each subpopulation of interest, whether or not the sampling is conducted separately. But stratification prior to execution of the sample is often advantageous since it enables the planner to customize the sampling design to the needs and features of each stratum. For example, among several states, one state may require greater precision in the estimation of its natural resources, so the intensity of sampling may need to be greater within that state. Moreover, because sampling occurs independently within each stratum, the sampling design may vary among strata to accommodate varying expectations of the survey results.

Another motivating reason for conducting stratified sampling is administrative convenience: survey personnel might be trained and supervised by different agencies in the various strata, and it might be cost effective to administer the overall survey by ceding the management of sampling within each stratum to the responsible agency.

From a statistical standpoint, stratification can be a very effective tool to increase the precision with which population parameters are estimated. Increased precision results when the homogeneity of y within a stratum is greater than that in the unstratified population. In other words, when it is possible to assign units into strata, such that the variation of y within strata is less than σ_y^2, then it is possible to estimate τ_y more precisely with a stratified sampling design than with an unstratified design. The degree to which one benefits from stratification depends, inter alia, on other aspects of the sampling design within the strata. Examples presented later will provide some empirical evidence of the statistical advantages that can be realized by stratification.

A stratified design is more efficient than an unstratified design if the stratified design provides greater precision in estimation for a given overall cost. Or, to put it another way, a stratified design is more efficient if it achieves a desired level of precision for a lower cost. However, stratification may involve an overhead cost—time or money to gather sufficient information to effect the stratification. This is time or money that otherwise could be spent to improve precision by procuring a larger sample under an unstratified design. Thus, deciding whether to stratify a population usually reduces to deciding whether an investment to set up the stratification will pay sufficient dividends in the form of increased precision.

Example 5.1

In Example 3.6, a simple random sample of $n = 140$ houses was selected and the concentration of radon gas was measured in each house. Although the overall sample average concentration was $9.04\,\mathrm{pCi\,L}^{-1}$, there was quite a distinct difference in concentrations among houses with basements vs. those lacking basements. Of the 79 houses with basements in the sample, the average concentration of radon was $4.79\,\mathrm{pCi\,L}^{-1}$ and the sample standard deviation was $5.18\,\mathrm{pCi\,L}^{-1}$. Of the 61 sampled houses without basements, the average concentration was $14.55\,\mathrm{pCi\,L}^{-1}$ and the sample standard deviation was $14.81\,\mathrm{pCi\,L}^{-1}$. Should a follow-up survey be conducted, these results suggest that the average concentration of radon per house may be estimated more precisely if the population of houses in Blueridge were divided into two strata: 1) houses with basements; and 2) houses without basements. This stratification could involve an overhead cost, viz., the cost of determining which houses in the sampling frame have basements.

Example 5.2

Barrett & Nutt (1979) discussed a design in which a pond was stratified by depth. The rationale was that the comparatively warm waters (epilimnion) near the surface during the summer season are prevented from mixing with the colder, deeper waters (hyperlimnion) by a thermocline layer. Typically, these three layers have distinctively different biological and chemical features, and hence it is reasonable to treat each pond as having three strata and to sample each stratum separately.

When multiresource surveys are undertaken for the purpose of precisely estimating multiple attributes, it is doubtful that any single stratification of the population will improve the precision with which all attributes can be estimated.

5.3 Estimation with stratified sampling

5.3.1 Notation

Abiding by notation found in Cochran (1977) and other standard texts on sampling methods, we use L to denote the number of strata. As implied in the chapter's introduction, each element, $\mathcal{U}_k, k = 1, \ldots, N$, of the population, $\mathcal{P}$, must be placed into one and only one stratum. For the sake of reference, $\mathcal{P}_h$ will be used to symbolize the subpopulation in stratum h, where $h = 1, \ldots, L$. For a continuously distributed population, each point within the continuum of $\mathcal{P}$ must belong to one and only one stratum. We defer discussion of stratified sampling of a continuous population until §5.8.

The criterion that is used to stratify the population will depend on context, but invariably stratification implies that auxiliary information is available. In the radon example, i.e., Example 5.1, the auxiliary information would be a record of whether or not each house in Blueridge had a basement. In Example 5.2, water temperature by depth was the auxiliary information used for stratification. In regionally stratified surveys, the stratification criterion might be defined by political boundary, so that the auxiliary information would be knowledge of the political unit (state, province, county, township, etc.) in which $\mathcal{U}_k$ occurs. Aerial photography or remotely sensed data from satellite imagery commonly is used in resource surveys to stratify the landscape by land cover or vegetative cover class. In such cases, cover class serves as the stratification criterion and knowledge of the boundaries between the classes serves as the auxiliary information. In these examples and in practice, both the specification of the stratification criterion and the determination of L, the number of strata, is a subjective decision of the sample designer or planner.

Having stratified the population into L non-overlapping strata that collectively include all of $\mathcal{P}$, the definition of a probability sample requires that a sample be selected from each stratum. Failing that, design-unbiased estimation of τ_y or any other population parameter is, with one exception, impossible. The exceptional case occurs when the omitted stratum is censused rather than sampled, and the value of the stratum parameter is appropriately included in the estimator of the corresponding parameter of $\mathcal{P}$, as discussed by Brewer (2002, p. 32).

Just as N represents the number of units in $\mathcal{P}$, let N_h represent the number of units in $\mathcal{P}_h$. Stratification of $\mathcal{P}$ implies that

$$N = N_1 + N_2 + \cdots + N_L$$

$$= \sum_{h=1}^{L} N_h$$

The total amount of attribute, y, in stratum h is

$$\tau_{y,h} = \sum_{\mathcal{U}_k \in \mathcal{P}_h} y_k,$$

where $\mathcal{U}_k \in \mathcal{P}_h$ indicates that the summation extends over all N_h units in $\mathcal{P}_h$. This implies that

$$\tau_y = \sum_{h=1}^{L} \tau_{y,h}. \tag{5.1}$$

If auxiliary information is available for all elements in $\mathcal{P}$, then we denote the total amount of the attribute, x, in stratum h by

$$\tau_{x,h} = \sum_{\mathcal{U}_k \in \mathcal{P}_h} x_k,$$

and the stratified population total by

$$\tau_x = \sum_{h=1}^{L} \tau_{x,h}. \tag{5.2}$$

5.3.2 HT estimation

Initially we consider a sampling strategy within each stratum that involves estimation of each $\tau_{y,h}$ by a HT estimator, denoted by $\hat{\tau}_{y\pi,h}$. We purposely do not specify the sampling design within each stratum in order to emphasize the point that the sampling design need not be identical in the L strata. We will consider examples of specific designs later.

Suppose n_h is the size of the sample stipulated by the sampling design for $\mathcal{P}_h$. Evidently $n_h \leq N_h$, and the overall size of the stratified sample is

$$n = \sum_{h=1}^{L} n_h. \tag{5.3}$$

In accordance with notation established in Chapters 2 and 3, let Ω_h denote the possible number of distinct samples under the sampling design for $\mathcal{P}_h$. The possible number of distinct samples under the stratified design is

$$\Omega = \prod_{h=1}^{L} \Omega_h.$$

Example 5.3

A population of size $N = 20$ yields $\Omega = 15{,}504$ distinct SRSwoR samples of size $n = 5$. Suppose the same population is stratified into $L = 2$ strata with $N_1 = 12$ and $N_2 = 8$ units in $\mathcal{P}_1$ and $\mathcal{P}_2$, respectively. Suppose further that

SRSwoR is employed in each stratum to select a sample of size $n_1 = 3$ in stratum 1 and of size $n_2 = 2$ in stratum 2. Then

$$\Omega_1 = \frac{12!}{3!\,9!} = 220$$

and

$$\Omega_2 = \frac{8!}{2!\,6!} = 28$$

Thus, there are $\Omega = \Omega_1 \times \Omega_2 = 6160$ distinctly different stratified random samples of overall size $n = 5$ units.

The reduced number of stratified random samples of size $n = 5$ results from the requirement that n be subdivided among the L strata. As a consequence, we can not select any samples that combine n units from a single stratum, though these same combinations of n units are possible samples in the absence of stratification.

In contrast to the above example, stratified systematic sampling may result in an increase in the number of possible samples.

Example 5.4

Consider a 1-in-a systematic sampling design. When $a = 4$ and $N = 20$, there are $\Omega = 4$ possible systematic samples, each of size $n = 5$ from an unstratified population. Using the same stratification as in the preceding example and 1-in-4 systematic sampling in each stratum, there are $a = 4$ possible systematic samples of size $n_1 = 3$ in stratum 1, and another 4 possible systematic samples, each of size $n_2 = 2$, from stratum 2. Therefore, there are a total of $\Omega = 16$ stratified systematic samples of size $n = 5$.

In this case, stratification gives rise to many more samples of five elements each than can be systematically selected from the unstratified population.

Because the HT estimator, $\hat{\tau}_{y\pi,h}$, unbiasedly estimates the stratum total, $\tau_{y,h}$, the overall population total, τ_y, is unbiasedly estimated by

$$\hat{\tau}_{y\pi,\text{st}} = \sum_{h=1}^{L} \hat{\tau}_{y\pi,h}. \tag{5.4}$$

Because sampling is conducted independently among the L strata, the variance of $\hat{\tau}_{y\pi,\text{st}}$ is the sum of the variances across the L strata, i.e.,

$$V\left[\hat{\tau}_{y\pi,\text{st}}\right] = \sum_{h=1}^{L} V\left[\hat{\tau}_{y\pi,h}\right]. \tag{5.5}$$

Likewise,

$$\hat{v}\left[\hat{\tau}_{y\pi,\text{st}}\right] = \sum_{h=1}^{L} \hat{v}\left[\hat{\tau}_{y\pi,h}\right] \tag{5.6}$$

ASIDE: $V[\hat{\tau}_{y\pi,\text{st}}]$ is a measure of spread of the sampling distribution of $\hat{\tau}_{y\pi,\text{st}}$, i.e., the distribution of all possible estimates that may result after the population has been partitioned into the L strata. It is not the variance of the distribution of all possible estimates over all possible stratifications of the population, only over the stratification for which this particular sampling is carried out. It is important to understand this distinction because the two sampling distributions are quite different. The variance $V[\hat{\tau}_{y\pi,\text{st}}]$, which is conventionally considered in probability sampling, is the variance of the distribution of estimates that obtains from all possible samples from the stipulated stratification of $\mathcal{P}$. An issue related to this arises in §5.7 when we discuss poststratification.

is a natural estimator of $V[\hat{\tau}_{y\pi,\text{st}}]$. If $\hat{v}[\hat{\tau}_{y\pi,h}]$ unbiasedly estimates $V[\hat{\tau}_{y\pi,h}]$, then $\hat{v}[\hat{\tau}_{y\pi,\text{st}}]$ unbiasedly estimates $V[\hat{\tau}_{y\pi,\text{st}}]$.

Providing that n_h is not too small, an approximate $100(1-\alpha)\%$ confidence interval for $\tau_{y,h}$ is given by

$$\hat{\tau}_{y\pi,h} \pm t_{n_h-1} \sqrt{\hat{v}[\hat{\tau}_{y\pi,h}]} \tag{5.7}$$

where t_{n_h-1} is the $1-(\alpha/2)$ percentile of the Student t distribution with n_h-1 degrees of freedom. When all the n_h are reasonably large, an approximate $100(1-\alpha)\%$ confidence interval for τ_y, based on t with $n-L$ degrees of freedom, is

$$\hat{\tau}_{y\pi,\text{st}} \pm t_{n-L} \sqrt{\hat{v}[\hat{\tau}_{y\pi,\text{st}}]}. \tag{5.8}$$

In (5.7) and (5.8), the margin of error, i.e., the $\pm$ part of the interval, has the same units of measure as the attribute of interest, y. Both of these $100(1-\alpha)\%$ intervals may be expressed equivalently by substituting the percentage margin of error:

$$\hat{\tau}_{y\pi,h} \pm t_{n_h-1} \frac{100\sqrt{\hat{v}[\hat{\tau}_{y\pi,h}]}}{\hat{\tau}_{y\pi,h}}\% \tag{5.9}$$

and

$$\hat{\tau}_{y\pi,\text{st}} \pm t_{n-L} \frac{100\sqrt{\hat{v}[\hat{\tau}_{y\pi,\text{st}}]}}{\hat{\tau}_{y\pi,\text{st}}}\%. \tag{5.10}$$

ASIDE: When n is large and none of the n_h are very small, the exact number of degrees of freedom to use to determine the critical value of t to use in constructing a confidence interval for τ_y is rather inconsequential: t_{n-L} will be inconsequentially different from the corresponding quantile of the standard normal distribution, z. In this situation, one may use z instead of t_{n-L} in (5.8).

The mean value per unit in $\mathcal{P}_h$ is

$$\mu_{y,h} = \frac{\tau_{y,h}}{N_h}, \tag{5.11}$$

which can be estimated unbiasedly by

$$\hat{\mu}_{y\pi,h} = \frac{\hat{\tau}_{y\pi,h}}{N_h}. \tag{5.12}$$

Because $V[\hat{\mu}_{y\pi,h}] = V[\hat{\tau}_{y\pi,h}]/N_h^2$, a reasonable $100(1-\alpha)\%$ interval estimate for $\mu_{y,h}$ is

$$\hat{\mu}_{y\pi,h} \pm t_{n_h-1} \sqrt{\hat{v}[\hat{\mu}_{y\pi,h}]}, \tag{5.13}$$

where $\hat{v}[\hat{\mu}_{y\pi,h}] = \hat{v}[\hat{\tau}_{y\pi,h}]/N_h^2$. Expressing the margin of error in percentage terms yields

$$\hat{\mu}_{y\pi,h} \pm t_{n_h-1} \frac{100\sqrt{\hat{v}[\hat{\mu}_{y\pi,h}]}}{\hat{\mu}_{y\pi,h}}\% \tag{5.14}$$

When there is a need to estimate the mean value per element in the population, $\mu_y = \tau_y/N$, a natural estimator is

$$\hat{\mu}_{y\pi,\text{st}} = \frac{\hat{\tau}_{y\pi,\text{st}}}{N}, \tag{5.15}$$

with variance

$$V[\hat{\mu}_{y\pi,\text{st}}] = \frac{V[\hat{\tau}_{y\pi,\text{st}}]}{N^2}. \tag{5.16}$$

An expression equivalent to (5.15) for $\hat{\mu}_{y\pi,\text{st}}$ is derived with the identity

$$\tau_y = N_1\mu_{y,1} + N_2\mu_{y,2} + \cdots + N_L\mu_{y,L}. \tag{5.17}$$

Substituting into $\mu_y = \tau_y/N$ yields

$$\begin{aligned}
\mu_y &= \frac{N_1\mu_{y,1} + N_2\mu_{y,2} + \cdots + N_L\mu_{y,L}}{N} \\
&= \frac{N_1}{N}\mu_{y,1} + \frac{N_2}{N}\mu_{y,2} + \cdots + \frac{N_L}{N}\mu_{y,L} \\
&= W_1\mu_{y,1} + W_2\mu_{y,2} + \cdots + W_L\mu_{y,L},
\end{aligned} \tag{5.18}$$

where $W_h = N_h/N$ customarily is called the 'stratum weight' for stratum h. Each stratum weight, W_h, is the proportion of units in $\mathcal{P}$ that are in stratum h. Consequently, $0 < W_h < 1$, and $\sum_{h=1}^{L} W_h = 1$. Comparing (5.17) and (5.18), we see that τ_y is simply the sum of the L strata totals; by contrast, μ_y is a weighted average of the strata averages.

ASIDE: In areal surveys of natural resources, the stratification criterion oftentimes is a land classification by type of predominant vegetative cover. In such
surveys, the strata weights, $W_h, h = 1, \ldots, L$, are regarded as the proportional
land area in each stratum, rather than the proportional number of discrete population units, which may remain unknown even at the conclusion of sampling.
That is, if A_h is the amount of land area in stratum h, a particular land cover
class, and $A = \sum_{h=1}^{L} A_h$, then $W_h = A_h/A$.

In view of (5.18), an alternative expression of $\hat{\mu}_{y\pi,\text{st}}$ is

$$\hat{\mu}_{y\pi,\text{st}} = W_1\hat{\mu}_{y\pi,1} + W_2\hat{\mu}_{y\pi,2} + \cdots + W_L\hat{\mu}_{y\pi,L}$$

$$= \sum_{h=1}^{L} W_h\hat{\mu}_{y\pi,h}. \tag{5.19}$$

In similar fashion, an alternative expression for (5.16) is

$$V\left[\hat{\mu}_{y\pi,\text{st}}\right] = \sum_{h=1}^{L} W_h^2\, V\left[\hat{\mu}_{y\pi,h}\right], \tag{5.20}$$

which can be estimated by

$$\hat{v}\left[\hat{\mu}_{y\pi,\text{st}}\right] = \sum_{h=1}^{L} W_h^2\, \hat{v}\left[\hat{\mu}_{y\pi,h}\right]. \tag{5.21}$$

The multiplicity of formulas in the preceding paragraphs is summarized in
Table 5.1.

5.3.3 More general estimation

Strata totals, $\tau_{y,h}, h = 1, \ldots, L$, need not be estimated by the HT estimator, nor is it
necessary to estimate each stratum total with the same estimator. Indeed, because
different sampling designs may be employed in different strata, it may be quite
counterproductive to precise estimation to use the HT or any other estimator in all
strata.

Adopting a more general notation than in the previous subsection, let $\hat{\tau}_{y,h}$ denote
an estimator of $\tau_{y,h}$. Depending on context, $\hat{\tau}_{y,h}$ might be any of the estimators
explicated thus far, or it might be the generalized ratio, regression, or difference
estimators, or others, to be introduced in later chapters. Irrespective of which
specific estimators are used to estimate $\tau_{y,h}, h = 1, 2, \ldots, L$, an estimator of
$\tau_y = \sum_{h=1}^{L} \tau_{y,h}$ is

$$\hat{\tau}_{y,\text{st}} = \sum_{h=1}^{L} \hat{\tau}_{y,h}. \tag{5.22}$$

Table 5.1 *Horvitz–Thompson estimation following stratified sampling*

	Stratum total	Population total
Parameter	$\tau_{y,h} = \displaystyle\sum_{u_k \in \mathcal{P}_h} y_k$	$\tau_y = \displaystyle\sum_{h=1}^{L} \tau_{y,h}$
Estimator	$\hat{\tau}_{y\pi,h} = \displaystyle\sum_{u_k \in \mathcal{S}_h} \frac{y_k}{\pi_k}$	$\hat{\tau}_{y\pi,\text{st}} = \displaystyle\sum_{h=1}^{L} \hat{\tau}_{y\pi,h}$
Variance of estimator	$V\left[\hat{\tau}_{y\pi,h}\right]$	$V\left[\hat{\tau}_{y\pi,\text{st}}\right] = \displaystyle\sum_{h=1}^{L} V\left[\hat{\tau}_{y\pi,h}\right]$
Estimated variance	$\hat{v}\left[\hat{\tau}_{y\pi,h}\right]$	$\hat{v}\left[\hat{\tau}_{y\pi,\text{st}}\right] = \displaystyle\sum_{h=1}^{L} \hat{v}\left[\hat{\tau}_{y\pi,h}\right]$
Confidence interval	$\hat{\tau}_{y\pi,h} \pm t_{n_h-1}\sqrt{\hat{v}\left[\hat{\tau}_{y\pi,h}\right]}$	$\hat{\tau}_{y\pi,\text{st}} \pm t_{n-L}\sqrt{\hat{v}\left[\hat{\tau}_{y\pi,\text{st}}\right]}$
Percentage margin of error	$\hat{\tau}_{y\pi,h} \pm \dfrac{t_{n_h-1}\sqrt{\hat{v}\left[\hat{\tau}_{y\pi,h}\right]}}{\hat{\tau}_{y\pi,h}/100\%}$	$\hat{\tau}_{y\pi,\text{st}} \pm \dfrac{t_{n-L}\sqrt{\hat{v}\left[\hat{\tau}_{y\pi,\text{st}}\right]}}{\hat{\tau}_{y\pi,\text{st}}/100\%}$

	Stratum mean	Population mean
Parameter	$\mu_{y,h} = \dfrac{1}{N_h}\displaystyle\sum_{u_k \in \mathcal{P}_h} y_k$	$\mu_y = \displaystyle\sum_{h=1}^{L} W_h \mu_{y,h}$
Estimator	$\hat{\mu}_{y\pi,h} = \dfrac{\hat{\tau}_{y\pi,h}}{N_h}$	$\hat{\mu}_{y\pi,\text{st}} = \displaystyle\sum_{h=1}^{L} W_h \hat{\mu}_{y\pi,h}$
Variance of estimator	$V\left[\hat{\mu}_{y\pi,h}\right]$	$V\left[\hat{\mu}_{y\pi,\text{st}}\right] = \displaystyle\sum_{h=1}^{L} W_h^2\, V\left[\hat{\mu}_{y\pi,h}\right]$
Estimated variance	$\hat{v}\left[\hat{\mu}_{y\pi,h}\right]$	$\hat{v}\left[\hat{\mu}_{y\pi,\text{st}}\right] = \displaystyle\sum_{h=1}^{L} W_h^2\, \hat{v}\left[\hat{\mu}_{y\pi,h}\right]$
Confidence interval	$\hat{\mu}_{y\pi,h} \pm t_{n_h-1}\sqrt{\hat{v}\left[\hat{\mu}_{y\pi,h}\right]}$	$\hat{\mu}_{y\pi,\text{st}} \pm t_{n-L}\sqrt{\hat{v}\left[\hat{\mu}_{y\pi,\text{st}}\right]}$
Percentage margin of error	$\hat{\mu}_{y\pi,h} \pm \dfrac{t_{n_h-1}\sqrt{\hat{v}\left[\hat{\mu}_{y\pi,h}\right]}}{\hat{\mu}_{y\pi,h}/100\%}$	$\hat{\mu}_{y\pi,\text{st}} \pm \dfrac{t_{n-L}\sqrt{\hat{v}\left[\hat{\mu}_{y\pi,\text{st}}\right]}}{\hat{\mu}_{y\pi,\text{st}}/100\%}$

Table 5.2 *General estimation following stratified sampling*

	Stratum total	Population total
Parameter	$\tau_{y,h} = \sum_{\mathcal{U}_k \in \mathscr{P}_h} y_k$	$\tau_y = \sum_{h=1}^{L} \tau_{y,h}$
Estimator	$\hat{\tau}_{y,h}$	$\hat{\tau}_{y,\text{st}} = \sum_{h=1}^{L} \hat{\tau}_{y,h}$
Variance of estimator	$V\left[\hat{\tau}_{y,h}\right]$	$V\left[\hat{\tau}_{y,\text{st}}\right] = \sum_{h=1}^{L} V\left[\hat{\tau}_{y,h}\right]$
Estimated variance	$\hat{v}\left[\hat{\tau}_{y,h}\right]$	$\hat{v}\left[\hat{\tau}_{y,\text{st}}\right] = \sum_{h=1}^{L} \hat{v}\left[\hat{\tau}_{y,h}\right]$
Confidence interval	$\hat{\tau}_{y,h} \pm t_{n_h-1}\sqrt{\hat{v}\left[\hat{\tau}_{y,h}\right]}$	$\hat{\tau}_{y,\text{st}} \pm t_{n-L}\sqrt{\hat{v}\left[\hat{\tau}_{y,\text{st}}\right]}$
Percentage margin of error	$\hat{\tau}_{y,h} \pm \dfrac{t_{n_h-1}\sqrt{\hat{v}\left[\hat{\tau}_{y,h}\right]}}{\hat{\tau}_{y,h}/100\%}$	$\hat{\tau}_{y,\text{st}} \pm \dfrac{t_{n-L}\sqrt{\hat{v}\left[\hat{\tau}_{y,\text{st}}\right]}}{\hat{\tau}_{y,\text{st}}/100\%}$
	Stratum mean	Population mean
Parameter	$\mu_{y,h} = \dfrac{1}{N_h}\sum_{\mathcal{U}_k \in \mathscr{P}_h} y_k$	$\mu_y = \sum_{h=1}^{L} W_h \mu_{y,h}$
Estimator	$\hat{\mu}_{y,h} = \dfrac{\hat{\tau}_{y,h}}{N_h}$	$\hat{\mu}_{y,\text{st}} = \sum_{h=1}^{L} W_h \hat{\mu}_{y,h}$
Variance of estimator	$V\left[\hat{\mu}_{y,h}\right]$	$V\left[\hat{\mu}_{y,\text{st}}\right] = \sum_{h=1}^{L} W_h^2 V\left[\hat{\mu}_{y,h}\right]$
Estimated variance	$\hat{v}\left[\hat{\mu}_{y,h}\right]$	$\hat{v}\left[\hat{\mu}_{y,\text{st}}\right] = \sum_{h=1}^{L} W_h^2 \hat{v}\left[\hat{\mu}_{y,h}\right]$
Confidence interval	$\hat{\mu}_{y,h} \pm t_{n_h-1}\sqrt{\hat{v}\left[\hat{\mu}_{y,h}\right]}$	$\hat{\mu}_{y,\text{st}} \pm t_{n-L}\sqrt{\hat{v}\left[\hat{\mu}_{y,\text{st}}\right]}$
Percentage margin of error	$\hat{\mu}_{y,h} \pm \dfrac{t_{n_h-1}\sqrt{\hat{v}\left[\hat{\mu}_{y,h}\right]}}{\hat{\mu}_{y,h}/100\%}$	$\hat{\mu}_{y,\text{st}} \pm \dfrac{t_{n-L}\sqrt{\hat{v}\left[\hat{\mu}_{y,\text{st}}\right]}}{\hat{\mu}_{y,\text{st}}/100\%}$

If any of the $\hat{\tau}_{y,h}$ in (5.22) biasedly estimate the corresponding $\tau_{y,h}$, then $\hat{\tau}_{y,\text{st}}$ will be a biased estimator of τ_y. The variance of $\hat{\tau}_{y,\text{st}}$ parallels that of $\hat{\tau}_{y\pi,\text{st}}$ in that

$$V\left[\hat{\tau}_{y,\text{st}}\right] = \sum_{h=1}^{L} V\left[\hat{\tau}_{y,h}\right]. \qquad (5.23)$$

Moreover, an estimator of μ_y is derived similarly to that of $\hat{\mu}_{y\pi,\text{st}}$:

$$\hat{\mu}_{y,\text{st}} = \frac{\hat{\tau}_{y,\text{st}}}{N}$$

$$= \sum_{h=1}^{L} W_h \hat{\mu}_{y,h} \qquad (5.24)$$

where $\hat{\mu}_{y,h} = \hat{\tau}_{y,h}/N_h$.

Just as Table 5.1 displays the formulas based on HT estimation following stratified sampling, Table 5.2 displays the more general formulas. In these tables the recurring theme is that estimates of strata totals add together simply to constitute an estimate of τ_y, and that estimates of their variances likewise are summed to provide an estimate of $V[\hat{\tau}_{y,\text{st}}]$. By contrast, strata means must be appropriately weighted when estimating the population mean.

5.3.4 Stratified random sampling

Stratified random sampling is the design wherein simple random sampling is used within each stratum. Thus, it is a special case of stratified sampling in general, which allows for any design within a stratum, not just SRS. It is worthwhile to emphasize, again, that in the general case there is no necessary implication that the same design be used in all strata.

Table 5.3 provides the formulas for HT estimation following stratified random sampling. Note that the variances of the estimators of the stratum totals and means are expressed in terms of $\sigma_{y,h}^2$, which is the variance of the y_k's within $\mathscr{P}_h$, i.e.,

$$\sigma_{y,h}^2 = \frac{1}{N_h - 1} \sum_{\mathcal{U}_k \in \mathscr{P}_h} \left(y_k - \mu_{y,h}\right)^2. \qquad (5.25)$$

Letting $\mathscr{S}_h$ denote the sample of n_h elements selected from $\mathscr{P}_h$, a design-unbiased estimator of $\sigma_{y,h}^2$ following SRSwoR is:

$$s_{y,h}^2 = \frac{1}{n_h - 1} \sum_{\mathcal{U}_k \in \mathscr{S}_h} (y_k - \bar{y}_h)^2, \qquad (5.26)$$

where $\bar{y}_h$ is the sample mean in $\mathscr{P}_h$,

$$\bar{y}_h = \frac{1}{n_h} \sum_{\mathcal{U}_k \in \mathscr{S}_h} y_k. \qquad (5.27)$$

Table 5.3 *Horvitz–Thompson estimation: stratified random sampling without replacement*

	Stratum total	Population total
Parameter	$\tau_{y,h} = \sum_{u_k \in \mathcal{P}_h} y_k$	$\tau_y = \sum_{h=1}^{L} \tau_{y,h}$
Estimator	$\hat{\tau}_{y\pi,h} = N_h \bar{y}_h$	$\hat{\tau}_{y\pi,\mathrm{st}} = \sum_{h=1}^{L} N_h \bar{y}_h$
Variance of estimator	$V\left[\hat{\tau}_{y\pi,h}\right] = \left[\dfrac{N_h - n_h}{n_h/N_h}\right] \sigma_{y,h}^2$	$V\left[\hat{\tau}_{y\pi,\mathrm{st}}\right] = \sum_{h=1}^{L} V\left[\hat{\tau}_{y\pi,h}\right]$
Estimated variance	$\hat{v}\left[\hat{\tau}_{y\pi,h}\right] = \left[\dfrac{N_h - n_h}{n_h/N_h}\right] s_{y,h}^2$	$\hat{v}\left[\hat{\tau}_{y\pi,\mathrm{st}}\right] = \sum_{h=1}^{L} \hat{v}\left[\hat{\tau}_{y\pi,h}\right]$
Confidence interval	$\hat{\tau}_{y\pi,h} \pm t_{n_h-1} \sqrt{\hat{v}\left[\hat{\tau}_{y\pi,h}\right]}$	$\hat{\tau}_{y\pi,\mathrm{st}} \pm t_{n-L} \sqrt{\hat{v}\left[\hat{\tau}_{y\pi,\mathrm{st}}\right]}$
Percentage margin of error	$\hat{\tau}_{y\pi,h} \pm \dfrac{t_{n_h-1} \sqrt{\hat{v}\left[\hat{\tau}_{y\pi,h}\right]}}{\hat{\tau}_{y\pi,h}/100\%}$	$\hat{\tau}_{y\pi,\mathrm{st}} \pm \dfrac{t_{n-L} \sqrt{\hat{v}\left[\hat{\tau}_{y\pi,\mathrm{st}}\right]}}{\hat{\tau}_{y\pi,\mathrm{st}}/100\%}$

	Stratum mean	Population mean
Parameter	$\mu_{y,h} = \dfrac{1}{N_h} \sum_{u_k \in \mathcal{P}_h} y_k$	$\mu_y = \sum_{h=1}^{L} W_h \mu_{y,h}$
Estimator	$\hat{\mu}_{y\pi,h} = \bar{y}_h$	$\hat{\mu}_{y\pi,\mathrm{st}} = \sum_{h=1}^{L} W_h \bar{y}_h$
Variance of estimator	$V\left[\hat{\mu}_{y\pi,h}\right] = \left[\dfrac{N_h - n_h}{n_h N_h}\right] \sigma_{y,h}^2$	$V\left[\hat{\mu}_{y\pi,\mathrm{st}}\right] = \sum_{h=1}^{L} W_h^2 V\left[\hat{\mu}_{y\pi,h}\right]$
Estimated variance	$\hat{v}\left[\hat{\mu}_{y\pi,h}\right] = \left[\dfrac{N_h - n_h}{n_h N_h}\right] s_{y,h}^2$	$\hat{v}\left[\hat{\mu}_{y\pi,\mathrm{st}}\right] = \sum_{h=1}^{L} W_h^2 \hat{v}\left[\hat{\mu}_{y\pi,h}\right]$
Confidence interval	$\hat{\mu}_{y\pi,h} \pm t_{n_h-1} \sqrt{\hat{v}\left[\hat{\mu}_{y\pi,h}\right]}$	$\hat{\mu}_{y\pi,\mathrm{st}} \pm t_{n-L} \sqrt{\hat{v}\left[\hat{\mu}_{y\pi,h}\right]}$
Percentage margin of error	$\hat{\mu}_{y\pi,h} \pm \dfrac{t_{n_h-1} \sqrt{\hat{v}\left[\hat{\mu}_{y\pi,h}\right]}}{\hat{\mu}_{y\pi,h}/100\%}$	$\hat{\mu}_{y\pi,\mathrm{st}} \pm \dfrac{t_{n-L} \sqrt{\hat{v}\left[\hat{\mu}_{y\pi,h}\right]}}{\hat{\mu}_{y\pi,\mathrm{st}}/100\%}$

The inclusion probability of $\mathcal{U}_k \in \mathcal{P}_h$ is $\pi_k = n_h/N_h$ under stratified random sampling. In general, stratified random sampling is not an equal probability sampling design: although the inclusion probability is identical for each and every unit within a given stratum, units in other strata may have different inclusion probabilities, except for the case when the sample is allocated proportionally across the L strata.

Example 5.5

Wilk et al. (1977) provide a wealth of data collected during a 13-month stratified random sampling of the New York Bight, the portion of the Atlantic Ocean between Long Island, New York and Cape May, New Jersey. Ocean depth was divided into $L = 7$ horizontal strata: $0-10$, $11-19$, $20-28$, $29-55$, $56-110$, $111-183$, and $184-366$ m. Trawl surveys were conducted in each stratum, and the weight of fish caught in each trawl was recorded. The mean catch weight in the seven strata are displayed in the following table. The strata weights are the proportional volume of water in each stratum. Since water is a continuous medium, N_h is infinite (unless water volume is considered to be a sum of discrete volumetric units as in Example 5.6). These data provide a 90% confidence interval extending from 55.4 kg to 98.1 kg for the average weight of fish per trawl.

Stratum	Depth (m)	Stratum weight W_h	Sample size n_h	Mean catch weight (kg) $\hat{\mu}_{y,h}$	Estimated variance $\hat{v}[\hat{\mu}_{y,h}]$
1	$0-10$	0.098	163	16.3	13.0
2	$11-19$	0.080	132	117.8	159.8
3	$20-28$	0.075	112	63.6	109.0
4	$29-55$	0.186	114	101.9	169.7
5	$56-110$	0.216	86	143.8	422.3
6	$111-183$	0.255	29	25.4	43.3
7	$184-366$	0.090	26	50.1	132.2

Example 5.6

The pond sample mentioned in Example 5.2 was a stratified random sample from $L = 3$ strata of lake water. Based on proportional volume, the stratum weight for the epilimnion layer was $W_1 = 1/7$; for the thermocline layer, $W_2 = 2/7$; and for the hyperlimnion layer, $W_3 = 4/7$. The sample units were liter containers of lake water, of which there were $n_1 = 20$ selected from the epilimnion layer at depths and locations selected by simple random sampling. In addition, there were $n_2 = 10$ samples taken from the thermocline layer, and $n_3 = 20$ from the hyperlimnion layer.

The following results were obtained, expressed in number of daphnia per liter:

$$\hat{\mu}_{y\pi,1} = 19.5, \qquad \hat{\mu}_{y\pi,2} = 11.3, \qquad \hat{\mu}_{y\pi,3} = 1.7\overline{3}$$

Therefore from (5.24),

$$\hat{\mu}_{y,st} = \left(\frac{1}{7} \times 19.5\right) + \left(\frac{2}{7} \times 11.3\right) + \left(\frac{4}{7} \times 1.7\bar{3}\right) = 7.0 \text{ daphnia } l^{-1}.$$

This estimate, when multiplied by the total liters of pond water, would estimate the total number of daphnia in the pond.

5.4 Sample allocation among strata

When planning a stratified sample with an overall sample size of n units or sampling locations, it is necessary to determine the size of the sample, $n_h, h = 1, \ldots, L$, to select in each stratum, such that (5.3) is satisfied. There are many ways in which the overall sampling effort among L strata can be allocated. A few of the more commonly considered *sample allocation rules* are described in this section.

5.4.1 Equal allocation

Perhaps the simplest procedure is to sample the same number of units from each stratum, i.e., to make $n_h = n/L$ in each $\mathcal{P}_h$. This equal allocation of the sample among the L strata may be particularly worthwhile when the strata sizes, as measured by N_h, are all approximately identical: $N_h \approx N/L$.

When the strata sizes, N_h, vary substantially, the larger strata will be sampled less intensely than smaller strata following equal allocation of the samples. As a consequence, strata totals and other parameters of possible interest may be estimated considerably less precisely for large strata than for small strata.

Example 5.7

The $N = 1{,}058$ tree population sampled in Example 3.8 was stratified by species into $L = 4$ strata. The distribution of total aboveground biomass in each species is depicted in Figure 5.1. A SRSwoR of size $n = 13$ trees was selected from each stratum. The sampling fraction of balsam fir was 4.4%; the fraction sampled from the black spruce, white birch, and white spruce strata were 4.1%, 4.7%, and 7.7%, respectively. The sample provided an estimate of average total aboveground biomass of 68.6 kg with an estimated standard error of 10.4 kg, both of which are smaller than the estimates provided by the $n = 52$ SRSwoR from the unstratified population in Example 3.8. The estimated standard errors of average total aboveground biomass were 12.6, 18.5, 28.8, and 17.6 kg, respectively.

If n is large, equal allocation may create the situation in which n_h is larger than N_h in a small stratum. When this unusual situation arises, one can include all N_h units in the sample, and then distribute the remaining sample draws, originally allocated to $\mathcal{P}_h$, evenly among the remaining strata. Alternatively, one could opt to combine

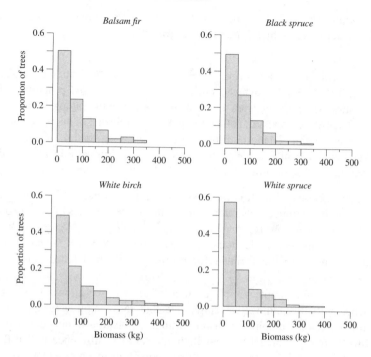

Figure 5.1 *Total aboveground biomass (kg) of* 296 *balsam fir,* 318 *black spruce,* 275 *white birch, and* 169 *white spruce trees.*

the smaller strata into larger but fewer strata in an effort to ensure that $n_h < N_h$ for all newly combined strata.

Equal allocation may not yield much gain in precision when stratum variances, $\sigma_{y,h}^2$, vary considerably in magnitude.

5.4.2 Proportional allocation

An alternative to the equal allocation rule is one which strives to select a constant fraction of each stratum. Specifically, if $f_h = n_h/N_h$ is the sampling fraction in $\mathcal{P}_h$, then the *proportional allocation* rule aims to select samples comprising the same proportion of each stratum, such that $f_1 = f_2 = \cdots = f_L$. This can be achieved by making

$$n_h = \left(\frac{n}{N}\right) N_h.$$

Note that the sampling fraction in each stratum, $f_h = n_h/N_h$, is the same as overall sampling fraction, $f = n/N$.

Example 5.8

Suppose $L = 2$, $N_1 = 1500$, and $N_2 = 2700$. An overall sampling intensity of

1% implies $f = n/N = 0.01$, and thus $n_1 = 15$ elements are sampled from $\mathcal{P}_1$ and $n_2 = 27$ elements are sampled from $\mathcal{P}_2$.

The proportional allocation rule, $n_h = (n/N)N_h$ can be written equivalently as $n_h = n(N_h/N) = nW_h$, which indicates that whatever fixed proportion, W_h, of the population occurs in $\mathcal{P}_h$, then that same proportion of the sample will be selected from $\mathcal{P}_h$. Thus, the composition of the sample will mimic that of the population based on strata membership. Many find this to be an appealing feature of this allocation plan. Another practical and appealing feature of proportional allocation is that n_h can never exceed N_h, in contrast to the equal allocation rule and others.

Under proportional allocation of the sample to the L strata followed by stratified random sampling within each stratum, the inclusion probabilities of units in different strata are equal: for any $\mathcal{U}_k$, $\pi_k = n_h/N_h = n(N_h/N)/N_h = n/N$. These inclusion probabilities are identical to those under SWSwoR in an unstratified population. Therefore the formula for estimating the population total given in Table 5.3, namely $\hat{\tau}_{y\pi,\text{st}} = \sum_{h=1}^{L} N_h \bar{y}_h$, simplifies to the familiar $\hat{\tau}_{y\pi,\text{st}} = N\bar{y}$, where $\bar{y}$ is the simple sample average of all n observations. This notational equivalence notwithstanding, we emphasize that the sampling design consisting of stratified random sampling with proportional allocation is not identical to that of SRSwoR from an unstratified population. A crucial difference is that the joint inclusion probability of any two units differs in the two designs, leading to different sampling distributions of $N\bar{y}$.

In practice, it is unlikely that n_h as computed by either formula will be integer valued, making it necessary to round up, or down, to the closest natural number. In this circumstance the equality of the strata sampling fractions will be only approximate.

Example 5.9

By rounding up to the next higher integer value, a nominal 5% proportionally allocated sample from the population used in Example 5.7 resulted in a sample of size $n_1 = 15$ in the balsam fir stratum, $n_2 = 16$ in the black spruce stratum, $n_3 = 14$ in the white birch stratum, and $n_4 = 9$ in the white spruce stratum. A stratified random sample with $n = 54$ trees allocated in this fashion provided an estimate of average total aboveground biomass of 74.7 kg with an estimated standard error of 9.8 kg. The estimated standard errors of average total aboveground biomass were 25.6, 10.3, 14.1, and 29.2 kg, respectively.

Oderwald (1993) presents a very understandable analysis of the gain resulting from proportional allocation in a stratified forest inventory when compared to the accuracy achievable in an unstratified design entailing equivalent sampling effort.

ASIDE: In Example 5.7 the estimated standard error of average total above-ground biomass in the white spruce stratum is less than the estimated standard error for the same stratum in Example 5.9. This results in part from the smaller proportionally allocated sample in this stratum than the 13 tree sample resulting from the equal allocation of the $n = 52$ sample elements in the earlier example. However, the balsam fir sample in Example 5.9 had two more trees than that with equal allocation, yet its estimated standard error is more than twice the magnitude observed in Example 5.7. This is a specific example of sampling variation: different samples yield different estimates because they contain different population elements. In Example 5.7, the minimum and maximum biomass values were 4.5 and 153 kg, respectively, whereas in Example 5.9 they were 1 and 327 kg. The greater within-sample variation is reflected in the greater estimate of standard error of the estimated average total aboveground biomass.

5.4.3 x-Proportional allocation

When auxiliary information, x_k, of the sort introduced in §3.3 is available for every unit of the population, the sample sizes in the $\mathcal{P}_h$ can be calculated proportionally to $\tau_{x,h}$, the strata total of the auxiliary variate in $\mathcal{P}_h$, i.e.,

$$n_h = n \left(\frac{\tau_{x,h}}{\tau_x} \right).$$

The efficacy of *x-proportional allocation* depends on auxiliary information that is well and positively correlated with the y-variate of interest. It is particularly propitious when sampling skew populations in which the attributes of a small number of the population elements constitute a large proportion of τ_y (Raj 1968).

5.4.4 Optimal allocation, equal sampling costs

The variance of the y_k values within $\mathcal{P}_h$ is expressed as $\sigma_{y,h}^2$, and it is a measure of the within-stratum heterogeneity of the y_k. Consider two strata, $\mathcal{P}_h$ and $\mathcal{P}_{h'}$, of roughly equivalent sizes, $N_h \approx N_{h'}$; the stratum with the greater variance will need to be sampled more intensely in order to estimate τ_h and $\tau_{h'}$ equally well. When the two strata also differ in size, then both sources of between-strata disparities interact to affect the precision with which the population total, τ_y can be estimated. The allocation rule presented in this section, due to Tschuprow (1922) and Neyman (1934), and often termed *Neyman allocation*, aims to determine the sample sizes in the L strata that minimize $V[\hat{\tau}_{y\pi,\text{st}}]$. It pertains only for the specific design of stratified random sampling. Equal-cost optimal allocation rules have not been discerned for the plethora of other sampling designs that could be used, owing to the difficulty of the optimization problem to be solved.

The sample size in $\mathcal{P}_h$ for stratified random sampling and equal sampling costs is

$$n_h = n \left(\frac{N_h \sigma_{y,h}}{\sum_{i=1}^{L} N_i \sigma_{y,i}} \right). \tag{5.28}$$

Details on the derivation of (5.28) can be found in Cochran (1977) and Särndal et al. (1992).

In (5.28) the fraction of n that is allocated to $\mathcal{P}_h$ depends both on the size of the stratum, N_h, and the within-stratum standard deviation, $\sigma_{y,h}$. In effect, equal-cost optimal allocation is allocation proportional to $N_h \sigma_{y,h}$. The stratum sample size n_h increases as stratum size increases and as $\sigma_{y,h}$ increases, whereas n_h decreases for smaller and less variable strata. But if $N_h > N_{h'}$, yet $\sigma_{y,h} < \sigma_{y,h'}$, then the relationship of n_h to $n_{h'}$ depends in a more complicated fashion on the product of strata size and variance. The equal-cost optimal allocation rule balances the need for smaller samples in smaller strata with the need for larger samples in more heterogeneous strata.

Example 5.10

Suppose the within-stratum stand deviations of $\mathcal{P}_1$ and $\mathcal{P}_2$ of Example 5.8 were $\sigma_{y,1} = 100$ and $\sigma_{y,2} = 300$, respectively. Then, $N_1 \sigma_{y,1} = 150,000$ and $N_2 \sigma_{y,2} = 810,000$. By the equal-cost optimal allocation rule, $n_1 = 7$ and $n_2 = 35$.

Example 5.11

Suppose the within-stratum stand deviations of $\mathcal{P}_1$ and $\mathcal{P}_2$ of Example 5.8 were $\sigma_{y,1} = 300$ and $\sigma_{y,2} = 100$, respectively. By the equal-cost optimal allocation rule, $n_1 = 26$ and $n_2 = 16$.

As with the equal allocation rule, it is possible that n_h calculated by (5.28) may exceed N_h. Remedies for this have already been discussed.

In principle, stratified random sampling with equal-cost optimal allocation will provide more precise estimation of τ_y than stratified random sampling with equal or proportional allocation. A difficulty with this equal-cost optimal allocation rule, however, is the need to know each and every $\sigma_{y,h}$ in order to calculate n_h. These values are unlikely to be known without error even at the conclusion of sampling. One must resort to substituting an estimate of each $\sigma_{y,h}$. Such estimates may be available from prior surveys of the population of current interest or from populations of similar character and composition. Another alternative is to conduct a small pilot survey of the population of current interest in order to obtain an approximate value for each $\sigma_{y,h}$.

Naturally, the use of estimates of the $\sigma_{y,h}$ in (5.28) vitiates the optimality of the calculated n_hs. The rationale for using the equal-cost optimal allocation rule with estimates of the $\sigma_{y,h}$, rather than the less demanding proportional allocation rule, is

> ASIDE: Sukhatme et al. (1984, p. 126) mention an alternative tactic for near
> optimal allocation when auxiliary information, x_k, is available for every unit of
> the population. Presuming the the auxiliary information is well and positively
> correlated with the y-variate of interest, their suggestion for *modified Neyman
> allocation* is to use the known $\sigma_{x,h}$ in place of the unknown $\sigma_{y,h}$ in (5.28).
> Särndal et al. (1992, p. 107) adopt the name 'x-optimal allocation.'

that $V[\hat{\tau}_{y\pi,\text{st}}]$ will be less with the former, even though it may not be the minimum
(optimal) variance achievable when the n_h are based on known values of the $\sigma_{y,h}$.

5.4.5 Optimal allocation, unequal sampling costs

When the cost of sampling a unit depends consequentially on the stratum in which
it resides, a rational approach to allocate sampling effort among the L strata is one
that seeks to estimate τ_y with greatest precision (minimum variance) for a stipulated
cost, C. The optimal allocation with unequal sampling costs is based on a linear cost
function:

$$C = c_0 + \sum_{h=1}^{L} n_h c_h, \tag{5.29}$$

where c_h represents the cost of sampling a population unit in the $\mathscr{P}_h$, and c_0
represents the combined overhead and administrative costs associated with the
sampling effort. With this cost function the strata sample sizes that minimize
$V[\hat{\tau}_{y\pi,\text{st}}]$ for a stipulated cost, C are computed as

$$n_h = (C - c_0) \left(\frac{N_h \sigma_{y,h}/\sqrt{c_h}}{\sum_{i=1}^{L} N_i \sigma_{y,i} \sqrt{c_i}} \right). \tag{5.30}$$

Note that in (5.30), the $\sqrt{c_h}$ is divided into σ_{yh} in the numerator term, but not in the
denominator term; this is not a typographical error, but rather a consequence of the
optimization. For details, consult Cochran (1977, p. 97–98), who derives the optimal
allocation when one wishes to minimize the cost of sampling for a stipulated value
of $V[\hat{\tau}_{y\pi,\text{st}}]$. With this optimal allocation scheme, the need for larger samples from
more heterogeneous strata is counterbalanced by the practical need to obtain the most
precise information for the money expended.

   Jessen (1978, p. 190) advised that when cost differences are small and poorly
estimated, the benefits accruing from this allocation rule may be more illusory than
real.

5.4.6 Power allocation

In Chapter 1 the coefficient of variation, γ_y, was introduced as the standard deviation
of the attribute y divided by μ_y. In similar fashion, an estimator's coefficient of
variation, CV $(\hat{\tau}_y)$ or CV $(\hat{\mu}_y)$, is its standard error divided by the parameter being

estimated, τ_y or μ_y. Bankier (1988) noted that while Neyman allocation minimizes the CV of the estimated population total or mean, CV for individual strata may be very much greater than that for the entire population. The power allocation method he proposed offers a compromise between Neyman allocation and one which achieves nearly equal CV for the individual strata. As with the optimal allocation rules presented above, a stratified random sampling design is presumed.

Let τ_z be some measure of size of stratum h. The power allocation scheme sets the size of the SRSwoR in stratum h to be

$$n_h = n \left(\frac{\tau_{z,h}^q \, \sigma_{y,h}/\mu_{y,h}}{\sum_{i=1}^{L} \tau_{z,i}^q \, \sigma_{y,i}/\mu_{y,i}} \right), \tag{5.31}$$

for some constant exponent q, $0 \leq q \leq 1$. Bankier (1988) shows that choosing $q = 1$ and $\tau_z = \tau_y$, (5.31) is identical to Neyman allocation, whereas when $q = 0$, CV $(\hat{\tau}_{y,h})$ are approximately equal for the L strata. As q is decreased from 1, CV $(\hat{\tau}_y)$ necessarily increases, whereas the CV $(\hat{\tau}_{y,h})$ will become more homogeneous, as illustrated convincingly in Bankier's example involving the estimation of migration into Canada at both national and provincial levels based in a single stratified random sample.

The subjective choice of the power, q, used in (5.31) for this example, involved weighing the importance of (a) having roughly the same level of precision for estimates at both the provincial and national levels against (b) the increased sampling burden needed to achieve the desired level of precision in the smaller, more variable provinces. Although the context will differ from one application to the next, this sort of weighing of competing benefits will always be confronted when selecting the value of q to use with the power allocation of sample sizes among the L strata.

5.4.7 Allocation for multiresource surveys

Many surveys of natural and environmental resources are multivariate in nature, i.e., there are two or more characteristics of interest, and the objective of sampling is to estimate the population totals or means of multiple resources from a single sample. Even when the population is not stratified prior to sampling, the determination of an adequate sample size for precise estimation of multiple resources is more problematic than for single resource surveys. One simplistic approach involves identifying the most important resource, and allocating the sample to maximize the precision of its estimation, without explicit consideration of the precision and costs associated with sampling and estimating the remaining resource parameters. For surveys with multiple stakeholders, this solution merely shifts the crux of the problem to one of identification of the most important resource, which may be a contentious task.

When the population is stratified prior to sampling, any attempt to optimally allocate the sampling effort among the L strata becomes considerably more demanding. Generally speaking, the approaches to determining the sample size needed in each stratum in a multivariate setting have been aimed at finding some compromise allocation of the optimal univariate results. The mathematical complexity of optimal allocation in multivariate surveys precludes even a brief description here. Those interested in pursuing this topic are referred to Bankier's (1988, p. 176) discussion of alternate techniques and the sources cited therein. Bethel (1989) not only presents a computing algorithm for a particular solution to the multivariate allocation problem, he also traces a rich literature in this general area.

5.4.8 Comparison of allocation rules

The equal allocation rule is easy to understand and communicate, and when the strata are all roughly identical in size, it approximates the proportional allocation rule. The chief complaint against this rule is that strata rarely are even approximately the same size, and it often makes little sense to expend equal effort and resources in all strata.

Proportional allocation, equal-cost optimal allocation, and optimal allocation with unequal costs successively account for meaningful differences among strata due to size, heterogeneity, and sampling costs. But lacking knowledge of strata standard deviations and consequently resorting to estimates of these parameters to calculate the allocation formulas may seriously jeopardize the intended feature the two optimality rules. Using known standard deviations of an auxiliary variate circumvents this problem, but then one is left wondering whether stratified random sampling, upon which the optimal allocation is based, constitutes the best use of this auxiliary information: there may be a more efficient way to incorporate auxiliary information in the sample design. Moreover, in multiresource surveys from which a number of population attributes will be estimated based on a single stratified sample, it is quite unlikely that any allocation which purports to be optimal for one attribute will also be optimal for all others. Indeed, Stevens & Olsen (1991), at the conclusion of a study to examine the potential advantages and liabilities of optimal allocation in an environmental monitoring program, asserted that "[W]e see no justification for stratifying the sample to optimize the estimate of a proportion of the population in some particular class." Stehman & Overton (2002, p. 1919) opined similarly: "... most environmental surveys have multiple goals, and optimizing for a single attribute is seldom justified."

Proportional allocation thus emerges as the allocation rule that many find conceptually and intuitively appealing, especially when insufficient information exists to permit an optimal allocation (see, e.g., Deming 1950; Hansen et al. 1953; Cochran 1977, p. 103). Swindel & Yandle (1972) used a game-theoretic approach to show that proportional allocation is uniquely good when the strata variances are unknown. In multiresource surveys, proportional allocation constitutes a 'middle of the road approach,' wherein one knows that no single resource will be optimally estimated, yet it is hoped that none will be estimated with unacceptably poor precision either. Moreover, when the strata coefficients of variation, $\gamma_{y,h}$ are fairly homogeneous, proportional allocation approximates Neyman allocation (see, e.g., Smith & Gavaris 1993; Hansen et al. 1953, p. 215). Lastly, with proportional allocation of the sample, n_h will never exceed N_h, as can happen under the optimal allocation rules.

Related to the issue of allocating the sample among the strata is the matter of the number of strata, L, to establish. Raj (1968, §4.7) asserted that stratification can be carried to the point where only one unit is selected from each stratum. A

ASIDE: Equal probability systematic sampling may be regarded as stratified sampling with $n_h = 1$ element sampled from each stratum. For those familiar with analysis of variance—in which variation is partitioned among units within a group separately from the variation among groups—this view of systematic sampling is very natural. The strata consist of each consecutive set of a units in the sampling frame. There will either be $L = [N/a]_{giv}$ or $L = [N/a]_{giv} + 1$ such strata, each with a units except possibly for the last stratum which may contain fewer. A sample of size $n_h = 1$ is selected from each, giving a total sample of size L.

point of diminishing returns generally exists, beyond which any increased precision from augmenting the number of strata will be trivially small. Determining an optimal number of strata is a complicated matter and, as shown by Cochran (1977, §5A.8) and Murthy (1967, §7.11), usually requires the specification of a population structure and a cost function, where the costs may be quite uncertain.

5.5 Incorrect assignment of population elements into strata

Even with a clearly defined criterion for stratification, it is almost inevitable that some elements of the population are assigned to strata incorrectly. Oftentimes these incorrectly stratified units will only be detected if they are included in the sample. Although it is tempting to try to correct the error by reassigning the misstratified units to their correct strata, doing so introduces bias into the estimation of the strata and population parameters. By contrast, leaving all units in the strata to which they were assigned, even if incorrectly for some, may make estimation less precise than it otherwise could have been, but no bias is introduced.

To understand how reassignment causes bias, recall that the population total is $\tau_y = \sum_{h=1}^{L} \tau_{y,h}$, irrespective of how the population is stratified into the L strata. This identity holds, always, without any presumption that the units assigned in each stratum meet the criteria that had been planned. After stratification and prior to sampling there are N_h units assigned to $\mathcal{P}_h, h = 1, \ldots, L$, regardless how accurately or truly the stratification of the N population units among the L strata is conducted. The inclusion probability of each unit in a particular stratum will differ from what it would have been, had it been placed into a different stratum (with the possible exception of when proportional allocation is carried out exactly). Moving a sample unit to its correct stratum, after it is drawn from its incorrect stratum, causes the sample unit's actual probability of inclusion in the sample to differ from what is used in estimation of strata totals. In addition, under a post-sampling reassignment protocol, the strata sizes become random variables rather than known constants, a fact which further precludes unbiased estimation.

Example 5.12

Imagine a population with two strata; $\mathcal{P}_1$ has $N_1 = 100$ units assigned to it and $\mathcal{P}_2$ has $N_2 = 200$ units assigned. Imagine further that a stratified random sample is selected, and that $n_1 = 10$ units are drawn from $\mathcal{P}_1$ and $n_2 = 20$ units are drawn from $\mathcal{P}_2$. After selecting the sample it is discovered that five of the elements selected from $\mathcal{P}_2$ actually ought to have been assigned to $\mathcal{P}_1$. Yielding to the temptation to correct the mistake, these five units are thus considered to be part of the first stratum, thereby inflating the size of $\mathcal{P}_1$ to 105 units and decreasing N_2 to 195 units. Proceeding with apparent HT estimation of τ_y following a simple random sample from $\mathcal{P}_1$ one would compute $N_1 \bar{y}_1$, which does not coincide with the actual HT estimator $\hat{\tau}_{y\pi,1} = \sum_{k=1}^{15} y_k / \pi_k = 10 \sum_{k=1}^{15} y_k$, (note, $\pi_k = 10/100 = 20/200 = 0.1$). Moreover, it is doubtful whether either estimator is unbiasedly estimating anything of interest. This adaptive reassignment of stratum membership after having observed the sample leads also to the disquieting realization that had a different sample of 20 been selected from $\mathcal{P}_2$ that did not contain any of these five mistakenly stratified units, others might have been found and reassigned. Not only are the strata sizes, N_h, random, but the sample sizes, n_h, are random, also, with this sampling strategy.

A preferable strategy is to accept the mistaken stratification of some units with grace, but not to reassign stratum membership once the sample has been selected. By keeping strata sizes and sample sizes fixed, not random, the HT estimator of strata and population parameters retain their unbiasedness. Inasmuch as the mistakenly assigned units make the strata more heterogeneous than they would otherwise have been, these estimators may also be more variable than they would have been had the stratification been conducted without error. Most people find the loss in precision due to the acceptance of misstratified units to be more acceptable than the introduction of bias due to reassigning strata membership after sampling has concluded.

5.6 Double sampling for stratification

Stratified sampling is possible even in situations where the strata weights are unknown a priori by conducting a two-phase or double sample. With this method, introduced by Neyman (1938), the first phase of sampling is intended to gather data inexpensively in order to permit estimation of the L strata weights, W_h. The rationale behind this sampling design is that one may hope to gain the advantage in precision of estimation normally expected from stratified sampling when the strata weights can be estimated accurately. In land cover, natural resource, and agricultural surveys, aerial photography often has been used to provide an areal frame for the first phase of sampling. A large number of points are sampled from the photographs and the stratum to which each point belongs is identified according the land cover or resource that occurs at the point. Each stratum weight is estimated as the proportion of first-phase sample points in the stratum: $\hat{W}_h = m_h/m$, where m is the size of the first phase of sampling and m_h is the number of these sampling units in $\mathcal{P}_h$.

The second phase of sampling consists of the selection of a subsample of the first phase sample and measuring the variate of interest, y, at each point in the subsample. Almost always, the first phase of sampling is much larger than the second. The population mean, μ_y is then estimated by

$$\hat{\mu}_{y,\text{dst}} = \sum_{h=1}^{L} \hat{W}_h \hat{\mu}_{y,h}, \tag{5.32}$$

in which the estimated strata weights are used in place of the unknown strata weights. This leads directly to

$$\hat{\tau}_{y,\text{dst}} = N \hat{\mu}_{y,\text{dst}}. \tag{5.33}$$

The distribution of $\hat{\mu}_{y,\text{dst}}$ is affected by sampling variation in both phases of the double sample. Not only does it reflect variation in the estimated value of μ_y among the possible second-phase samples for a given first phase-sample, but also among the variation in $\hat{W}_h$ among the possible first-phase samples. Not surprisingly, the variance of $\hat{\mu}_{y,\text{dst}}$ following double sampling for stratification is greater than $V[\hat{\mu}_{y,\text{st}}]$ following stratified sampling with known strata weights.

An expression for $V[\hat{\mu}_{y,\text{dst}}]$ depends, inter alia, on the sampling designs used in both phases of the double sample. If SRSwoR is used in both phases, and HT estimation furnishes the strata means, then $\hat{\mu}_{y,\text{dst}}$ is a design-unbiased estimator of μ_y. Cochran (1977, p. 330) and Sukhatme et al. (1984, p. 139) provide an approximation to the variance (see, also, Schreuder et al. (1993, p. 168–170), who discuss the derivation of this approximation and others.):

$$V[\hat{\mu}_{y\pi,\text{dst}}] \approx V[\hat{\mu}_{y,\text{st}}] + \left(\frac{N-m}{N-1}\right) \frac{1}{m} \sum_{h=1}^{L} W_h \left(\mu_{y,h} - \mu_y\right)^2, \tag{5.34}$$

where m denotes the size of the first phase sample. The second term of (5.34) is the inflation in variance incurred by double sampling for stratification. It diminishes as the size of the first phase of sampling, m, is increased, hence it is usually important that phase one data be inexpensive to acquire so that a large sample can be selected. An estimator of $V[\hat{\mu}_{y\pi,\text{dst}}]$ is given by

$$\hat{v}[\hat{\mu}_{y,\text{dst}}] = \frac{N-1}{N} \sum_{h=1}^{L} \left(\frac{m_h-1}{m-1} - \frac{n_h-1}{N-1}\right) \frac{\hat{W}_h s_{y,h}^2}{n_h}$$

$$+ \frac{N-m}{N(m-1)} \sum_{h=1}^{L} \hat{W}_h \left(\hat{\mu}_{y,h} - \hat{\mu}_{y,\text{dst}}\right)^2, \tag{5.35}$$

where n_h is the size of the second phase sample in $\mathcal{P}_h$. If N is very much greater than the first phase sample, and if the first-phase sample itself is large, then $\hat{v}[\hat{\mu}_{y,\text{dst}}]$

in (5.35) simplifies to

$$\hat{v}\left[\hat{\mu}_{y,\text{dst}}\right] \approx \sum_{h=1}^{L} \frac{\hat{W}_h^2 s_{y,h}^2}{n_h} + \frac{1}{m-1} \sum_{h=1}^{L} \hat{W}_h \left(\hat{\mu}_{y,h} - \hat{\mu}_{y,\text{dst}}\right)^2. \qquad (5.36)$$

From (5.34) and (5.35) are obtained analogous results for estimating the population total with double sampling for stratification:

$$V\left[\hat{\tau}_{y,\text{dst}}\right] \approx N^2 V\left[\hat{\mu}_{y,\text{dst}}\right], \qquad (5.37)$$

and

$$\hat{v}\left[\hat{\tau}_{y,\text{dst}}\right] = N^2 \hat{v}\left[\hat{\mu}_{y,\text{dst}}\right]. \qquad (5.38)$$

Särndal et al. (1992, p. 351) discuss double sampling for stratification with arbitrary probability sampling designs in phase one and phase two, as well as the HT-like estimator of τ_y, its variance, and an unbiased estimator of variance. Using $\pi_{k(1)}$ to denote the inclusion probability of $\mathcal{U}_k$ in phase 1, and $\pi_{k(2)}$ to denote the conditional inclusion probability in phase 2 (this will vary from one first phase sample to another), the estimator of $\tau_{y,h}$ they propose is

$$\hat{\tau}_{y\pi,h}^* = \sum_{\mathcal{U}_k \in \mathcal{S}_h} \frac{y_k}{\pi_{k(1)}\pi_{k(2)}}.$$

Thus to estimate τ_y:

$$\hat{\tau}_y^* = \sum_{h=1}^{L} \hat{\tau}_{y\pi,h}^*$$

and

$$\hat{\mu}_y^* = \frac{\hat{\tau}_y^*}{N}.$$

Särndal et al. (1992, p. 352) also present the relevant estimation formulae when the probability sampling design at phase one is arbitrary and the second phase design is simple random sampling of the phase one units. Their mathematical notation appears more complicated initially than that used in this book, but their treatment is quite thorough.

Example 5.13

Forest surveys, or 'forest inventories' in the vernacular of forestry, often consist of a simple random sample of fixed area plots within the forest, the measurement of tree characteristics found thereon, and HT estimation of the abundance or density of one or more characteristics. MacLean (1972a,b) reports on a double sampling forest inventory in which aerial photography was used to stratify land in northwest portion of the USA into nine strata in a first phase of sampling. The photography was generally less than 4 years old, but one region relied on 15-year old images. A systematic grid of 18, 548 'photo plots' was overlaid on the images, and the proportion of these photo plots that fell into each stratum was computed. This served as the estimated stratum weight, W_h, because the proportional number of photo plots in each stratum provides an estimate of the fractional amount of land area in the stratum, i.e., $W_h = A_h/A$, where A_h was the land area in the stratum, and A was the total land area of the survey. A total of $L = 9$ strata were established: nonforest, noncommercial forest, and seven commercial forest strata. The second phase sample was allocated proportionally to the area of each stratum by selecting one-sixteenth of the photo points as the location of a ground plot. A measurement or field crew visited each field plot to take the actual measurements required by the survey. As a result of this double sampling with proportional allocation in the second phase, MacLean reported that the estimator of total timber volume was twice as precise as the estimator from an unstratified inventory using the same number of field plots.

5.7 Poststratification

5.7.1 Preliminary details about poststratification

It may be possible to take advantage of opportunities to increase the precision of estimation, which accrue from stratifying the population prior to sampling, even when the sample design is not a stratified design. The strategy of stratifying the sample after it has been selected and then estimating strata and population parameters in the usual stratified fashion is known as poststratification. The term 'poststratification' is used by many authors indiscriminately to mean the act of partitioning the selected sample into L discrete strata as well as the estimation of parameters of interest. Zhang (2000), however, did distinguish between these two meanings in order to emphasize that any sample may be poststratified, however, poststratified estimation is possible only when the strata sizes, N_h, or weights, W_h, are known.

One situation in which poststratification may be considered arises when information to identify homogeneous subgroups becomes available only after the sample has been selected. For example, satellite imagery, which heretofore had been unaffordable, is acquired following field sampling for natural resources, thereby permitting poststratification by land cover class. In another situation, poststratification may be useful when the sampling frame does not contain the information needed to assign the $\mathcal{U}_k$ to desired strata before the selection of the sample. As an example of this, imagine that civil authorities have a list frame of all single-family houses on the tax roster of a municipality. This roster serves as a frame from which a SRSwoR of dwellings is to be drawn in order to sample for the presence of arsenic-treated wood in exterior decks. A different municipal office keeps records of construction permits issued each year, thereby enabling the stratification of all exterior decks by year of construction. However, the recordkeeping of construction permits is separate from that of maintaining the tax roster, and merging of the two is administratively inefficient, thereby precluding an a priori stratification of dwellings in town. Yet another scenario for poststratification, mentioned by Holt & Smith (1979, p. 33) and Särndal et al. (1992, p. 268), occurs following multiresource surveys, in which a priori stratification works well for precise estimation of the principal resources of interest,

ASIDE: When using photography or other remotely sensed images, the assignment of photo plots or points in strata will not be utterly accurate. As hinted in Example 5.13, the available imagery may be so old that the present land-cover or vegetation class differs from what is revealed in the image. Also, classification may be erroneous because the apparent class, when looking down on a point from above, may differ from what is revealed on the ground. Moreover, some points will be difficult to classify into a discrete stratum, as described in a report by Frayer (1978) which, because of its insightfulness, deserves to be more widely known: "Most of the photo plots were probably easy to classify. The ones which took the most time on the part of the photo interpreters (and consequently were the most expensive to handle) were the ones which were doubtful. You've seen it before: an interpreter looks at a photo plot, is unsure how to classify it and after looking at it some more he asks advice of other interpreters. Finally, after spending much more time on this than the 'easy' ones, it's put into one class or another."

His clever solution is to create an additional stratum that consists of all the 'I do not know' photo plots. The first phase of sampling is made more efficient by separating these troublesome photo plots from the majority of others that are comparatively easy to classify. By way of example involving $m = 80,000$ photo plots in the first phase of double sampling for stratification, he shows how this simple tactic enabled a reduction in standard error of estimate of almost 50%. He concluded, "We were able to do this by absorbing most of the photo interpretation error into one class which represents a small segment of the population. At the same time we have reduced photo interpretation costs."

Frayer's example is included in section 5.11.2 in the Appendix.

but poorly for a large number of secondary resources. Therefore, poststrata can be formed after sample selection, which groups these secondary resources into more uniform subgroups than do the original strata. Similarly, when pre-sample strata are based on geographic regions or political subdivisions to enable separate reporting by region, the precision of estimation at the population level may be improved by using post-strata criteria more closely correlated with resource variates of interest.

Following poststratification of the sample of fixed overall size n, the strata sample sizes, $n_h, h = 1, 2, \ldots, L$, are random variables, because one cannot determine the number of units in each poststrata in advance of sampling. That is, the size of the sample in each stratum will depend on the particular sample that is selected, and the sample, itself, is a realization of a random, probabilistic process.

5.7.2 Poststratification of a SRSwoR sample

Although postratification may follow sample selection under any probability sampling design, we initially consider the situation where the unstratified population is sampled under the SRSwoR design with a sample size of n units. The expected value of n_h, i.e., the expected sample size in poststratum $\mathcal{P}_h$, is

$$E[n_h] = n\frac{N_h}{N}$$

$$= nW_h$$

(5.39)

Thus, poststratification of a SRSwoR sample of fixed size n yields strata samples that will vary in size from one sample to another, but that are, on average, identical in size to those one would choose by a priori stratification with proportional allocation.

An estimator of the stratum total, $\tau_{y,h}$ is

$$\hat{\tau}_{y,\text{pst},h} = \sum_{u_k \in \mathcal{s}_h} \frac{y_k}{n_h/N_h}$$

$$= N_h \bar{y}_h.$$

(5.40)

Despite the superficial similarity of (5.40) to $\hat{\tau}_{y\pi,h}$ in Tables 5.1 and 5.3, the statistical properties of $\hat{\tau}_{y,\text{pst},h}$ differ from those of $\hat{\tau}_{y\pi,h}$ because $\bar{y}_h$ in $\hat{\tau}_{y,\text{pst},h}$ is a ratio of two random variables. Nonetheless, provided the population has been sampled by SRSwoR or SRSwR, $\hat{\tau}_{y,\text{pst},h}$ is a design-unbiased estimator of $\tau_{y,h}$ (see Chapter 5 Appendix).

The estimators of strata totals add together to estimate τ_y, i.e.,

$$\hat{\tau}_{y,\text{pst}} = \sum_{h=1}^{L} \hat{\tau}_{y,\text{pst},h}.$$

(5.41)

Likewise, $\hat{\tau}_{y,\text{pst}}$ is a design-unbiased estimator of τ_y (see Chapter 5 Appendix), provided that the strata sizes are known without error. When the stratum weight is incorrectly presumed to be W_h', the bias of $\hat{\tau}_{y,\text{pst}}$ following poststratification of a SRS sample is $N \sum_{h=1}^{L} (W_h' - W_h) \mu_{y,h}$ (Smith 1991). Consequently, the bias due to inaccurately determined strata sizes or weights may outweigh any potential gains in precision of estimation made possible by poststratifying. Pfeffermann & Krieger (1991) present an alternative, regression-type estimator for situations where information on poststrata sizes is missing or in error.

The variance of the conditional distribution of $\hat{\tau}_{y,\text{pst}}$, given the set of observed strata sample sizes, is

$$V\left[\hat{\tau}_{y,\text{pst}} \mid n_h, h = 1, \ldots, L\right] = \sum_{h=1}^{L} N_h^2 \left(\frac{1}{n_h} - \frac{1}{N_h}\right) \sigma_{y,h}^2.$$

(5.42)

This expression of variance is a quantitative measure of the average squared distance between estimates of τ_y over all samples with the same set of realized strata sample sizes, n_h. That is, it accounts for the variable set of y values one would get among the subset of $_NC_n$ (see page 36 to review the meaning of the $_NC_n$ notation) samples with the same set of realized strata sample sizes. This is not the same as the variance of the distribution of all possible estimates possible under the design, because the latter distribution necessarily accounts for the variable set of n_h values that occur. This latter, unconditional variance is given by the expression:

$$V\left[\hat{\tau}_{y,\text{pst}}\right] = \sum_{h=1}^{L} N_h^2 \left(E\left[\frac{1}{n_h}\right] - \frac{1}{N_h}\right)\sigma_{y,h}^2, \tag{5.43}$$

an approximation of which is derived by expanding $1/n_h$ in a Taylor series and which yields the expression

$$V\left[\hat{\tau}_{y,\text{pst}}\right] \approx N\left(\frac{1}{n} - \frac{1}{N}\right)\sum_{h=1}^{L} N_h\sigma_{y,h}^2 + \frac{N^2}{n^2}\sum_{h=1}^{L}\left(1 - \frac{N_h}{N}\right)\sigma_{y,h}^2. \tag{5.44}$$

Some sampling texts, e.g., Thompson (2002, p. 124), show the second term of (5.44) multiplied by $(N - n)/(N - 1)$, the effect of which becomes inconsequential when the population is very much greater in size than the sample.

An unbiased estimator of $V\left[\hat{\tau}_{y,\text{pst}} \mid n_h, h = 1, \ldots, L\right]$ given in (5.42) is

$$\hat{v}\left[\hat{\tau}_{y,\text{pst}} \mid n_h, h = 1, \ldots, L\right] = \sum_{h=1}^{L} N_h^2\left(\frac{1}{n_h} - \frac{1}{N_h}\right)s_{y,h}^2. \tag{5.45}$$

In a similar fashion, an unbiased estimator of $V\left[\hat{\tau}_{y,\text{pst}}\right]$ given in (5.44) is

$$\hat{v}\left[\hat{\tau}_{y,\text{pst}}\right] = N\left(\frac{1}{n} - \frac{1}{N}\right)\sum_{h=1}^{L} N_h s_{y,h}^2 + \frac{N^2}{n^2}\sum_{h=1}^{L}\left(1 - \frac{N_h}{N}\right)s_{y,h}^2. \tag{5.46}$$

Whether the variance of $\hat{\tau}_{y,\text{pst}}$ should be assessed conditionally or unconditionally is a contentious issue among statisticians. A conditional assessment is based on the distribution of all possible sample estimates with the observed poststrata sample sizes. This conditional distribution reflects variation among estimates that might have been realized had the sample sizes been allocated a priori. By contrast, an unconditional assessment is based on the distribution of all estimates from all possible samples with all realizable poststrata sample sizes (including cases where $n_h = 0$). This unconditional distribution reflects the variation in estimates from all possible allocations of poststrata sample sizes.

Many statisticians concur that the unconditional variance is appropriate for survey planning purposes, but that the conditional variance is more appropriate for inferring the reliability of the estimator based on the observed sample. An informative, although somewhat advanced discussion of these issues is found in Holt & Smith (1979) and Smith (1991). We support the view that the conditional variance and its estimator are most appropriate when reporting the results of estimation from poststratification.

5.7.3 Poststratification of samples other than SRSwoR

Poststratification may be used even when the population has been stratified a priori and sampled according to that stratified design. For example, suppose the original strata were defined by political boundaries, which may not be well correlated with

ASIDE: A brief explanation for preferring the conditional variance is that by
using the realized poststrata sample sizes, $n_h, h = 1, \ldots, L$, the variance is
sensitive to the actual sample allocation, whereas the unconditional variance, by
its use of the expected poststrata sample sizes, is not. One poststratified sample
of overall size n might result in realized poststrata sample sizes that are nearly
identical to those from a proportional allocation of the n units, whereas another
might be far from it. One would rightfully expect the precision from the former
sample to be greater than that from the latter, and this is the result one gets from
the conditional variance, (5.42). In contrast the unconditional variance, (5.44), is
identical for both samples.

subdivisions within those boundaries of environmental variables such as concentrations of toxins in the air. Therefore, the sample from each politically based stratum
could be poststratified by other, more relevant criteria to permit more precise estimation of air pollution. For example, provinces of Canada vary greatly in land area and
there is reason to have estimates of air pollution effects for each province. This need
argues for a stratified sample wherein each province is a stratum. Environmental pro-
tection administrators, however, might be more interested in knowing the magnitude
of difference in air pollution levels between urban and rural areas. Therefore, it would
be sensible to poststratify each provincial stratum into two poststrata, a rural subdivision and an urban subdivision. In this situation, the poststrata weights within each
geographic stratum must be known, and this may limit the applicability of poststratification in this setting. When they are known, Särndal et al. (1992, p. 268) provide an
estimator of τ_y. Smith (1991) addresses the bias that occurs when poststrata weights
are incorrect.

We are unaware of any applications of poststratification following the selection of
a systematic sample.

For general unequal probability sampling, an estimator discussed by Smith (1991)
and Zhang (2000) replaces $\hat{\tau}_{y,\text{pst},h}$ in (5.41) with

$$\hat{\tau}_{y,\text{genpst},h} = N_h \frac{\sum_{u_k \in \mathscr{P}_h} y_k / \pi_k}{\sum_{u_k \in \mathscr{P}_h} 1 / \pi_k}. \tag{5.47}$$

Essentially $\hat{\tau}_{y,\text{genpst},h}$ replaces $\bar{y}_h$ in (5.40) with an alternative ratio estimator of
the stratum mean, $\mu_{y,h}$. An expanded discussion of ratio-type estimators is given in
Chapter 6.

5.8 Stratified sampling of a continuous population

We have defined a continuous population in terms of a domain of integration, $\mathscr{D}$,
comprising infinitely many points (see §1.5.2). Stratified sampling of a continuous
population first involves dividing the domain, $\mathscr{D}$, into strata in the form of two
or more mutually exclusive and completely exhaustive subdomains, $\mathscr{D}_h, h =$

$1, 2, \ldots, L$. These strata, or subdomains, are then sampled independently using the methods of Chapter 4.

Recall that the total amount of attribute of interest in $\mathcal{D}$ is

$$\tau_\rho = \int_{\mathcal{D}} \rho(\mathbf{x})\, d\mathbf{x}.$$

where $\rho(\mathbf{x})$ is the density of the attribute at a point $\mathbf{x}$ within $\mathcal{D}$. In the stratified population, the amount of attribute in stratum h is

$$\tau_{\rho,h} = \int_{\mathcal{D}_h} \rho(\mathbf{x})\, d\mathbf{x}.$$

Hence,

$$\tau_\rho = \sum_{h=1}^{L} \tau_{\rho,h}.$$

The mean attribute density in stratum h is

$$\mu_{\rho,h} = \frac{\tau_{\rho,h}}{D_h} \tag{5.48}$$

where D_h is the size—the length, area, or volume—of stratum h, i.e.,

$$D_h = \int_{\mathcal{D}_h} d\mathbf{x}$$

The domain $\mathcal{D}$ divides into the subdomains without gaps or overlaps, so the size of $\mathcal{D}$ is $D = D_1 + D_2 + \cdots + D_L$.

The mean attribute density in the total population is a weighted average of the stratum means

$$\mu_\rho = \frac{\tau_\rho}{D}$$

$$= \frac{1}{D} \sum_{h=1}^{L} \tau_{\rho,h}$$

$$= \frac{1}{D} \sum_{h=1}^{L} D_h \left(\frac{\tau_{\rho,h}}{D_h} \right)$$

$$= \frac{1}{D} \sum_{h=1}^{L} D_h \mu_{\rho,h}.$$

We can also express μ_ρ in terms of strata weights, i.e.,

$$\mu_\rho = \sum_{h=1}^{L} W_h \mu_{\rho,h},$$

where W_h is the weight for stratum h, i.e.,

$$W_h = \frac{D_h}{D}.$$

5.8.1 Probability densities

In order to sample the stratified population, we require a probability density function, $f_h(\mathbf{x})$, for selecting the sample points from each stratum. The sampling design may vary among strata, so the probability density may be constant across one stratum and vary across another. For example, the crude Monte Carlo design could be used in, say, stratum 1, in which case the probability density, $f_1(\mathbf{x})$, would be constant or uniform for all $\mathbf{x}$ in the subdomain $\mathcal{D}_1$. By contrast, importance sampling might be used in stratum 2, in which case the probability density would vary across the subdomain $\mathcal{D}_2$.

Regardless of the shape of $f_h(\mathbf{x})$, two constraints must be met, i.e.,

$$f_h(\mathbf{x}) > 0 \ \text{ for all } \ \mathbf{x} \in \mathcal{D}_h$$

and

$$\int_{\mathcal{D}_h} f_h(\mathbf{x}) \, d\mathbf{x} = 1.$$

If $f(\mathbf{x})$ has been specified for $\mathcal{D}$, we may integrate over the subdomain $\mathcal{D}_h$

$$F_h = \int_{\mathcal{D}_h} f(\mathbf{x}) \, d\mathbf{x}, \quad h = 1, 2, \ldots, L$$

and define

$$f_h(\mathbf{x}) = \frac{f(\mathbf{x})}{F_h} \ \text{ for all } \ \mathbf{x} \in \mathcal{D}_h$$

in which case,

$$\int_{\mathcal{D}_h} f_h(\mathbf{x}) \, d\mathbf{x} = \frac{1}{F_h} \int_{\mathcal{D}_h} f(\mathbf{x}) \, d\mathbf{x} = \frac{F_h}{F_h} = 1$$

as required.

5.8.2 Estimation

We assume, in this section, that one of the designs described in Chapter 4 is used in stratum h. Estimation formulae are summarized in Table 5.4.

Let $\hat{\tau}_{\rho,h_s}$ be an estimator of $\tau_{\rho,h}$ based on the sth of n_h sample points in stratum h. Combining the results from $n_h > 1$ selections,

$$\hat{\tau}_{\rho,h} = \frac{1}{n_h} \sum_{s=1}^{n_h} \hat{\tau}_{\rho,h_s},$$

Table 5.4 *General estimation following stratified sampling of a continuum*

	Stratum total	Population total
Parameter	$\tau_{\rho,h} = \int_{\mathcal{D}_h} \rho(\mathbf{x})\,d\mathbf{x}$	$\tau_\rho = \sum_{h=1}^{L} \tau_{\rho,h}$
Estimator	$\hat{\tau}_{\rho,h} = \dfrac{1}{n_h} \sum_{s=1}^{n_h} \hat{\tau}_{\rho,hs}$	$\hat{\tau}_{\rho,\mathrm{st}} = \sum_{h=1}^{L} \hat{\tau}_{\rho,h}$
Variance of estimator	$V\left[\hat{\tau}_{\rho,h}\right]$	$V\left[\hat{\tau}_{\rho,\mathrm{st}}\right] = \sum_{h=1}^{L} V\left[\hat{\tau}_{\rho,h}\right]$
Estimated variance	$\hat{v}\left[\hat{\tau}_{\rho,h}\right] = \dfrac{\sum_s^{n_h} \left(\hat{\tau}_{\rho,hs} - \hat{\tau}_{\rho,h}\right)^2}{n_h(n_h - 1)}$	$\hat{v}\left[\hat{\tau}_{\rho,\mathrm{st}}\right] = \sum_{h=1}^{L} \hat{v}\left[\hat{\tau}_{\rho,h}\right]$
Confidence interval	$\hat{\tau}_{\rho,h} \pm t_{n_h-1} \sqrt{\hat{v}\left[\hat{\tau}_{\rho,h}\right]}$	$\hat{\tau}_{\rho,\mathrm{st}} \pm t_{n-L} \sqrt{\hat{v}\left[\hat{\tau}_{\rho,\mathrm{st}}\right]}$
Percentage margin of error	$\hat{\tau}_{\rho,h} \pm \dfrac{t_{n_h-1} \sqrt{\hat{v}\left[\hat{\tau}_{\rho,h}\right]}}{\hat{\tau}_{\rho,h}/100\%}$	$\hat{\tau}_{\rho,\mathrm{st}} \pm \dfrac{t_{n-L} \sqrt{\hat{v}\left[\hat{\tau}_{\rho,\mathrm{st}}\right]}}{\hat{\tau}_{\rho,\mathrm{st}}/100\%}$

	Stratum mean	Population mean
Parameter	$\mu_{\rho,h} = \int_{\mathcal{D}_h} \dfrac{\rho(\mathbf{x})}{D_h}\,d\mathbf{x}$	$\mu_\rho = \sum_{h=1}^{L} \dfrac{D_h}{D} \mu_{\rho,h}$
Estimator	$\hat{\mu}_{\rho,h} = \dfrac{\hat{\tau}_{\rho,h}}{D_h}$	$\hat{\mu}_{\rho,\mathrm{st}} = \dfrac{\hat{\tau}_{\rho,\mathrm{st}}}{D}$
Variance of estimator	$V\left[\hat{\mu}_{\rho,h}\right] = \dfrac{V\left[\hat{\tau}_{\rho,h}\right]}{D_h^2}$	$V\left[\hat{\mu}_{\rho,\mathrm{st}}\right] = \dfrac{V\left[\hat{\tau}_{\rho,\mathrm{st}}\right]}{D^2}$
Estimated variance	$\hat{v}\left[\hat{\mu}_{\rho,h}\right] = \dfrac{\hat{v}\left[\hat{\tau}_{\rho,h}\right]}{D_h^2}$	$\hat{v}\left[\hat{\mu}_{\rho,\mathrm{st}}\right] = \dfrac{\hat{v}\left[\hat{\tau}_{\rho,\mathrm{st}}\right]}{D^2}$
Confidence interval	$\hat{\mu}_{\rho,h} \pm t_{n_h-1} \sqrt{\hat{v}\left[\hat{\mu}_{\rho,h}\right]}$	$\hat{\mu}_{\rho,\mathrm{st}} \pm t_{n-L} \sqrt{\hat{v}\left[\hat{\mu}_{\rho,\mathrm{st}}\right]}$
Percentage margin of error	$\hat{\mu}_{\rho,h} \pm \dfrac{t_{n_h-1} \sqrt{\hat{v}\left[\hat{\mu}_{\rho,h}\right]}}{\hat{\mu}_{\rho,h}/100\%}$	$\hat{\mu}_{\rho,\mathrm{st}} \pm \dfrac{t_{n-L} \sqrt{\hat{v}\left[\hat{\mu}_{\rho,\mathrm{st}}\right]}}{\hat{\mu}_{\rho,\mathrm{st}}/100\%}$

which also estimates $\tau_{\rho,h}$. The variance of $\hat{\tau}_{\rho,h}$, i.e., $V\left[\hat{\tau}_{\rho,h}\right]$, is estimated by

$$\hat{v}\left[\hat{\tau}_{\rho,h}\right] = \frac{1}{n_h(n_h-1)} \sum_{s=1}^{n_h} \left(\hat{\tau}_{\rho,h_s} - \hat{\tau}_{\rho,h}\right)^2, \qquad n_h \geq 2.$$

The population total, τ_ρ, is estimated by the sum of the strata estimates, i.e.,

$$\hat{\tau}_{\rho,\text{st}} = \sum_{h=1}^{L} \hat{\tau}_{\rho,h}.$$

The variance of $\hat{\tau}_{\rho,\text{st}}$ is

$$V\left[\hat{\tau}_{\rho,\text{st}}\right] = \sum_{h=1}^{L} V\left[\hat{\tau}_{\rho,h}\right],$$

which is estimated by

$$\hat{v}\left[\hat{\tau}_{\rho,\text{st}}\right] = \sum_{h=1}^{L} \hat{v}\left[\hat{\tau}_{\rho,h}\right].$$

The mean attribute density in stratum h, $h = 1, 2, \ldots, L$, is estimated by

$$\hat{\mu}_{\rho,h} = \frac{\hat{\tau}_{\rho,h}}{D_h}.$$

The variance of $\hat{\mu}_{\rho,h}$ is

$$V\left[\hat{\mu}_{\rho,h}\right] = \frac{V\left[\hat{\tau}_{\rho,h}\right]}{D_h^2},$$

which is estimated by

$$\hat{v}\left[\hat{\mu}_{\rho,h}\right] = \frac{\hat{v}\left[\hat{\tau}_{\rho,h}\right]}{D_h^2}.$$

The mean attribute density in the population, μ_ρ, is estimated by

$$\hat{\mu}_{\rho,\text{st}} = \sum_{h=1}^{L} W_h \hat{\mu}_{\rho,h}$$

$$= \frac{1}{D} \sum_{h=1}^{L} D_h \hat{\mu}_{\rho,h}$$

$$= \frac{1}{D} \sum_{h=1}^{L} \hat{\tau}_{\rho,h}$$

$$= \frac{\hat{\tau}_{\rho,\text{st}}}{D}.$$

The variance of $\hat{\mu}_{\rho,\text{st}}$ is

$$V\left[\hat{\mu}_{\rho,\text{st}}\right] = \frac{V\left[\hat{\tau}_{\rho,\text{st}}\right]}{D^2},$$

which is estimated by

$$\hat{v}\left[\hat{\mu}_{\rho,\text{st}}\right] = \frac{\hat{v}\left[\hat{\tau}_{\rho,\text{st}}\right]}{D^2}.$$

5.8.3 Sample allocation

The sample allocation methods for stratified sampling of discrete populations have continuous analogs. Equal allocation remains just that: the allocation of an equal number of sample points to each stratum. Under proportional allocation, the number of sample points in stratum h is proportional to F_h, the integral of the probability density, $f(\mathbf{x})$, over the subdomain $\mathcal{D}_h$ i.e.,

$$n_h = n\left(\frac{F_h}{F}\right) = nF_h$$

where $n = \sum_{h=1}^{L} n_h$ and $F = \sum_{h=1}^{L} F_h = 1$ (see, e.g., Rubinstein (1981, p. 133) or Evans & Swartz (2000, p. 186)). If $f(\mathbf{x})$ is uniform and equal across all subdomains and, therefore, uniform over $\mathcal{D}$, then

$$n_h = n\left(\frac{D_h}{D}\right)$$

The need to round to a whole number of sample points in each stratum nearly always precludes exact proportionality.

Optimal allocation generally is not possible in stratifications of natural and environmental continuums. As in stratifications of discrete populations, the optimization requires knowledge of the within-stratum variances. Since continuums comprise populations of infinitely many points, the prospect of ascertaining the requisite variances is hopeless.

An allocation is optimal for the estimation of τ_ρ, i.e., $V[\hat{\tau}_{\rho,\text{st}}]$ is minimal, if

$$n_h = n\left(\frac{\sqrt{V\left[\hat{\tau}_{\rho,h_s}\right]}}{\sum_{i=1}^{L}\sqrt{V\left[\hat{\tau}_{\rho,i_s}\right]}}\right)$$

See Evans and Swartz (2000, p. 185) for a proof. If importance sampling is the sampling design in stratum h, then

$$V\left[\hat{\tau}_{\rho,h_s}\right] = \int_{\mathcal{D}_h} \frac{\rho^2(\mathbf{x})}{f_h(\mathbf{x})}\,d\mathbf{x} - \tau_{\rho,h}^2$$

Of course, if we could calculate $V[\hat{\tau}_{\rho,h_s}] = n_h V[\hat{\tau}_{\rho,h}]$ for each stratum, then we would also be able to calculate τ_ρ, thus obviating any need of stratified sampling. Any attempt at optimal allocation necessarily involves the use of estimates of the variances. Formulae that incorporate costs are analogous to the formulae for discrete populations.

5.9 Terms to Remember

Allocation rules	Sample allocation rules
Double sampling for stratification	Strata
Equal allocation	Stratified sampling
Power allocation	Stratified random sampling
Proportional allocation	Stratum
Optimal allocation	Stratum weight
Poststratification	x-Proportional allocation

5.10 Exercises

1. Retrieve the daphnia data and verify the result reported in Example 5.6, and also estimate the standard error of the estimated population total number of daphnia.

2. Retrieve the daphnia data used in Example 5.6 and compute a 90% confidence interval for the mean number of daphnia per liter in each stratum and for the population.

3. With the aboveground biomass population used in Example 5.7, compare the strata sample sizes when allocating the sample proportionally with that needed for optimal allocation (equal costs). In both cases, the overall sample size should be set to $n = 52$.

4. How does the proportional allocation in the preceding Exercise compare to x-Proportional allocation when using tree diameter as the auxiliary variate? Again, set overall sample size to $n = 52$.

5. Select stratified random samples from the biomass population according to the allocations in the previous two exercises.

6. Use the population of biomass data to compute $\hat{\tau}_{y\pi}$ from a simple random sample of $n = 52$ units. Compare its magnitude to $V[\hat{\tau}_{y\pi,\text{st}}]$ for a proportionally-allocated stratified random sample of $n = 52$ units (as given in Table 5.3). By how much does the stratification increase precision of estimation of τ_y?

7. Use the population of biomass data to compute $V[\hat{\tau}_{y,\text{pst}}]$ in (5.44) for a SRSwoR sample of $n = 52$ units. How does this compare to the results of the preceding exercise?

5.11 Appendix

5.11.1 Proof of design-unbiasedness of $\hat{\tau}_{y,\text{pst},h}$.

Conditionally on the selected sample sizes $n_h, h = 1, \ldots, L$, $\hat{\tau}_{y,\text{pst},h}$ is a design-unbiased estimator of $\tau_{y,h}$ because under this condition, it is identical to the HT estimator of $\tau_{y,h}$. To be specific, suppose that the realized sample size in $\mathcal{P}_h$ is

$n_h = 1$. Then,

$$E\left[\hat{\tau}_{y,\text{pst},h} \mid n_h = 1\right] = E\left[\sum_{u_k \in \mathcal{S}_h} \frac{y_k}{n_h/N_h} \mid n_h = 1\right]$$

$$= N_h \mu_{y,h}$$

$$= \tau_{y,h},$$

(5.49)

where $\mu_{y,h} = N_h^{-1} \sum_{u_k \in \mathcal{P}_h} y_k$. This same result, namely $\tau_{y,h}$, occurs when $n_h = 2$, $n_h = 3$, or any other value, v, up through $v = N_h$. Therefore, since the expected value of $\hat{\tau}_{y,\text{pst},h}$ conditionally on any realized value of n_h is just the constant value of the target parameter $\tau_{y,h}$, its unconditional expected value is also $\tau_{y,h}$. Formally, the unconditional expected value of $\hat{\tau}_{y,\text{pst},h}$ is

$$E\left[\hat{\tau}_{y,\text{pst},h}\right] = E_v\left\{E\left[\hat{\tau}_{y,\text{pst},h} \mid n_h = v\right]\right\}$$

$$= \sum_{v=1}^{N_h} \tau_{y,h}\, p(n_h = v)$$

$$= \tau_{y,h} \sum_{v=1}^{N_h} p(n_h = v)$$

$$= \tau_{y,h},$$

(5.50)

since $\sum_{v=1}^{N_h} p(n_h = v) = 1$. Consequently,

$$E\left[\hat{\tau}_{y,\text{pst}}\right] = E\left[\sum_{h=1}^{L} \hat{\tau}_{y,\text{pst},h}\right]$$

$$= \sum_{h=1}^{L} E\left[\hat{\tau}_{y,\text{pst},h}\right]$$

$$= \sum_{h=1}^{L} \tau_{y,h}$$

$$= \tau_y$$

(5.51)

5.11.2 Double sampling for stratification from Resource Inventory Notes (Frayer 1978)

The first-phase sample consists of 80,000 photo plots, i.e., point locations of aerial photographs over a region $\mathcal{A}$ with land area $A = 20{,}580{,}000$ ha. A trained photo interpreter classifies each photo plot into one of two strata: forest ($h = 1$) or non-forest ($h = 2$). There were $m_1 = 64{,}526$ forested photo plots and

$m_2 = 15{,}474$ non-forested photo plots, thereby yielding estimated strata weights of $\hat{W}_1 = 64{,}526/80{,}000 = 0.806575$ and $\hat{W}_2 = 15{,}474/80{,}000 = 0.193425$. Of the $m_1 = 64{,}526$ forested photo plots, $n_1 = 403$ were located and measured by a field crew in the second phase of sampling, and of the $m_2 = 15{,}474$ nonforested plots, $n_2 = 97$ were visited likewise in phase two. The target parameter was the total number of hectares in forest:

$$\tau_y = A \times (\text{proportion of area under forest cover})$$
$$= A\,\mu_{y,\text{st}}$$
$$= A\left(W_1\mu_{y,1} + W_2\mu_{y,2}\right),$$

where $W_h, h = 1, 2$, is the proportion of land area in $\mathcal{A}$ in $\mathcal{P}_h$, and $\mu_{y,h}, h = 1, 2$, is the proportion of forested land in $\mathcal{P}_h$.

The second phase of sampling, the ground phase, revealed that 386 of the $n_1 = 403$ field plots in $\mathcal{P}_1$ were truly under forest cover, as were 10 of the $n_2 = 97$ field plots in $\mathcal{P}_2$. Hence, $\mu_{y,1}$ was estimated as $\hat{\mu}_{y,1} = 386/403 = 0.957816$ and $\mu_{y,2}$ as $\hat{\mu}_{y,2} = 10/97 = 0.103093$. With these data, the proportion of $\mathcal{A}$ in forest is estimated to be $\hat{\mu}_{y,\text{st}} = 0.792491$, yielding an estimate of total forested area of

$$\hat{\tau}_{y,\text{dst}} = 16{,}309{,}474 \text{ ha.}$$

The estimated standard error of $\hat{\tau}_{y,\text{dst}}$ using (5.36) is

$$\sqrt{\hat{v}\left[\hat{\tau}_{y,\text{dst}}\right]} = 208{,}707 \text{ ha.}$$

When a third 'I do not know' stratum was established in the first (photo-interpretation) phase, the following results were obtained: $m_1 = 64{,}101$ forested photo plots yielding an estimated stratum weight of $\hat{W}_1 = 64{,}101/80{,}000 = 0.801263$; $m_2 = 15{,}104$ non-forested photo plots yielding an estimated stratum weight of $\hat{W}_2 = 15{,}104/80{,}000 = 0.188800$; and $m_3 = 795$ 'I do not know' photo plots yielding an estimated stratum weight of $\hat{W}_3 = 795/80{,}000 = 0.009938$.

The second phase of sampling provided 396 forested plots among the $n_1 = 401$ second-phase ground plots in the forest stratum; a single forested plot among the $n_2 = 94$ second-phase ground plots in the non-forest stratum; and two forested plots among the $n_3 = 5$ ground plot selected from the 'I do not know' stratum. Hence,

$$\hat{\tau}_{y,\text{dst}} = A \times \left[\hat{W}_1\hat{\mu}_{y,1} + \hat{W}_2\hat{\mu}_{y,2} + \hat{W}_3\hat{\mu}_{y,3}\right]$$

$$= 20{,}580{,}000\left[0.801263\left(\frac{396}{401}\right) + 0.188800\left(\frac{1}{94}\right) + 0.009938\left(\frac{2}{5}\right)\right]$$

$$= 20{,}580{,}000 \times 0.797255$$

$$= 16{,}407{,}512 \text{ ha.}$$

While the estimate of forested area is not much changed by the establishment of the third, uncertain stratum in the photo-interpretation phase, the estimated standard error is reduced substantially to

$$\sqrt{\hat{v}\left[\hat{\tau}_{y,\text{dst}}\right]} = 113{,}229 \text{ ha.}$$

CHAPTER 6

Using Auxiliary Information to Improve Estimation

The use of auxiliary information was introduced in Chapter 3 in connection with the selection of samples. In Example 3.20, an easily measured tree attribute, diameter (x), was known for all elements in the population of red oak trees, whereas the attribute of interest, the aggregate volume in the boles of the trees (y), was estimated from a systematic sample. The trees were ordered from smallest to largest for the systematic selection. This proved efficient because bole volume, y, is correlated with diameter, x, so any sample which tends to span the range of diameters also tends to span the range of volumes. Elsewhere in Chapter 3 we described (a) how to select units with replacement with probability proportional to an auxiliary variate, x, and (b) how to use an auxiliary variate to select units systematically with unequal probability.

In this chapter we consider the use auxiliary information for the purpose of improving the estimation of τ_y, regardless of how samples are selected. To effectively serve that purpose, the auxiliary attribute, x, must be well, and usually positively, correlated with the attribute of interest, y.

The question of how well correlated the two variates must be is a difficult one to answer, in general, because the auxiliary information usually costs something (labor, financial resources, effort) to acquire. Hence, when deciding whether to use auxiliary information, one must consider the potential reduction in variance achievable by measuring both x and y relative to the reduction that would be realized by investing in a larger sample and measuring y only. The most common measure of the strength of correlation between two variables is the *linear correlation coefficient*, ρ_{xy}, which we discuss in the appendix to this chapter. In some situations it may be possible to establish a minimum value of ρ_{xy} needed in order to make the use of auxiliary information worthwhile. The appended discussion is intended mainly to impart an appreciation of linear correlation via graphical examination of the relationship between two variates. Indeed, we make relatively little direct use of ρ_{xy} or estimates of ρ_{xy} when estimating τ_y or other population parameters, while nonetheless appealing repeatedly to the notion of correlation between two variates. In a loose sense, if y and x are correlated, there is information about y that can be gleaned from a knowledge of x. Ratio and regression estimators presented in this chapter exploit this feature. These estimators are practical and feasible when information about x can be obtained for lesser cost than information about y; if not, it may be better to sample y solely and use estimators that do not depend on auxiliary information.

An excellent overview of ratio and regression estimation following SRSwoR is provided by Rao (1988). For a more general treatment, Särndal et al. (1992) is recommended. Regression estimation is surveyed nicely by Fuller (2002) but at a somewhat challenging level of mathematical detail.

6.1 Generalized ratio estimator

The generalized ratio estimator was mentioned briefly in Chapter 3 in the discussion of the Poisson sampling design, yet it has far wider applicability than was indicated there. To motivate the generalized ratio estimator of τ_y, note that τ_y can be written as a product:

$$\tau_y = R_{y|x}\tau_x, \tag{6.1}$$

where the population parameter $R_{y|x} = \tau_y/\tau_x$, as defined in Chapter 1. This suggests that an alternative estimator of τ_y can be constructed as the product of an estimator of $R_{y|x}$ and the known population total, τ_x. For example, one could estimate $R_{y|x}$ by the ratio of the HT estimators of τ_y and τ_x:

$$\hat{R}_{y|x} = \frac{\hat{\tau}_{y\pi}}{\hat{\tau}_{x\pi}}. \tag{6.2}$$

A ratio estimator of τ_y is

$$\hat{\tau}_{y\pi,\mathrm{rat}} = \hat{R}_{y|x}\tau_x, \tag{6.3}$$

which can be expressed equivalently as

$$\hat{\tau}_{y\pi,\mathrm{rat}} = \hat{\tau}_{y\pi}\left(\frac{\tau_x}{\hat{\tau}_{x\pi}}\right). \tag{6.4}$$

The latter expression reveals that $\hat{\tau}_{y\pi,\mathrm{rat}}$ constitutes a multiplicative adjustment of the HT estimator of τ_y: when the sample provides an estimate $\hat{\tau}_{x\pi}$ which is smaller in magnitude than the known τ_x, then $\tau_x/\hat{\tau}_{x\pi} > 1$ and $\hat{\tau}_{y\pi}$ is adjusted upwards; in samples where $\tau_x/\hat{\tau}_{x\pi} < 1$, $\hat{\tau}_{y\pi}$ is adjusted downwards. The efficacy of this adjustment of $\hat{\tau}_{y\pi}$ derives from the presumed positive correlation between x and y, which implies that when $\hat{\tau}_{x\pi}$ is too small as an estimate of τ_x, then it is likely that $\hat{\tau}_{y\pi}$ will be too small as an estimate of τ_y, and hence an upwards adjustment will tend to bring it closer in value to τ_y. Conversely, when $\hat{\tau}_{x\pi}$ is too large, $\hat{\tau}_{y\pi}$ will tend to be larger than τ_y, and the ensuing downward multiplicative adjustment will bring it closer to τ_y. As is evident from (6.3), $\hat{\tau}_{y\pi,\mathrm{rat}}$ presumes knowledge of the aggregate value, τ_x. It is not necessary to have a census of all $x_k, k = 1, \ldots N$, however.

When x and y are well and positively correlated, a consequence of this ratio adjustment to $\hat{\tau}_{y\pi}$ is that the randomization distribution of possible estimates with the chosen sampling design will be clustered more tightly and hence $V\left[\hat{\tau}_{y\pi,\mathrm{rat}}\right]$ can be expected to be less than $V\left[\hat{\tau}_{y\pi}\right]$. In other words $\hat{\tau}_{y\pi,\mathrm{rat}}$ will be a more precise estimator of τ_y than $\hat{\tau}_{y\pi}$. As mentioned in Chapter 3, if y and x are perfectly correlated, then the variance of $\hat{\tau}_{y\pi,\mathrm{rat}}$ is identically zero, a characteristic that Brewer and Hanif (1982, p. 7) term the *ratio estimator property*. In a practical setting where perfect correlation will never hold, the magnitude of the gain in precision depends, also, on the shape of the relationship between y and x: the closer that it is to being

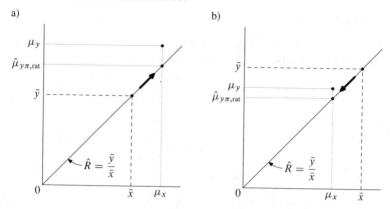

Figure 6.1 *A graphical display to accompany Example 6.1 illustrating ratio estimation following a SRSwoR design.*

a straightline trend, the better. As we discuss later in Example 6.9 it is necessary for the linear trend to pass at least approximately through the origin $y = 0, x = 0$ for $\hat{\tau}_{y\pi,\text{rat}}$ to be more efficient than $\hat{\tau}_{y\pi}$ or other alternative estimators that use auxiliary information. Relationships that intersect the vertical axis at a point far removed from $y = 0$ can be exploited better by an estimator that makes an additive, rather than a multiplicative, adjustment to $\hat{\tau}_{y\pi}$, as will be explained in §6.8.

Example 6.1

Suppose the sampling design is SRSwoR so that $\hat{\tau}_{y\pi} = N\bar{y}$, $\hat{\tau}_{x\pi} = N\bar{x}$, and hence

$$\hat{R}_{y|x} = \frac{\bar{y}}{\bar{x}}. \tag{6.5}$$

Substitution into (6.3) yields

$$\hat{\tau}_{y\pi,\text{rat}} = \frac{\bar{y}}{\bar{x}} \tau_x$$

$$= \frac{\bar{y}}{\bar{x}} N\mu_x$$

$$= N\bar{y} \frac{\mu_x}{\bar{x}}$$

$$= \hat{\tau}_{y\pi} \frac{\mu_x}{\bar{x}}. \tag{6.6}$$

Example 6.2

Suppose the sampling design is Poisson sampling with expected sample size $E[n]$ and $\pi_k = E[n]x_k/\tau_x$, as presented in §3.3.2. Hence,

$$\hat{\tau}_{y\pi} = \frac{\tau_x}{E[n]} \sum_{\mathcal{U}_k \in s} \frac{y_k}{x_k}$$

and

$$\hat{\tau}_{x\pi} = \left(\frac{\tau_x}{E[n]}\right) n$$

so that $\hat{R}_{y|x} = \hat{\tau}_{y\pi}/\hat{\tau}_{x\pi}$ is equivalent to

$$\hat{R}_{y|x} = \frac{1}{n} \sum_{\mathcal{U}_k \in s} \frac{y_k}{x_k} \tag{6.7}$$

and

$$\hat{\tau}_{y\pi,\text{rat}} = \frac{\tau_x}{n} \sum_{\mathcal{U}_k \in s} \frac{y_k}{x_k}, \tag{6.8}$$

which also can be expressed in the form of (6.4), i.e.,

$$\hat{\tau}_{y\pi,\text{rat}} = \hat{\tau}_{y\pi} \frac{\tau_x}{n\tau_x/E[n]}$$

$$= \frac{E[n]}{n} \hat{\tau}_{y\pi}.$$

Example 6.3

With the unequal probability systematic sampling design of §3.3, the inclusion probability of $\mathcal{U}_k$ is proportional to x_k, viz., $\pi_k = nx_k/\tau_x$, providing that all $x_k < \tau_x/n$. Therefore, $\hat{\tau}_{x\pi} = \sum_{\mathcal{U}_k \in s} x_k/\pi_k = \tau_x$, and, as a consequence, not only does $\hat{R}_{y|x}$ simplify to $\hat{R}_{y|x} = (1/n) \sum_{\mathcal{U}_k \in s} y_k/x_k$, but also, $\hat{\tau}_{y\pi,\text{rat}}$ is identical to $\hat{\tau}_{y\pi}$. This result implies that all the properties of the HT estimator apply to $\hat{\tau}_{y\pi,\text{rat}}$ following systematic sampling with inclusion probability proportional to x_k.

Example 6.4

An important use of ratio estimation arises when there is interest in subgroups of unknown size in the population. This was mentioned in a side comment

ASIDE: The result displayed in (6.5) for $\hat{R}_{y|x}$ following SRSwoR has given rise to the name *ratio of means* estimator for $\hat{\tau}_{y\pi,\text{rat}}$ when used with this design.

> ASIDE: Ratio estimation has a long history, which antedates the formal compilation of sample survey methods by many decades. Hald's (2003) historical excursus devotes an entire chapter to Laplace's use of (6.5) to estimate the population of France in 1786.

following Example 3.7, where the interest was in estimating the average radon concentration in houses with basements separately from that of houses lacking basements. For this situation, define the variate of interest, y, as follows:

$$y_k = \begin{cases} \text{radon concentration,} & \text{if house } \mathcal{U}_k \text{ has a basement;} \\ 0, & \text{otherwise.} \end{cases}$$

Let the auxiliary variate be a binary valued indicator:

$$x_k = \begin{cases} 1, & \text{if house } \mathcal{U}_k \text{ has a basement;} \\ 0, & \text{otherwise.} \end{cases}$$

Assuming n houses are selected by SRSwoR, $\hat{\tau}_{y\pi} = N\bar{y} = (N/n)\sum_{\mathcal{U}_k \in s_1} y_k$ and $\hat{\tau}_{x\pi} = N\bar{x} = (N/n)n_1$, where n_1 is the number of houses in the sample with basements. Consequently, $\hat{R}_{y|x} = \hat{\tau}_{y\pi}/\hat{\tau}_{x\pi}$ is identical to the estimator, $\bar{y}_1$, given in (3.14):

$$\bar{y}_1 = \frac{1}{n_1} \sum_{\mathcal{U}_k \in s_1} y_k,$$

where s_1 indicates the subset of the sample s consisting of the n_1 houses with basements. In other words, the mean value of y for a subgroup of the SRSwoR sample is just a special case of $\hat{R}_{y|x}$.

If the sampling design is unequal probability sampling with replacement (list sampling of §3.3.1), with selection probabilities $p_k = x_k/\tau_x$, an alternative estimator of $R_{y|x}$ is

$$\hat{R}'_{y|x} = \frac{\hat{\tau}_{yp}}{\hat{\tau}_{xp}}. \tag{6.9}$$

In contrast to $\hat{R}_{y|x}$, $\hat{R}'_{y|x}$ following list sampling is an unbiased estimator of $R_{y|x}$. A demonstration of this is left as an exercise at the end of the chapter.

6.2 Bias of the generalized ratio estimator

Evidently, $R_{y|x} = \tau_y/\tau_x = E[\hat{\tau}_{y\pi}]/E[\hat{\tau}_{x\pi}]$. This identity does not ensure that $E[\hat{R}_{y|x}] = E[\hat{\tau}_{y\pi}/\hat{\tau}_{x\pi}]$ coincides with $E[\hat{\tau}_{y\pi}]/E[\hat{\tau}_{x\pi}]$, however. Lahiri (1951) apparently was the first to recognize that $\hat{R}_{y|x}$ unbiasedly estimates $R_{y|x}$ for those sampling designs which result in $p(s) \propto \sum_{\mathcal{U}_k \in s} x_k$. Failing this, $E[\hat{R}_{y|x}] \neq R_{y|x}$, and hence $B[\hat{R}_{y|x} : R_{y|x}] \neq 0$. Whenever $B[\hat{R}_{y|x} : R_{y|x}] \neq 0$, $\hat{\tau}_{y\pi,\mathrm{rat}}$ is a design-biased estimator of τ_y: $B[\hat{\tau}_{y\pi,\text{rat}} : \tau_y] \neq 0$. In particular, $\hat{\tau}_{y\pi,\text{rat}}$ is a design-biased estimator of τ_y when the design is SRSwoR.

Its bias under some designs notwithstanding, $\hat{\tau}_{y\pi,\text{rat}}$ has proved to be a very accurate estimator of τ_y in situations where x and y are positively correlated. Indeed Särndal et al. (1992, p. 248) asserted, "Over the history of survey sampling, most of the considerable experience gathered with this estimator points to excellent performance under a variety of conditions."

Various bounds on $B[\hat{R}_{y|x} : R_{y|x}]$ have been derived, as Särndal et al. (1992) have summarized well. These results show that $B[\hat{R}_{y|x} : R_{y|x}]$ is sufficiently minor to obviate concern when the size of the sample is large enough to meet requirements of precision: both $B[\hat{R}_{y|x} : R_{y|x}]$ and $V[\hat{\tau}_{y\pi,\text{rat}}]$ decrease with increasing sample size, the former at a faster rate than the latter.

There are a number of ways to modify $\hat{R}_{y|x}$ to reduce or remove its bias as an estimator of $R_{y|x}$. None are universally better than $\hat{R}_{y|x}$ when using mean square error as the criterion by which to compare performance, and almost all presume a SRSwoR sampling design. Sukhatme et al. (1984, §5.9), provides a nice review. For situations where bias of $\hat{R}_{y|x}$ is a primary concern, any of these alternative estimators can be used. For reasonably large samples for which the bias of $\hat{R}_{y|x}$ can be expected to be negligible, the simplicity of $\hat{R}_{y|x}$ is attractive.

Example 6.5

To illustrate the diminution of the bias of $\hat{R}_{y|x}$ with increasing sample size, we drew every possible SRSwoR sample of $n = 2$ trees from the sugar maple population presented in Table 2.1, and then evaluated $E[\hat{R}_{y|x}]$. We repeated this exercise for samples of size $3, 4, \ldots, 29$. The magnitude of $B[\hat{R}_{y|x} : R_{y|x}]$, expressed as a percentage of $R_{y|x}$, is plotted against n in Figure 6.2. For samples of $n > 5$, the relative bias is less than 0.5%.

Example 6.6

In Example 3.32, results were reported from a simulated sampling experiment involving samples of size $n = 12$ from the $N = 236$ population of red oak trees. As part of that experiment, 100,000 samples were selected according to a SRSwoR design. In this experiment, y_k was the volume of tree $\mathcal{U}_k$ and x_k was its basal area. The average $\hat{R}_{y|x}$ value among the 100,000 estimates was 1.04% larger than the parameter value $R_{y|x} = 9.2635 \, \text{m}^3 \, \text{m}^{-2}$. That is, the relative bias of $\hat{R}_{y|x}$ as an estimator of $R_{y|x}$ was observed to be 1.04% for this particular design and population and choice of y and x attributes. Therefore, the bias in $\hat{\tau}_{y\pi,\text{rat}}$ as an estimator of τ_y is also 1.04% under these circumstances.

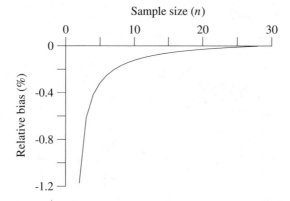

Figure 6.2 *The relative bias of $\hat{R}_{y|x}$ versus n when sampling the $N = 29$ tree population of sugar maples by SRSwoR.*

6.3 Variance of the generalized ratio estimator

Using principles of the calculus omitted here, $\hat{R}_{y|x}$ can be expanded as a linear series, leading to the approximation

$$\hat{R}_{y|x} - R_{y|x} \approx \frac{1}{\tau_x} \sum_{u_k \in s} \frac{y_k - R_{y|x} x_k}{\pi_k}.$$

The right side of this approximation is nothing more than a scaled version of the HT estimator of $\sum_{k=1}^{N} r_k$, where

$$r_k = y_k - R_{y|x} x_k.$$

In a regression context, r_k commonly is termed a residual value, a nomenclature adopted here as well.

The variance of the HT estimator, τ_y, can be expressed generally as

$$V\left[\hat{\tau}_{y\pi}\right] = \sum_{k=1}^{N} y_k^2 \left(\frac{1 - \pi_k}{\pi_k}\right) + \sum_{k=1}^{N} \sum_{\substack{k'\neq k \\ k'=1}}^{N} y_k y_{k'} \left(\frac{\pi_{kk'} - \pi_k \pi_{k'}}{\pi_k \pi_{k'}}\right). \tag{6.10}$$

By analogy to this result,

$$V\left[\hat{R}_{y|x}\right] = V\left[\hat{R}_{y|x} - R_{y|x}\right]$$

$$\approx V\left[\frac{1}{\tau_x}\sum_{\mathcal{U}_k \in s}\frac{r_k}{\pi_k}\right]$$

$$= \frac{1}{\tau_x^2}\left[\sum_{k=1}^{N}r_k^2\left(\frac{1 - \pi_k}{\pi_k}\right) + \sum_{k=1}^{N}\sum_{\substack{k'\neq k \\ k'=1}}^{N}r_k r_{k'}\left(\frac{\pi_{kk'} - \pi_k \pi_{k'}}{\pi_k \pi_{k'}}\right)\right]. \quad (6.11)$$

We introduce the symbol $V_a\left[\hat{R}_{y|x}\right]$ for the right side of (6.11), viz.,

$$V_a\left[\hat{R}_{y|x}\right] = \frac{1}{\tau_x^2}\left[\sum_{k=1}^{N}r_k^2\left(\frac{1 - \pi_k}{\pi_k}\right) + \sum_{k=1}^{N}\sum_{\substack{k'\neq k \\ k'=1}}^{N}r_k r_{k'}\left(\frac{\pi_{kk'} - \pi_k \pi_{k'}}{\pi_k \pi_{k'}}\right)\right] \quad (6.12)$$

and hence

$$V_a\left[\hat{\tau}_{y\pi,\mathrm{rat}}\right] = \tau_x^2 V_a\left[\hat{R}_{y|x}\right]$$

$$= \sum_{k=1}^{N}r_k^2\left(\frac{1 - \pi_k}{\pi_k}\right) + \sum_{k=1}^{N}\sum_{\substack{k'\neq k \\ k'=1}}^{N}r_k r_{k'}\left(\frac{\pi_{kk'} - \pi_k \pi_{k'}}{\pi_k \pi_{k'}}\right), \quad (6.13)$$

which is usually a satisfactory approximation to $V\left[\hat{\tau}_{y\pi,\mathrm{rat}}\right]$.

With a SRSwoR design, $V_a\left[\hat{R}_{y|x}\right]$ simplifies to

$$V_a\left[\hat{R}_{y|x}\right] = \frac{1}{\mu_x^2}\left(\frac{1}{n} - \frac{1}{N}\right)\sigma_r^2, \quad (6.14)$$

where σ_r^2 is the residual variance:

$$\sigma_r^2 = \frac{1}{N-1}\sum_{k=1}^{N}r_k^2. \quad (6.15)$$

This leads to

$$V_a\left[\hat{\tau}_{y\pi,\mathrm{rat}}\right] = N^2\left(\frac{1}{n} - \frac{1}{N}\right)\sigma_r^2 \quad (6.16)$$

as an approximate variance of $\hat{\tau}_{y\pi,\mathrm{rat}}$ under a SRSwoR strategy.

An alternative expression for (6.14) is given by

$$V_a\left[\hat{R}_{y|x}\right] = \frac{1}{\mu_x^2}\left(\frac{1}{n} - \frac{1}{N}\right)\left(\sigma_y^2 + R_{y|x}^2\sigma_x^2 - 2R_{y|x}\rho_{xy}\sigma_x\sigma_y\right). \quad (6.17)$$

The latter expression is revealing: it makes evident the reduction in variance resulting from a large, positive linear correlation between y and x.

We emphasize that (6.13) is the general result for any design and that (6.14) and (6.17) apply only for SRSwoR.

The following simple example serves to illustrate the formulae of the previous three sections. We emphasize that it is illustrative only, and that in practice ratio estimation with such a small sample as the one presented is ill-advised.

Example 6.7

The data used in this example appear in the following table.

$\mathcal{U}_k$	y_k	x_k
$\mathcal{U}_1$	1	1
$\mathcal{U}_2$	3	2
$\mathcal{U}_3$	5	4

For these data,

$$R_{y|x} = \frac{9}{7} \approx 1.286$$

$$\sigma_y^2 = \frac{1}{2}\left[(1-3)^2 + (3-3)^2 + (5-3)^2\right] = 4$$

$$\sigma_x^2 = \frac{1}{2}\left[\left(1-\frac{7}{3}\right)^2 + \left(2-\frac{7}{3}\right)^2 + \left(4-\frac{7}{3}\right)^2\right] = \frac{7}{3}$$

$$\sigma_{xy}^2 = \frac{1}{2}\left[(1-3)\left(1-\frac{7}{3}\right) + (3-3)\left(2-\frac{7}{3}\right) + (5-3)\left(4-\frac{7}{3}\right)\right] = 3$$

$$\rho_{xy} \approx 0.98$$

The $\Omega = 3$ samples of size $n = 2$ are $\{\mathcal{U}_1, \mathcal{U}_2\}$, $\{\mathcal{U}_1, \mathcal{U}_3\}$, and $\{\mathcal{U}_2, \mathcal{U}_3\}$, which yield the following estimates of $R_{y|x}$: $4/3, 6/5,$ and $4/3$. Presuming a SRSwoR design so that $p(s) = 1/3$ for each of the $\Omega = 3$ samples,

$$E\left[\hat{R}_{y|x}\right] = \frac{1}{3}\left(\frac{4}{3} + \frac{6}{5} + \frac{4}{3}\right) \approx 1.289$$

with bias of magnitude $B\left[\hat{R}_{y|x} : R_{y|x}\right] \approx 0.003$, or 0.24% in relative terms.

Because $\pi_k = 2/3$ and $\pi_{kk'} = 1/3$, the variance of $\hat{\tau}_{y\pi}$ from (6.10) evaluates to

$$V\left[\hat{\tau}_{y\pi}\right] = (1 + 9 + 25)\left(\frac{1}{2}\right) + 2(3 + 5 + 15)\left(-\frac{1}{4}\right) = 6.$$

The ratio residuals are $r_1 = y_1 - R_{y|x}x_1 = -2/7, r_2 = 3/7,$ and $r_3 = -1/7$. The variance of these residuals, from (6.15), is

$$\sigma_r^2 = \frac{1}{2}\left(\frac{4 + 9 + 1}{49}\right) = 0.142857.$$

ASIDE: With the fabricated data of Example 6.7, the variance of $\hat{R}_{y|x}$ was 0.003951 so that in this case $V_a[\hat{R}_{y|x}]$ exceeds $V[\hat{R}_{y|x}]$ by about 11%. While this example is intended to be illustrative of the degree of approximation introduced by the linear expansion of $\hat{R}_{y|x}$ to derive $V_a[\hat{R}_{y|x}]$, there are many other factors that influence how close in value $V_a[\hat{R}_{y|x}]$ is to $V[\hat{R}_{y|x}]$.

Using $\mu_x = 7/3$, the approximate variance of $\hat{R}_{y|x}$ from (6.14) is

$$V_a\left[\hat{R}_{y|x}\right] = \frac{1}{(7/3)^2}\left(\frac{1}{2} - \frac{1}{3}\right)0.142857 = 0.004373.$$

The approximate variance of $\hat{\tau}_{y\pi,\text{rat}}$ from (6.16) is

$$V_a\left[\hat{\tau}_{y\pi,\text{rat}}\right] = 0.214286,$$

which is only 3.6% of the value of $V[\hat{\tau}_{y\pi}]$.

Example 6.8

There are $\Omega = 635{,}376$ distinct samples of size $n = 4$ that can be selected from the $N = 64$ leaf population shown in Figure 3.4. For this population, $R_{y|x} = 200\,\text{cm}^2/\text{g}$, and for the SRSwoR design with $n = 4$, it is possible to compute the variance of $\hat{R}_{y|x}$ and to compare it to the usual approximation. In this case, the difference between $V_a[\hat{R}_{y|x}]$ and $V[\hat{R}_{y|x}]$ is -8.3% of the latter. For the $\Omega = 74{,}974{,}368$ samples of size $n = 6$, the difference between $V_a[\hat{R}_{y|x}]$ and $V[\hat{R}_{y|x}]$ is -5.9% of the latter.

Example 6.9

Consider the data used in Example 6.7, but modified by the addition of 5 to each value of y. So doing does not change the value of $p_{xy} \approx 0.98$, however, the straightline trend now intercepts the vertical y-axis at a point closer to five than zero. Due to this, $V_a[\hat{R}_{y|x}] = 0.332362$, which is a huge increase over the approximate variance of $\hat{R}_{y|x}$ in Example 6.7. Because of this, $V_a[\hat{\tau}_{y\pi,\text{rat}}] = 16.283$, which is more than double the variance of $\hat{\tau}_{y\pi}$. Moreover, the bias of $\hat{R}_{y|x}$ is 21.851%, and the difference between $V_a[\hat{R}_{y|x}]$ and $V[\hat{R}_{y|x}]$, as computed from (2.2), -30.5%. While overly simplistic, this example serves to underscore the deterioration of $\hat{\tau}_{y\pi,\text{rat}}$ when the trend between y and x is far removed from the origin, despite a very strong positive correlation between the two variates. In this situation, the generalized regression estimator presented in §6.8 is preferable.

The usual procedure is used to estimate μ_y, namely

$$\hat{\mu}_{y\pi,\text{rat}} = \frac{\hat{\tau}_{y\pi,\text{rat}}}{N},$$

the approximate variance of which is

$$V_a\left[\hat{\mu}_{y\pi,\text{rat}}\right] = \frac{V_a\left[\hat{\tau}_{y\pi,\text{rat}}\right]}{N^2}$$

$$= \frac{1}{N^2}\left[\sum_{k=1}^{N} r_k^2\left(\frac{1-\pi_k}{\pi_k}\right) + \sum_{k=1}^{N}\sum_{\substack{k'\neq k \\ k'=1}}^{N} r_k r_{k'}\left(\frac{\pi_{kk'}-\pi_k\pi_{k'}}{\pi_k\pi_{k'}}\right)\right].$$

(6.18)

Under a SRSwoR design, this simplifies to

$$V_a\left[\hat{\mu}_{y\pi,\text{rat}}\right] = \frac{V_a\left[\hat{\tau}_{y\pi,\text{rat}}\right]}{N^2}$$

$$= \left(\frac{1}{n} - \frac{1}{N}\right)\sigma_r^2.$$

(6.19)

Example 6.10

The ratio estimator was introduced explicitly in Chapter 3 in conjunction with the Bernoulli sampling design. In this context, $x_k = 1$ for all $\mathcal{U}_k$, and therefore $\tau_x = N$. Consequently, $R_{y|x} = \tau_y/\tau_x = \tau_y/N = \mu_y$, and $\hat{\tau}_{x\pi} = \hat{N}_\pi = n/\pi = nN/E[n]$. This leads to

$$\hat{R}_{y|x} = \frac{\hat{\tau}_{y\pi}}{\hat{N}_\pi}$$

$$= \frac{E[n]}{nN}\hat{\tau}_{y\pi}$$

$$= \frac{E[n]}{n}\hat{\mu}_{y\pi}$$

$$= \tilde{\mu}_{y\pi,\text{rat}},$$

as presented in (3.41). Using this result one gets

$$\hat{\tau}_{y\pi,\text{rat}} = \hat{R}_{y|x}\tau_x$$

$$= \frac{E[n]}{n}\hat{\mu}_{y\pi}N$$

$$= \frac{E[n]}{n}\hat{\tau}_{y\pi},$$

as presented in (3.39c).

Since $r_k = y_k - R_{y|x} x_k = y_k - \mu_y$ and $\pi_{kk'} = \pi^2$ with Bernoulli sampling, $V_a[\hat{\tau}_{y\pi,\text{rat}}]$ in (6.18) reduces to

$$V_a[\hat{\tau}_{y\pi,\text{rat}}] = \frac{1-\pi}{\pi} \sum_{k=1}^{N} (y_k - \mu_y)^2 \tag{6.20}$$

as in (3.42a). Upon substituting $E[n]/N$ for π one gets

$$V_a[\hat{\tau}_{y\pi,\text{rat}}] = N^2 \left(\frac{1}{E[n]} - \frac{1}{N} \right) \frac{N-1}{N} \sigma_y^2$$

$$\approx N^2 \left(\frac{1}{E[n]} - \frac{1}{N} \right) \sigma_y^2, \tag{6.21}$$

as in (3.42b).

To check the closeness of the above approximations, a simulation experiment was run in which 100,000 Bernoulli samples were drawn from the red oak population using an inclusion probability of $\pi = 0.05$ for each element in the population. Based on a population size of $N = 236$, the expected sample size was $E[n] = N\pi \approx 12$. The variance observed among the 100,000 estimates $\hat{\tau}_{y\pi,\text{rat}}$ was 4863.3 m^6 (30.3% relative standard error). The approximate variance, $V_a[\hat{\tau}_{y\pi,\text{rat}}]$, computed by (6.20) was 4391.0 m^6 (28.8%), and that computed by (6.21) was 4409.7 m^6 (28.9%).

Example 6.11

In the same SRSwoR simulation experiment reported in Example 6.6, the relative standard error among the 100,000 estimates, $\hat{\tau}_{y\pi,\text{rat}}$, was 6.5%. The relative standard error computed by using the variance approximation, $V_a[\hat{\tau}_{y\pi,\text{rat}}]$, was very close, 6.4%.

6.4 Estimated variance of the generalized ratio estimator

Consider, first, the estimation of the variance of $\hat{\tau}_{y\pi,\text{rat}}$ in conjunction with fixed sample size designs. Two alternative estimators of $V_a[\hat{R}_{y|x}]$ commonly are used:

$$\hat{v}_1[\hat{R}_{y|x}] = \frac{1}{\hat{\tau}_{x\pi}^2} \left[\sum_{u_k \in s} \hat{r}_k^2 \left(\frac{1-\pi_k}{\pi_k^2} \right) + \sum_{u_k \in s} \sum_{\substack{k' \neq k \\ u_{k'} \in s}} \hat{r}_k \hat{r}_{k'} \left(\frac{\pi_{kk'} - \pi_k \pi_{k'}}{\pi_k \pi_{k'} \pi_{kk'}} \right) \right] \tag{6.22}$$

and

$$\hat{v}_2\left[\hat{R}_{y|x}\right] = \frac{1}{\tau_x^2}\left[\sum_{\mathcal{U}_k \in s}\hat{r}_k^2\left(\frac{1-\pi_k}{\pi_k^2}\right) + \sum_{\mathcal{U}_k \in s}\sum_{\substack{k'\neq k\\ \mathcal{U}_{k'} \in s}}\hat{r}_k\hat{r}_{k'}\left(\frac{\pi_{kk'}-\pi_k\pi_{k'}}{\pi_k\pi_{k'}\pi_{kk'}}\right)\right],$$

(6.23)

where $\hat{r}_k = y_k - \hat{R}_{y|x}x_k$. These two estimators differ only in the divisor of $\hat{\tau}_{x\pi}^2$ or τ_x^2, where the latter is invariant to the particular sample that is selected. For that reason, $\hat{v}_1\,[\,\hat{R}_{y|x}\,]$ is preferred by many.

Using (6.22) or (6.23), the variance of $\hat{\tau}_{y\pi,\text{rat}}$ may be estimated as

$$\hat{v}_1\left[\hat{\tau}_{y\pi,\text{rat}}\right] = \tau_x^2\,\hat{v}_1\left[\hat{R}_{y|x}\right]$$

(6.24)

or

$$\hat{v}_2\left[\hat{\tau}_{y\pi,\text{rat}}\right] = \tau_x^2\,\hat{v}_2\left[\hat{R}_{y|x}\right].$$

(6.25)

It follows that the variance of $\hat{\mu}_{y\pi,\text{rat}}$ may be estimated as

$$\hat{v}_1\left[\hat{\mu}_{y\pi,\text{rat}}\right] = \frac{\tau_x^2}{N^2}\,\hat{v}_1\left[\hat{R}_{y|x}\right]$$

$$= \frac{1}{N^2}\,\hat{v}_1\left[\hat{\tau}_{y\pi,\text{rat}}\right]$$

(6.26)

or

$$\hat{v}_2\left[\hat{\mu}_{y\pi,\text{rat}}\right] = \frac{\tau_x^2}{N^2}\,\hat{v}_2\left[\hat{R}_{y|x}\right]$$

$$= \frac{1}{N^2}\,\hat{v}_2\left[\hat{\tau}_{y\pi,\text{rat}}\right].$$

(6.27)

Under SRSwoR, the joint inclusion probability has the straightforward form $\pi_{kk'} = [n(n-1)]/[N(N-1)]$. Using this result, (6.22) simplifies to

$$\hat{v}_1\left[\hat{R}_{y|x}\right] = \frac{1}{\bar{x}^2}\left(\frac{1}{n}-\frac{1}{N}\right)s_r^2,$$

(6.28)

and (6.23) to

$$\hat{v}_2\left[\hat{R}_{y|x}\right] = \frac{1}{\mu_x^2}\left(\frac{1}{n}-\frac{1}{N}\right)s_r^2,$$

(6.29)

where

$$s_r^2 = \frac{1}{n-1}\sum_{\mathcal{U}_k \in s}\left(y_k - \hat{R}_{y|x}x_k\right)^2.$$

(6.30)

For designs other than SRSwoR, the estimators, $\hat{v}_1[\hat{R}_{y|x}]$ and $\hat{v}_2[\hat{R}_{y|x}]$, may simplify further than is shown, depending on the joint inclusion probabilities, $\pi_{kk'}$. For with-replacement designs, $\pi_{kk'} = \pi_k\pi_{k'}$, and hence the second term in each of these variance estimators vanishes. Hartley & Rao (1962) provided an approximation of $\pi_{kk'}$ for unequal probability systematic sampling, which we reproduce for sake of convenience in this chapter's Appendix (see p. 204).

Example 6.12

From the exhaustive sampling of the sugar maple tree population which provided the results displayed in Figure 6.2, both $\hat{v}_1 [\hat{R}_{y|x}]$ and $\hat{v}_2 [\hat{R}_{y|x}]$ were negatively biased estimators of $V[\hat{R}_{y|x}]$. While the absolute bias of $\hat{v}_1 [\hat{R}_{y|x}]$ exceeded that of $\hat{v}_2 [\hat{R}_{y|x}]$ for all sample sizes, for $n \geq 10$ the expected values of two estimators were virtually identical. In contrast, the variance of $v_1 [\hat{R}_{y|x}]$ was always sufficiently smaller than the variance of $v_2 [\hat{R}_{y|x}]$, so the mean square error of $v_1 [\hat{R}_{y|x}]$ was uniformly smaller than that of $v_2 [\hat{R}_{y|x}]$.

In the 100,000 sample simulation experiment from the red oak population (see Examples 6.6 and 6.11), the 100,000 estimates, $v_1 [\hat{R}_{y|x}]$ and $v_2 [\hat{R}_{y|x}]$, provided average values that were nearly identical. Both slightly underestimated the observed variation in $\hat{\tau}_{y\pi,\text{rat}}$ among samples, on average. In terms of relative standard error of estimation, they were 5.90% and 5.91%, respectively.

The comparative performance of $\hat{v}_1 [\hat{R}_{y|x}]$ versus $\hat{v}_2 [\hat{R}_{y|x}]$ has been studied too little to know how well these limited results generalize to other sampling designs and populations.

From (6.28) and (6.29) it follows that

$$\hat{v}_1 \left[\hat{\tau}_{y\pi,\text{rat}} \right] = \frac{N^2 \mu_x^2}{\bar{x}^2} \left(\frac{1}{n} - \frac{1}{N} \right) s_r^2, \tag{6.31}$$

and

$$\hat{v}_2 \left[\hat{\tau}_{y\pi,\text{rat}} \right] = N^2 \left(\frac{1}{n} - \frac{1}{N} \right) s_r^2, \tag{6.32}$$

under SRSwoR. For $\hat{\mu}_{y\pi,\text{rat}}$ the analogous estimators are

$$\hat{v}_1 \left[\hat{\mu}_{y\pi,\text{rat}} \right] = \frac{\mu_x^2}{\bar{x}^2} \left(\frac{1}{n} - \frac{1}{N} \right) s_r^2, \tag{6.33}$$

and

$$\hat{v}_2 \left[\hat{\mu}_{y\pi,\text{rat}} \right] = \left(\frac{1}{n} - \frac{1}{N} \right) s_r^2, \tag{6.34}$$

A technique introduced by Quenouille (1956) can be used both to reduce the bias of $\hat{R}_{y|x}$ and to estimate its variance robustly. Later dubbed the jackknife owing to its utility, Gregoire (1984) demonstrated its application to the ratio estimator with a SRSwoR design, as explained in the Appendix (§6.12).

For random n designs such as the Bernoulli and Poisson, the estimation of variance must also account for the variation introduced by the randomness of the size of the sample itself. Grosenbaugh (1976, p. 174) proposed an estimator that he denoted as vS. Written in our notation, this estimator is

$$\hat{v}_3 \left[\hat{\tau}_{y\pi,\text{rat}} \right] = \left(\frac{1}{E[n]} - \frac{1}{N} \right) \frac{1}{E[n](n-1)} \sum_{u_k \in s} \left(\frac{y_k}{\pi_k} - \hat{\tau}_{y\pi,\text{rat}} \right)^2. \tag{6.35}$$

Brewer & Gregoire (2000) proposed the following alternative to (6.35):

$$\hat{v}_4\left[\hat{\tau}_{y\pi,\text{rat}}\right] = \sum_{\mathcal{U}_k \in s} \left(\frac{n - \sum_{k=1}^{N} \pi_k^2/n}{n-1} - \frac{\pi_k n}{E[n]}\right)\left(\frac{y_k E[n]}{\pi_k n} - \frac{\hat{\tau}_{y\pi,\text{rat}}}{n}\right)^2. \quad (6.36)$$

6.5 Confidence interval estimation

Using $v_1\left[\hat{R}_{y|x}\right]$, an approximate $100(1-\alpha)\%$ confidence interval for $\hat{R}_{y|x}$ is given by

$$\hat{R}_{y|x} \pm t_{n-1}\sqrt{\hat{v}_1\left[\hat{R}_{y|x}\right]}. \quad (6.37)$$

Similarly,

$$\hat{\tau}_{y\pi,\text{rat}} \pm t_{n-1}\sqrt{\hat{v}_1\left[\hat{\tau}_{y\pi,\text{rat}}\right]}, \quad (6.38)$$

and

$$\hat{\mu}_{y\pi,\text{rat}} \pm t_{n-1}\sqrt{\hat{v}_1\left[\hat{\mu}_{y\pi,\text{rat}}\right]}. \quad (6.39)$$

Intervals based on the other estimators of variance follow in an analogous fashion.

6.6 Ratio estimation with systematic sampling design

As presented in Chapter 3, well-correlated auxiliary information can be used to great advantage when sampling systematically by ordering the sampling frame in order of increasing (or decreasing) magnitude of x. A further gain in precision may be possible by using $\hat{\tau}_{y\pi,\text{rat}}$.

Example 6.13

The eight possible equal probability systematic samples of size $n = 8$ were selected from the $N = 64$ population of leaves used in Example 6.8. The variate of interest, y, was the total leaf surface area, which totaled $\tau_y = 1{,}582\,\text{cm}^2$. The sampling frame was arranged in order of increasing leaf weight. Drawing all possible systematic samples from the ordered frame, the standard error of $\hat{\tau}_{y\pi}$, expressed relatively as a percentage of τ_y, was 4.3% ($V[\hat{\tau}_{y\pi}] = 4{,}540\,\text{cm}^4$). In contrast, systematic selection from the randomly ordered frame using the square root of leaf weight as the auxiliary variate resulted in a relative standard error of $\hat{\tau}_{y\pi,\text{rat}}$ that was 3.0% ($V[\hat{\tau}_{y\pi,\text{rat}}] = 2{,}204\,\text{cm}^4$). When using the logarithm of leaf weight as the auxiliary variate, the relative standard error of $\hat{\tau}_{y\pi,\text{rat}}$ when sampling from the randomly ordered frame was 2.8% ($V[\hat{\tau}_{y\pi,\text{rat}}] = 1{,}987\,\text{cm}^4$).

In this case, a further gain in precision is realized by the sampling strategy which combines ordering of the sampling frame with the generalized ratio estimator. When sampling systematically from the frame ordered by leaf weight the relative standard error of $\hat{\tau}_{y\pi,\text{rat}}$ was 2.3% with the square root of leaf weight as the auxiliary variate, and it was 2.4% with the logarithm of leaf weight as the auxiliary variate.

Example 6.14

All possible equal probability systematic samples were selected from the $N = 1{,}058$ unstratified population of trees shown in Figure 5.1 using a sampling interval of $a = 50$. The variate of interest, y, was total aboveground biomass, which totaled $\tau_y = 76{,}396.1$ kg. Using tree basal area as the auxiliary variate, x, the sampling frame was arranged in order of increasing x. Drawing all possible systematic samples from the ordered frame, the standard error of $\hat{\tau}_{y\pi}$, expressed relatively as a percentage of τ_y was 8.7%, whereas the relative standard error of $\hat{\tau}_{y\pi,\text{rat}}$ was 5.6%. In fact the relative standard error of $\hat{\tau}_{y\pi,\text{rat}}$ was 7.5% even when the sampling frame was not deliberately ordered.

When the sample has been selected systematically, it is not possible to unbiasedly estimate the variance of $\hat{R}_{y|x}$, $\hat{\tau}_{y\pi,\text{rat}}$, or $\hat{\mu}_{y\pi,\text{rat}}$.

6.7 Generalized ratio estimation with stratified sampling

When a stratified sampling design is used, at least two options generally are considered for ratio estimation of population and strata parameters.

With the first option the stratum ratio, $R_{y|x,\text{st},h} = \tau_{y,h}/\tau_{x,h}$ is estimated separately for each of the L strata as

$$\hat{R}_{y|x,\text{st},h} = \frac{\hat{\tau}_{y\pi,h}}{\hat{\tau}_{x\pi,h}}. \tag{6.40}$$

With the second option the population ratio $R_{y|x} = \tau_y/\tau_x$ is estimated as the ratio of the estimator of τ_y from a stratified population to the analogous estimator of τ_x, namely

$$\hat{R}_{y|x,\text{st},c} = \frac{\hat{\tau}_{y\pi,\text{st}}}{\hat{\tau}_{x\pi,\text{st}}}, \tag{6.41}$$

where $\hat{\tau}_{y\pi,\text{st}}$ was defined in (5.4) of Chapter 5, and $\hat{\tau}_{x\pi,\text{st}}$ is the corresponding estimator of τ_x.

The subscribed h in $\hat{R}_{y|x,\text{st},h}$ is a necessary index, inasmuch as a ratio is computed for each stratum. In contrast, the subscribed 'c' in $\hat{R}_{y|x,\text{st},c}$ is used as a visual cue that all L strata are *combined* when computing this single ratio for the entire stratified population.

6.7.1 Ratio estimation of population and strata totals

When $R_{y|x}$ is estimated separately for each stratum, $\tau_{y,h}$ is estimated by

$$\hat{\tau}_{y\pi,h,\text{srat}} = \hat{R}_{y|x,\text{st},h}\tau_{x,h} \tag{6.42}$$

which leads naturally to an estimator of the stratified population total, τ_y, as

$$\hat{\tau}_{y\pi,\text{st},\text{srat}} = \sum_{h=1}^{L} \hat{\tau}_{y\pi,h,\text{srat}}. \tag{6.43}$$

With the combined estimator of $R_{y|x}$, the stratum total $\tau_{y,h}$ can be estimated by

$$\hat{\tau}_{y\pi,h,\text{crat}} = \hat{R}_{y|x,\text{st,c}}\tau_{x,h}, \tag{6.44}$$

and τ_y, is estimated as

$$\hat{\tau}_{y\pi,\text{st,crat}} = \hat{R}_{y|x,\text{st,c}}\tau_x$$

$$= \sum_{h=1}^{L} \hat{\tau}_{y\pi,h,\text{crat}} \tag{6.45}$$

The subscribed 'srat' in $\hat{\tau}_{y\pi,\text{st,srat}}$ is a mnemonic device to remind that this estimator uses an estimate of $R_{y|x}$ for each stratum *separately*, whereas $\hat{\tau}_{y\pi,\text{st,crat}}$ uses a single estimate for all L strata *combined*. Both $\hat{\tau}_{y\pi,\text{st,srat}}$ and $\hat{\tau}_{y\pi,\text{st,crat}}$ are biased estimators of τ_y, but the bias of the estimators is expected to be small provided that the linear correlation between y and x is strong, and the within-strata samples are not too small.

Because (6.44) uses information from other strata to estimate the $\tau_{y,h}$, it is known as a *synthetic* or indirect estimator (Rao 2003).

The approximate variance of the separate ratio estimator, $\hat{R}_{y|x,\text{st},h}$, is given by

$$V_a\left[\hat{R}_{y|x,\text{st},h}\right] = \frac{1}{\tau_{x,h}^2}$$

$$\times \left[\sum_{u_k \in \mathcal{P}_h} r_k^2 \left(\frac{1-\pi_k}{\pi_k}\right) + \sum_{u_k \in \mathcal{P}_h} \sum_{\substack{k' \neq k \\ u_{k'} \in \mathcal{P}_h}} r_k r_{k'} \left(\frac{\pi_{kk'} - \pi_k \pi_{k'}}{\pi_k \pi_{k'}}\right) \right] \tag{6.46}$$

in which r_k is the separate-ratio residual: $r_k = y_k - R_{y|x,\text{st},h}x_k$. Therefore,

$$V_a\left[\hat{\tau}_{y\pi,h,\text{srat}}\right] = \tau_{x,h}^2 V_a\left[\hat{R}_{y|x,\text{st},h}\right]$$

$$= \sum_{u_k \in \mathcal{P}_h} r_k^2 \left(\frac{1-\pi_k}{\pi_k}\right) + \sum_{u_k \in \mathcal{P}_h} \sum_{\substack{k' \neq k \\ u_{k'} \in \mathcal{P}_h}} r_k r_{k'} \left(\frac{\pi_{kk'} - \pi_k \pi_{k'}}{\pi_k \pi_{k'}}\right) \tag{6.47}$$

which leads to

$$V_a\left[\hat{\tau}_{y\pi,\text{st,srat}}\right] = \sum_{h=1}^{L} V_a\left[\hat{\tau}_{y\pi,h,\text{srat}}\right], \tag{6.48}$$

because the estimated strata totals, $\hat{\tau}_{y\pi,h,\text{srat}}, h = 1, \cdots L$ are independent.

For the special case of stratified random sampling, (6.46) simplifies to

$$V_a\left[\hat{R}_{y|x,\text{st},h}\right] = \frac{1}{\mu_{x,h}^2} \sum_{u_k \in \mathcal{P}_h} \left(\frac{1}{n_h} - \frac{1}{N_h}\right) \sigma_{rh}^2 \tag{6.49}$$

where σ_{rh}^2 is the residual variance:

$$\sigma_{rh}^2 = \frac{1}{N_h - 1} \sum_{\mathcal{U}_k \in \mathcal{P}_h} r_k^2. \tag{6.50}$$

Hence

$$V_a \left[\hat{\tau}_{y\pi,h,\mathrm{srat}} \right] = N_h^2 \left(\frac{1}{n_h} - \frac{1}{N_h} \right) \sigma_{rh}^2 \tag{6.51}$$

and

$$V_a \left[\hat{\tau}_{y\pi,\mathrm{st,srat}} \right] = \sum_{h=1}^{L} N_h^2 \left(\frac{1}{n_h} - \frac{1}{N_h} \right) \sigma_{rh}^2. \tag{6.52}$$

Turning to the combined ratio estimator, the approximate variance of $\hat{R}_{y|x,\mathrm{st,c}}$ is given by (6.12) provided that the residuals in that expression are $r_k = y_k - R_{y|x} x_k$. With this same provision on the ratio residuals, (6.14) is the approximate variance of $\hat{R}_{y|x,\mathrm{st,c}}$ with a stratified random sampling design.

The variance of the combined estimator of the stratum total is approximated by

$$V_a \left[\hat{\tau}_{y\pi,h,\mathrm{crat}} \right] = \tau_{x,h}^2 V_a \left[\hat{R}_{y|x,\mathrm{st,c}} \right]$$

$$= \frac{\tau_{x,h}^2}{\tau_x^2} \left[\sum_{\mathcal{U}_k \in \mathcal{P}_h} r_k^2 \left(\frac{1 - \pi_k}{\pi_k} \right) + \sum_{\mathcal{U}_k \in \mathcal{P}_h} \sum_{\substack{k' \neq k \\ \mathcal{U}_{k'} \in \mathcal{P}_h}} r_k r_{k'} \left(\frac{\pi_{kk'} - \pi_k \pi_{k'}}{\pi_k \pi_{k'}} \right) \right], \tag{6.53}$$

whereas the variance of the stratified population total is

$$V_a \left[\hat{\tau}_{y\pi,\mathrm{st,crat}} \right] = \sum_{\mathcal{U}_k \in \mathcal{P}_h} r_k^2 \left(\frac{1 - \pi_k}{\pi_k} \right) + \sum_{\mathcal{U}_k \in \mathcal{P}_h} \sum_{\substack{k' \neq k \\ \mathcal{U}_{k'} \in \mathcal{P}_h}} r_k r_{k'} \left(\frac{\pi_{kk'} - \pi_k \pi_{k'}}{\pi_k \pi_{k'}} \right). \tag{6.54}$$

We note that

$$V_a \left[\hat{\tau}_{y\pi,\mathrm{st,crat}} \right] \neq \sum_{h=1}^{L} V_a \left[\hat{\tau}_{y\pi,h,\mathrm{crat}} \right], \tag{6.55}$$

because positive covariance is induced among estimated strata totals owing to the common use of $\hat{R}_{y|x,\mathrm{st,c}}$ in all strata.

For the SRSwoR design, (6.54) simplifies to

$$V_a \left[\hat{\tau}_{y\pi,\mathrm{st,crat}} \right] = \sum_{h=1}^{L} N_h^2 \left(\frac{1}{n_h} - \frac{1}{N_h} \right) \sigma_{rh}^2 \tag{6.56}$$

where σ_{rh}^2 is the residual variance:

$$\sigma_{rh}^2 = \frac{1}{N_h - 1} \sum_{\mathcal{U}_k \in \mathcal{P}_h} \left(y_k - R_{y|x} x_k \right)^2. \tag{6.57}$$

Regarding ratio estimation following stratified sampling, Cochran (1977, §6.12) asserted that $\hat{\tau}_{y\pi,\text{st,srat}}$ is likely to be more precise if (a) the sample in each stratum is large enough that $V_a[\hat{\tau}_{y\pi,\text{st,srat}}]$ provides a satisfactory approximation of the actual variance of $\hat{\tau}_{y\pi,\text{st,srat}}$ and (b) its bias is small. Failing those conditions, the use of $\hat{\tau}_{y\pi,\text{st,crat}}$ is recommended. Jessen (1978, §7.9) concurs with this recommendation, yet Sukhatme & Sukhatme (1970, §4.11) do not, at least when the strata ratios do not vary considerably. Rao & Ramachandran (1974) used a model to relate y to x and then studied conditions that must hold for separate ratio estimation to be more precise than combined ratio estimation. Their results, which are too complicated to present here, provide quantitative guidelines on within-stratum sample sizes following proportional allocation that are needed to ensure the superiority of separate ratio estimation.

Example 6.15

In Example 5.9 we reported the results of a single, proportionally allocated stratified random sample of size $n = 54$ trees comprising a population of $N = 1058$ from four different species. The stratification variable was species, and the variable of interest, y, was total aboveground biomass. This sample consisted of $n_1 = 15$ balsam fir trees, $n_2 = 16$ black spruce trees, $n_3 = 14$ white birch trees, and $n_4 = 9$ white spruce trees.

The results from that particular sample notwithstanding, in the present example we used data from the entire population with (5.5) to compute the variance of $\hat{\tau}_{y\pi,\text{st}}$. For this sampling design, the relative standard error of $\hat{\tau}_{y\pi,\text{st}}$ is 14.3%.

In contrast, the relative standard error of $\hat{\tau}_{y\pi,\text{st,crat}}$ in (6.45) is reduced to 8.6% when bole diameter was used as the auxiliary variate for combined ratio estimation of τ_y. When the basal area of the bole was used as the auxiliary variate, the relative standard error was further reduced to 4.5%. As expected, ratio estimation results in a substantial increase in the precision of estimation in this context.

For this population, the further gain in precision by using ratio estimation separately in each stratum and then estimating τ_y with (6.43) is modest: 8.3% and 3.7% when diameter and basal area, respectively, are used as the auxiliary variate. Evidently, the relationship between aboveground biomass and bole diameter or basal area is so similar in these four species, that there is little additional gain by estimating $R_{y|x}$ separately by stratum.

6.7.2 Estimating the variances of $\hat{\tau}_{y\pi,\text{st,srat}}$ and $\hat{\tau}_{y\pi,\text{st,crat}}$

The usual estimator of $V_a[\hat{\tau}_{y\pi,\text{st,srat}}]$ following stratified random sampling is

$$\hat{v}_1[\hat{\tau}_{y\pi,\text{st,srat}}] = \sum_{h=1}^{L} \frac{N_h^2 \hat{\tau}_{x,h}^2}{\hat{\tau}_{x\pi,h}^2} \left(\frac{1}{n_h} - \frac{1}{N_h}\right) s_{rh}^2, \tag{6.58}$$

where $\hat{r}_k$ in s_{rh}^2 is computed as $r_k = y_k - \hat{R}_{y|x,\text{st},h} x_k$. This is a special case of the following expression which may be used with any sampling design:

$$\hat{v}_1\left[\hat{\tau}_{y\pi,\text{st},\text{srat}}\right] = \sum_{h=1}^{L}\left[\sum_{u_k\in\mathscr{S}_h} g_k^2\hat{r}_k^2\left(\frac{1-\pi_k}{\pi_k^2}\right)\right]$$

$$+\sum_{h=1}^{L}\left[\sum_{u_k\in\mathscr{S}_h}\sum_{\substack{k'\neq k\\u_{k'}\in\mathscr{S}_h}} g_k\hat{r}_k g_{k'}\hat{r}_{k'}\left(\frac{\pi_{kk'}-\pi_k\pi_{k'}}{\pi_{kk'}\pi_k\pi_{k'}}\right)\right] \quad (6.59)$$

where

$$g_k = 1 + \frac{(\tau_{x,h}-\hat{\tau}_{x\pi,h})\,x_k}{\sum_{u_k\in\mathscr{S}_h} x_{k\pi}^2/\pi_k}.$$

The variance estimators of $V_a\left[\hat{\tau}_{y\pi,\text{st},\text{crat}}\right]$ following stratified random sampling use $\hat{r}_k = y_k - \hat{R}_{y|x,\text{st},c} x_k$. For the SRSwoR design, the variance estimator for the combined-ratio estimator is

$$\hat{v}_1\left[\hat{\tau}_{y\pi,\text{st},\text{crat}}\right] = \frac{\tau_x^2}{\hat{\tau}_{x\pi}^2}\sum_{h=1}^{L} N_h^2\left(\frac{1}{n_h}-\frac{1}{N_h}\right) s_{rh}^2, \quad (6.60)$$

and for any design it is

$$\hat{v}_1\left[\hat{\tau}_{y\pi,\text{st},\text{srat}}\right] = \sum_{h=1}^{L}\left[\sum_{u_k\in\mathscr{S}_h} g_k^2\hat{r}_k^2\left(\frac{1-\pi_k}{\pi_k^2}\right)\right]$$

$$+\sum_{h=1}^{L}\left[\sum_{u_k\in\mathscr{S}_h}\sum_{\substack{k'\neq k\\u_{k'}\in\mathscr{S}_h}} g_k\hat{r}_k g_{k'}\hat{r}_{k'}\left(\frac{\pi_{kk'}-\pi_k\pi_{k'}}{\pi_{kk'}\pi_k\pi_{k'}}\right)\right], \quad (6.61)$$

where

$$g_k = 1 + \frac{(\tau_x-\hat{\tau}_{x\pi})\,x_k}{\sum_{k=1}^{N} x_{k\pi}^2/\pi_k}.$$

6.7.3 Ratio estimation of population and strata means

To estimate the mean of stratum h with the separate ratio estimator, divide (6.42) by N_h to get

$$\hat{\mu}_{y\pi,h,\text{srat}} = \hat{R}_{y|x,\text{st},h}\hat{\mu}_{y\pi,h}, \quad (6.62)$$

The population mean is estimated by dividing $\hat{\tau}_{y\pi,\text{st},\text{crat}}$ by N, i.e.,

$$\hat{\mu}_{y\pi,\text{st},\text{srat}} = \frac{\hat{\tau}_{y\pi,\text{st},\text{srat}}}{N}, \quad (6.63)$$

which is identical to

$$\hat{\mu}_{y\pi,\text{st,srat}} = \sum_{h=1}^{L} W_h \hat{\mu}_{y\pi,h,\text{srat}}. \tag{6.64}$$

The variance of $\hat{\mu}_{y\pi,h,\text{srat}}$ is approximately

$$V_a\left[\hat{\mu}_{y\pi,h,\text{srat}}\right] = \frac{V_a\left[\hat{\tau}_{y\pi,h,\text{srat}}\right]}{N^2} \tag{6.65}$$

$$= \mu_{x,h}^2 V_a\left[\hat{R}_{y|x,\text{st},h}\right].$$

Estimators of these parameters with the combined ratio estimator for each stratum are derived similarly:

$$\hat{\mu}_{y\pi,h,\text{crat}} = \hat{R}_{y|x,\text{st,c}}\mu_{x,h} \tag{6.66}$$

and

$$\hat{\mu}_{y\pi,\text{st,crat}} = \sum_{h=1}^{L} W_h \hat{\mu}_{y\pi,h,\text{crat}}. \tag{6.67}$$

The expressions for the approximate variance of $\hat{\mu}_{y\pi,h,\text{srat}}$, $\hat{\mu}_{y\pi,\text{st,srat}}$, $\hat{\mu}_{y\pi,h,\text{crat}}$, and $\hat{\mu}_{y\pi,\text{st,crat}}$ can be derived by dividing the expressions given earlier for $\hat{\tau}_{y\pi,h,\text{srat}}$, $\hat{\tau}_{y\pi,\text{st,srat}}$, $\hat{\tau}_{y\pi,h,\text{crat}}$, and $\hat{\tau}_{y\pi,\text{st,crat}}$ by N^2. Variance estimators of strata or population means are obtained similarly.

6.8 Generalized regression estimator

One can view the generalized ratio estimator in (6.3) as the ordinate of a straight line evaluated at τ_x, where the line passes both through the points $(\hat{\mu}_{y\pi}, \hat{\mu}_{x\pi})$ and $(\hat{\tau}_{y\pi}, \hat{\tau}_{x\pi})$, and which extrapolates to the origin $(0,0)$. When y and x are well correlated but do not follow the trend implied by this line, the regression estimator is a possible alternative estimator of τ_y. Useful when the data are either positively or negatively correlated, regression estimation exploits a linear relationship of the form

$$y \approx A + Bx, \tag{6.68}$$

where B is the slope of the line and A is the intercept—the ordinate of the line at the point $x = 0$. If the relationship shown in (6.68) held exactly, then $A = \mu_y - B\mu_x$ for any parametric value of B. As is customary, we define the slope B as the following parametric function:

$$B = \frac{\sigma_{xy}}{\sigma_x^2} = \frac{\sum_{k=1}^{N} x_k y_k - \tau_x \tau_y/N}{\sum_{k=1}^{N} x_k^2 - \tau_x^2/N}. \tag{6.69}$$

The slope, B, as defined in (6.69) is unique for any universe of the population values $\{y_k, x_k, k = 1, \ldots, N\}$, provided that $\sigma_x^2 \neq 0$ (Jönrup & Rennermalm 1976). Furthermore, from a sample of x and y values, it is possible to estimate A and B in (6.68) consistently. Equation 6.68 is a *regression estimator* of τ_y if it is evaluted at $x = \tau_x$ with consistent estimates of A and B.

The generalized ratio estimator of (6.3) is a special case of (6.68) in which the

ASIDE: In multiresource surveys one may wish to estimate many attributes, though the auxiliary information at one's disposal may be positively correlated with some attributes but not all. When the auxiliary variate is negatively correlated with the variate of interest, the analog to the generalized ratio estimator is the generalized product estimator:

$$\hat{\tau}_{y\pi,\mathrm{pr}} = \frac{\hat{\tau}_{y\pi}\,\hat{\tau}_{x\pi}}{\tau_x}.$$

This estimator was introduced by Robson (1957) and Murthy (1964). For use following SRSwoR, it simplifies to

$$\hat{\tau}_{y\pi,\mathrm{pr}} = \frac{\hat{\tau}_{y\pi}\,\bar{x}}{\mu_x}.$$

Gupta (1972) and Singh & Horn (1998) have examined the product estimator under general unequal probability sampling designs. The latter authors have proposed a clever composite estimator that uses the linear correlation coefficient between the variate of interest and the auxiliary variate in such a way that the estimator reverts automatically to the generalized ratio estimator when the linear correlation is positive and to the product estimator under negative linear correlation. Singh & Espejo (2003) have proposed a mixture of the two. Much empirical work remains to be done to determine the practical utility of these estimators in multiresource surveys.

magnitude of the intercept, A, is predetermined as zero and the slope is estimated by $\hat{R}_{y|x}$. The utility of regression estimation results from having a nonzero estimate of the intercept term and different estimator of the slope.

6.8.1 Regression estimation following SRSwoR

To ease the notational burden, we start by considering regression estimation following a SRSwoR sample which selects n units from the discrete population of N units. The conventional estimator of the slope of the line is

$$\hat{B} = \frac{s_{xy}}{s_x^2}, \tag{6.70}$$

where s_{xy} is the sample covariance between y and x, i.e.,

$$s_{xy} = \frac{1}{n-1} \sum_{\mathcal{U}_k \in s} (y_k - \bar{y})(x_k - \bar{x}), \tag{6.71}$$

and s_x^2 is the sample variance of x,

$$s_x^2 = \frac{1}{n-1} \sum_{\mathcal{U}_k \in s} (x_k - \bar{x})^2. \tag{6.72}$$

Following SRSwoR, s_{xy} is a design-unbiased estimator of the population covariance, σ_{xy} (see the Appendix, §6.12) and s_x^2 estimates σ_x^2 unbiasedly. Like $\hat{R}_{y|x}$, $\hat{B}$ is a ratio of random variables, and therefore it is a biased estimator of the population parameter $B = \sigma_{xy}/\sigma_x^2$.

Following SRSwoR, the estimator of the intercept of the line is

$$\hat{A} = \frac{\hat{\tau}_{y\pi} - \hat{B}\,\hat{\tau}_{x\pi}}{N} \tag{6.73}$$

$$= \bar{y} - \hat{B}\bar{x},$$

which can be regarded as an estimator of the population parameter $A = \mu_y - B\mu_x$. Both $\hat{A}$ and $\hat{B}$ consistently estimate A and B, respectively, because when $n = N$, $\hat{A} = A$ and $\hat{B} = B$.

The conventional regression estimator of τ_y following SRSwoR is

$$\hat{\tau}_{y\pi,\text{reg}} = N\hat{A} + \hat{B}\tau_x \tag{6.74a}$$

$$= N\left[\bar{y} + \hat{B}\left(\mu_x - \bar{x}\right)\right] \tag{6.74b}$$

which can be expressed equivalently as

$$\hat{\tau}_{y\pi,\text{reg}} = \hat{\tau}_{y\pi} + \hat{B}\left(\tau_x - \hat{\tau}_{x\pi}\right). \tag{6.75}$$

The latter expression makes it evident that the $\hat{\tau}_{y\pi,\text{reg}}$ constitutes an additive adjustment of the HT estimator of τ_y, in contrast to the multiplicative adjustment implicit in $\hat{\tau}_{y\pi,\text{rat}}$. Like the generalized ratio estimator, $\hat{\tau}_{y\pi,\text{reg}}$ is a biased estimator of τ_y, but the bias is usually negligible if (6.74) reasonably portrays the trend in the data and the sample is not too small.

The variance of $\hat{\tau}_{y\pi,\text{reg}}$ is closely approximated by

$$V_a\left[\hat{\tau}_{y\pi,\text{reg}}\right] = N^2\left(\frac{1}{n} - \frac{1}{N}\right)\sigma_{\text{reg}}^2 \tag{6.76}$$

where σ_{reg}^2 is the residual variance:

$$\sigma_{\text{reg}}^2 = \frac{1}{N-2}\sum_{k=1}^{N}(y_k - A - Bx_k)^2. \tag{6.77}$$

The usual estimator of (6.76) is

$$\hat{v}\left[\hat{\tau}_{y\pi,\text{reg}}\right] = N^2\left(\frac{1}{n} - \frac{1}{N}\right)s_{\text{reg}}^2 \tag{6.78}$$

where s_{reg}^2 is the residual variance:

$$s_{\text{reg}}^2 = \frac{1}{n-2}\sum_{u_k \in s} r_k^2 \tag{6.79}$$

and

$$r_k = y_k - \hat{A} - \hat{B}x_k.$$

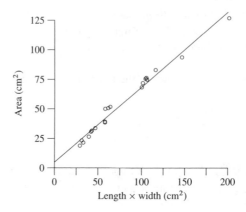

Figure 6.3 *Leaf surface area versus leaf length × width of 20 Eucalyptus nitens leaves.*

Example 6.16

A simple random sample of 20 leaves was selected without replacement from a population of 744 leaves of shining gum (*Eucalyptus nitens*). The area, length, and width of each leaf was measured, with the objective of estimating the foliar area of all 744 leaves. The sample of leaves is shown in Figure 6.3 and listed in Table 6.1. In the figure the rectangular area, computed as the product of leaf length, and width is shown on the horizontal axis. Measuring the length and width of each leaf is far less time-consuming than measuring its area, and the strong linear relationship between leaf area and this rectangular area is evident in Figure 6.3 suggests that the regression estimator of total area might work well.

The average leaf area in the sample is $\bar{y} = 52.12\,\text{cm}^2$ and the average rectangular area is $\bar{x} = 74.39\,\text{cm}^2$. The sample covariance is $s_{xy} = 1262.4\,\text{cm}^2$, and the sample variance of the auxiliary variate is $s_x^2 = 1990.0\,\text{cm}^2$. The aggregate rectangular area for all 744 leaves was $\tau_x = 57{,}266.6\,\text{cm}^2$.

Using these values, the slope, B, is estimated as $\hat{B} = 1262.4/1990.0 = 0.6344$ and the intercept, A, is estimated as $\hat{A} = 52.12 - 0.6344(74.39)\,\text{cm}^2 = 4.95\,\text{cm}^2$. Therefore, $\hat{\tau}_{y\pi,\text{reg}} = 3679.8 + 0.6344(57{,}266.6) = 40{,}009.8\,\text{cm}^2$ is the estimate of total surface area. From (6.79), $s_{\text{reg}}^2 = 18.06\,\text{cm}^4$, and with this result the estimated standard error of $\hat{\tau}_{y\pi,\text{reg}}$ is $697.7\,\text{cm}^2$. This is considerably smaller than the estimated standard error of $\hat{\tau}_{y\pi}$ which is $4{,}696.2\,\text{cm}^2$.

Example 6.17

In the same simulated sampling trial reported in Example 3.32, the regression estimator, $\hat{\tau}_{y\pi,\text{reg}}$, was computed for each of 100,000 SRSwoR samples from the red oak tree population. Each sample included $n = 12$ trees. Tree basal area serves as the covariate in the regression estimator.

Table 6.1 *A simple random sample of 20 Eucalyptus nitens leaves from a population of 744 leaves.*

Leaf	Area	Length × Width	Leaf	Area	Length × Width
	(cm^2)	(cm^2)		(cm^2)	(cm^2)
1	80.7	113.08	11	28.5	39.15
2	69.7	98.40	12	47.8	55.35
3	66.1	97.17	13	73.4	101.43
4	124.6	198.40	14	48.4	58.80
5	72.6	103.20	15	74.1	102.50
6	36.3	55.10	16	24.3	36.57
7	37.0	55.10	17	16.7	26.28
8	31.5	43.96	18	19.3	30.20
9	21.1	28.35	19	91.4	143.63
10	49.5	61.10	20	29.4	40.00

The average value among the 100,000 estimates was within -0.2% of the target value, τ_y. In other words, the observed bias of $\hat{\tau}_{y\pi,\text{reg}}$ in this case is negligibly small.

Compared to the relative standard error of 28.6% for the 100,000 HT estimates of τ_y, the relative standard error observed for $\hat{\tau}_{y\pi,\text{reg}}$ was 4.9%. The gain in precision which was realized by including this auxiliary information in the regression estimator is sizeable.

We also investigated the closeness of $V_a[\hat{\tau}_{y\pi,\text{reg}}]$ as given in (6.76) to the variance observed in the simulation. In this case, the relative standard error of $\hat{\tau}_{y\pi,\text{reg}}$ computed on the basis of $V_a[\hat{\tau}_{y\pi,\text{reg}}]$ was 4.98%, compared to the 4.92% actually observed.

The performance of the variance estimator, $\hat{v}_1[\hat{\tau}_{y\pi,\text{reg}}]$ given in (6.78), was monitored, also, by computing its average value among the 100,000 samples. The relative standard error of $\hat{\tau}_{y\pi,\text{reg}}$ computed on the basis of this average value was 3.86%, which understates the actual relative standard error of $\hat{\tau}_{y\pi,\text{reg}}$. The variance estimator, $\hat{v}_2[\hat{\tau}_{y\pi,\text{reg}}]$, yielded results nearly identical to $\hat{v}_1[\hat{\tau}_{y\pi,\text{reg}}]$.

6.8.2 Regression estimation following any sampling design

For sampling designs more general than SRSwoR it is possible to consistently estimate B by combining separate HT estimators of each of its terms, as in (6.69). To be explicit, a consistent estimator of B is

$$\hat{B}_\pi = \frac{\hat{t}_{xy\pi} - \hat{t}_{x\pi}\hat{t}_{y\pi}/\hat{N}_\pi}{\hat{t}_{x^2\pi} - \hat{t}_{x\pi}^2/\hat{N}_\pi} \tag{6.80}$$

ASIDE: Estimation of τ_y or μ_y following poststratification of the sample (see §5.7) can be cast as a specific case of regression estimation, as explicated by Bethlehem & Keller (1987). Smith (1991) indicates that, in this framework, conditional estimation of variance is not possible.

where

$$\hat{\tau}_{xy\pi} = \sum_{u_k \in s} \frac{x_k y_k}{\pi_k} \qquad (6.81)$$

and

$$\hat{\tau}_{x^2\pi} = \sum_{u_k \in s} \frac{x_{k\pi}^2}{\pi_k}, \qquad (6.82)$$

which may be expressed equivalently (see the Appendix, p. 204) as

$$\hat{B}_\pi = \frac{\sum_{u_k \in s}(x_k - \hat{\tau}_{x\pi}/\hat{N}_\pi)(y_k - \hat{\tau}_{y\pi}/\hat{N}_\pi)/\pi_k}{\sum_{u_k \in s}(x_k - \hat{\tau}_{x\pi}/\hat{N}_\pi)^2/\pi_k}. \qquad (6.83)$$

The corresponding estimator of A is

$$\hat{A}_\pi = \frac{\hat{\tau}_{y\pi} - \hat{B}_\pi \hat{\tau}_{x\pi}}{\hat{N}_\pi}. \qquad (6.84)$$

As before

$$\hat{\tau}_{y\pi,\text{reg}} = \sum_{k=1}^{N} \hat{y}_k \qquad (6.85a)$$

$$= N\hat{A}_\pi + \hat{B}_\pi \tau_x \qquad (6.85b)$$

$$= \frac{N}{\hat{N}} \hat{\tau}_{y\pi} + \hat{B}_\pi \left(\tau_x - \frac{N}{\hat{N}_\pi} \hat{\tau}_{x\pi} \right), \qquad (6.85c)$$

where $\hat{y}_k = \hat{A} + \hat{B}x_k$, is a biased estimator of τ_y which may be very precise if the linear correlation between y and x is strong.

The variance of $\hat{\tau}_{y\pi,\text{reg}}$ following a general sampling design is approximated by

$$V_a \left[\hat{\tau}_{y\pi,\text{reg}} \right] = \sum_{k=1}^{N} r_k^2 \left(\frac{1 - \pi_k}{\pi_k} \right) + \sum_{k=1}^{N} \sum_{\substack{k'\neq k \\ k'=1}}^{N} r_k r_{k'} \left(\frac{\pi_{kk'} - \pi_k \pi_{k'}}{\pi_k \pi_{k'}} \right), \qquad (6.86)$$

where the regression residual is defined as $r_k = y_k - A - Bx_k$. Except for the difference in the definition of r_k, $V_a \left[\hat{\tau}_{y\pi,\text{reg}} \right]$ is identical to $V_a \left[\hat{\tau}_{y\pi,\text{rat}} \right]$.

For fixed sample size designs, the customary estimator of $V_a\left[\hat{\tau}_{y\pi,\text{reg}}\right]$ is

$$\hat{v}_1\left[\hat{\tau}_{y\pi,\text{reg}}\right] = \sum_{\mathcal{U}_k \in s} \hat{r}_k^2\left(\frac{1-\pi_k}{\pi_k^2}\right) + \sum_{\mathcal{U}_k \in s}\sum_{\substack{k' \neq k \\ \mathcal{U}_{k'} \in s}} \hat{r}_k\hat{r}_{k'}\left(\frac{\pi_{kk'} - \pi_k\pi_{k'}}{\pi_k\pi_{k'}\pi_{kk'}}\right), \quad (6.87)$$

where $\hat{r}_k = y_k - \hat{A}_\pi - \hat{B}_\pi x_k$.

It is possible to express $\hat{\tau}_{y\pi,\text{reg}}$ as a linear combination of the sample y_k values. Borrowing the notation of Särndal et al. (1992), $\hat{\tau}_{y\pi,\text{reg}} = \sum_{\mathcal{U}_k \in s}(g_k/\pi_k)y_k$, where g_k is a sample-dependent function of x_k and quantities other than y_k (see (6.5.12), p. 233 of Särndal et al. (1992)). This led Särndal (1982) to propose and advocate an alternative estimator of the variance of $\hat{\tau}_{y\pi,\text{reg}}$ as

$$\hat{v}_2\left[\hat{\tau}_{y\pi,\text{reg}}\right] = \sum_{\mathcal{U}_k \in s} g_k^2\hat{r}_k^2\left(\frac{1-\pi_k}{\pi_k^2}\right)$$

$$+ \sum_{\mathcal{U}_k \in s}\sum_{\substack{k' \neq k \\ \mathcal{U}_{k'} \in s}} g_k\hat{r}_k g_{k'}\hat{r}_{k'}\left(\frac{\pi_{kk'} - \pi_k\pi_{k'}}{\pi_k\pi_{k'}\pi_{kk'}}\right), \quad (6.88)$$

where

$$g_k = 1 + \frac{\left(N - \hat{N}_\pi\right)\left(\hat{\tau}_{x^2\pi} - x_k\hat{\tau}_{x\pi}\right) + \left(\tau_x - \hat{\tau}_{x\pi}\right)\left(x_k\hat{N}_\pi - \hat{\tau}_{x\pi}\right)}{\hat{N}_\pi\hat{\tau}_{x^2\pi} - \hat{\tau}_{x\pi}^2}.$$

Following SRSwoR, the preceding expression for g_k collapses to

$$g_k = 1 + \frac{n\left(\mu_x - \bar{x}\right)\left(x_k - \bar{x}\right)}{\sum_{\mathcal{U}_k \in s}\left(x_k - \bar{x}\right)^2}.$$

Following unequal probability systematic sampling with $\pi_k = nx_k/\tau_x$, the preceding expression for g_k collapses to

$$g_k = 1 + \frac{\left(N - \hat{N}_\pi\right)\left(\hat{\tau}_{x^2\pi} - x_k\hat{\tau}_{x\pi}\right)}{\hat{N}_\pi\hat{\tau}_{x^2\pi} - \hat{\tau}_{x\pi}^2}.$$

Example 6.18

In the same simulated sampling trial reported in Example 3.32, the regression estimator, $\hat{\tau}_{y\pi,\text{reg}}$, was computed for each of the 100,000 unequal probability systematic samples. Not only did tree basal area serve as the auxiliary variate to affix the inclusion probability $\pi_k = nx_k/\tau_x$, it was also used as the covariate in the regression estimator. As reported above, each sample included $n = 12$ trees. The average value among the 100,000 estimates was within 0.4% of the target value, τ_y. In other words, the observed bias of $\hat{\tau}_{y\pi,\text{reg}}$ is negligibly small in this case.

Compared to the relative standard error of 5.25% for the 100,000 HT estimates of τ_y under this probability proportional to size ($\pi_k \propto x_k$) systematic sampling design, the relative standard error observed for $\hat{\tau}_{y\pi,\text{reg}}$ was 4.0%. The gain in precision beyond that which was realized by including elements with probability proportional to basal area was modest, yet it is provided at no marginal increase in cost.

We also investigated the closeness of $V_a[\hat{\tau}_{y\pi,\text{reg}}]$ as given in (6.86) to the variance observed in the simulation. The requisite joint inclusion probabilities in (6.86) were computed using the results provided by Hartley & Rao (1962) and reproduced in this chapter's Appendix on page 204. In this case, the relative standard error of $\hat{\tau}_{y\pi,\text{reg}}$ computed on the basis of $V_a[\hat{\tau}_{y\pi,\text{reg}}]$ is 3.4%, compared to the 4.0% actually observed. The discrepancy between the variance observed in the $\hat{\tau}_{y\pi,\text{reg}}$ estimates and that computed from the formula for approximate variance is noticeably larger than in Example 6.17.

Moreover, the variance estimator, $\hat{v}_1[\hat{\tau}_{y\pi,\text{reg}}]$ given in (6.87) was 3.3% on average among the 100,000 samples, which is much closer to the discrepant approximate variance result than it is to the observed variance.

The regression estimator may be extended to include covariates in addition to x_k. For a thorough treatment of a multiple linear regression estimator of τ_y, consult Särndal et al. (1992, Chapter 6).

Regression estimation may also be used following stratified sampling. The coefficients of the regression equation can be estimated separately by strata or for all strata combined, in direct analogy to the situation discussed in §6.7 for ratio estimation.

Example 6.19

Following an extensive simulation study, Valiant (1990) concluded that systematic sampling within strata from a frame ordered by x performed very well compared to stratified random sampling. Based on results from stratified systematic sampling of six artificial populations having a variety of straight-line and curvilinear trends with both uniform and nonuniform variation of observations around the trend lines, he concluded that a jackknife variance estimator successfully estimated the variance of the estimated population mean even in systematic samples, as long as the size of the sample was not too small.

6.9 Double sampling with ratio and regression estimation

Double sampling was introduced in §5.6, for the purpose of estimating strata weights for a stratified sampling design. Another common application of double sampling occurs in conjunction with ratio or regression estimation.

When double sampling for stratification, a large first-phase sample is conducted for the purpose of estimating the strata weights, $W_h, h = 1, \ldots, L$. By contrast, when employing double sampling with ratio and regression estimation, the objective of the first phase of sampling is to provide a precise estimate of τ_x, thereby avoiding the need to know the value of this population parameter. The rationale for double sampling in this context is that it may be infeasible to know τ_x yet feasible to estimate its value precisely without suffering too great a loss in precision in estimating τ_y by so doing.

Let s_1 denote the first phase sample comprising n_1 elements, $\mathcal{U}_k$, each of which is included in s_1 with probability $\pi_{k(1)}$. For each $\mathcal{U}_k \in s_1$, x_k is measured and recorded, but y_k is not, unless $\mathcal{U}_k$ has also been included into the second phase sample. Based on s_1 alone, τ_x is estimated unbiasedly by $\hat{\tau}_{x\pi(1)}$:

$$\hat{\tau}_{x\pi(1)} = \sum_{\mathcal{U}_k \in s_1} \frac{x_k}{\pi_{k(1)}}, \tag{6.89}$$

where $\mathcal{U}_k \in s_1$ indicates summation over all elements that are included into the first-phase sample.

In many applications of double sampling for ratio and regression estimation, the second-phase sample, s_2, is a subsample of s_1. The results which follow presume this second-phase subsampling design in which $n_2 < n_1$ elements, $\mathcal{U}_k$, are selected from s_1 with inclusion probability $\pi_{k(2)}$. When s_2 is not a subset of s_1, the properties of the double sampling ratio and regression estimators differ from those presented below.

A pseudo-HT estimator of τ_y is $\hat{\tau}'_{y\pi}$:

$$\hat{\tau}'_{y\pi} = \sum_{\mathcal{U}_k \in s_2} \frac{y_k}{\pi_{k(1)}\pi_{k(2)}}. \tag{6.90}$$

The reason that $\hat{\tau}'_{y\pi}$ in (6.90) is not the HT estimator is that the actual inclusion probability of $\mathcal{U}_k$, π_k, is not identical to $\pi_{k(1)}\pi_{k(2)}$, for reasons explained in Särndal et al. (1992, §9.1). We defer to these authors for a more comprehensive discussion of $\hat{\tau}'_{y\pi}$ and its properties. A corresponding estimator of τ_x, which we use below, is

$$\hat{\tau}'_{x\pi} = \sum_{\mathcal{U}_k \in s_2} \frac{x_k}{\pi_{k(1)}\pi_{k(2)}}. \tag{6.91}$$

If $\mathcal{U}_k \in s_2$, then y_k is measured. The selection of s_2 and the subsequent measurement of y_k constitutes the second phase of sampling. Evidently, $\sum_{\mathcal{U}_k \in s_1} y_k$ can be unbiasedly estimated from this second-phase information by

$$\hat{\tau}_{y\pi(2)} = \sum_{\mathcal{U}_k \in s_2} \frac{y_k}{\pi_{k(2)}}, \tag{6.92}$$

and $\sum_{\mathcal{U}_k \in s_1} x_k$ is unbiasedly estimated by

$$\hat{\tau}_{x\pi(2)} = \sum_{\mathcal{U}_k \in s_2} \frac{x_k}{\pi_{k(2)}}. \tag{6.93}$$

This suggests that

$$\hat{R}_{y|x(2)} = \frac{\hat{\tau}_{y\pi(2)}}{\hat{\tau}_{x\pi(2)}} \tag{6.94}$$

may serve usefully as an estimator of $R_{y|x}$, leading to the double sampling ratio estimator as

$$\hat{\tau}_{y\pi,\text{rat,ds}} = \hat{R}_{y|x(2)}\hat{\tau}_{x\pi(1)} + \left(\hat{\tau}'_{y\pi} - \hat{R}_{y|x(2)}\hat{\tau}'_{x\pi}\right). \tag{6.95}$$

In (6.95), one may interpret the parenthesized term as an adjustment to account for the use of $\hat{\tau}_{x\pi(1)}$ rather than τ_x in the term preceding it. In the case of SRSwoR at both phases of sampling, $\hat{\tau}_{y\pi,\text{rat,ds}}$ simplifies to

$$\hat{\tau}_{y\pi,\text{rat,ds}} = N\hat{R}_{y|x(2)}\bar{x}_1, \tag{6.96}$$

where $\bar{x}_1$ is the sample mean of x in the first-phase sample. To those familiar with traditional books of sampling methods, such as Cochran (1977), (6.96) will seem familiar, whereas (6.95) will seem foreign, yet it is the latter which permits use of ratio estimation following a broad suite of (fixed sample size) double sampling designs other than SRSwoR.

As derived by Särndal et al. (1992, §9.7, eqn. 9.7.27), the variance of $\hat{\tau}_{y\pi,\text{rat,ds}}$ is approximately

$$V_a\left[\hat{\tau}_{y\pi,\text{rat,ds}}\right]$$

$$\approx E_{(1)}\left[\sum_{\mathcal{U}_k \in s_1} r_k^2\left(\frac{1-\pi_{k(2)}}{\pi_{k(1)}^2\pi_{k(2)}}\right) + \sum_{\mathcal{U}_k \in s_1}\sum_{\substack{k' \neq k \\ \mathcal{U}_{k'} \in s_1}} r_k r_{k'}\left(\frac{\pi_{kk'(2)} - \pi_{k(2)}\pi_{k'(2)}}{\pi_{k(1)}\pi_{k(2)}\pi_{k'(1)}\pi_{k'(2)}}\right)\right]$$

$$+ \sum_{k=1}^{N} y_k^2\left(\frac{1-\pi_{k(1)}}{\pi_{k(1)}}\right) + \sum_{k=1}^{N}\sum_{\substack{k' \neq k \\ k'=1}}^{N} y_k y_{k'}\left(\frac{\pi_{kk'(1)} - \pi_{k(1)}\pi_{k'(1)}}{\pi_{k(1)}\pi_{k'(1)}}\right), \tag{6.97}$$

where $E_{(1)}$ indicates expected value over all first-phase samples; r_k in this context is the residual value $r_k = y_k - R_{y|x(1)}x_k$, with $R_{y|x(1)} = \tau_{y(1)}/\tau_{x(1)}$. $R_{y|x(1)}$ obviously will vary from one first-phase sample to another.

Again relying on Särndal et al. (1992, §9.7, eqn. 9.7.28), $V_a\left[\hat{\tau}_{y\pi,\text{rat,ds}}\right]$ is estimated by

$$\hat{v}_1[\hat{\tau}_{y\pi,\text{rat,ds}}] = \sum_{\mathcal{U}_k \in s_2} \hat{g}_k^2\hat{r}_k^2\left(\frac{1-\pi_{k(2)}}{\pi_{k(1)}^2\pi_{k(2)}^2}\right) + \sum_{\mathcal{U}_k \in s_2} y_k^2\left(\frac{1-\pi_{k(1)}}{\pi_{k(1)}^2\pi_{k(2)}}\right)$$

$$+ \sum_{\mathcal{U}_k \in s_2}\sum_{\substack{k' \neq k \\ \mathcal{U}_{k'} \in s_2}} g_k\hat{r}_k g_{k'}\hat{r}_{k'}\left(\frac{\pi_{kk'(2)} - \pi_{k(2)}\pi_{k'(2)}}{\pi_{kk'(2)}\pi_{k(1)}\pi_{k(2)}\pi_{k'(1)}\pi_{k'(2)}}\right)$$

$$+ \sum_{\mathcal{U}_k \in s_2}\sum_{\substack{k' \neq k \\ \mathcal{U}_{k'} \in s_2}} y_k y_{k'}\left(\frac{\pi_{kk'(1)} - \pi_{k(1)}\pi_{k'(1)}}{\pi_{kk'(1)}\pi_{kk'(2)}\pi_{k(1)}\pi_{k'(1)}}\right), \tag{6.98}$$

where g_k in this context is

$$g_k = \frac{\hat{\tau}_{x\pi(1)}}{\hat{\tau}_{x\pi(2)}}, \tag{6.99}$$

and where

$$\hat{r}_k = y_k - \hat{R}_{y|x(2)} x_k. \tag{6.100}$$

When the sampling design for both phases is SRSwoR, $\hat{v}_1[\hat{\tau}_{y\pi,\mathrm{rat,ds}}]$ simplifies (greatly) to

$$\hat{v}_1[\hat{\tau}_{y\pi,\mathrm{rat,ds}}] = N^2 \left(\frac{1}{n_1} - \frac{1}{N} \right) s_y^2 + N^2 \left(\frac{1}{n_2} - \frac{1}{n_1} \right) \left(\frac{\bar{x}_1}{\bar{x}_2} \right)^2 s_r^2, \tag{6.101}$$

where s_r^2 is the sample variance of the $\hat{r}_k$ values in the second phase sample, and s_y^2 is the sample variance of the y_k values in the second phase, i.e.,

$$s_r^2 = \frac{1}{n_2 - 1} \sum_{u_k \in s_2} \hat{r}_k^2 \tag{6.102}$$

and

$$s_y^2 = \frac{1}{n_2 - 1} \sum_{u_k \in s_2} (y_k - \bar{y}_2)^2 . \tag{6.103}$$

The first term in (6.101) estimates the variance of the HT estimator of τ_y when drawing a simple random sample of n_1 units without replacement, whereas the second term constitutes the increased variability incurred by the second phase of sampling n_2 units from the n_1 units of the first phase. The simplification of $\hat{v}_1[\hat{\tau}_{y\pi,\mathrm{rat,ds}}]$ to the expression in (6.101) is left as an exercise at the end of the chapter.

Example 6.20

In Example 3.8 we reported an estimate of $\hat{\mu}_y = \bar{y} = 81.0$ kg following a SRSwoR of size $n = 52$ trees from the $N = 1058$ population of trees shown in Figure 5.1. The estimated standard error of $\hat{\mu}_y$ was 12.9 kg. This population was also sampled systematically with equal probability from an ordered frame in Example 6.14.

Here we report the results from a double sample of the population using SRSwoR at both phases. The size of the first-phase sample was set at $n_1 = 104$, while the size of the second-phase sample was $n_2 = 26$. The auxiliary variate was the basal area of each tree, which implies that the only measurement of trees in the first-phase sample was its diameter. The variate of interest was aboveground biomass, as in the previous examples.

The sample mean biomass from phase 1 was $\bar{x}_1 = 0.0178228\,\mathrm{m}^2$, hence $\hat{\tau}_{x\pi(1)} = 18.9\,\mathrm{m}^2$. From the second-phase sample, $\hat{\tau}_{x\pi(2)} = 21.7\,\mathrm{m}^2$ and $\hat{\tau}_{y\pi(2)} = 88,400$ kg and $s_y^2 = 7682.29\,\mathrm{kg}^2$. With these results, the population ratio was estimated as $\hat{R}_{y|x(2)} = 4063.7\,\mathrm{kg\,m}^{-1}$, and $s_r^2 = 1760.22\,\mathrm{kg}^2\,\mathrm{m}^{-4}$. Consequently, the estimator of the total biomass from (6.96) is $\hat{\tau}_{y\pi,\mathrm{rat,ds}} = 76,798.6$ kg, in other words, the estimated average biomass per tree is $\hat{\mu}_{y\pi,\mathrm{rat,ds}} = 72.6$ kg.

Appealing to (6.101), we obtain

$$\hat{v}_1\left[\hat{\tau}_{y\pi,\text{rat,ds}}\right] = 1058^2 \left(\frac{1}{104} - \frac{1}{1058}\right) 7682.29$$

$$+ 1058^2 \left(\frac{1}{26} - \frac{1}{104}\right) \left(\frac{0.0178228}{0.020561}\right)^2 1760.22$$

$$= 11{,}745{,}670.5\,\text{kg}^2.$$

Consequently, $\hat{v}_1\left[\hat{\mu}_{y\pi,\text{rat,ds}}\right] = 104.9\,\text{kg}^2$, in other words the estimated standard error of $\hat{\mu}_{y\pi,\text{rat,ds}}$ is estimated at 10.2 kg.

In other words, a double sample coupled with ratio estimation resulted in a about a 20% kg reduction in standard error compared to the one-phase SRSwoR results from Example 3.8. Here the diameter of twice as many trees were measured, but the biomass of only half as were measured in the earlier example.

Turning now to the regression estimator of τ_y following double sampling, the regression coefficients, A and B, of §6.8 are estimated with the n_2 pairs of $x - y$ values obtained from the units sampled in the second phase. For a general sampling design, the double-sampling estimator of B is

$$\hat{B}_{\pi(2)} = \frac{\hat{\tau}'_{xy\pi} - \hat{\tau}'_{x\pi}\hat{\tau}'_{y\pi}/\hat{N}'_{\pi}}{\hat{\tau}'_{x^2\pi} - (\hat{\tau}'_{x\pi})^2/\hat{N}'_{\pi}} \qquad (6.104a)$$

$$= \frac{\sum_{u_k \in s_2}(x_k - \hat{\tau}'_{x\pi}/\hat{N}'_{\pi})(y_k - \hat{\tau}'_{y\pi}/\hat{N}'_{\pi})/\pi_{k(1)}\pi_{k(2)}}{\sum_{u_k \in s_2}(x_k - \hat{\tau}'_{x\pi}/\hat{N}'_{\pi})^2/\pi_{k(1)}\pi_{k(2)}}, \qquad (6.104b)$$

where the terms not previously defined are

$$\hat{\tau}'_{xy\pi} = \sum_{u_k \in s_2} \frac{x_k y_k}{\pi_{k(1)}\pi_{k(2)}}, \qquad (6.105)$$

$$\hat{\tau}'_{x^2\pi} = \sum_{u_k \in s_2} \frac{x_k^2}{\pi_{k(1)}\pi_{k(2)}}, \qquad (6.106)$$

and

$$\hat{N}'_{\pi} = \sum_{u_k \in s_2} \frac{1}{\pi_{k(1)}\pi_{k(2)}}. \qquad (6.107)$$

The corresponding estimator of A is

$$\hat{A}_{\pi(2)} = \frac{\hat{\tau}'_{y\pi} - \hat{B}_{\pi(2)}\tau'_{x\pi}}{\hat{N}'_k}. \qquad (6.108)$$

The double sampling regression estimator has a form that is similar to that of the double sampling ratio estimator $\hat{\tau}_{y\pi,\text{rat,ds}}$ in (6.95):

$$\hat{\tau}_{y\pi,\text{reg,ds}} = \hat{N}_{\pi(1)} \hat{A}_{\pi(2)} + \hat{B}_{\pi(2)} \hat{\tau}_{x\pi(1)}$$
$$+ \left(\hat{\tau}'_{y\pi} - \hat{N}'_{\pi} \hat{A}_{\pi(2)} - \hat{B}_{\pi(2)} \hat{\tau}'_{x\pi} \right). \qquad (6.109)$$

Here, too, the parenthesized term may be interpreted as the adjustment that is needed to account for the use of $\hat{N}_{\pi(1)}$ and $\hat{\tau}_{x\pi(1)}$ rather than N and τ_x (compare (6.109) to (6.85)b).

The approximate variance of $\hat{\tau}_{y\pi,\text{reg,ds}}$ is given by the same expression as used for the approximate variance of $\hat{\tau}_{y\pi,\text{rat,ds}}$ in (6.97), but with the following regression residuals:

$$r_k = y_k - A - B x_k. \qquad (6.110)$$

Furthermore, the expression given for $\hat{v}_1 \left[\hat{\tau}_{y\pi,\text{rat,ds}} \right]$ in (6.98) can be used to estimate $V_a \left[\hat{\tau}_{y\pi,\text{reg,ds}} \right]$ provided that the following values of g_k and $\hat{r}_k$ are substituted into it:

$$g_k = \frac{\hat{N}_{\pi(1)}}{\hat{N}'_\pi} \frac{\hat{\tau}'_{x^2\pi} - \hat{\mu}_{x\pi(1)} \hat{\tau}'_{x\pi}}{\hat{\tau}'_{x^2\pi} - \hat{\mu}'_{x\pi} \hat{\tau}'_{x\pi}} + \frac{\left(\hat{\tau}_{x\pi(1)} - \hat{\mu}'_{x\pi} \hat{N}_{\pi(1)} \right) x_k}{\hat{\tau}'_{x^2\pi} - \hat{\mu}'_{x\pi} \hat{\tau}'_{x\pi}} \qquad (6.111)$$

and

$$\hat{r}_k = y_k - \hat{A}_{\pi(2)} - \hat{B}_{\pi(2)} x_k. \qquad (6.112)$$

In (6.111), $\hat{\mu}_{x\pi(1)} = \hat{\tau}_{x\pi(1)} \hat{N}_{\pi(1)}$ and $\hat{\mu}'_{x\pi} = \hat{\tau}'_{x\pi} \hat{N}'_\pi$.

Särndal et al. (1992) assert that the variance estimator can be simplified by putting $g_k = 1$ rather than evaluating (6.111), but did not indicate how this simplification affects the performance of $\hat{v}_1 \left[\hat{\tau}_{y\pi,\text{reg,ds}} \right]$ as an estimator of $V_a \left[\hat{\tau}_{y\pi,\text{reg,ds}} \right]$.

For SRSwoR at both phases, $\hat{B}_{\pi(2)}$ simplifies to

$$\hat{B}_{\pi(2)} = s_{xy(2)} / s^2_{x(2)}, \qquad (6.113)$$

where $s_{xy(2)}$ and $s^2_{x(2)}$ are the sample covariance between y and x and the sample variance of x, respectively, both computed from the second-phase sample values. Moreover, $\hat{\tau}_{y\pi,\text{reg,ds}}$ simplifies to

$$\hat{\tau}_{y\pi,\text{reg,ds}} = N \bar{y}_2 + \hat{B}_{\pi(2)} \left(\bar{x}_1 - \bar{x}_2 \right), \qquad (6.114)$$

and $\hat{v}_1 \left[\hat{\tau}_{y\pi,\text{reg,ds}} \right]$ simplifies to the same expression as given for $\hat{v}_1 \left[\hat{\tau}_{y\pi,\text{rat,ds}} \right]$ in (6.101) provided that the regression residuals given by (6.112) are used to compute s^2_r.

Example 6.21

Gilbert & Eberhardt (1976) recount an application of double sampling with regression estimation for the purpose of estimating the average amount of plutonium in surface soil. In this application, y was the measure of plutonium (in curies) by an accurate but expensive procedure, and x was the measure of plutonium by a more fallible but less expensive device. They concluded that double sampling with regression estimation reduced the standard error of estimated plutonium by 35%, and that a reduction in the cost of sampling by 20% to 30% was possible without any sacrifice in the level of precision.

6.10 Terms to Remember

Combined ratio estimator	Ratio estimator
Covariance	Ratio estimator property
Double sampling ratio estimator	Ratio of means
Double sampling regression estimator	Regression estimator
Linear correlation coefficient	Separate ratio estimator

6.11 Exercises

1. Use the technique shown on page 85 in the Appendix to Chapter 3 to derive the expected value and variance of $\hat{\tau}_{yp}$ to show that $\hat{R}'_{y|x}$ is an unbiased estimator of $R_{y|x}$.

2. Derive a simplified expression for (6.13) when the sampling design is Bernoulli with constant inclusion probability, π.

3. Derive a simplified expression for (6.13) when the sampling design is the Poisson design explained in Example 6.2.

4. Using the Eucalypt leaf data of Example 6.16, compute the HT estimate of total leaf area and verify that the estimated standard error is $4,696.2 \text{ cm}^2$.

5. Using the Eucalypt leaf data of Example 6.16, compute the ratio estimate of total leaf area and estimate its standard error.

6. Show that $\hat{N} = \sum_{u_k \in s} 1/\pi_k$ is identical to N under SRSwoR. Use this result to show, further, that $\hat{A}_\pi$ in (6.84) and $\hat{B}_\pi$ in (6.80) simplify to $\hat{A}$ in (6.73) and $\hat{B}$ in (6.70), respectively.

7. Show that $\hat{v}_2 [\hat{\tau}_{y\pi,\text{rat}}]$ in (6.25) simplifies to (3.2.3) under Bernoulli sampling.

8. Derive $\hat{v}_1 [\hat{R}_{y|x}]$ as expressed in (6.28) from its more general expression in (6.22). Use this result to derive (6.29) from (6.22).

9. The data shown in Table 6.2 were obtained from a sample of $n = 20$ 1-meter wide belt transects. The purpose of the study was to estimate the total number of pale white larkspur plants in this grassy region between a river and the edge of

Table 6.2 *Pale white larkspur abundance on a sample of 1 m wide belt transects in Stehman & Salzer (2000). Area of all N = 150 transects was* $\tau_x = 5897\ m^2$.

Transect	No. plants	Transect area (m^2)	Transect	No. plants	Transect area (m^2)
1	0	22.0	11	19	37.0
2	22	27.0	12	63	42.0
3	1	33.0	13	55	30.5
4	12	38.4	14	39	30.0
5	4	41.0	15	17	27.0
6	21	45.0	16	7	3.0
7	77	34.0	17	18	13.0
8	27	34.0	18	3	12.0
9	4	41.0	19	12	14.0
10	23	31.0	20	5	13.0

the forest. Because neither the forest edge nor the river were straight, the lengths of the $N = 157$ non-overlapping belt transects varied. With the area of transect as the x-variable, use $\hat{\tau}_{y\pi,\text{rat}}$ to compute a 90% confidence interval for the total number of pale white larkspur on the property.

10. Barrett & Nutt (1979) provide data derived from aerial photography and field examination of the number of dead trees in a 80 ha parcel of forested land where a disease had inflicted considerable mortality. As they aptly describe, the field counts are much more accurate than the counts of dead trees assessed from the photos, yet it is also considerably more costly. The relationship between field (y) and photo (x) counts of dead trees is shown in Figure 6.4 for the complete census of 80 'photoplots.'

As an exercise, select a SRSwoR of $n = 10$ of these photos and with the paired $x - y$ data, estimate the total number of dead trees on this parcel of land. Do these data suggest that the ratio or the regression estimator would be preferred? Compute a 90% confidence interval for τ_y using a) the ratio estimator, and b) the regression estimator. Which of the two intervals is narrower? Explain the reason for the result you obtained.

11. Using the biomass data displayed in Figure 3.1 and discussed in Examples 6.16 and Example 6.15, among others, select a stratified random sample of overall sample size $n = 28$ trees. Allocate the sample equally among the four strata. Compute a 90% confidence interval for total foliar biomass for each species and for the population as a whole using a) the HT estimator, b) the combined ratio estimator using basal area as the auxiliary variate, and c) the separate ratio estimator using basal area as the auxiliary variate. Is the improvement offered by ratio estimation observed in Example 6.15 apparent in these results, also? Is the

Table 6.3 *Counts of dead trees from aerial photography and from field assessment. Data provided in Barrett & Nutt (1979, p. 175).*

Photo no.	Photo count	Field count	Photo no.	Photo count	Field count
1	9	10	41	18	16
2	11	6	42	16	10
3	3	4	43	13	7
4	8	10	44	14	13
5	17	13	45	17	12
6	2	4	46	12	11
7	16	10	47	14	9
8	13	9	48	8	6
9	12	10	49	10	13
10	6	7	50	15	15
11	14	15	51	12	9
12	17	17	52	14	11
13	9	8	53	14	9
14	15	13	54	13	8
15	16	14	55	18	11
16	6	8	56	5	6
17	12	8	57	19	15
18	7	5	58	19	14
19	9	9	59	9	7
20	14	12	60	10	6
21	16	12	61	1	2
22	15	9	62	7	8
23	11	9	63	8	9
24	16	11	64	12	7
25	11	7	65	3	8
26	10	8	66	6	6
27	9	6	67	7	9
28	8	7	68	10	8
29	18	12	69	12	10
30	16	12	70	13	10
31	3	6	71	12	12
32	10	9	72	2	3
33	15	11	73	8	11
34	17	13	74	11	10
35	1	5	75	4	7
36	15	10	76	19	14
37	5	5	77	18	13
38	19	12	78	7	7
39	17	15	79	19	13
40	10	11	80	11	8

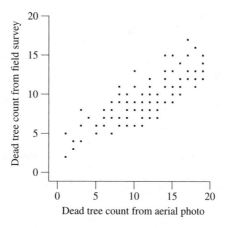

Figure 6.4 *Dead tree counts from ground plots and from photos of the ground plots.*

similarity between the combined ratio estimator and the separate ratio estimator observed in Example 6.15 also observed when estimating aggregate foliar biomass for each species?

12. Repeat the previous exercise but allocate the sample of $n = 28$ trees proportionally to the sizes of the strata, i.e., $n_h = n(N_h/N)$.

13. Repeat the previous exercise but allocate the sample of $n = 28$ trees proportionally to the aggregate basal area in the strata, i.e., $n_h = n(\tau_{x,h}/\tau_x)$, as discussed in §5.4.3.

14. Derive g_k as given on page 191 for SRSwoR.

15. Derive g_k as given on page 191 for unequal probability systematic sampling with $\pi_k = nx_k/\tau_x$.

16. Starting with the general expression for $\hat{v}_1\left[\hat{\tau}_{y\pi,\mathrm{rat,ds}}\right]$ given in (6.98), derive the result for SRSwoR given in (6.101).

6.12 Appendix

6.12.1 Linear correlation coefficient

The strength of the linear correlation between two variates, x and y, is quantified by the linear correlation coefficient. For a population of N distinct elements, this coefficient is

$$\rho_{xy} = \frac{\sigma_{xy}}{\sqrt{\sigma_y^2 \sigma_x^2}} \tag{6.115}$$

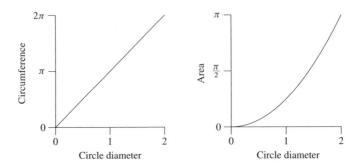

Figure 6.5 *Two functional relationships, the left of which is linear.*

where

$$\sigma_{xy} = \frac{1}{N-1} \sum_{k=1}^{N} \left(y_k - \mu_y\right)\left(x_k - \mu_x\right) \tag{6.116}$$

is the covariance between y and x. Two features of ρ_{xy} are noteworthy:

1. It has no units of measure, therefore its magnitude does not depend on the unit of measure associated with x or y

2. Its value is bounded by the interval $-1 \leq \rho_{xy} \leq 1$, with $\rho_{xy} = 1$ indicating perfect positive correlation, $\rho_{xy} = -1$ indicating perfect negative correlation, and $\rho_{xy} = 0$ indicating that x and y are uncorrelated.

When $\rho_{xy} = 1$, a graph of y_k versus x_k would reveal a straight line with positive slope. That is, for some constant value, a, and positive constant, b, $y_k = a + bx_k, k = 1, \ldots, N$. When $\rho_{xy} = -1$, the same relationship holds but with b now being a negative constant. For sake of illustration, suppose we had a collection of N circles and y_k was the circumference and x_k the diameter of the kth circle. Because diameter multiplied by the mathematical constant, π, yields circumference (i.e., $y_k = 0 + \pi x_k$), any collection of circle diameters and circumferences would yield $\rho_{xy} = 1$. Perfect linear correlation, either positive or negative, means that any value of y_k can be exactly calculated from x_k, provided that a and b were known. In other words, a linear correlation of ρ_{xy} means that there is a linear functional relationship between x and y.

Two variates may be functionally related, but not linearly so, in which case $\rho_{xy} < 1$. For example, let y_k be the area of a circle and x_k be diameter. Geometry informs us that $y_k = \pi x_k^2/4$, which is a relationship that can not be expressed as $y_k = a + bx_k$ for any constant b. These two functional relationships are displayed in Figure 6.5.

A value of $\rho_{xy} = 0$ indicates that the two variates have no linear association, which does not mean that x and y are unassociated or independent. This point is demonstrated in Figure 6.6. In each frame of the figure, the linear correlation coefficient is identically zero. In the left frame there is no association between the y variate on the vertical axis and the x variate on the horizontal axis. Yet there is a clear pattern among the x and y values in the right frame, despite $\rho_{xy} = 0$. It would be a

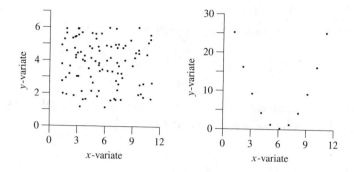

Figure 6.6 *Two relationships, each of which with* $\rho_{xy} = 0$.

mistake, therefore, to conclude that two variates are unrelated whenever $\rho_{xy} = 0$, as this coefficient is a measure of linear association only.

It is rare that attributes of biological, ecological, and environmental phenomena are perfectly linearly correlated or completely uncorrelated. A positive correlation between two variates, whether linear or not, will be revealed graphically as a trend showing increasing value of y for increasing value of x, as, for example, in Figure 3.2 which displays bole volume (y) versus bole diameter (x) of 236 red oak trees. At any value of diameter on the horizontal axis, there is a spectrum of volumes that could correspond to it. The smaller the range of y at each x, the more highly correlated are x and y. A positive trend bespeaks a positive correlation, and a positive linear correlation coefficient, ρ_{xy}, even if the trend is curvilinear. A positive value of ρ_{xy} does not imply that x and y share a straight-line relationship.

The population parameter, ρ_{xy}, commonly is estimated from a SRSwoR by

$$\hat{\rho}_{xy} = \frac{\sum_{u_k \in s} (y_k - \bar{y})(x_k - \bar{x})}{(n-1)\sqrt{s_y^2 s_x^2}}. \tag{6.117}$$

6.12.2 The jackknife estimator of $R_{y|x}$.

Let $\hat{R}_{y|x,-k}$ be the estimate of $R_{y|x}$ using all n observations but the kth. In other words,

$$\hat{R}_{y|x,-k} = \frac{\hat{\tau}_{y\pi} - y_k/\pi_k}{\hat{\tau}_{x\pi} - x_k/\pi_k}. \tag{6.118}$$

The 'omit-one' jackknife estimator of $R_{y|x}$ is

$$\hat{R}_{y|x,J} = n\hat{R}_{y|x} - \frac{n-1}{n} \sum_{k=1}^{n} \hat{R}_{y|x,-k} \tag{6.119}$$

$$= \frac{1}{n} \sum_{k=1}^{n} \hat{R}_{y|x,\bullet}, \tag{6.120}$$

where $\hat{R}_{y|x,\bullet} = \frac{1}{n} \sum_{k=1}^{n} \hat{R}_{y|x,-k}$.

The approximate variance of $\hat{R}_{y|x,J}$ is unbiasedly estimated by

$$\hat{v}_J \left[\hat{R}_{y|x,J} \right] = \frac{1}{n} \sum_{k=1}^{n} \left(\hat{R}_{y|x,-k} - \hat{R}_{y|x,\bullet} \right)^2. \tag{6.121}$$

Efron (1982) showed that $\hat{v}_J \left[\hat{R}_{y|x,J} \right]$ has a slight positive bias, and he asserted that $\hat{v}_J \left[\hat{R}_{y|x,J} \right]$ is appropriate as an estimator of $V \left[\hat{R}_{y|x} \right]$, also.

6.12.3 Equivalent expression of B_π

To show the equivalence of (6.80) to (6.83), begin by expanding the product inside the numerator of (6.83):

$$\sum_{\mathcal{U}_k \in s} \frac{\left(x_k - \hat{\tau}_{x\pi}/\hat{N}_\pi \right) \left(y_k - \hat{\tau}_{y\pi}/\hat{N}_\pi \right)}{\pi_k}$$

$$= \sum_{\mathcal{U}_k \in s} \frac{x_k y_k}{\pi_k} - \frac{\hat{\tau}_{y\pi}}{\hat{N}_\pi} \sum_{\mathcal{U}_k \in s} \frac{x_k}{\pi_k} - \frac{\hat{\tau}_{x\pi}}{\hat{N}_\pi} \sum_{\mathcal{U}_k \in s} \frac{y_k}{\pi_k} + \frac{\hat{\tau}_{x\pi} \hat{\tau}_{y\pi}}{\hat{N}_\pi^2} \sum_{\mathcal{U}_k \in s} \frac{1}{\pi_k}$$

$$= \hat{\tau}_{xy\pi} - \frac{\hat{\tau}_{y\pi}}{\hat{N}_\pi} \hat{\tau}_{x\pi} - \frac{\hat{\tau}_{x\pi}}{\hat{N}_\pi} \hat{\tau}_{y\pi} + \frac{\hat{\tau}_{x\pi} \hat{\tau}_{y\pi}}{\hat{N}_\pi^2} \hat{N}_\pi$$

$$= \hat{\tau}_{xy\pi} - \frac{\hat{\tau}_{x\pi} \hat{\tau}_{y\pi}}{\hat{N}_\pi},$$

which is identical to the numerator in (6.80).

The equivalence of the denominator of (6.83) to that of (6.80) follows a similar progression:

$$\sum_{\mathcal{U}_k \in s} \frac{\left(x_k - \hat{\tau}_{x\pi}/\hat{N}_\pi \right)^2}{\pi_k} = \sum_{\mathcal{U}_k \in s} \frac{x_k^2}{\pi_k} - 2\frac{\hat{\tau}_{x\pi}}{\hat{N}_\pi} \sum_{\mathcal{U}_k \in s} \frac{x_k}{\pi_k} + \frac{\hat{\tau}_{x\pi}^2}{\hat{N}_\pi^2} \sum_{\mathcal{U}_k \in s} \frac{1}{\pi_k}$$

$$= \hat{\tau}_{x^2\pi} - 2\frac{\hat{\tau}_{x\pi}}{\hat{N}_\pi} \hat{\tau}_{x\pi} + \frac{\hat{\tau}_{x\pi}^2}{\hat{N}_\pi^2} \hat{N}_\pi$$

$$= \hat{\tau}_{xy\pi} - \frac{\hat{\tau}_{x\pi}^2}{\hat{N}_\pi},$$

which is identical to the denominator in (6.80).

6.12.4 Joint inclusion probability under unequal probability systematic sampling

The following expression was derived by Hartley & Rao (1962) as an approximation to the joint inclusion probability of $\mathcal{U}_k$ and $\mathcal{U}_{k'}$. With appropriate adjustment to the differences in notation, it appears as their expression (5.15).

$$\pi_{kk'} = \frac{n-1}{n}\pi_k\pi_{k'} + \frac{n-1}{n^2}\pi_k\pi_{k'}\left(\pi_k + \pi_{k'}\right)$$

$$- \frac{n-1}{n^3}\pi_k\pi_{k'}\phi + 2\frac{n-1}{n^3}\pi_k\pi_{k'}\left(\pi_k^2 + \pi_k\pi_{k'} + \pi_{k'}^2\right)$$

$$- 3\frac{n-1}{n^4}\pi_k\pi_{k'}\left(\pi_k + \pi_{k'}\right)\phi + 3\frac{n-1}{n^5}\pi_k\pi_{k'}\phi^2$$

$$- 2\frac{n-1}{n^4}\pi_k\pi_{k'}\varphi \qquad\qquad (6.122)$$

In the above expression, $\phi = \sum_{k=1}^{N}\pi_k^2$ and $\varphi = \sum_{k=1}^{N}\pi_k^3$.

CHAPTER 7

Sampling with Fixed Area Plots

7.1 Introduction

Plot sampling often is used where the populations of interest comprise elements that are distributed spatially over a landscape, e.g., plants, ant hills, wildlife dens, etc. Sampling over the landscape usually relies on an areal sampling frame because of the infeasibility of compiling a list frame of individual elements or clusters of elements. By an areal sampling frame we mean a device, such as a map or a GIS, that permits the selection of any point location within a region, denoted by $\mathcal{A}$, on which all the discrete elements $\mathcal{U}_k$, $k = 1, \ldots, N$, are situated. Sampling locations are point locations, which ordinarily are selected uniformly at random from the continuous areal frame. A sampling location may serve as the center point of a circular plot, the centroid or a corner point of a rectangular plot (quadrat), or it may be distinct from the plot, where the sampling protocol indicates the distance and direction from the sampling location to a plot with a prescribed shape and size. Plots of any shape are permissible, but shapes other than circular or rectangular are rarely employed in practice. The elements that occur within a plot constitute a probability sample of the population of discrete elements.

In addition to the sampling design that establishes a single plot at each sampling location, there are a number of variants. Frequently plots of the same shape, but of different sizes, will be located at a single sampling location. Often a nested-plot design is employed to permit smaller but more frequently occurring elements to be sampled on smaller plots in order to allocate the sampling effort more equably among the larger elements of the population. For example, trees exceeding some threshold diameter will be sampled on 0.1 ha plots, whereas trees smaller than the threshold diameter will be sampled only if they occur on a 0.05 ha plot located at the same place in $\mathcal{A}$. Another variant is the establishment of not one but a cluster of plots equidistant from a central sampling location. An example of a plot cluster appears in Figure 7.5.

These variants notwithstanding, a distinguishing feature of plot sampling is that, with some exceptions, each element of the population is sampled with a probability proportional to the area of the plot in which it occurs. The exceptional elements are close to the boundary of $\mathcal{A}$, where, it turns out, their inclusion probabilities are somewhat diminished. This interesting complication will be explained in detail in section 7.5. With the exception of these elements close to the boundary of $\mathcal{A}$, plot sampling is an example of equal probability sampling with an areal frame.

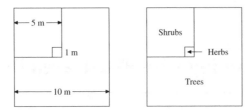

Figure 7.1 *The layout of the nested plot from Ponce-Hernandez (2004). As described in Example 7.1, successively smaller vegetation, organisms, and features are measured on successively smaller plots.*

Example 7.1

In a manual prepared by the Food and Agriculture Organization of the United Nations (Ponce-Hernandez 2004), a nest of square plots, shown in Figure 7.1, was described for the sampling of aboveground biomass and assessment of land degradation.

Morphometric measurements of the trees and large woody detritus are gathered over the entire 10 m × 10 m quadrat. On this plot, too, tree species and individual organisms within a species are recorded for the purpose of biodiversity assessment, as well as site measurements and observations for land degradation assessment.

In addition to the measurements just noted, the shrub layer is also measured in the 5 m × 5 m quadrat. At this level the stem and canopy of shrubs are measured, plus small deadwood, and shrub species and individual shrub organisms are identified and recorded.

Information about the herbaceous species is added to the mix in the 1 m^2 plot. Litterfall, fine debris, and stems and roots of herbaceous species and grasses are sampled for the determination of live and dead biomass. The number of herbaceous species and number of individuals within species are counted.

7.2 Notation

As already noted, $\mathcal{A}$ indicates the region on which the population of interest is situated. The region often is demarcated by political or property boundaries. For example, $\mathcal{A}$ may be a parcel of forested property owned by an individual; or a national or district park; or a conservation area or landfill; or an industrial plantation of commercially harvestable trees or agricultural crop; or a riparian zone surrounding a recreation area; and so on. Let the horizontal land area of $\mathcal{A}$ be denoted by A.

The number of sampling locations at which a plot or cluster of plots is established on $\mathcal{A}$ is denoted by m. With plot sampling it is convenient to regard the number of sampling locations as the size of the sample, rather than the number, n, of population elements that are included in the sample: the number of plots to establish is a design parameter to be stipulated by the survey planner, whereas n is not. The number of elements sampled from the population at each sampling location can not be known in advance of sampling.

Let $\mathbb{P}_s$ denote sample plot s, where $s = 1, \ldots, m$. With respect to the origin of orthogonal axes encompassing the horizontal plane of $\mathcal{A}$, let (x_s, z_s) denote the location coordinates of $\mathbb{P}_s$. For example, the x-axis may point East, the z-axis may point North, and $(x, z) = (0, 0)$ may be the southwest corner of a rectangle which completely encloses the horizontal projection of $\mathcal{A}$ onto a map or GIS.

Let a denote the area of each plot. A circular plot of radius R m implies that $a = 10^{-4} \pi R^2$ ha. A rectangular plot of dimension L m by W m implies that $a = 10^{-4} LW$ ha. Although plots of other shapes are rarely used in practice, it is not uncommon to find plots in clusters arranged in a geometric pattern such as a star or an ell or a hexagon.

7.2.1 Selection and installation

Sampling locations ordinarily are selected by the acceptance-rejection method. Let X and Z be the length and width of a rectangle that encompasses $\mathcal{A}$. Let u_1 and u_2 be two uniform random numbers drawn from U[0, 1]. Calculate $x = u_1 X$ and $z = u_2 Z$. If the coordinates (x, z) occur in $\mathcal{A}$, then this point location is selected as a sampling location; otherwise, this point location is rejected and the selection procedure is repeated with two new random numbers.

All distances and dimensions of plots are measured in the horizontal plane. On steeply sloping land, this requirement may cause circular plots to appear as ellipses, if dimensions are measured on the slant. Likewise, a square plot may appear as a rectangle or a distorted rectangle, depending on the orientation of the plot with respect to the aspect of the landscape.

7.3 Sampling protocol

All population elements situated inside a plot are included in the sample for that plot—provided all these elements are within $\mathcal{A}$. If a plot overlaps the boundary of $\mathcal{A}$, then it may contain some elements that are not in $\mathcal{A}$; these elements are not part of the sample. We regard location of an element as a point property. For this purpose, we let (x_k, z_k) denote the location of $\mathcal{U}_k, k = 1, \ldots, N$ on $\mathcal{A}$. In contrast to the random nature of the sampling locations, $(x_s, z_s), s = 1, \ldots, m$, the locations of the elements are regarded as fixed. With a circular plot of radius R, the protocol implies that an element, $\mathcal{U}_k$, is tallied if (x_k, z_k) is closer than R to (x_s, z_s). Regardless of plot shape, there is a need to determine the location of $\mathcal{U}_k$ unambiguously. If the population of interest is one of single-stemmed plants, the location of $\mathcal{U}_k$ is normally taken to be the center of the base of the stem. For other types of populations, the point location of each element may be less obvious. For example, if the population consists of woody detritus or coarse woody debris (CWD), part of $\mathcal{U}_k$ may lie outside the plot.

Part of the sampling protocol must be an a priori definition of the point on each $\mathcal{U}_k$ that determines its location. In the case of log-shaped CWD, this point might be defined as the furthermost extreme of the small end of the log; or the midpoint of

ASIDE: We do not envision the situation where the the region $\mathcal{A}$ is tessellated into a finite number of non-overlapping cells, each of approximate area a, from which m will be selected at random. In the design we have proposed, plots may well overlap if randomly located over $\mathcal{A}$. To prevent overlap, plots often will be established on a systematic grid. Whether randomly or systematically established, there are an infinite number of potential plot locations, (x_s, z_s), in the design we present. Except for very small land areas, we do not view the tessellation of $\mathcal{A}$ as a realistic or practical design for sampling with an areal frame. For that reason, we do not consider it further.

An alternative design, having elements of both the tessellation design above and systematic sampling, has been termed 'unaligned systematic sampling.' In this design, $\mathcal{A}$ is tesselated into a grid of m rectangular cells, each of which will be considerably larger in area than a. Within each cell, one plot is located at random. Analytical and empirical results from simulation studies reported by Barabesi & Pisani (2004) indicate that this design warrants further attention. Despite its apparent merits, however, we do not consider this design for the location of plots on $\mathcal{A}$ further, either.

the log; or any other point that can be unequivocally identified on any piece of CWD that might be encountered. How the location of $\mathcal{U}_k$ is defined is less important than (*i*) the ease with which it can be determined in the field, and (*ii*) the consistent use of this definition for all plots in a particular survey. Failure of the latter is a form of measurement error which may make estimation of the population parameters less precise than otherwise. Errors from the inconsistent or sloppy determination of the location of population elements within a plot may also result in biased estimation of population parameter values. An element is either in a plot, or not; and the errors that accrue from incorrect determinations rarely average to zero. The risk of measurement error of this sort usually is lessened by working in a systematic manner. For example, tally elements within a circular plot by starting from a fixed direction, e.g., due north, and working in a clockwise direction.

7.4 Estimation

7.4.1 Inclusion zone of an element

By definition (see §1.3.2), the inclusion probability of $\mathcal{U}_k$ is the probability of including it in the sample. For present purposes, we focus on the probability of including $\mathcal{U}_k$ in the sample from a sample plot, $\mathbb{P}_s$. Whether or not $\mathcal{U}_k \in \mathbb{P}_s$ depends on the location of $\mathcal{U}_k$ relative to (x_s, z_s). It is instructive to visualize the locus of points on $\mathcal{A}$ where (x_s, z_s) could be randomly located such that $\mathcal{U}_k \in \mathbb{P}_s$. We do so with a series of examples for plots of various shapes and orientations. In the figures accompanying these examples, the open circle $\circ$ indicates the point (x_s, z_s)

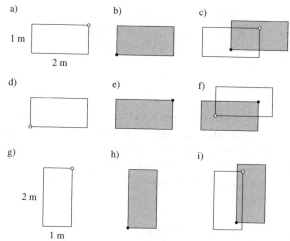

Figure 7.2 *Three differently located 1 m × 2 m plots and the inclusion zones implied by them. The ○ indicates the location, (x_s, z_s) of a typical plot, and • indicates the location, (x_k, z_k) of a typical population element, $\mathcal{U}_k$. The rectangular plot described in Example 7.2 is shown in frame (a); the inclusion zone of $\mathcal{U}_k$ corresponding to plots of this size, shape, orientation in frame (b); an example plot in the inclusion zone is shown in (c). Frames (d), (e), and (f) display the sample plot described in Example 7.3, with the corresponding inclusion zone and specific example. Frames (g), (h), and (i) display the sample plot described in Example 7.4, with the corresponding inclusion zone and specific example.*

by which the plot is located. The filled dot • indicates the location of an element of the population, $\mathcal{U}_k$.

Example 7.2

Consider a rectangular plot of dimensions 1 m by 2 m, oriented with its longer side in an East-West direction, and with (x_s, z_s) as its northeast corner, as displayed in Figure 7.2a. Only those points within the shaded region of Figure 7.2b could serve as the location of $\mathbb{P}_s$ such that $\mathcal{U}_k \in \mathbb{P}_s$. One such point location of a plot is shown in frame (c), from which it is clear that this sample plot would include $\mathcal{U}_k$. We term this shaded region the *inclusion zone* for $\mathcal{U}_k$. Any (x_s, z_s) located outside the shaded region could not include that element in a plot of this size, shape, orientation, and located by its northeast corner.

Example 7.3

Consider the same size plot as in the preceding example, but with its southeastern corner as the point which serves to locate the plot on $\mathcal{A}$, as shown in Figure 7.2d.

The inclusion zone of $\mathcal{U}_k$ for a plot of this dimension and orientation is shown in frame (e). Evidently it is the same size as the inclusion zone in Example 7.2, but its placement relative to (x_k, z_k) has shifted.

Example 7.4

Suppose the same plot dimensions as in Example 7.2, but with the longer side of the plot aligned in a North-South direction. The plot is located by its northeastern corner, as in shown in Figure 7.2f. The inclusion zone of $\mathcal{U}_k$ for a plot of this dimension and orientation is shown in frame (g). The change in orientation of the plot causes a similar change in the orientation of the inclusion zone with respect to (x_k, z_k).

The inclusion zone of an element is a subregion of $\mathcal{A}$. The shape and areal size of the inclusion zone is determined by, and is identical to, the shape and size of the intended field plot. The location of the inclusion zone relative to the point location of $\mathcal{U}_k$ is determined by the placement of a plot relative to the sampling location (x_s, z_s). As illustrated in Examples 7.2 and 7.3, any change in the position of a plot relative to the sampling location causes a predictable change in the position of the inclusion zone relative to the location of the element. Examples 7.2 and 7.4 illustrate the effect of changing the orientation of the plot on the position of the inclusion zone relative to the location of $\mathcal{U}_k$.

The following examples illustrate further the relationship between the shape and orientation of the plot around its point of location to the shape, orientation, and location of the inclusion zone of $\mathcal{U}_k$.

Example 7.5

A triangular plot rarely is used in practice, however it is instructive to consider the use of one. We suppose that each side of the plot is 2 m long, that its base runs in an East-West direction, and the location point (x_s, z_s) is its upper vertex, as shown in frame (a) of Figure 7.3. Obviously for any such plot to include $\mathcal{U}_k$, (x_s, z_s) of the plot must be north of $\mathcal{U}_k$, leading to an inclusion zone whose orientation is inverted from that of the triangular plot as described, as shown in frame (b) of the figure. In frame (c) is a typical plot that is within the element's inclusion zone and hence includes the element within it.

Example 7.6

Suppose one elects to use a plot in the shape of an L, and the inner corner of the L serves to locate the plot at (x_s, z_s). Suppose that the short leg runs in a North-South direction and is 3 m long and 2 m wide. The long side of the L-shaped plot is 5 m long and 1 m wide, as shown in frame (a) of Figure 7.4. The

Figure 7.3 *An example of a triangular plot: a) the plot located at ○ described in Example 7.5; b) the inclusion zone for $\mathcal{U}_k$ when sampling with such a triangular plot; c) a typical plot in the inclusion zone of $\mathcal{U}_k$.*

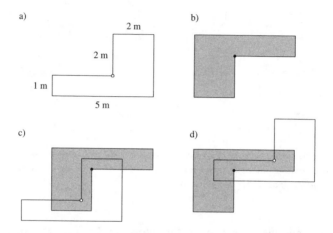

Figure 7.4 *The inclusion zone for $\mathcal{U}_k$ when sampling with an L-shaped plot described in Example 7.6. The layout of the plot located at ○ is shown in (a); the corresponding inclusion zone of $\mathcal{U}_k$ is shown in frame (b); typical plots within the inclusion zone of $\mathcal{U}_k$ are shown in frames (c) and (d).*

corresponding inclusion zone for $\mathcal{U}_k$ is shown in frame (b). Two typical plots which include $\mathcal{U}_k$ are shown in frames (c) and (d).

Example 7.7

A circular plot of radius R located at (x_s, z_s) implies a circular inclusion zone for each element of the population. Any plot located within R of (x_k, z_k) will include $\mathcal{U}_k$.

Example 7.8

Suppose the sampling location (x_s, z_s) is the center of a triangular cluster of circular plots, each of radius R. One plot is located 20 m directly north of

ASIDE: An image of the inclusion zone corresponding to any size and shape plot can be generated by making the location point, (x_s, z_s), of the plot coincident with the location, (x_k, z_k), of $\mathcal{U}_k$, and then rotating the plot $180°$ around (x_k, z_k). The subregion mapped out by this rotated 'plot' is the locus of points where the (x_s, z_s) of any plot of the prescribed size and shape can be located and include $\mathcal{U}_k$.

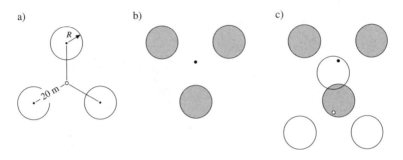

Figure 7.5 *The triangular cluster of circular plots described in Example 7.8 is shown in frame (a). The corresponding inclusion zone for $\mathcal{U}_k$ appears in frame (b). A typical plot cluster within the inclusion zone of $\mathcal{U}_k$ is shown in frame (c).*

(x_s, z_s), while the other two satellite plots are 20 m distant from (x_s, z_s) at $120°$, and $240°$, respectively. The orientation of a typical plot cluster is shown in frame (a) of Figure 7.5. The inclusion zone for $\mathcal{U}_k$ is shown in frame (b). With this type of plot cluster, the sampling location (x_s, z_s) is not within any of the plots of the cluster.

If the sampling protocol prescribes that all (x_s, z_s) must be located within $\mathcal{A}$, then it is possible that some units of the population will have an inclusion zone smaller than the size of the plot or plot cluster. This occurs for those $\mathcal{U}_k$ sufficiently close to the edge of $\mathcal{A}$ that part of its nominal inclusion zone lays outside of $\mathcal{A}$, and hence gets truncated by the boundary. This situation is illustrated in Figure 7.6, in which we presume that a single circular plot is established at each (x_s, z_s). In this figure, $\mathcal{U}_1$ is located a distance greater than R from any boundary of $\mathcal{A}$, and hence has an inclusion zone with an area equal to the area of a sample plot. In contrast, $\mathcal{U}_2$ is less than R from one boundary of $\mathcal{A}$; because its inclusion zone is truncated by the boundary, its area is less than that of a sample plot. In the same figure, the inclusion zone of $\mathcal{U}_3$ is truncated by adjoining boundaries. We discuss methods to deal with edge units with truncated inclusion zones in §7.5. For the present, we denote the area of the inclusion zone of $\mathcal{U}_k$ by a_k. When a single plot of area a is established at each sampling location, $a_k \leq a$; when a cluster of plots is established at each sampling location, $a_k \leq ca$, where c denotes the number of plots in each cluster; when different size plots are nested at each sampling location, the area of the inclusion zone for $\mathcal{U}_k$ is

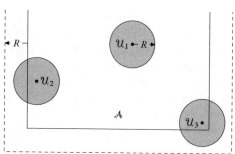

Figure 7.6 *When sampling with circular plots, each having radius R, element $\mathcal{U}_1$ is sufficiently into the interior of $\mathcal{A}$ that its inclusion zone is the same size and shape as the plot. In contrast, elements $\mathcal{U}_2$ and $\mathcal{U}_3$ are so close to one or more boundaries of $\mathcal{A}$ that their inclusion zones are truncated.*

less than or equal to the area of the nested plot appropriate for elements of its size class.

7.4.2 Estimation following plot sampling

The notion of inclusion zone is integral to understanding the inclusion probability of $\mathcal{U}_k$ by a randomly located sampling unit, i.e., a plot or plot cluster. The latter is the horizontal area where a sampling unit can be located such that it would include $\mathcal{U}_k$, expressed as a proportion of the area where a sampling unit can be located. When (x_s, z_s) is restricted to $\mathcal{A}$, the above implies that

$$\pi_k = \frac{a_k}{A} \tag{7.1}$$

The probability of including each $\mathcal{U}_k$ at each sampling location enables HT estimation of τ_y for whatever characteristic y is measured on those $\mathcal{U}_k$ sampled. That is, the estimate of τ_y from $\mathbb{P}_s$ is

$$\hat{\tau}_{y\pi s} = \sum_{\mathcal{U}_k \in \mathbb{P}_s} \frac{y_k}{\pi_k} \tag{7.2a}$$

$$= A \sum_{\mathcal{U}_k \in \mathbb{P}_s} \frac{y_k}{a_k} \tag{7.2b}$$

$$= A \sum_{\mathcal{U}_k \in \mathbb{P}_s} \rho_k \tag{7.2c}$$

$$= A\rho_s, \tag{7.2d}$$

where $\rho_k = y_k/a_k$ is the value of y_k prorated to a unit area basis, and ρ_s is the sum of all such prorated values for plot s. If plot s contains no elements, then $\rho_s = 0$ and, therefore, $\hat{\tau}_{y\pi s} = 0$.

ASIDE: The quantity $\rho_k = y_k/a_k$ can be interpreted as an attribute density of $\mathcal{U}_k$ within its inclusion zone, i.e., the amount of attribute per unit land area. Likewise, ρ_s is the total attribute density—the total amount of attribute per unit land area—at the sampling location (x_s, z_s). These quantities are fundamental to the formulation of plot sampling as an application of Monte Carlo integration, which is described in Chapter 10.

The installation of multiple, independent sample plots on $\mathcal{A}$, followed by HT estimation of τ_y with the data from each plot, has been called replicated sampling; e.g., see Barabesi & Fattorini (1998), and Barabesi & Pisani (2004, §4). The customary estimator of τ_y based on a replicated sample of m plots is

$$\hat{\tau}_{y\pi,\text{rep}} = \frac{1}{m} \sum_{s=1}^{m} \hat{\tau}_{y\pi s} \tag{7.3a}$$

$$= \frac{A}{m} \sum_{s=1}^{m} \rho_s \tag{7.3b}$$

$$= A\bar{\rho}, \tag{7.3c}$$

where $\bar{\rho}$ is the average ρ_s value among the m plots.

The design-based variance of $\hat{\tau}_{y\pi s}$ is the same as given in (6.10), namely

$$V\left[\hat{\tau}_{y\pi s}\right] = \sum_{k=1}^{N} y_k^2 \left(\frac{1-\pi_k}{\pi_k}\right) + \sum_{k=1}^{N} \sum_{\substack{k' \neq k \\ k'=1}}^{N} y_k y_{k'} \left(\frac{\pi_{kk'} - \pi_k \pi_{k'}}{\pi_k \pi_{k'}}\right), \tag{7.4}$$

where $\pi_{kk'} = a_{kk'}/A$ and $a_{kk'}$ is the area of the joint inclusion zone for $\mathcal{U}_k$ and $\mathcal{U}_{k'}$. The joint inclusion zone for any two elements is the locus of points common to the inclusion zones of both $\mathcal{U}_k$ for $\mathcal{U}_{k'}$; see the Appendix (§7.10) for a discussion of joint inclusion zones.

If the m plots have been independently established on $\mathcal{A}$, the variance of $\hat{\tau}_{y\pi,\text{rep}}$ is

$$V\left[\hat{\tau}_{y\pi,\text{rep}}\right] = \frac{1}{m} V\left[\hat{\tau}_{y\pi s}\right]. \tag{7.5}$$

Moreover, $V\left[\hat{\tau}_{y\pi,\text{rep}}\right]$ can be estimated unbiasedly by

$$\hat{v}\left[\hat{\tau}_{y\pi,\text{rep}}\right] = \frac{1}{m(m-1)} \sum_{s=1}^{m} \left(\hat{\tau}_{y\pi s} - \hat{\tau}_{y\pi,\text{rep}}\right)^2 \tag{7.6a}$$

$$= \frac{A^2}{m(m-1)} \sum_{s=1}^{m} (\rho_s - \bar{\rho})^2. \tag{7.6b}$$

As the number, m, of plots grows large, theoretical results from mathematical

Table 7.1 *Tree diameters and numbers of Nectria cankers from a sample of 8 plots of size $a = 0.01$ ha on a tract of $A = 10$ ha.*

Plot	Dia.	No. cankers	Plot	Dia.	No. cankers
	(cm)			(cm)	
1	10		5	13	1
	12			15	
	15	2			
			6	9	1
2	13	2		10	
	14			11	1
	18	2		12	
	18			14	
	19	1		15	1
				17	1
3	11				
	11	1	7	10	
				12	
4	10	1		12	1
	10			14	2
	11	2		18	
	12			18	4
	14				
	16	1	8	10	1
				13	
5	9			15	2
	10			17	
				18	3

statistics indicate that the distribution of $\hat{\tau}_{y\pi,\text{rep}}$ is approximately normal or Gaussian. A confidence interval for τ_y is constructed using a t-value with $m - 1$ degrees of freedom i.e.,

$$\hat{\tau}_{y\pi,\text{rep}} \pm t_{m-1} \sqrt{\hat{v}\left[\hat{\tau}_{y\pi,\text{rep}}\right]}. \tag{7.7}$$

Example 7.9

Barrett & Nutt (1979) report the results of a plot sample with $m = 8$ plots, each of area $a = 0.01$ hectare. The diameter of each tree on a sample plot was recorded, as well as the number of Nectria cankers on each tree, as shown in Table 7.1. Because none of the plots were near the edge of the tract, the inclusion zone of each of the trees, and hence cankers, on these plots had an inclusion probability of $\pi_k = 0.01/10 = 0.001$. From the first plot alone, the total number of cankers on the 10 hectare tract is estimated to be

$$\hat{\tau}_{y\pi 1} = \frac{2}{0.001} = 2000 \text{ cankers.}$$

From the second plot, the number of cankers on the entire tract is estimated to be

$$\hat{\tau}_{y\pi 2} = \frac{2}{0.001} + \frac{2}{0.001} + \frac{1}{0.001} = \frac{5}{0.001} = 5000 \text{ cankers.}$$

Continuing in this fashion and averaging the $m = 8$ plot estimates together provides a replicated sampling estimate of

$$\hat{\tau}_{y\pi,\text{rep}} = 3750 \text{ cankers.}$$

The plot-to-plot variance was $s_y = 5,071,429$, leading to an estimated standard error of $\hat{\tau}_{y\pi,\text{rep}}$ of

$$\sqrt{\hat{v}\left[\hat{\tau}_{y\pi,\text{rep}}\right]} = 796,$$

or 21%. A 90% interval estimate of the number of cankers is $3750 \pm 40\%$, or from 2242 to 5258 cankers.

Example 7.10

Often there is interest in estimating the size of the population, N. This is simply τ_y when $y_k = 1$ (the count of an individual element) for all $\mathcal{U}_k$ on $\mathcal{A}$. With a plot sample consisting of a single plot of area a at each sampling location, $\hat{\tau}_{y\pi s}$ simplifies to

$$\hat{\tau}_{y\pi s} = \sum_{\mathcal{U}_k \in \mathbb{P}_s} \frac{1}{\pi_k} \tag{7.8a}$$

$$= A \sum_{\mathcal{U}_k \in \mathbb{P}_s} \frac{1}{a_k}. \tag{7.8b}$$

If there are no elements in $\mathbb{P}_s$ close to the edge of $\mathcal{A}$, then the above result further simplifies to $\hat{\tau}_{y\pi s} = n_s A/a$, where n_s is the count of the number of elements sampled on $\mathbb{P}_s$.

Combining the estimates from all m plots provides

$$\hat{\tau}_{y\pi,\text{rep}} = \frac{1}{m} \sum_{s=1}^{m} \hat{\tau}_{y\pi s} \qquad (7.9a)$$

$$= \frac{1}{m} \sum_{s=1}^{m} \sum_{u_k \in \mathbb{P}_s} \frac{1}{\pi_k} \qquad (7.9b)$$

$$= \frac{A}{m} \sum_{s=1}^{m} \sum_{u_k \in \mathbb{P}_s} \frac{1}{a_k} \qquad (7.9c)$$

$$= \hat{N}_{\pi,\text{rep}}, \text{ say.} \qquad (7.9d)$$

Barring the occurrence of edge elements in the sample, this simplifies to

$$\hat{\tau}_{y\pi,\text{rep}} = \frac{A}{ma} \sum_{s=1}^{m} n_s \qquad (7.10a)$$

$$= \frac{nA}{ma}, \qquad (7.10b)$$

where $n = \sum_{s=1}^{m} n_s$ is the total number of elements sampled on all m plots.

7.4.3 Prorating estimates to unit area values

The average amount of y per unit area, say λ_y, is a population parameter functionally related to τ_y: $\lambda_y = \tau_y / A$. Similarly, λ_y may be estimated unbiasedly with the data sampled at $\mathbb{P}_s$ by

$$\hat{\lambda}_{y\pi s} = \frac{\hat{\tau}_{y\pi s}}{A}. \qquad (7.11)$$

The estimator of λ_y from the replicated sample of m plots is

$$\hat{\lambda}_{y\pi,\text{rep}} = \frac{1}{m} \sum_{s=1}^{m} \hat{\lambda}_{y\pi s} \qquad (7.12a)$$

$$= \frac{\hat{\tau}_{y\pi,\text{rep}}}{A}. \qquad (7.12b)$$

This estimator can also be expressed in terms of the ρ_s introduced earlier:

$$\hat{\lambda}_{y\pi,\text{rep}} = \frac{1}{Am} \sum_{s=1}^{m} \sum_{\mathcal{U}_k \in \mathbb{P}_s} \frac{y_k}{a_k/A} \qquad (7.13a)$$

$$= \frac{1}{m} \sum_{s=1}^{m} \sum_{\mathcal{U}_k \in \mathbb{P}_s} \frac{y_k}{a_k} \qquad (7.13b)$$

$$= \frac{1}{m} \sum_{s=1}^{m} \rho_s \qquad (7.13c)$$

$$= \bar{\rho}. \qquad (7.13d)$$

Result (7.13b) makes it clear that estimates prorated to a unit area basis do not require knowledge of the area of $\mathcal{A}$ nor the explicit determination of inclusion probabilities.

The variance of $\hat{\lambda}_{y\pi,\text{rep}}$ is

$$V\left[\hat{\lambda}_{y\pi,\text{rep}}\right] = \frac{1}{A^2} V\left[\hat{\tau}_{y\pi,\text{rep}}\right], \qquad (7.14)$$

which is unbiasedly estimated by

$$\hat{v}\left[\lambda_{y\pi,\text{rep}}\right] = \frac{1}{m(m-1)} \sum_{s=1}^{m} \left(\hat{\lambda}_{y\pi s} - \hat{\lambda}_{y\pi,\text{rep}}\right)^2. \qquad (7.15)$$

Alternatively, one can use $\hat{v}\left[\hat{\tau}_{y\pi,\text{rep}}\right]$ in

$$\hat{v}\left[\lambda_{y\pi,\text{rep}}\right] = \frac{1}{A^2} \hat{v}\left[\hat{\tau}_{y\pi,\text{rep}}\right]. \qquad (7.16)$$

to obtain the estimated variance.

Example 7.11

The result, n_s/a, in Example 7.10 represents the number of elements in sample plot s, prorated to a unit area basis. When multiplied by A, the area of $\mathcal{A}$, as in (7.9c), it estimates the size of the population in the entire region.

To convert this to an estimate of frequency per unit area, calculate

$$\hat{\lambda}_N = \frac{\hat{N}_{\pi,\text{rep}}}{A}, \qquad (7.17)$$

whose standard error is computed analogously:

$$\sqrt{\hat{v}\left[\hat{\lambda}_N\right]} = \frac{1}{A} \sqrt{\hat{v}\left[\hat{N}_{\pi,\text{rep}}\right]}, \qquad (7.18)$$

where $\hat{v}\left[\hat{N}_{\pi,\text{rep}}\right]$ is given by (7.6b).

Table 7.2 *Biomass (g) of* Agropyron smithii *Rydb. and all grasses and forbs sampled from 12 quadrats on a grassland prairie of A* = 8 *ac* = 3.24 *ha.*

| Plot | Grasses and forbs | |
	A. smithii	All
	(g)	(g)
1	27.5	76.0
2	37.4	138.6
3	1.2	54.5
4	8.7	79.8
5	43.2	98.0
6	40.3	106.0
7	30.1	93.1
8	20.4	81.2
9	25.7	81.1
10	34.3	131.1
11	36.3	118.1
12	17.5	82.6

Example 7.12

The grass *Agropyron smithii* Rydb. is native to the prairies of North Dakota (USA). Data from a sample of 12 rectangular quadrats were presented in Hanson (1934). A summary of these data appear in Table 7.2, and the complete listing of data can be obtained at the book's website at http://www.crcpress.com.

Each quadrat had an area of $a = 0.8\,\text{m}^2$, so that the inclusion probability of each grassland plant was approximately 0.000,024. From (7.3c) we get $\hat{\tau}_{y\pi,\text{rep}} = 1088.8$ Mg, with an estimated standard error of $\sqrt{\hat{v}\left[\hat{\tau}_{y\pi,\text{rep}}\right]} = 150.9$ Mg, which is 13.9% of $\hat{\tau}_{y\pi,\text{rep}}$. In percentage terms a 90% confidence interval for the total biomass of *A. smithii* on this prairie is 1088.8 Mg $\pm 25\%$, or 817.7 to 1359.8 Mg.

Example 7.13

The results from the preceding example can be expressed as a density by means of (7.13) by dividing 1088.8 Mg by 3.24 ha to yield an interval estimate of $\hat{\lambda}_y = 336\,\text{Mg ha}^{-1} \pm 25\%$.

Example 7.14

Table 7.2 also shows the amount of sampled biomass for all grasses and forbs on each plot. The biomass per hectare is estimated to be 1187.5 Mg, and the estimated standard error in percentage terms is 7.5%. The latter is lower than the

13.9% when estimating the biomass of only *A. smithii*. It is almost always true that an estimate of a subset of a population is less precise than the corresponding estimate for the whole, regardless of whether the criterion for subsetting is species, sex, age, or other grouping variable. In this case, it is testimony to the fact that grass vegetation is more uniformly distributed on the prairie than is *A. smithii*.

7.4.4 Estimating the mean attribute per element

When sampling with an areal frame the size of the population, N, usually is unknown. Because of this, estimation of $\mu_y = \tau_y/N$ by $\hat{\mu}_y = \hat{\tau}_y/N$ evidently is impossible with data from a plot sample. However, Example 7.10 shows how one can estimate N by treating y_k as a count of $\mathcal{U}_k$. This suggests that when y is some additional characteristic of interest, μ_y can be estimated by dividing (7.3c) by (7.9d):

$$\hat{\mu}_{y\pi,\mathrm{rat}} = \frac{\hat{\tau}_{y\pi,\mathrm{rep}}}{\hat{N}_{\pi,\mathrm{rep}}}. \tag{7.19}$$

This estimator is reminiscent of the estimator of the population ratio, $\hat{R}_{y|x}$ in (6.2), presented in Chapter 6. In $\hat{\mu}_{y\pi,\mathrm{rat}}$, however, we are using the replicated sampling estimators of τ_y and N, rather than the HT estimators of these parameters. Although they are not quite the same, we do not expect their sampling distributions to be markedly different in large samples of plots (in the case of $\hat{\mu}_{y\pi,\mathrm{rat}}$) and elements (in the case of $\hat{R}_{y|x}$).

7.4.5 Stratification

It often is possible to stratify an areal sampling frame of $\mathcal{A}$ to great advantage in order to simplify the sampling effort in the field, to obtain separate estimates of the population parameters by strata for administrative or management purposes, or to increase the precision of estimation. As explained in Chapter 5, the criterion used to stratify $\mathcal{A}$ will vary from one type of survey to the next, but common stratification variables are land cover class, ownership, land use, and size or age class of the population elements. For surveys of small land areas, the benefits of stratification often will justify the cost of an initial effort to traverse the property for the purpose of deciding on an apt stratification criterion (e.g., cover type, drainage, past land use), and rough delineation of strata boundaries. For extensive surveys covering large land areas, satellite images or other forms of remotely sensed data are often used for this purpose.

   Sampling and estimation following stratification of $\mathcal{A}$ proceeds along the lines presented in Chapter 5, using the estimators presented earlier in this section of each stratum's parameters of interest.

7.4.6 Sampling intensity

It is generally true that precision is directly related to sampling intensity (SI), which is the proportion of land area included in the sample:

$$SI = \frac{mca}{A}. \tag{7.20}$$

In the forestry vernacular, SI expressed as a percentage is known as the percent cruise.

Although SI is not identical to the sampling fraction, n/N, it generally is the case that by increasing SI one effectively increases the sampling fraction, and hence the precision of estimation, too. Evidently SI increases as the number of plots increases, and for a fixed number of plots, m, SI increases as the area, a, of each plot increases. If plots are allowed to overlap, then it is possible for SI to exceed unity. In practice SI is a small proportion of the land area of $\mathcal{A}$.

7.4.7 Elements near the plot border

One of the mensurational burdens which accompany plot sampling is the determination of which of the population elements are contained within the plot located at (x_s, z_s). For many elements it will be obvious whether they are inside or outside of the plot, but there will some elements that are so near to the border, or margin, of the plot, that an optical determination of their inclusion in the plot will be difficult. Errors in this determination introduce bias into the estimators of τ_y and λ_y presented above. A commonly employed tactic is to include the first 'borderline element' into the sample, and to exclude the next one encountered, and to keep alternating inclusion into the sample with this pairwise strategy. This is bad practice for at least two reasons. First, for elements near the boundary of the plot, optical determination of presence within the plot by human eyesight is faulty and inconsistent, and will yield results that are not repeatable if different observers were to conduct the sampling. Second, errors of inclusion or exclusion by this method will rarely even out or sum to zero on average.

Measurement errors of this sort are never entirely avoidable, but a modicum of good field procedure will reduce their prevalence and impact considerably: use a mechanical instrument (e.g., a tape measure and compass) or an electronic instrument (e.g., a laser range finder) of verified tolerance, in order to determine whether the element is in the plot, or not. If the burden to take careful measurements is too great for the budget allocated to the sample, then the SI should be lessened. A diminished SI will decrease the precision of estimation, but lessened precision usually is preferable to introducing a bias of indeterminant and possibly sizable magnitude by failing to measure carefully.

7.5 Edge effect

The boundary overlap problem, as it is known in the forestry literature, arises where an element, $\mathcal{U}_k$, occurs so close to the boundary of $\mathcal{A}$ that its inclusion zone overlaps the boundary. However, boundary overlap is not restricted to sampling in forestry; indeed, it attends all applications of sampling with an areal frame.

Boundary overlap effectively truncates the element's inclusion zone, reducing the area of the inclusion zone by the area of the overlapping portion. Were it not for elements near the edge of A with truncated inclusion zones, the inclusion probability of all elements would be a constant value ca/A, where a is the area of each plot in a cluster of $c \geq 1$ plots established at any (x_s, z_s) in A. The estimator $\hat{\tau}_{y\pi s}$ in (7.2d) requires knowledge of the inclusion probability of each element in the sample, which means that the inclusion areas of any edge units need to be determined. For example, additional measurements would be required in the field to determine the truncated circular inclusion area of U_2 and U_3 in Figure 7.6. Determination of the inclusion area of each element provides a general solution to the boundary overlap problem for both single-plot and cluster-plot designs that use plots of any shape. We call this the 'measure π method.'

Other, less labor intensive, solutions have been advanced to solve the boundary overlap problem, and most of them apply to designs that prescribe independent circular plots ($c = 1$). An exception is the buffer method, which applies to both single-plot and cluster-plot designs.

7.5.1 External peripheral zone

This method—commonly called the buffer method—solves the boundary overlap problem allowing sampling locations to fall both within A and within a tract buffer, i.e., an external zone that surrounds A. In essence, this method 'un-truncates' the inclusion zones of the elements near the boundary. The external buffer zone must be wide enough that no population element within A can have an inclusion zone which overlaps the exterior boundary of the peripheral zone; otherwise, some inclusion zones may still be truncated, though to a lesser extent. For example in Figure 7.6, the peripheral zone must be at least as wide as the radius of the plot or plot cluster. With this method, first suggested by Masuyama (1954), the population of interest remains the N elements within A: similar elements located in the peripheral zone are not included in the tally from any sample plot even if the plot itself is within the peripheral zone. The effect of enlarging the region where (x_s, z_s) may be located is twofold: for all U_k, the area of the inclusion zone is ca (for single plots, $c = 1$), and the constant inclusion probability is $\pi_k = ca/A^*$, where $A^* = A + A_{pz}$ and A_{pz} is the horizontal land area of the peripheral zone. With this method, some of the field effort is reduced because there is no need to check or otherwise be concerned with elements near the edge of A. Estimation is simplified, too, because $\hat{\tau}_{y\pi s}$ in (7.2d) reduces to

$$\hat{\tau}_{y\pi s} = \frac{A^*}{ca} \sum_{U_k \in P_s} y_k \tag{7.21a}$$

$$= A^* \rho_s, \tag{7.21b}$$

where $\rho_s = \sum_{u_k \in P_s} y_k/ca$ is the aggregate sum of the y-attribute, prorated to a unit area basis, measured on P_s. The price to be paid for the convenience of obviating the boundary overlap problem in this manner is a likely increase in the variance of $\hat{\tau}_{y\pi,\mathrm{rep}}$ and the difficulties that may be introduced in the field work to locate plots outside $\mathcal{A}$.

7.5.2 Pullback method

If it is infeasible to locate plots outside the boundaries of $\mathcal{A}$ and also infeasible to take the requisite measurements to calculate the inclusion area of edge units, a number of alternative tactics can be used. One, which applies to single-plot designs, is to alter the sampling protocol: any P_s that overlaps the boundary of $\mathcal{A}$ is relocated orthogonally back from the edge until the overlap is nil. This tactic was called the *Move-to-R* method in Gregoire & Scott (1990) and the *pullback* method in Gregoire & Scott (2003), who studied the method for the situation where boundaries are straight and plots are circular. This method does not really solve the boundary overlap problem, because it ignores the fact that it is the overlap of inclusion zones of population elements that is the source of the problem, not the overlap of a plot cluster with the edge. Nonetheless, this method has been widely used in conjunction with the estimator

$$\check{\tau}_{y\pi s} = A\rho_s. \tag{7.22}$$

Because $\pi_k \neq ca/A$ for all elements of the population, $\check{\tau}_{y\pi s}$ is biased. Figure 7.7 provides a detailed illustration of the alteration to the inclusion zone of edge elements caused by this pullback protocol.

When using a single plot at each sampling location, an alternative tactic is to use $\check{\tau}_{y\pi s}$ but to leave the sample plots where they were initially located, regardless of whether some overlap the boundary of $\mathcal{A}$. Again, $\check{\tau}_{y\pi s}$ is biased under this *no-correction* plot sampling method, although the magnitude of the bias will differ from its magnitude under the pullback method.

For the above two methods, the magnitude of the bias is affected by the size of the sampling units, because smaller plot clusters provide correspondingly smaller inclusion areas. There will be fewer population elements with truncated inclusion areas when smaller plots are used. The ratio of the length of the edge to area of the region being sampled also affects bias, because the smaller the ratio, the smaller the bias. Advice that bias is ignorable when this ratio is sufficiently small, however, ought to be heeded with caution. If the region is stratified into distinct land use classes, the length of edge in a stratum in relation to the stratum area is the relevant metric to examine in this regard, rather than the length of edge along the external boundary of the unstratified region. Also, if plots are not permitted to be located in roads, bodies of water, or other subregions of $\mathcal{A}$, then the length of edge along all these areas, which are excluded from the areal sampling frame, also must be taken into account. In other words, many more elements in $\mathcal{A}$ may have truncated inclusion areas than would be inferred from a simple calculation of the ratio of external edge length to A. Finally, the composition of the population may change as one recedes from the edge into the interior of $\mathcal{A}$. This almost always is true when sampling trees

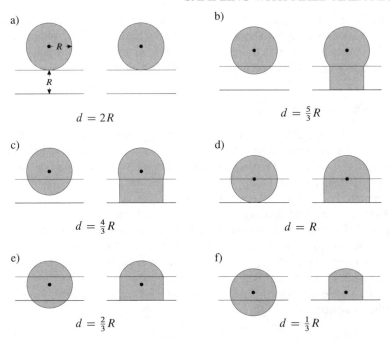

Figure 7.7 *Inclusion zones without and with alteration by the pullback method; d denotes the distance separating $\mathcal{U}_k$ from the edge.*

or other vegetation and where there is a distinct difference in light conditions near the edge compared to the interior. In many vegetation surveys, estimates of frequency, biomass, or other attributes are desired for each separate species or species group. To the extent that some species occur only along the edge, or with much greater frequency along the edge compared to the interior, the pullback or no-correction sampling strategies presented above may result in estimates with consequential bias. By the same reasoning, estimates indexed by size-class of vegetation may have unacceptably large bias.

7.5.3 Grosenbaugh's method

Grosenbaugh (1958) suggested an alteration to the sampling protocol when sampling with circular plots of radius R that permits an easy determination of the inclusion area of edge units. He suggested that any $\mathcal{P}_s$ that are located within an internal peripheral zone of width R be semicircular in shape of radius R, such that the flat edge of the semicircle is oriented towards the outside of $\mathcal{A}$ with its flat side parallel to the boundary, as shown in Figure 7.8a. This has the effect of halving the area of the inclusion zone of a element, $\mathcal{U}_k$, in the peripheral zone, and orienting the semicircular inclusion zone towards the interior, as shown Figure 7.8b. Consequently its inclusion probability is $\pi_k = a/2A$. For any $\mathcal{P}_s$ located in a right-angled corner

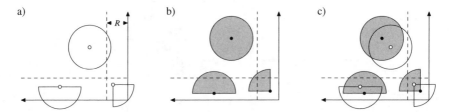

Figure 7.8 *The Grosenbaugh (1958) method to deal with boundary overlap. Half- and quarter-circle plots prescribed by the Grosenbaugh (1958) method alter the plot sampling protocol for edge elements. a) Half-circle and quarter-circle plots in the interior peripheral zone of width R; (b) the corresponding inclusion areas for trees within the interior and peripheral zones; (c) typical plots within the inclusion zones of the trees shown in frame (b).*

zone, a quarter-circle plot is used, resulting in an inclusion probability of $a/4A$. With this method of sampling near the edge, as long as those elements that are in the interior peripheral zone are noted, $\hat{\tau}_{y\pi s}$ is easy to compute, and remains an unbiased estimator of τ_y.

Certain limitations of the above procedure are evident. For regions with boundaries that are not straight lines, the inclusion zone corresponding to a semicircular plot will deviate from a semicircle in a way that will make the actual inclusion area difficult to determine exactly. For regions that are not rectangular, the inclusion zone for a tree near a corner will not be exactly a quarter-circle in shape or size. Therefore, if $a/2A$ and $a/4A$ are used as the inclusion probabilities for edge and corner elements, then $\hat{\tau}_{y\pi s}$ will be biased. For very irregularly shaped regions, field implementation of this method may become altogether too unwieldy. Moreover, the method is not very amenable to plot clusters where $c \geq 2$ plots are established at each sampling location.

7.5.4 Mirage method

The mirage correction procedure solves the boundary overlap problem by both altering the sampling protocol in the field and using an estimator similar to $\hat{\tau}_{y\pi s}$, but with each y_k measured in $\mathbb{P}_s$ multiplied by a weight, t_k. The method provides for unbiased estimates if the boundaries are straight with square corners. Also called the reflection method, the mirage method was introduced into the forestry literature by Schmid (1969). Regardless of where $\mathbb{P}_s$ is located, one tallies all $\mathcal{U}_k$ within the plot of radius R. Then if $\mathbb{P}_s$ is within an internal peripheral zone of width R, the plot location is reflected across the boundary, as shown in Figure 7.9a. To do this in the field, the distance of (x_s, z_s) orthogonally to the boundary is measured. Keeping the same bearing, an identical distance is measured exterior to $\mathcal{A}$, at which point the 'mirage plot' (or 'reflected plot') is established.

An additional tally is taken from the miraged plot, i.e., any $\mathcal{U}_k$ within $\mathcal{A}$ that also is within the bounds of the miraged plot is included with the sample from the original plot location. Gregoire (1982) showed, for circular plots, that it is impossible

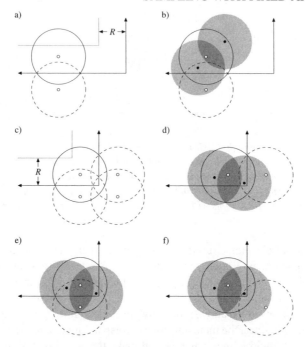

Figure 7.9 *The mirage correction method for boundary overlap. In (a), a sample plot at (x_s, z_s), and its reflection across the boundary; in (b), an example showing the inclusion zones of two trees. Both trees are selected by the original plot, and the tree closer to the edge is selected again from the miraged plot; in (c) the plot is miraged across both boundaries and reflected $180°$ across the square corner; the tree closer to the corner is selected by the original plot and the mirage plot across the right boundary (d), the bottom boundary (e), and the corner-mirage plot (f), but the other tree is selected only by the original plot and the mirage plot across the bottom boundary.*

to include any element from the mirage plot that had not already been tallied in the original plot. A 'multi-tally estimator' of τ_y from the sample at $\mathbb{P}_s$, which includes any elements also tallied from the miraged location of (x_s, z_s), is

$$\hat{\tau}_{yms} = \frac{A}{a} \sum_{\mathcal{U}_k \in \mathbb{P}_s} t_k y_k, \qquad (7.23)$$

where $t_k = 1$ if $\mathcal{U}_k$ was tallied from (x_s, z_s) only, or $t_k = 2$ if $\mathcal{U}_k$ was tallied from both (x_s, z_s) and its miraged location. In Figure 7.9b the tree closer to the edge of the tract is tallied on both the original and the miraged plot ($t = 2$), but the tree farther to the interior is tallied only on the original plot ($t = 1$). Gregoire (1982) showed that $E[t_k] = a/A$, and thereby established the unbiasedness of $\hat{\tau}_{yms}$ as an estimator of τ_y. In right-angled corners, three reflections of (x_s, z_s) are required, as shown in Figure 7.9c. Naturally, the multi-tally estimate, $\hat{\tau}_{yms}$, substitutes for $\hat{\tau}_{y\pi s}$ in the estimator of the variance.

Table 7.3 *Performance of estimators of N and total basal area for Example 7.15. All results are expressed as a percentage of N* = 4676 *trees or of the total basal area,* 109.7 m^2.

Method	Estimator	Bias		Root MSE	
		N	Basal area	N	Basal area
		$\cdots\cdots\cdots\cdots$ (%) $\cdots\cdots\cdots\cdots$			
No-correction	$\breve{\tau}_{y\pi s}$	−4.6	−4.0	57.1	46.6
Pullback	$\breve{\tau}_{y\pi s}$	−1.4	1.4	57.7	46.6
Grosenbaugh	$\hat{\tau}_{y\pi s}$	0	0	66.0	53.6
Mirage	$\hat{\tau}_{yms}$	0	0	58.9	46.1
Measure π	$\hat{\tau}_{y\pi s}$	0	0	58.9	46.6

Similar to the Grosenbaugh method, the mirage method works exactly to remove bias that otherwise would be incurred from boundary overlap only when $\mathcal{A}$ is rectangular. For slightly curved boundaries, the bias of $\hat{\tau}_{yms}$ is likely to be small.

Example 7.15

In order to compare the performance of the pullback, no-correction, Grosenbaugh, and mirage strategies of dealing with boundary overlap, Gregoire & Scott (1990) computed the bias and root MSE of the estimators associated with these methods. In their case study, they used the measurements of size and location of 4676 sapling and sawtimber trees in a 5.2 ha tract. The upper half of the tract contained a stand of 3396 saplings, whereas the lower half contained a stand of 1280 sawtimber-size trees. The sampling unit was a 0.04 ha circular plot. As a percentage of the N = 4767 trees on the plot, the bias of $\breve{\tau}_{y\pi s}$ as an estimator of N for the pullback method was −1.4% and for the no-correction method it was −4.6%.

They also computed $\sqrt{\text{MSE}\left(\hat{\tau}_{y\pi s}\right)}$ when using the actual inclusion area to compute the inclusion probability of each tree under the measure-π method, and compared this to $\sqrt{\text{MSE}\left(\breve{\tau}_{y\pi s}\right)}$ under the no-correction method. Despite the bias of the latter, it was slightly more accurate, based on a comparison of root mean square errors. These results and other comparisons appear in Table 7.3.

7.5.5 Walkthrough method

The walkthrough method (Ducey et al. 2004) may be used to correct for edge effect if the plot shape is radially symmetric about (x_s, z_s), i.e., if both the plot and a 180° rotation of the plot about (x_s, z_s) cover the same ground. Designs that meet this requirement include, but are not limited to, those that prescribe a circular, square, or rectangular plot centered about (x_s, z_s). Square or rectangular plots that use (x_s, z_s) as a corner point do not meet this requirement.

If a plot is symmetric about (x_s, z_s), then the inclusion zone of each element is

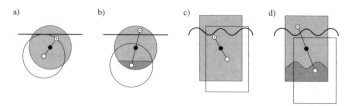

Figure 7.10 *The walkthrough method: walk from plot center (○) to element $\mathcal{U}_k$ (●), and beyond the element an equal distance to the reflection point (⊙). If the reflection point is inside the tract, as in (a) and (c), then $\mathcal{U}_k$ is tallied once ($t_k = 1$). If the reflection point is outside the tract, as in (b) and (d), then $\mathcal{U}_k$ is tallied twice ($t_k = 2$). If the plot center were to fall anywhere in the dark-shaded region in (b) or (d), then $\mathcal{U}_k$ would be tallied twice. Note that the area of the dark-shaded region equals the area of the region of the inclusion zone that is outside the tract. Note also that the method works with curved or straight boundaries.*

symmetric about the element's center point. Hence, if (x_s, z_s) falls in the inclusion zone of $\mathcal{U}_k$, then one may walk from (x_s, z_s) to the center point of $\mathcal{U}_k$ and then beyond the center point an equal distance to a reflection point of (x_s, z_s). In effect, the walkthrough procedure checks to see, for each $\mathcal{U}_k \in \mathbb{P}_s$, whether the reflection point of (x_s, z_s) is inside or outside of $\mathcal{A}$: if inside, the element is tallied once; if outside, the element is tallied twice.

Operationally, the walkthrough method is applied element by element on each plot (Figure 7.10). One walks straight from the center point of the plot, (x_s, z_s), to the center point of $\mathcal{U}_k$ and then straight through and beyond $\mathcal{U}_k$ an equal distance. Ordinarily, if the boundary is encountered before the walk is completed, then the reflection point is outside of $\mathcal{A}$ and $\mathcal{U}_k$ is tallied twice ($t_k = 2$). Conversely, if the reflection point is reached before encountering the boundary, then $\mathcal{U}_k$ is tallied once ($t_k = 1$). In some cases it might be possible to cross a curved or zig-zag boundary twice, first leaving and then re-entering $\mathcal{A}$, in which case the reflection point of (x_s, z_s) is inside of $\mathcal{A}$, so $\mathcal{U}_k$ is tallied once. Usually, however, any element that is closer to the boundary than to (x_s, z_s) will be tallied twice. And, if circular or rectangular plots are used, then only those elements that appear about equidistant between the boundary and (x_s, z_s) actually require walkthrough, the tallies for other elements ($t_k = 1$ or $t_k = 2$) being obvious.

The walkthrough method is more generally applicable than Grosenbaugh's method or the mirage method, since it can be applied with straight or curved boundaries. Moreover, the method is applicable where work outside of $\mathcal{A}$ is prohibited or infeasible, for example, where a boundary is marked by a natural feature such as a cliff or the shore of a river or lake. The method provides for unbiased estimates if, for all k, every point in the inclusion zone of $\mathcal{U}_k$, but outside of $\mathcal{A}$, has a reflection point inside of $\mathcal{A}$. Some amount of bias will persist if more than half of any inclusion zone is outside of $\mathcal{A}$.

Ordinarily, the walkthrough method is expected to greatly reduce edge bias, not necessarily eliminate it entirely. Either way, τ is estimated with the multi-tally

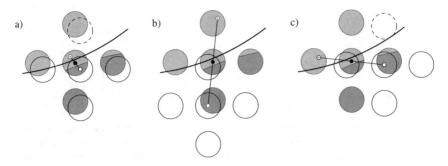

Figure 7.11 *The walkthrough method is applicable to radially symmetric plot clusters—each satellite plot is matched by another equidistant from the central plot in the opposite direction. If the plot center were to fall anywhere in the dark-shaded region in (a), (b) or (c), then $\mathcal{U}_k$ would be tallied twice. Plots that fall completely outside of $\mathcal{A}$ (dashed circles) are not installed. That plots may overlap the boundary is not a problem, but elements that occur in these plots are not tallied unless they occur within $\mathcal{A}$.*

estimator (7.23). Ducey et al. (2004) reported that, in simulations, the walkthrough method reduced underestimation of tree volume on a 46.5 ha tract in New Hampshire from 6.7% to 0.11%, when the tract was sampled with fixed-radius plots.

7.5.6 Edge corrections for plot clusters

The boundary overlap problem for plot clusters is somewhat complicated owing to the dispersed and disjoint inclusion zones of the population elements. Mandallaz (1991) devised a design where sample locations are allowed to fall in an external peripheral zone. In each cluster, circular plots are installed only if their center points fall within $\mathcal{A}$, so the number of plots in a cluster is a random variable. Boundary overlap is corrected on a plot-by-plot basis within each cluster, possibly with the walkthrough method or the mirage method. The estimate of τ is then calculated by treating all the plots of all clusters as independent plots. The estimate that results is asymptotically unbiased.

Valentine et al. (2006) described, in detail, the theory and operation of four methods that correct for boundary overlap, where sampling locations are not allowed to fall outside of $\mathcal{A}$. Here we provide brief sketches of the protocols. All four methods were formulated for clusters of circular plots, with one plot centered about the sampling location (x_s, z_s). However, squares, rectangles, and some other shapes could substitute for circles, provided all the plots in a cluster have the same directional orientation, e.g., the long edges of rectangles all point in the same direction. Boundaries may be straight or curved, and corrections usually can be carried out without leaving $\mathcal{A}$. The authors noted that radially symmetric cluster designs seem to afford the easiest options for correction of edge bias. The multi-tally

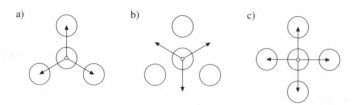

Figure 7.12 *Direction vectors indicate the direction and distance from the sampling location* (○) *to the center points of satellite plots (a). Inverse vectors are direction vectors rotated 180°* (b). *In radial symmetric clusters, each direction vector is coincident with in an inverse vector* (c).

estimator is used to estimate τ with each of the methods, i.e.,

$$\hat{\tau}_{yms} = \frac{A}{ac} \sum_{u_k \in \mathbb{P}_s} t_k y_k, \qquad (7.24)$$

where c is the usual number of plots in the cluster and the fixed number of plot-shaped regions in the inclusion zone of each element.

The walkthrough method, which applies to single plots that are radially symmetric about the sampling location, also applies to radially symmetric plot clusters (Figure 7.11). The protocols are the same as for single plots, though the walks are longer, since the inclusion zones are disjoint. Plots that are entirely outside $\mathcal{A}$ are not installed. Nevertheless, the value of c is not altered in the estimator, as ac is the area of each inclusion zone, whether c plots are installed or not. A closely related 'walkabout method' applies to some radially asymmetric cluster designs (see Valentine et al. 2006).

Two other methods, the vectorwalk and reflection methods, are most easily described in terms of direction vectors and inverse vectors (Figure 7.12). Direction vectors indicate the direction and distance from the sampling location, (x_s, z_s), to the center points of a cluster's satellite plots. The inverse vectors are simply the direction vectors rotated 180°. In radially symmetric clusters, each direction vector is coincident with an inverse vector (Figure 7.12c).

Vectorwalk method

Under the protocols of the vectorwalk method, each satellite plot is installed only if the center point of the plot is inside $\mathcal{A}$ (Figure 7.13). Edge correction is performed in each individual plot within the cluster by the walkthrough method. If the boundary is straight, the mirage method may substitute for walkthrough. The plot-by-plot application of walkthrough or mirage provides a final tally ($t_k = 1$ or $t_k = 2$) for each element, $\mathcal{U}_k$, that occurs in a satellite plot, but it provides only a preliminary tally ($t'_k = 1$ or $t'_k = 2$) for an element, $\mathcal{U}_k$, that occurs in the central plot (Figure 7.13b). For a final tally we start at the sampling location, (x_s, y_s), in the central plot and walk along each of the $c - 1$ inverse vectors, keeping count of the total number of inverse vectors, v, that intersect the boundary (Figure 7.13c). The

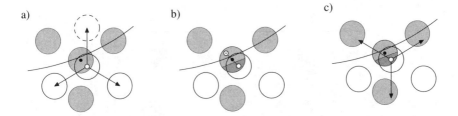

Figure 7.13 *a) An element, which is centered in its nominal inclusion zone (grey), occurs in the central plot of a plot cluster, which is shown with its direction vectors. The 'north plot' is not installed because its center point is outside $\mathcal{A}$. b) Application of the walkthrough method indicates that the element is tallied twice, but this is a preliminary tally, since the element occurs in the central plot. c) Vector walks would indicate that only the 'northwest inverse vector' crosses the boundary, so the final tally for the element is $t_k = 4$.*

final tally for the element $\mathcal{U}_k$ in the central plot is $t_k = (1 + v)t'_k$. The vector walks are not necessary if radially symmetric clusters are used, because each inverse vector is coincident with a direction vector, and it turns out that v is simply the number of satellite plots that were not installed. Regardless of whether or not c plots are installed in the cluster, the value of c is not altered in the estimator.

Reflection method

The reflection method for symmetric plot clusters begins with installation: each satellite plot is installed in the usual way unless the direction vector intersects the boundary. In that case, the direction vector is folded back over itself at the boundary and the 'reflected plot' is installed where the vector terminates inside of $\mathcal{A}$. If need be, the walkthrough method is applied in each installed plot, which provides the edge-corrected tally for each element in that plot (the mirage method may substitute for walkthrough if the boundary is straight). It is possible for plots to overlap, so the same element may have different edge-corrected tallies in two or more different plots.

The method is successful if every location point in a section of an inclusion zone that falls outside of $\mathcal{A}$ has a reflection point inside of $\mathcal{A}$ (Figure 7.14). These reflection points coalesce into reflected sections of inclusion zone. In general, if the sampling location, (x_s, z_s), is within a reflected section of an element's inclusion zone, then the element is tallied in a reflected plot.

The reflection method for radially symmetric plot clusters also applies to radially asymmetric plot clusters, provided the orientation of each cluster, i.e., the direction of the first direction vector, is selected uniformly at random. If the orientation of each asymmetric cluster is fixed, then the reflection method is slightly more complicated. Each satellite plot is installed only if its center point falls inside of $\mathcal{A}$. Direction vectors are not folded at the boundary. Instead, we check whether any inverse vectors intersect the boundary. If, and only if, an inverse vector intersects the boundary, then that inverse vector is folded back over itself at the boundary and a plot is installed

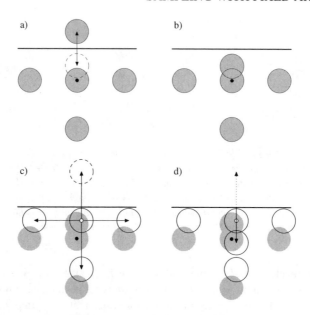

Figure 7.14 *The reflection method: a) The 'north section' of the inclusion zone of an element is outside* A. *Each point in the north section has a reflection point inside* A. *These reflection points coalesce into a 'reflected north section' (dashed circle). The reflected section plus the sections of the inclusion zone that are already inside of* A *constitute a 'reflected inclusion zone.' c) Under the protocols of the reflection method for radially symmetric clusters, a satellite plot is installed as usual unless its direction vector intersects the boundary of* A. *In that case, the direction vector is folded back at the boundary and the 'reflected plot' is installed where the vector terminates inside* A *(d). If the sampling location falls in a reflected section of the inclusion zone, then, in general, the element is tallied by a reflected plot.*

where the inverse vector terminates inside of A. Walkthrough is applied on a plot-by-plot basis, which provides the tally for each element in each plot. Again, the same element may occur in more than one plot and have a different tally in each plot. The reflection method always results in the installation of c plots in each radially symmetric plot cluster. However, in an asymmetric plot cluster with fixed orientation, the number of installed plots may turn out to be c, less than c, or more than c. Again, regardless of whether or not c plots are installed in the cluster, the value of c is not altered in the estimator, as the area of each inclusion zone, ac, is unaffected by the edge-corrected plot count.

7.6 Plot size and shape

An issue of longstanding interest is the most appropriate, or best, size of the plot to use when sampling natural, ecological, agricultural, entomological, or environmental resources. Related to this issue is the optimal shape of the plot and its effect on the precision and cost of sampling. There is a voluminous literature on these two topics, with an array of different and partly contradictory results.

The search for an optimal plot size and shape—the 'perennial quest' according to Daubenmire (1959)—has been motivated by a desire to minimize the cost and time of sampling yet still achieve the objectives of the survey, experiment, or monitoring plan. The best size/shape of plot is likely to differ depending on the resource of interest. Sampling grasslands for aboveground biomass is a far different task than sampling trees in a mature forest for the same purpose; an ideal plot size for trees is usually far too large for sampling grasses, herbs, and forbs. Even for a single resource, the optimal plot size/shape for one purpose may not be best for other purposes; for example, estimating the number of species may require larger plots than is needed for estimation with the same precision of the number of individuals in all species. Moreover, the criterion to determine optimality is likely to influence the result. For example, Wiegert (1962) observed that optimal quadrat size when sampling grass and forb biomass is sensitive to the pattern and dispersion of vegetation in the field. He searched for the optimal quadrat size that provided the smallest confidence limits for estimating μ_y for a predetermined cost. There is no reason to believe that this will indicate the same optimal size when a different constraint is imposed.

In the context of forest inventory, Mesavage & Grosenbaugh (1956) realized that optimal plot size and arrangement of plots on a forested tract is unique to each tract. Freese (1961) also intoned that the relationship between plot size and variability can not be generalized, and that for exactness a special study of the population of interest must be undertaken. Because of this lack of generality, many investigations have been launched to determine the ideal plot size. In the following, we summarize the results of a few of these studies, only. Additional results for grassland studies can be found in Bonham (1989, §5.3.2).

Clapham (1932) concluded that long narrow quadrats (with length 16 times greater than width) were best for sampling herbaceous vegetation. Such plots tended to minimize the variance of plant counts among plots, compared to other shaped plots. For phytosociological sampling, for purposes relevant to forest ecology, Bormann (1953) also concluded that long narrow plots were best. Bormann examined aggregate basal area of trees > 2.5 cm in a climax oak-hickory forest. Long, narrow strips plots were recommended by Meyer (1948) and Freese (1961)

When investigating the heterogeneity of yield of agricultural crops, Smith (1938) derived an empirical relationship between the variance of yield among plots and the size of the plot. Namely, he discerned that the sample variance decreases according to a power law:

$$s_y^2 \propto a^{-b}, \tag{7.25}$$

for some real-valued exponent, b. The value of b varied depending both on the crop and on the season of year. He combined this empirical law with a linear cost function to derive an optimal plot size. He did not find any consistent change of variability related to plot shape.

Both Mahalanobis (1946) and Sukhatme (1947) found that the measurement error

> ASIDE: The empirical relationship propounded by Smith (1938) relating vari-
> ability to plot size is unrelated to another empirical relationship that relates vari-
> ability to density. Taylor's power law, as it has come to be known, asserts that
> $s_y^2 \propto \lambda_y^b$. Clark & Perry (1994) report that Bliss (1941) apparently was the first
> to notice this relationship, but it attracted widespread attention of ecologists only
> after publication by Taylor (1971) in *Nature*.

increased as plot size decreased, thereby biasing the estimate of rice yield per
unit area. This increasing overestimation as the size of sample plots shrank was
attributed to the tendency to include plants that were actually outside the plot. The
greater edge to area ratio in small plots compared to larger plots exacerbates this
effect. Wiegert (1962) echoed this warning about sizeable measurement error when
counting elements near the border of the plot.

For sampling timber volume in old-growth Douglas-fir forests of the Pacific
northwest (USA), Johnson & Hixon (1952) investigated the efficiency of circular
plots with 5 different radii and rectangular plots with different sizes and shapes.
Defining efficiency to mean smallest variance for a given amount of work, they
concluded that rectangular plots were more efficient than circular plots. Further,
when comparing the efficiency of rectangular plots of the same area but different
shapes, the less elongated plot was better, and its ideal size was approximately 0.1 ha.
This size plot is appreciably larger than the 0.04 ha plot recommended by Mesavage
& Grosenbaugh (1956) for sampling in shortleaf pine and mixed hardwood forests
of Arkansas, U.S.A. These authors also found plots located on a square lattice to be
the most efficient arrangement.

The results of Mesavage & Grosenbaugh (1956) contrast with that of Daubenmire
(1959) who concluded that the elongate plot is superior, but are attended by two
practical limitations: 1) they are difficult to lay out, and 2) the frequency of borderline
plants, i.e., those near the border of the plot which require close decisions as to
whether a plant is inside the plot or not, increases. Because variability is greatly
influenced by the clumpiness of the population and the space between clumps, long
and narrow plots are less likely to entirely miss or wholly include a clump.

Kulow (1966) examined the efficiency of circular, triangular, square, and rectangular plots of various sizes for the purpose of estimating basal area per hectare. He
concluded that the shape of the plots did not appreciably affect sampling efficiency,
and that larger plots were more efficient.

In an aptly named article on "Plot size optimization," Zeide (1980) proposed
a methodology to determine optimal size of plot for sampling. It is based on an
empirical relationship, apparently first discussed by Freese (1961), namely

$$\frac{s_{y1}^2}{s_{y2}^2} = \sqrt{\frac{a_2}{a_1}}, \qquad (7.26)$$

where the subscribed numbers index plots of different sizes. By relating the time needed to measure a plot and the time to travel between plots, an expression for the optimal plot size to minimize the total time requirement of the survey can be derived.

Spetich & Parker (1998) deduced that the most efficient plot size was directly related to the patchiness of mortality, with smaller plots being more efficient when mortality was less evenly distributed. They were interested in aboveground biomass of live trees greater than 10 cm in diameter.

An excellent discussion of the size, shape, and configuration of plots for the purpose of silvicultural research is contained in Curtis & Marshall (2005, pp. 9–22).

7.7 Estimating change

Interest in the value of one or more parameters of the population being sampled may extend to the change in these values since some prior survey. Surveys conducted over time, known as longitudinal surveys, commonly are conducted by government agencies charged with oversight of natural and environmental resources. But non-governmental stakeholders may also be concerned with the changing value(s) of the attribute(s) of interest, and also may rely on longitudinal studies. Repeated surveys may be conducted on a periodic basis or on an irregular schedule. The statistical issues surrounding repeated surveys are summarized well in Fuller (1990) and Fuller (1999), whereas more descriptive discussions of specific longitudinal surveys for various natural resources can be found in Olsen et al. (1999), Nusser & Goebel (1997), and Scott et al. (1999).

In an extension of prior notation, let $\tau_y(t_1)$ denote the the population or stratum total of y at the occasion of the first survey, and let $\tau_y(t_2)$ denote the total at the second occasion. The change parameter of interest is thus

$$\Delta_y = \tau_y(t_2) - \tau_y(t_1). \tag{7.27}$$

If (7.3c) is used to estimate both $\tau_y(t_2)$ and $\tau_y(t_1)$, then a natural estimator of Δ_y is

$$\hat{\Delta}_y = \hat{\tau}_{y\pi,\text{rep}}(t_2) - \hat{\tau}_{y\pi,\text{rep}}(t_1). \tag{7.28}$$

This estimator of change is very general: there is no presumption that the same plots used at occasion t_1 are also used a occasion t_2, and there is no presumption that the same number of plots are used on both occasions, nor that the same population elements are measured on both occasions. However, if the plots sampled at t_1 are again visited at t_2, then $\hat{\Delta}_y$ in (7.28) can be expressed and computed alternatively as

$$\hat{\Delta}_y = \frac{1}{m} \sum_{s=1}^{m} \left(\hat{\tau}_{y\pi s}(t_2) - \hat{\tau}_{y\pi s}(t_1) \right). \tag{7.29}$$

If both $\hat{\tau}_{y\pi,\text{rep}}(t_2)$ and $\hat{\tau}_{y\pi,\text{rep}}(t_1)$ are unbiased, then $\hat{\Delta}_y$ is also. Its variance is

$$V[\hat{\Delta}_y] = V\left[\hat{\tau}_{y\pi,\text{rep}}(t_2)\right] + V\left[\hat{\tau}_{y\pi,\text{rep}}(t_1)\right] - 2C\left[\hat{\tau}_{y\pi,\text{rep}}(t_2), \hat{\tau}_{y\pi,\text{rep}}(t_1)\right], \tag{7.30}$$

where $C\left[\hat{\tau}_{y\pi,\text{rep}}(t_2), \hat{\tau}_{y\pi,\text{rep}}(t_1)\right]$ is the covariance between the two estimators. If the sample plots are independently selected at t_1 and t_2, then this covariance will be identically zero, leading to the well known result that the variance between two independent estimators is the sum of their respective variances. When the same plots are used on both occasions, $\hat{\tau}_{y\pi s}(t_2)$ generally is positively correlated with $\hat{\tau}_{y\pi s}(t_1)$, though the strength of the correlation usually diminishes as the interval of time between t_1 and t_2 lengthens. However, any positive correlation implies a positive covariance, so the variance of $\hat{\Delta}_y$ is expected to be smaller if the same plots, rather than different plots, are used on the two occasions. To put it another way, $\hat{\Delta}_y$ is expected to be a more precise estimator of Δ_y if the same plots are measured at both t_2 and at t_1.

A natural estimator of $V\left[\hat{\Delta}_y\right]$ is

$$\hat{v}\left[\hat{\Delta}_y\right] = \hat{v}\left[\hat{\tau}_{y\pi,\text{rep}}(t_2)\right] + \hat{v}\left[\hat{\tau}_{y\pi,\text{rep}}(t_1)\right] - 2\hat{c}\left[\hat{\tau}_{y\pi,\text{rep}}(t_2), \hat{\tau}_{y\pi,\text{rep}}(t_1)\right], \quad (7.31)$$

where $\hat{v}\left[\hat{\tau}_{y\pi,\text{rep}}(t_2)\right]$ and $\hat{v}\left[\hat{\tau}_{y\pi,\text{rep}}(t_1)\right]$ are given in (7.6b), and where

$$\hat{c}\left[\hat{\tau}_{y\pi,\text{rep}}(t_2), \hat{\tau}_{y\pi,\text{rep}}(t_1)\right]$$

$$= \frac{1}{m(m-1)} \sum_{s=1}^{m} \left[\hat{\tau}_{y\pi s}(t_2) - \hat{\tau}_{y\pi,\text{rep}}(t_2)\right]\left[\hat{\tau}_{y\pi s}(t_1) - \hat{\tau}_{y\pi,\text{rep}}(t_1)\right] \quad (7.32)$$

is a design-unbiased estimator of $C\left[\hat{\tau}_{y\pi,\text{rep}}(t_2), \hat{\tau}_{y\pi,\text{rep}}(t_1)\right]$.

If the same population elements are measured on both occasions, and the sampled elements at t_1 are labeled in a manner which permits these same elements to be identified at t_2, then the change in y_k can be computed and used to estimate Δ_y. Provided that no new elements have become members of the population in between t_1 and t_2, then a second alternative expression for (7.28) is

$$\hat{\Delta}_y = \frac{1}{m} \sum_{s=1}^{m} \sum_{u_k \in \mathbb{P}_s} \frac{y_k(t_2) - y_k(t_1)}{\pi_k}. \quad (7.33)$$

If interest lies solely with the estimation of Δ_y, there is no advantage to computing $\hat{\Delta}_y$ by (7.33) rather than (7.29): both will yield identical estimates. By contrast, if there is interest in estimating the change in identifiable subsets within the population or within each stratum, then (7.33) permits this. Suppose, for example, that the population consists of trees in a forest, and that we are interested in estimating the change in aggregate aboveground biomass for trees that survived until t_2 separately from the change in those that died in the interim. For this purpose, it is necessary to be able to match up the biomass of each tree measured at t_2 with its biomass at t_1.

In the general statistics literature on survey sampling, special subsets of interest are known as domains of interest (see Särndal et al. 1992, p. 69). These domains differ from population strata, because elements in one domain may appear in more than one stratum. In addition, a priori information to permit stratification based on domain membership usually is lacking. Nor do domains constitute post-strata in the usual sense, inasmuch as the sizes of the different domains are unknown even after sampling has concluded. Indeed, estimation of the size of each domain may be an objective of the survey.

To provide a more general treatment of change in domains of interest, as well as for changes in the composition of the population between t_1 and t_2, we relax the constraint that no elements enter or exit the population between surveys. The turnover of elements is common in organic populations owing to birth and death processes, or death and decay processes. Or, if domains correspond to size classes, organisms move from one domain to another through growth or decay.

Let there be D domains of interest, and let $\mathcal{D}_d$ indicate the dth domain, where $d = 1, \ldots, D$. The population totals in the dth domain at t_1 and t_2 are $\tau_{yd}(t_1)$ and $\tau_{yd}(t_2)$, respectively. Let Δ_{yd} denote the change in the domain total, then

$$\Delta_{yd} = \tau_{yd}(t_2) - \tau_{yd}(t_1), \qquad d = 1, \ldots, D. \qquad (7.34)$$

If every $\mathcal{U}_k$ belongs to a domain of interest, and these domains do not overlap in composition, then

$$\Delta_y = \sum_{d=1}^{D} \Delta_{yd}, \qquad \mathcal{D}_d \cap \mathcal{D}_{d'} = \emptyset, \, d \neq d'. \qquad (7.35)$$

In order to estimate Δ_{yd}, we must specify the population of interest in domain d. For example, the population may comprise the elements present only at t_1, plus the elements present at both t_1 and t_2, plus the elements present only at t_2. Let δ_{y_k} denote the change in y_k from t_1 to t_2. Then,

$$\delta_{y_k} = \begin{cases} -y_k(t_1), & \mathcal{U}_k \text{ is present only at } t_1; \\ y_k(t_2) - y_k(t_1), & \mathcal{U}_k \text{ is present at } t_1 \text{ and } t_2; \\ y_k(t_2), & \mathcal{U}_k \text{ is present only at } t_2. \end{cases} \qquad (7.36)$$

We assume that the same size plot is used on both occasions, so that π_k is constant over time. The parameter Δ_{yd} is estimated from the elements in plot s by

$$\hat{\Delta}_{yds} = \sum_{\mathcal{U}_k \in \mathbb{P}_s} \frac{\delta_{y_k} \xi_k}{\pi_k}, \qquad (7.37)$$

where

$$\xi_k = \begin{cases} 1, & \text{if } \mathcal{U}_k \in \mathcal{D}_d; \\ 0, & \text{otherwise.} \end{cases} \qquad (7.38)$$

Alternatively, we may use changes in attribute densities, $\delta_{\rho_k}, k = 1, \ldots, N$, where

$$\delta_{\rho_k} = \frac{\delta_{y_k}}{ca} \qquad (7.39)$$

in which case,

$$\hat{\Delta}_{yds} = A \sum_{\mathcal{U}_k \in \mathbb{P}_s} \delta_{\rho_k} \xi_k \qquad (7.40\text{a})$$

$$= \frac{A}{ac} \sum_{\mathcal{U}_k \in \mathbb{P}_s} \delta_{y_k} \xi_k. \qquad (7.40\text{b})$$

In replicated sampling, Δ_{yd} can be estimated by

$$\hat{\Delta}_{yd} = \frac{1}{m} \sum_{s=1}^{m} \sum_{\mathcal{U}_k \in \mathbb{P}_s} \frac{\delta_{y_k} \xi_k}{\pi_k}, \tag{7.41a}$$

$$= \frac{A}{m} \sum_{s=1}^{m} \sum_{\mathcal{U}_k \in \mathbb{P}_s} \frac{\delta_{y_k} \xi_k}{ca}, \tag{7.41b}$$

$$= \frac{A}{m} \sum_{s=1}^{m} \sum_{\mathcal{U}_k \in \mathbb{P}_s} \delta_{\rho_k} \xi_k, \tag{7.41c}$$

$$= \frac{1}{m} \sum_{s=1}^{m} \hat{\Delta}_{yds} \tag{7.41d}$$

The variance of $\hat{\Delta}_{yd}$ is

$$V[\hat{\Delta}_{yd}] = \frac{1}{m} \left[\sum_{k=1}^{N} \delta_{y_k}^2 \xi_k \left(\frac{1 - \pi_k}{\pi_k} \right) + \sum_{k=1}^{N} \sum_{\substack{k' \neq k \\ k'=1}}^{N} \delta_{y_k} \delta_{y_{k'}} \xi_k \xi_{k'} \left(\frac{\pi_{kk'} - \pi_k \pi_{k'}}{\pi_k \pi_{k'}} \right) \right]. \tag{7.42}$$

Moreover, $V[\hat{\Delta}_{yd}]$ can be estimated unbiasedly by

$$\hat{v}[\Delta_{yd}] = \frac{1}{m(m-1)} \sum_{s=1}^{m} \left(\hat{\Delta}_{yds} - \hat{\Delta}_{yd} \right)^2. \tag{7.43}$$

Example 7.16

In forestry, changes in populations of trees often are estimated from periodic measurements of trees in permanent sample plots. Trees smaller than some minimal size ordinarily are not measured. Once a tree achieves the minimal size (usually minimal diameter), and can be measured for first time, it is called an 'ingrowth tree.' Trees that can be measured on two successive occasions are called 'survivor trees,' and those that die naturally (as opposed to being cut) between one occasion and the next are called 'mortality trees.' Each of these three subpopulations—ingrowth, survivors, and mortality—are domains of interest for forest managers and, by convention, are called components of change. Restricting our temporal interest to measurements at t_1 and t_2, trees measured on a sample plot only at t_1 are mortality, trees measured at t_1 and t_2 contribute to survivor growth, and trees measured only at t_2 contribute to ingrowth. Hence,

$$\delta_{y_k} = \begin{cases} -y_k(t_1), & \mathcal{U}_k \text{ is a mortality tree;} \\ y_k(t_2) - y_k(t_1), & \mathcal{U}_k \text{ is a survivor;} \\ y_k(t_2), & \mathcal{U}_k \text{ is ingrowth.} \end{cases}$$

Table 7.4 *Basal areas (in^2) of trees at t_1 and t_2 (min. tree dia. > 5.5 in) on a 60ft × 40ft rectangular plot.*

t_1	t_2	t_1	t_2	t_1	t_2
.... (in^2)					
27.34	72.38	26.42	81.71	28.27	78.54
	39.59		66.48	24.63	
27.34	56.74	29.22	49.02	25.52	
24.63	59.45		39.59		38.48
26.42	76.98	31.17	75.43	27.34	78.54
28.27	75.43	30.19	72.38		52.81
	69.40	27.34	69.40	32.17	78.54
28.27	44.18		39.59		

Table 7.4 contains basal areas (in^2) of trees on a 40 ft by 60 ft (0.0551 ac) rectangular plot in a loblolly pine plantation at ages $t_1 = 8$ and $t_2 = 20$ years. A tree was not measured unless its diameter exceeded 5.5 in. The plantation size, A, is unknown.

Basal area lost to mortality on the plot from t_1 to t_2 was

$$-24.63 - 25.52 = -50.15 \, \text{in}^2.$$

or $-0.35 \, \text{ft}^2$, and the change in basal area from the growth of survivors was

$$(72.38 - 27.34) + (56.74 - 27.34)$$
$$+ \cdots + (78.54 - 27.34) + (78.54 - 32.17) = 574.31 \, \text{in}^2$$

or $3.99 \, \text{ft}^2$. Ingrowth of basal area on the plot at time t_2 was

$$39.59 + 69.40 + 66.48 + 39.59 + 39.59 + 38.48 + 52.81 = 345.94 \, \text{in}^2$$

or $2.40 \, \text{ft}^2$. The total increase in basal area on the plot from t_1 to t_2 was $-0.35 + 3.99 + 2.40 = 6.04 \, \text{ft}^2$. Based on the information from the plot, total basal area growth in the plantation is estimated to be $\hat{A}_{ys}/A = 6.04 \, \text{ft}^2/0.0551 \, \text{ac} = 109.6 \, \text{ft}^2 \text{ac}^{-1}$ over the 12 year period. Breaking this total down into the components of change yields: $\hat{A}_{y1s}/A = 6.4 \, \text{ft}^2 \text{ac}^{-1}$ lost to mortality; $\hat{A}_{y2s}/A = 72.4 \, \text{ft}^2 \text{ac}^{-1}$ accrued from survivor growth; and $\hat{A}_{y3s}/A = 43.6 \, \text{ft}^2 \text{ac}^{-1}$ accrued from ingrowth.

7.8 Terms to remember

Areal frame	Inclusion zone	Plot cluster
Boundary overlap	Joint inclusion zone	Pullback method
Change estimation	Measure π method	Reflection method
Domains of interest	Mirage method	Replicated sampling
Edge effect	No-correction method	Vectorwalk method
Grosenbaugh method	Optimal plot size	Walkthrough method

7.9 Exercises

1. Use the data in Table 7.2 to derive the results presented in Example 7.12.

2. Use the data in Table 7.2 to derive the results presented in Example 7.13.

3. Use the data in Table 7.2 to derive the results presented in Example 7.14.

4. With the data used in Example 7.9 and tabulated in Table 7.1, compute a 90% confidence interval for the number of trees per hectare.

5. With the data used in Example 7.9 and tabulated in Table 7.1, compute a 90% confidence interval for the aggregate basal area per acre per hectare. The basal area of a tree stem is defined as the cross-sectional area of a tree, assuming a circular cross-sectional shape. With diameter, d, measured in cm, the basal area of an individual tree is computed as $\pi d^2/40,000\text{m}^2$.

6. With the data in Table 7.1, how would you estimate the average number of cankers per tree?

7. With the data in Table 7.5 from Monkevich (1994), estimate the number of fragments per hectare of petrified wood in each size class for each year. Also, estimate $V\left[\hat{\lambda}_y\right]$ of each of these estimates.

8. Use the petrified wood data of Table 7.5. Pool the counts for both size classes together. For example, the pooled count in 1993 for quadrat 1 is 330. For each year, estimate the number of fragments per hectare in this pooled size class, and estimate the variance of each.

9. Suppose one discounted all the quadrats with zero counts in the petrified wood data, and adjusted the sample size accordingly. Therefore, consider the sample of size class 1 in 1993 as consisting only of the 15 quadrats on which there was at least a single fragment of petrified wood. Would the estimator $\hat{\lambda}_y$ still be an unbiased estimator of the number of fragments per hectare in this size class? Would it be a more precise estimator?

10. Use the petrified wood data of Table 7.5 to estimate the change in abundance of fragments of petrified wood in size class 2 from 1993 to 1994.

11. A forest inventory was conducted with $a = 0.1$ ha circular plots. The plots were established independently in 3 strata: I-softwood cover type, II-mixed cover type,

Table 7.5 *Number of pieces of petrified wood from 1/24,000 hectare quadrats in the Petrified Wood National Park, Arizona, USA. Size class 1 included pieces smaller than 0.635 cm. Size class 2 included pieces larger than 0.635 cm, but smaller than 2.540 cm. Dashes indicate the absence of a count owing to destruction of the quadrat.*

Quadrat	Size class 1 1993	Size class 1 1994	Size class 2 1993	Size class 2 1994
1	120	0	110	8
2	220	0	84	4
3	4	0	46	5
4	0	124	172	324
5	0	14	2	68
6	2	1	22	8
7	11	34	69	45
8	0	0	4	10
9	6	0	7	11
10	0	0	1	6
11	120	–	110	–
12	220	–	84	–
13	4	–	46	–
14	0	–	172	–
15	1	12	7	71
16	4	19	15	66
17	0	4	10	37
18	22	5	23	85
19	31	23	114	21
20	0	12	7	70
21	0	76	49	128
22	0	25	12	20
23	12	68	46	180
24	0	36	53	332
25	0	0	5	11
26	0	0	0	0
27	0	0	0	0
28	0	0	0	0
29	0	0	0	0
30	0	0	0	0
31	0	0	0	0
32	0	0	6	6
33	0	0	0	0
34	2	5	12	12
35	0	0	0	0

Table 7.6 *Aggregate bole volume of wood in standing trees on a* = 0.1 *ha plots.*

Stratum I		Stratum II		Stratum III	
Plot	Volume	Plot	Volume	Plot	Volume
	(m^3)		(m^3)		(m^3)
1	8.1	1	2.5	1	0.0
2	8.9	2	8.5	2	0.1
3	13.7	3	7.2	3	0.3
4	22.3	4	6.5	4	0.0
5	7.0	5	6.3	5	1.9
6	19.4	6	5.8	6	1.3
7	20.4	7	6.2	7	2.2
8	28.1	8	4.5	8	3.7
9	10.6	9	9.1	9	0.0
10	27.6	10	4.1	10	4.4
11	14.3	11	10.6	11	0.0
12	23.3	12	3.1	12	0.4
13	12.3	13	2.9	13	1.1
14	31.9	14	8.7		
15	12.5	15	8.0		
16	18.5	16	9.3		
17	6.5	17	5.4		
18	9.9	18	6.9		
19	23.5	19	11.3		
20	17.7	20	12.3		
21	9.3	21	2.2		
22	33.1				
23	17.2				
24	4.1				
25	11.6				
26	7.8				

III-hardwood cover type. The volume of wood in the boles of trees on each plot are shown in Table 7.6. These data were presented in Cunia (1979, p.73).

The area of Stratum I was 42,541 ha; stratum II contained 17,250 ha; and stratum II had 32,435 ha. Compute 90% confidence interval estimates for the total volume of wood in each stratum, and for the entire tract.

12. Sample plots were randomly located at four different elevations on Watershed 5 of the Hubbard Brook Experimental Forest http://www.hubbardbrook. org/research/overview/hbguidebook.htm. The Lower elevation was centered at 525 m; the Mid at 585 m; the Upper at 725 m; and the High at 800 m. These plots served as littertraps for coarse woody debris (> 2 cm diameter), which was collected and weighed annually between the years of 1996 through 2002. These data appear in Table 7.7. Each plot had an area of a = 6.25 m^2. a)

Table 7.7 *Weight of coarse litterfall (> 2 cm diameter) at the Hubbard Brook Experimental Forest by year.*

Elevation	Plot	Biomass						
		1996	1997	1998	1999	2000	2001	2002
		··········	··········	··········	$(g\,m^{-2})$	··········	··········	··········
Lower	165	19.64	16.05	39.67	67.88	23.40	83.67	0.00
Lower	162	28.47	9.62	34.86	59.27	33.47	206.80	64.04
Lower	171	20.72	28.59	39.79	33.74	52.20	156.51	12.30
Lower	170	18.15	32.85	73.72	21.80	78.02	158.85	7.58
Lower	168	13.39	11.33	21.00	47.16	36.70	83.24	–
Lower	172	42.67	19.69	35.90	23.12	35.84	–	0.29
Mid	147	11.71	6.23	79.21	15.75	30.06	50.60	2.07
Mid	153	8.16	10.69	84.86	19.34	35.95	225.82	36.55
Mid	144	52.30	6.69	49.11	38.06	22.91	–	0.00
Mid	142	37.56	9.59	46.75	42.39	39.72	117.56	21.26
Mid	152	55.04	11.36	46.94	22.25	46.02	93.87	54.28
Mid	154	–	–	–	–	–	82.39	0.00
Upper	121	57.44	10.14					19.43
Upper	127/130	36.29	31.80					7.55
Upper	132	23.59	20.09					3.68
Upper	134/133	81.77	12.41	–	–	–	–	4.16
Upper	129	9.48	15.14					47.98
Upper	123	15.42	11.49					0.00
High	104	39.58	10.13					63.51
High	109	38.86	27.13	–	–	–	–	6.34
High	105	25.32	9.38					71.04
High	106	96.71	21.16					31.27
High	115	41.28	6.60					136.57
High	111	119.75	21.30	–	–	–	–	14.18

Compute a 90% confidence interval estimate of the dry weight per square meter in the Lower elevation for each year; b) Pool the Lower elevation data for all years, and then compute 90% confidence interval estimate of the dry weight per square meter; c) Explain why the width of the interval computed from the pooled data is shorter than that for the separate intervals computed for each year's data.

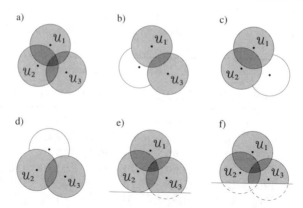

Figure 7.15 *a) Three circular inclusion zones, each overlapping the other two. The joint inclusion area of two elements is the shared area of their overlapping inclusion zones, as depicted by the darkly shaded areas in (b), (c), and (d). An inclusion zone is truncated by the boundary in (e), and two inclusion zones and their joint inclusion zone are truncated by the boundary in (f).*

7.10 Appendix

7.10.1 Joint inclusion zones

The inclusion zones of two or more elements in $\mathcal{A}$ may overlap. Indeed, two or more elements occur in the same plot only if their inclusion zones overlap. The region of the overlap between two elements is called joint inclusion zone. If the inclusion zones of three elements—$\mathcal{U}_1$, $\mathcal{U}_2$, and $\mathcal{U}_3$—overlap, as in Figure 7.15, then three joint inclusion zones are present: the region shared by $\mathcal{U}_1$ and $\mathcal{U}_2$, the region shared by $\mathcal{U}_1$ and $\mathcal{U}_3$, and the shared region of $\mathcal{U}_2$ and $\mathcal{U}_3$. If four elements occur in the same plot, there are six joint inclusion zones. Any two elements, $\mathcal{U}_k$ and $\mathcal{U}_{k'}$, with overlapping inclusion zones have a joint inclusion probability ($\pi_{kk'}$) equal to the area of their joint inclusion zone divided by the area of the tract. This joint inclusion probability is diminished if part of their joint inclusion zone is truncated by the boundary of $\mathcal{A}$.

CHAPTER 8

Bitterlich Sampling

8.1 Introduction

The sampling design described in this chapter is applicable chiefly to standing trees. It was first articulated by Walter Bitterlich (1949) under the German name *Winkelsählprobe*, which translates into English as 'angle-count sampling.' An Austrian forester, Bitterlich began as early as 1931 (see Bitterlich 1984) to think about distances between neighboring trees in the forest and the geometric interrelationships between them. From these musings sprang an ingenious probability-proportional-to-size sampling design that allows for the precise estimation of aggregate basal area per hectare merely by counting trees that are selected for the sample. Shortly after Bitterlich's initial publication of the method, an American forester, Lewis R. Grosenbaugh, discerned the probabilistic basis of angle-count sampling, deduced its applicability for estimating characteristics of the forest other than basal area, and he coined the term 'horizontal point sampling' (Grosenbaugh 1952, 1958). Other names that have appeared for this sampling design are Bitterlich sampling, variable radius plot sampling, Relaskop sampling, point sampling, plotless sampling, and prism sampling.

Stage & Rennie (1994) present a very readable discussion of the commonalities of Bitterlich sampling and plot sampling with circular plots of fixed radius. A Monte Carlo integration approach to Bitterlich sampling is described in Chapter 10.

8.2 Fundamental concepts

As in Chapter 7, the population of interest comprises N trees that are located within a region $\mathcal{A}$ with horizontal area A, and we assume that interest lies in estimating some parameter of this population such as the total number of trees, total aboveground biomass, basal area, carbon, and so on. There may be supplemental interest in estimating these parameter values for each species or group of related species, or compiling estimates by size class, or on a per unit area basis.

As in the previous chapter, we assume that a set of m sampling locations are established at random in $\mathcal{A}$. Without risk of confusion (we hope!), the sth sampling unit, or *sampling point* as it is known in the forest-sampling literature, is denoted by $\mathbb{P}_s$, and the location coordinates of $\mathbb{P}_s$ are denoted by (x_s, z_s). The kth tree of the population is denoted by $\mathcal{U}_k$ and its fixed location coordinates are denoted by (x_k, z_k).

With Bitterlich sampling the trees of interest may vary from one application to another. Almost always there is a minimum size requirement, stipulated in terms of the diameter of the bole of the tree, or its height, or possibly a combination of the two.

In Example 8.4, the population of interest comprises only those trees with diameters 20 cm or larger. The size threshold usually is dictated by the purpose of the forest inventory. Because inventories are conducted for a variety of different reasons, there is no universally accepted size threshold. Moreover, the Bitterlich sample may be restricted to certain species of trees, e.g., just the conifers of some minimum height, rather than all trees growing on the forested region.

8.2.1 Limiting distance for inclusion

There is a very close relationship between Bitterlich sampling and sampling with circular plots with fixed radius R, but there is an important difference, as well. The circular-plot design prescribes a circular inclusion zone with fixed radius R centered about each tree. If the sample point, $\mathbb{P}_s$, occurs within the inclusion zone of $\mathcal{U}_k$, then tree $\mathcal{U}_k$ is selected into the sample. As implied in Example 7.7, the decision to include $\mathcal{U}_k$ is solely a function of R. In this sense, R is a *limiting distance* because $\mathcal{U}_k$ would be excluded from the sample if the distance from (x_s, z_s) to (x_k, z_k) were greater than R.

Like the fixed-radius plot design, the Bitterlich design prescribes a circular inclusion zone for each tree, and the tree's limiting distance is equivalent to the radius of its inclusion zone. However, under the Bitterlich design, the radius of the inclusion zone for a particular tree is proportional to the radius of the tree's bole at breast height (1.37 m). Let r_k (m) and R_k (m), respectively, denote the bole radius and the inclusion zone radius of tree $\mathcal{U}_k$. Then

$$R_k = \alpha r_k \tag{8.1}$$

where α (m m^{-1}) is constant, independent of k. As will be explained, the value of α is selected indirectly as part of the sampling design, and it is closely related to Bitterlich's 'angle.' However, it is important to note that the values of α and r_k jointly determine the limiting distance, R_k, for tree $\mathcal{U}_k$. If the distance from $\mathbb{P}_s$ to tree $\mathcal{U}_k$ does not exceed R_k, then $\mathcal{U}_k$ is selected into the sample at $\mathbb{P}_s$.

By convention, the cross-sectional area of a tree bole at breast height is called *basal area*. Let b_k (m^2) denote the basal area of tree $\mathcal{U}_k$,

$$b_k = \pi r_k^2. \tag{8.2}$$

Let a_k (ha) denote the area of the inclusion zone of $\mathcal{U}_k$. Since 1 m^2 = 10^{-4} ha,

$$a_k = 10^{-4} \pi R_k^2 \tag{8.3a}$$

$$= 10^{-4} \alpha^2 b_k. \tag{8.3b}$$

Consequently, barring truncation by the edge of $\mathcal{A}$, the inclusion probability of $\mathcal{U}_k$ is

$$\pi_k = \frac{a_k}{A} \tag{8.4}$$

where A is measured in hectares.

ASIDE: The literature of forest sampling is replete with descriptions of Bitterlich sampling which mention 'imaginary circles surrounding trees,' 'imaginary zones,' 'π-circles,' and other apocrypha. These fanciful terms signify nothing more than the circular inclusion zones of trees when sampling with Bitterlich's method.

8.2.2 Basal area factor

The ratio $F = b_k/a_k$ (m^2 ha^{-1}) is tree basal area per unit land area of inclusion zone. Because α is constant, F is also constant, independent of k, i.e.,

$$F = \frac{10^4}{\alpha^2}. \tag{8.5}$$

In the Bitterlich sampling literature, F is called the *basal area factor*. In practice, the basal area factor is a design parameter in direct analogy to the choice of plot size, a, when designing a plot sample. The sampler chooses the value of F, and this determines the value of α, i.e.,

$$\alpha = \frac{100}{\sqrt{F}}. \tag{8.6}$$

Because F determines α and because $a_k = 10^{-4}\alpha^2 b_k = b_k/F$, we can also express the inclusion probability of $\mathcal{U}_k$ in terms of F, i.e.,

$$\pi_k = \frac{b_k}{FA}. \tag{8.7}$$

From (8.7) it is evident that the probability of including tree $\mathcal{U}_k$ in a sample is proportional to the tree's basal area. This is a defining characteristic of Bitterlich sampling, and one that makes it a method of sampling with probability proportional to size (pps).

8.2.3 Plot radius factor

Practitioners of Bitterlich sampling usually calculate the limiting distance, R_k (m), from diameter, d_k, measured in cm. The *plot radius factor*, κ (m cm^{-1}), is the ratio of R_k to d_k, so

$$R_k = \kappa d_k \tag{8.8}$$

independent of k. Note that d_k (cm) $= 2\,r_k$ (m) $\times 100$ (cm m^{-1}), therefore

$$\kappa = \frac{\alpha}{200}. \tag{8.9}$$

Substitution of (8.6) relates the plot radius factor to the basal area factor, i.e.,

$$\kappa = \frac{1}{2\sqrt{F}}. \tag{8.10}$$

A better name for κ would be the 'limiting distance factor,' since it relates a tree's limiting distance, R_k, to the tree's diameter. However, plot radius factor is the standard terminology in numerous forest mensuration texts.

Example 8.1

A Bitterlich sample with basal area factor $F = 4$ $m^2\,ha^{-1}$ provides $\kappa = 1/(2\sqrt{4}) = 0.25$ $m\,cm^{-1}$. Thus, a tree with diameter $d_k = 16$ cm has a limiting distance of $R_k = 0.25$ $m\,cm^{-1} \times 16\,cm = 4.0$ m, and its inclusion area is $a_k = 10^{-4}\pi(4.0)^2 = 0.005027$ ha. By contrast, a tree with a 24 cm diameter would have a limiting distance of 6.0 m, and an inclusion area of 0.011310 ha.

The inverse relationship between limiting distance and basal area factor implies that a tree may be excluded from the sample if sampling is conducted with $F = 4$ but included if sampling with $F = 3$. For example, the limiting distance of the tree with a 20 cm diameter, when sampling with a basal area factor of size $F = 3$ $m^2\,ha^{-1}$, is 5.77 m. Had this tree been 5.50 m from the sample point, then it would have been included in the latter case, but not in the former case.

8.2.4 Sampling protocol

As implied by (8.1) and the preceding example, $\mathcal{U}_k$ is included into the sample $\mathbb{P}_s$ if the distance from (x_k, z_k) to (x_s, z_s) is less than or equal to R_k. This could be discerned by measuring d_k, calculating the limiting distance, R_k, with the plot radius factor and comparing this limiting distance to the actual distance from (x_k, z_k) to (x_s, z_s). This would be repeated for all trees in the vicinity of (x_s, z_s). Heuristically, one can think of sampling with nested, concentric circular plots centered at (x_s, z_s), where the plot radius for $\mathcal{U}_k$ is R_k, which obtains from d_k and the plot radius factor, κ. Indeed, this construct has given rise to an alternative name for Bitterlich sampling: variable radius plot sampling. Fortunately, however, various devices can be used to determine optically whether or not the distance from (x_s, z_s) to (x_k, z_k) exceeds R_k, so, for the most part, the measurements of actual distances and the calculations of limiting distances are rendered unnecessary.

The devices collectively are known as *angle gauges* and their use relates directly to the name 'angle count sampling' that Bitterlich bestowed on this method of sampling standing trees. The angle to which Bitterlich refers can be derived by picturing a point exactly $R_k = \alpha r_k$ (m) away from (x_k, z_k). Such a point lies on the perimeter of the inclusion zone of tree $\mathcal{U}_k$ (Figure 8.1a). Rays emanating from this point, which are exactly tangent to the (circular) bole of $\mathcal{U}_k$ as shown in Figure 8.1b, form an angle ν. The length from each point of tangency to the center of the tree is simply the tree radius, r_k (m), so

$$\sin\left(\frac{\nu}{2}\right) = \frac{r_k}{R_k} = \frac{r_k}{\alpha r_k} = \frac{1}{\alpha}.$$

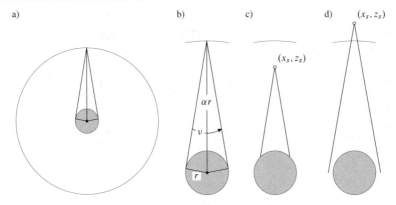

Figure 8.1 a,b) *The ratio $r/\alpha r$, i.e., the ratio of the cross section of a tree to the radius of the tree's inclusion zone, defines the angle, ν, of an angle gauge. c) The width of a tree fills the field of view of the angle gauge if the sample point falls inside the inclusion zone. d) Otherwise, the sample point is outside the inclusion zone.*

Therefore,

$$\nu = 2\arcsin\left(\frac{1}{\alpha}\right), \tag{8.11}$$

which is equivalent to

$$\nu = 2\arcsin\left(\frac{1}{200\kappa}\right), \tag{8.12}$$

Substituting (8.6) relates the angle ν to the design parameter F, i.e.,

$$\nu = 2\arcsin\left(\frac{\sqrt{F}}{100}\right). \tag{8.13}$$

Example 8.2

Bitterlich sampling with a basal area factor of size $F = 4\ \text{m}^2\,\text{ha}^{-1}$ implies that $\nu = 0.040$ radians, or $2.29°$. A basal area factor of size $F = 3\ \text{m}^2\,\text{ha}^{-1}$ implies that $\nu = 0.035$ radians, or $1.98°$. The smaller the basal area factor, F, the smaller the angle ν.

When conducting Bitterlich sampling at the sample point, (x_s, z_s), each tree in the vicinity is sighted with the angle gauge such that the rays propagated at angle ν emanate directly over the sample point. If a tree is closer than its limiting distance to (x_s, z_s), its diameter at breast height (1.37 m aboveground) will appear wider than the distance between the rays, as in Figure 8.1c. Conversely, it will appear narrower than the distance between the rays if its distance away from (x_s, z_s) exceeds its limiting distance, as in Figure 8.1d. This result suggests the procedure to follow at $\mathbb{P}_s$ to decide which trees to include in the sample at that point: trees included in the sample are those that appear larger than the projected angle when viewed at breast height

---

ASIDE: Equivalent basal area factors (BAF) and plot radius factors (PRF) can be couched in terms of the ratios α^2 and α, for example,

$$\text{BAF}: \quad \frac{10^4}{\alpha^2} \left(m^2\, ha^{-1} \right) = \frac{1}{\alpha^2} \left(m^2\, m^{-2} \right) = \frac{1}{\alpha^2} \left(ft^2\, ft^{-2} \right) = \frac{43560}{\alpha^2} \left(ft^2\, ac^{-1} \right)$$

$$\text{PRF}: \quad \frac{\alpha}{200} \left(m\, cm^{-1} \right) = \frac{\alpha}{2} \left(m\, m^{-1} \right) = \frac{\alpha}{2} \left(ft\, ft^{-1} \right) = \frac{\alpha}{24} \left(ft\, in^{-1} \right)$$

---

through the angle gauge. Notwithstanding the use of an angle gauge or other optical device, the decision to include $\mathcal{U}_k$ into $\mathbb{P}_s$ is based on distance only; the tree's width fills the field of view in Figure 8.1c only when its distance from (x_s, z_s) is R_k or less.

With this protocol for sampling standing trees in the forest, proper care must be exercised that no trees are inadvertently omitted from the tally at a sampling point. With Bitterlich sampling, very large trees can be quite some distance from (x_s, z_s) yet still be closer than their limiting distance. For example, suppose $\mathcal{U}_k$ has a diameter of $d_k = 107$ cm, and that sampling is conducted with a wedge prism that has been manufactured at an angle to ensure $F = 3\, m^2\, ha^{-1}$. This tree can be 30 m from (x_s, z_s), yet still be included in the sample at that point. The opposite phenomenon will also be encountered, namely the occurrence of small diameter trees close to the sampling point yet further away from it to be included in the tally from that point. When sampling with fixed-area circular plots, if $\mathcal{U}_k$ is closer to (x_s, z_s) than a neighboring tree, and if the neighbor is in the sample at (x_s, z_s), then $\mathcal{U}_k$ will be in it, also. This need not be the case with Bitterlich sampling, as the limiting distance and hence the inclusion probability is a function of tree diameter. It is a tree's relative proximity to the sample point that determines sample inclusion with Bitterlich sampling, not absolute proximity. In this regard the title of Bitterlich's book is revealing: *The Relascope Idea: Relative Measurements in Forestry*.

Although Bitterlich sampling can be conducted without the aid of an optical device, this is never done on a routine basis. Occasionally there will be trees that will be borderline when viewed through an angle gauge. This situation is similar to the one discussed in §7.4.7 in the context of sampling with fixed-area plots. A borderline tree in the context of Bitterlich sampling is one that appears to be exactly at its limiting distance when viewed through the optical device. In accordance with the recommendation of Iles (2003, p. 506), we believe that the horizontal distance of each borderline tree from the sample point should be measured and compared to the limiting distance appropriate for a tree of its diameter. In particular, the practice of including every second borderline tree in the sample is not to be trusted, owing to possible systematic errors of sighting which will inject bias of unknown magnitude into the estimation of population parameters.

8.2.5 *Bitterlich sampling with English units*

In North American forestry, practitioners often perform Bitterlich sampling in English units. Tree diameters are measured in inches (in), basal areas are calculated in square feet (ft^2), and land areas are measured in acres (ac).

An English basal area factor, F_E ($\text{ft}^2\,\text{ac}^{-1}$), obtains directly from a metric basal area factor, F ($\text{m}^2\,\text{ha}^{-1}$), with the conversion formula $4.356\,\text{ft}^2\,\text{ac}^{-1} = 1\,\text{m}^2\,\text{ha}^{-1}$, i.e.,

$$F_E = 4.356\,F. \tag{8.14}$$

Since $\alpha\,\left(\text{m}\,\text{m}^{-1}\right) = \alpha\,\left(\text{ft}\,\text{ft}^{-1}\right)$, equation (8.5) converts to

$$F_E = \frac{43560}{\alpha^2}. \tag{8.15}$$

Substituting (8.15) into (8.11) gives the angle, ν, appropriate for the design parameter F_E, i.e.,

$$\nu = 2\arcsin\sqrt{\frac{F_E}{43560}}. \tag{8.16}$$

The English plot radius factor, κ_E, has units $\text{ft}\,\text{in}^{-1}$, where $1\,\text{ft} = 12\,\text{in}$, so

$$\kappa_E = \frac{\alpha}{24}. \tag{8.17}$$

Equations (8.15) and (8.17) relate the plot radius factor to the basal area factor, i.e.,

$$\kappa_E = \sqrt{\frac{75.625}{F_E}}. \tag{8.18}$$

English basal area factors of 10, 20, and, 40 $\text{ft}^2\,\text{ac}^{-1}$ are popular choices. Thus, for example, a basal area factor of 20 $\text{ft}^2\,\text{ac}^{-1}$ provides $\kappa_E = 1.9445\,\text{ft}\,\text{in}^{-1}$, so the inclusion zone radius for a tree with a 10 in diameter is 19.45 ft. A metric basal area factor of $4\,\text{m}^2\,\text{ha}^{-1}$ converts to 17.424 $\text{ft}^2\,\text{ac}^{-1}$, which provides $\kappa_E = 2.0833\,\text{ft}\,\text{in}^{-1}$, so a 10 in tree has an inclusion zone radius of 20.83 ft.

8.2.6 *Noncircular tree boles*

When a tree's cross-sectional shape at breast height is not circular, then the tree's basal area will be computed inaccurately by (8.2). This has a crucial effect on the calculation of the limiting distance and inclusion probability of the tree. Grosenbaugh (1958) provided an extensive discussion about the consequences of elliptical boles on bias in estimation following Bitterlich sampling, and he provided a protocol to reduce it. Ellipticity of tree boles is ignored in most applications, inasmuch as the bias will be less than 1% if the ratio of the minor to major axes of the ellipse is not less than 0.9 (Grosenbaugh 1958, p. 25)

ASIDE: There are a number of fine texts on forest mensuration that describe
various angle gauges and their proper use in Bitterlich sampling. For example,
the section on Implementing Point Sampling, in Chapter 11 of Avery & Burkhart
(2002), explains the use of stick-type angle gauges, wedge prisms, and Bitterlich's relascope. In Chapter 7 of Schreuder et al. (1993, pp. 268–273) there is an
informative section on Instruments Used in VRP Sampling that also compares
the advantages and disadvantages of the wedge prism to the relascope.

8.2.7 Expected number of trees selected at a point

Let ξ_{ks} be a binary indicator of inclusion of $\mathcal{U}_k$ at the sth sampling location.
Specifically,

$$\xi_{ks} = \begin{cases} 1, & \text{if } \mathcal{U}_k \in \mathbb{P}_s; \\ 0, & \text{otherwise.} \end{cases} \tag{8.19}$$

Then n_s, number of trees selected at (x_s, z_s), can be expressed as $n_s = \sum_{k=1}^{N} \xi_{ks}$
which has expected value $E[n_s] = \sum_{k=1}^{N} E[\xi_{ks}] = \sum_{k=1}^{N} \pi_k$. When sampling
with fixed-size plots, each with area a, the expected number of trees selected on a
plot is approximately (that is, ignoring the truncated inclusion probabilities of trees
near the edge) $\lambda_N a$, where λ_N is the average number of trees per hectare. With
Bitterlich sampling it is approximately λ_b / F, where λ_b is the average basal area per
hectare. These results indicate that with fixed-size plots, the average number of trees
depends on the density as measured by numbers of trees but not their size, whereas
with Bitterlich sampling it depends on the density as measured by basal area. Since
the number of small trees in a forest usually outnumber the number of large trees,
fixed-size plot sampling will entail sampling more of the smaller trees of the forest
than will Bitterlich sampling.

Conventional wisdom from practical experience suggests that F should be chosen
so that n_s is in the range of 4 to 8.

8.3 Estimation following Bitterlich sampling

8.3.1 Estimating τ_y

In Chapter 7, the HT estimator, (7.2d), was presented as the estimator of τ_y based
on the sample from the sth plot. This also is suitable when trees are selected by the
Bitterlich design, namely

$$\hat{\tau}_{y\pi s} = \sum_{\mathcal{U}_k \in \mathbb{P}_s} \frac{y_k}{\pi_k} \tag{8.20a}$$

$$= FA \sum_{\mathcal{U}_k \in \mathbb{P}_s} \frac{y_k}{b_k}. \tag{8.20b}$$

If no trees are sampled on $\mathbb{P}_s$, then $\hat{\tau}_{y\pi s} = 0$.

Barring edge effects, an unbiased estimator of τ_y based on a Bitterlich sample of m points is

$$\hat{\tau}_{y\pi,\text{rep}} = \frac{1}{m} \sum_{s=1}^{m} \hat{\tau}_{y\pi s} \tag{8.21a}$$

$$= \frac{FA}{m} \sum_{s=1}^{m} \sum_{\mathcal{U}_k \in \mathbb{P}_s} \frac{y_k}{b_k}. \tag{8.21b}$$

The similarity of (8.21) to (7.3c) is evident. Indeed, all the estimators of population parameters presented in Chapter 7 are equally applicable to data acquired through Bitterlich sampling, provided the inclusion probabilities appropriate for Bitterlich sampling, i.e., $\pi_k = b_k/FA$, are used in place of the inclusion probabilities appropriate for fixed-area plot sampling. Therefore, we shall not repeat all the estimators presented in Chapter 7 again in this chapter.

An alternative to $\hat{\tau}_{b\pi,\text{rep}}$ was proposed by Flewelling & Iles (2004) which dispenses with the need to know A.

8.3.2 Estimating basal area

It is worthwhile, however, to consider estimation of total basal area of the population, namely τ_y when $y_k = b_k$, as in (8.2), for all $\mathcal{U}_k$. Because basal area has such a central role in Bitterlich sampling, we designate its total on $\mathcal{A}$ and its per hectare value as τ_b and λ_b, respectively. From the sample selected at $\mathbb{P}_s$, an unbiased estimator of τ_b is

$$\hat{\tau}_{b\pi s} = \sum_{\mathcal{U}_k \in \mathbb{P}_s} \frac{b_k}{\pi_k}$$

$$= FA \sum_{\mathcal{U}_k \in \mathbb{P}_s} \frac{b_k}{b_k} \tag{8.22a}$$

$$= n_s FA, \tag{8.22b}$$

where n_s is the number of trees that are selected at $\mathbb{P}_s$. The trees selected at a sample point, $\mathbb{P}_s$, are said to be 'in' according to forestry vernacular. We shall refer to such trees as 'in-trees.'

Based on a replicated sample of m points,

$$\hat{\tau}_{b\pi,\text{rep}} = \frac{1}{m} \sum_{s=1}^{m} \hat{\tau}_{b\pi s} \tag{8.23a}$$

$$= \frac{FA}{m} \sum_{s=1}^{m} n_s \tag{8.23b}$$

$$= \frac{FAn}{m} \tag{8.23c}$$

$$= \bar{n}FA \tag{8.23d}$$

is unbiased for τ_b. In (8.23c), n is the total number of in-trees selected in the m points, and in (8.23d), $\bar{n} = \sum_{s=1}^{m} n_s/m$ is the average number of in-trees at a point. On a per unit area basis,

$$\hat{\lambda}_{b\pi s} = \frac{\hat{\tau}_{b\pi s}}{A} \tag{8.24}$$

$$= n_s F \tag{8.25}$$

unbiasedly estimates basal area per hectare, as does

$$\hat{\lambda}_{b\pi,\text{rep}} = \frac{1}{m} \sum_{s=1}^{m} \hat{\lambda}_{b\pi s}$$

$$= \bar{n}F. \tag{8.26}$$

These important results indicate that one need only count the number of in-trees at a sample point in order to unbiasedly estimate aggregate basal area and basal area per land area. This is the origin of Bitterlich's moniker 'angle count sampling.'

Example 8.3

At $\mathbb{P}_s$ a total of $n_s = 7$ trees were selected by Bitterlich sampling using an angle gauge which provides a basal area factor of $F = 4\,\text{m}^2\,\text{ha}^{-1}$. This provides an estimate of $\hat{\lambda}_{bs} = 28\,\text{m}^2\,\text{ha}^{-1}$.

Aggregate basal area and basal area per hectare are very highly correlated with aboveground biomass, volume, carbon, foliar surface area, and other parameters of the forest that are of importance ecologically, and therefore of silvicultural interest for managers of forest resources. Being able to estimate basal area simply by counting trees with the probabilistic basis provided by Bitterlich sampling was a major breakthrough in the field of forest inventory.

In §3.3 we discussed the statistical implications of unequal probability sampling when elements are selected for the sample with probabilities that are proportional to some measure of the size of each element. Of principal importance, is that if

$\pi_k \propto x_k$, and y_k is strongly correlated with x_k, then we can expect that HT estimation of τ_y under this pps-sampling design will be more precise than under an equal probability sampling design. This is exactly the situation for which Bitterlich sampling is ideal, when the goals of a forest inventory are to estimate precisely the amount of biomass, volume, and carbon stored in the trees of the forest.

8.3.3 Estimating population size

Based on a Bitterlich sample at $\mathbb{P}_s$, an estimator of the total number of trees in the population is

$$\hat{N}_{\pi s} = \sum_{u_k \in \mathbb{P}_s} \frac{1}{\pi_k} \tag{8.27a}$$

$$= FA \sum_{u_k \in \mathbb{P}_s} \frac{1}{b_k}, \tag{8.27b}$$

which leads to

$$\hat{N}_{\pi,\text{rep}} = \frac{1}{m} \sum_{s=1}^{m} \hat{N}_{\pi s}. \tag{8.28}$$

As has been mentioned elsewhere in this book, $y_k = 1$ implicitly when estimating N, and $\hat{N}_{\pi,\text{rep}}$ is termed the estimator of tree frequency in the literature on forest inventory. Because the correlation between tree frequency and basal area is nil, there is no special advantage to Bitterlich sampling for estimation of population size.

8.3.4 Estimating μ_y

To estimate the average value of y per tree in the population of interest, i.e., $\mu_y = \tau_y / N$, one ratio-type estimator is

$$\hat{\mu}_{y\pi,\text{rat}} = \frac{\hat{\tau}_{y\pi,\text{rep}}}{\hat{N}_{\pi,\text{rep}}}, \tag{8.29}$$

as in §7.4.4.

An alternative estimator is the replicated estimator of μ_y, namely

$$\hat{\mu}_{y\pi,\text{rep}} = \frac{1}{m} \sum_{s=1}^{m} \sum_{u_k \in \mathbb{P}_s} \frac{\hat{\tau}_{y\pi s}}{\hat{N}_{\pi,s}}. \tag{8.30}$$

8.3.5 Variances and variance estimators

As in Chapter 7, $V\left[\hat{\tau}_{y\pi,\text{rep}}\right]$ is unbiasedly estimated by

$$\hat{v}\left[\hat{\tau}_{y\pi,\text{rep}}\right] = \frac{1}{m(m-1)} \sum_{s=1}^{m} \left(\hat{\tau}_{y\pi s} - \hat{\tau}_{y\pi,\text{rep}}\right)^2. \tag{8.31}$$

Similarly, $V\left[\hat{\lambda}_{y\pi,\text{rep}}\right]$ is unbiasedly estimated by

$$\hat{v}\left[\hat{\lambda}_{y\pi,\text{rep}}\right] = \frac{1}{m(m-1)} \sum_{s=1}^{m} \left(\hat{\lambda}_{y\pi s} - \hat{\lambda}_{y\pi,\text{rep}}\right)^2. \tag{8.32}$$

8.3.6 Product estimator

In an obvious fashion the replicated-sampling estimator of τ_y can be expressed as a product:

$$\hat{\tau}_{y\pi,\text{rep}} = \hat{\tau}_{b\pi,\text{rep}} \left(\frac{\hat{\tau}_{y\pi,\text{rep}}}{\hat{\tau}_{b\pi,\text{rep}}}\right). \tag{8.33}$$

Less obviously, the ratio term reduces to a simple average value:

$$\frac{\hat{\tau}_{y\pi,\text{rep}}}{\hat{\tau}_{b\pi,\text{rep}}} = \frac{(AF/m)\sum_{s=1}^{m}\sum_{u_k \in \mathbb{P}_s} y_k/b_k}{(AF/m)\,n} \tag{8.34a}$$

$$= \frac{1}{n}\sum_{s=1}^{m}\sum_{u_k \in \mathbb{P}_s} v_k \tag{8.34b}$$

$$= \bar{v}, \tag{8.34c}$$

where $v_k = y_k/b_k$, and $\bar{v}$ is the average v_k value across the n trees selected in the m replicated sample points. Substituting this result into (8.33) yields

$$\hat{\tau}_{y\pi,\text{rep}} = \hat{\tau}_{b\pi,\text{rep}}\bar{v} \tag{8.35a}$$

$$= \bar{n}FA\bar{v}. \tag{8.35b}$$

In forestry, v_k is called a 'VBAR' if y_k is tree volume. VBAR (pronounced vee bar) is an acronym originally for volume to basal area ratio for tree u_k. Iles (2003) suggested that the broader applicability of (8.35) might be better appreciated if VBAR were interpreted as variable of interest (i.e., attribute of interest) to basal area ratio.

Expressing $\hat{\tau}_{y\pi,\text{rep}}$ as a product suggests an alternative expression for the variance of $\hat{\tau}_{y\pi,\text{rep}}$ based on the derivations by Goodman (1960) for the variance of a product of random variables. He showed that when the two multiplicands, say a and b, are independent random variables, a design-unbiased estimator of the variance of their product is

$$\hat{v}\,[ab] = a^2\hat{v}\,[b] + b^2\hat{v}\,[a] - \hat{v}\,[a]\,\hat{v}\,[b]. \tag{8.36}$$

In the notation of this chapter, this result is

$$\hat{v}\left[\hat{\tau}_{b\pi,\text{rep}}\,\bar{v}\right] = \hat{\tau}_{b\pi,\text{rep}}^2\hat{v}\,[\bar{v}] + \bar{v}^2\hat{v}\left[\hat{\tau}_{b\pi,\text{rep}}\right] - \hat{v}\left[\hat{\tau}_{b\pi,\text{rep}}\right]\hat{v}\,[\bar{v}], \tag{8.37}$$

where $\hat{v}\,[\bar{v}]$ is an estimator of the variance of $\bar{v} = \hat{\tau}_{y\pi,\text{rep}}/\hat{\tau}_{b\pi,\text{rep}}$.

When the random variables are not independent, the variance of their product is more complicated. Inasmuch as $\hat{\tau}_{b\pi,\mathrm{rep}}$ appears in both terms of (8.33), obviously these terms are dependent random variables. However, their correlation is quite small, apparently, in many applications; Brooks (2006), however, observed positive correlations in the range 0.25–0.55.

The magnitude of the third term in (8.37) is almost always considerably smaller than that of the first two terms. Indeed, in the forestry literature, an alternative estimator (since Bell & Alexander 1957) of the variance of $\hat{\tau}_{y,\mathrm{rat,ds}}$ is commonly seen, and credited to Bruce (1961). It comprises the first two terms of (8.37), i.e.,

$$\tilde{v}\left[\hat{\tau}_{y\pi,\mathrm{rep}}\right] = \hat{\tau}_{b\pi,\mathrm{rep}}^2 \hat{v}\left[\bar{v}\right] + \bar{v}^2 \hat{v}\left[\hat{\tau}_{b\pi,\mathrm{rep}}\right], \tag{8.38}$$

where $\hat{v}[\bar{v}]$ is often computed as

$$\hat{v}[\bar{v}] = \frac{1}{n(n-1)} \sum_{s=1}^{m} \sum_{u_k \in \mathcal{P}_s} (v_k - \bar{v})^2. \tag{8.39}$$

This estimator would be design-unbiased if the n trees used in the calculation of $\bar{v}$ were chosen by SRSwR. However, the n trees constitute a pooled sample of trees selected at m sample points by Bitterlich sampling.

We find no compelling reason to use $\hat{v}[\hat{\tau}_{y\pi,\mathrm{rep}}]$ or $\hat{v}[\hat{\tau}_{b\pi,\mathrm{rep}}\bar{v}]$ as an estimator of $V[\hat{\tau}_{y\pi,\mathrm{rep}}]$, because the conventional estimator, $\hat{v}[\hat{\tau}_{y\pi,\mathrm{rep}}]$, provides unbiased estimates and it renders any concerns about correlation nugatory. However, the variance of $\bar{v}$ may be of interest in and of itself. We note that $\tilde{v}[\hat{\tau}_{y\pi,\mathrm{rep}}] = \hat{v}[\hat{\tau}_{y\pi,\mathrm{rep}}]$ if we estimate $V[\bar{v}]$ by

$$\hat{v}[\bar{v}] = \frac{1}{\hat{\tau}_{b\pi,\mathrm{rep}}^2} \left(\hat{v}\left[\hat{\tau}_{y\pi,\mathrm{rep}}\right] - \bar{v}^2\, \hat{v}\left[\hat{\tau}_{b\pi,\mathrm{rep}}\right] \right) \tag{8.40}$$

or, equivalently,

$$\hat{v}[\bar{v}] = \frac{1}{\hat{\lambda}_{b\pi,\mathrm{rep}}^2} \left(\hat{v}\left[\hat{\lambda}_{y\pi,\mathrm{rep}}\right] - \bar{v}^2\, \hat{v}\left[\hat{\lambda}_{b\pi,\mathrm{rep}}\right] \right). \tag{8.41}$$

Alternatively, because $\bar{v}$ is just a ratio estimator analogous to $\hat{R}_{y|x}$ of Chapter 6, the variance estimators presented in (6.28) and (6.29) are possible alternative estimators for the variance of $\bar{v}$. Specifically, the following estimator of the approximate variance of $\bar{v}$:

$$\hat{v}[\bar{v}] = \hat{v}\left[\frac{\hat{\tau}_{y\pi,\mathrm{rep}}}{\hat{\tau}_{b\pi,\mathrm{rep}}}\right] = \frac{1}{\hat{\tau}_{b\pi,\mathrm{rep}}^2} \frac{s_r^2}{m}, \tag{8.42}$$

where

$$s_r^2 = \frac{1}{m-1} \sum_{s=1}^{m} \left(\hat{\tau}_{y\pi s} - \bar{v}\hat{\tau}_{b\pi s}\right)^2. \tag{8.43}$$

An equivalent expression is

$$\hat{v}[\bar{v}] = \hat{v}\left[\frac{\hat{\lambda}_{y\pi,\mathrm{rep}}}{\hat{\lambda}_{b\pi,\mathrm{rep}}}\right] = \frac{1}{\hat{\lambda}_{b\pi,\mathrm{rep}}^2} \frac{s_{r'}^2}{m}, \tag{8.44}$$

Table 8.1 *Diameters (at breast height) and heights of trees from one Bitterlich sample location. Aboveground biomass values were calculated with equations from Jenkins et al. (2004).*

Species code	Diameter	Height	Biomass
	(cm)	(m)	(kg)
QUGA	49.5	13.8	1445.21
PSME	49.5	31.1	938.96
QUGA	24.6	16.3	316.25
PSME	69.3	31.3	2034.81
PSME	41.7	30.8	630.70
PSME	46.2	30.8	801.27
PSME	31.8	24.1	337.87
QUGA	23.4	17.8	281.85
PSME	49.8	29.3	950.07

where

$$s_{r'}^2 = \frac{1}{m-1} \sum_{s=1}^{m} \left(\hat{\lambda}_{y\pi s} - \bar{\hat{v}}\hat{\lambda}_{b\pi s} \right)^2. \tag{8.45}$$

Example 8.4

The data displayed in Table 8.1 were tallied on a single sampling location in the Pacific northwest region of the USA. A prism with factor $F_E = 20 \, \text{ft}^2 \, \text{ac}^{-1}$ was used to select the sample trees. All measurements that were recorded in English units have been converted to metric units. The equivalent metric basal area factor is $F = 4.59 \, \text{m}^2 \, \text{ha}^{-1}$.

On the basis of this sample point alone, the estimated basal area per hectare is, using (8.24),

$$\hat{\lambda}_{b\pi s} = 9(4.59) = 41.3 \, \text{m}^2 \, \text{ha}^{-1}.$$

The corresponding number of trees per acre is estimated as

$$\hat{\lambda}_{N\pi s} = \frac{\hat{N}_{\pi s}}{A} = 405.8 \, \text{trees ha}^{-1},$$

and the above-ground biomass is estimated as

$$\hat{\lambda}_{y\pi s} = \frac{\hat{t}_{y\pi s}}{A} = 227,352 \, \text{kg ha}^{-1}.$$

Estimates of basal area, tree frequency, and biomass of QUGA are obtained by using these same formulas restricted to the data from only the three QUGA trees sampled at this point. The corresponding estimates are 13.8 $\text{m}^2 \, \text{ha}^{-1}$, 227.1 trees ha^{-1}, and 95,068 kg ha^{-1}. For PSME, the estimates provided from this single point are 27.5 $\text{m}^2 \, \text{ha}^{-1}$, 178.6 trees ha^{-1}, and 132,284 kg ha^{-1}.

Evidently, the estimates for the individual species sum to the estimate obtained for all species combined. For example, the estimated 227.1 QUGA and the 178.6 PSME trees ha^{-1} sum to 405.7 trees ha^{-1}, which is identical, but for rounding error, to the estimate computed directly. This additivity will always obtain with linear, homogeneous estimators such as the HT estimators $\hat{\tau}_{y\pi s}$ and $\hat{\lambda}_{y\pi s}$ used here. Additivity is sacrificed when either the ratio or regression estimator is used. In particular, $\hat{\mu}_{y\pi,\text{rat}}$ will not be additive across species.

In addition to separate estimates for each species, estimates may also be computed separately for discrete size (e.g., diameter or height) classes of trees. Estimates that are summed across size classes will be additive, too, subject to the proviso of the preceding paragraph. In other words, if trees are grouped into successive 5 cm diameter classes, not only can the number of trees per hecare in each class be computed, but these will sum to the same estimate that obtains by pooling all trees together for all diameter classes.

Example 8.5

The sample data in Table 8.1 are part of a forest inventory comprising $m = 27$ sample locations. These data are stored in file OlympicNaturalResourcesCenter_bigtrees_phase2.dat.

On these 27 sample points, the average number of sample trees per point was $\bar{n} = 6.926$ trees per point. From (8.26), the estimated basal area per hectare from the replicated sample data is $\hat{\lambda}_{b\pi,\text{rep}} = 31.8$ $m^2\,ha^{-1}$. Using (8.32), the variance of $\hat{\lambda}_{b\pi,\text{rep}}$ is estimated to be 17.5 $m^4\,ha^{-2}$, which yields an estimated standard error, relative to $\hat{\lambda}_{b\pi,\text{rep}}$, of 13%. A 90% confidence interval for λ_b is 31.8 $m^2\,ha^{-1} \pm 22.5\%$, which extends from 24.7 to 38.9 $m^2\,ha^{-1}$.

Estimates of tree frequency and aggregate biomass per hectare is left as an exercise at the end of the chapter.

Example 8.6

To continue the preceding example, six species of trees were sampled on the $m = 27$ points. Table 8.2 summarizes the replicated sampling estimates of basal area per hectare, the standard error for each estimate as a percentage of estimated basal area, and the corresponding 90% confidence interval for λ_b. A column is also included which displays the number of sample points on which each species was found. This information is displayed as a implicit reminder that the size of the sample on which the estimates are based is $m = 27$, regardless of the number of points on which a species was actually found. Consequently, all confidence intervals shown in the rightmost column were computed with a t-value appropriate for 26 degrees of freedom.

A noteworthy aspect of these results is that individual species estimates are more variable than the estimate for all species combined. This is a common

Table 8.2 *Replicated sampling estimates of basal area per hectare for each species found in the m = 27 point sample. The column for 'No. points' displays the number of sampling points on which at least one tree of the species was selected.*

Species code	No. points	$\hat{\lambda}_{b\pi,\text{rep}}$	$\sqrt{\hat{v}\left[\lambda_{y\pi,\text{rep}}\right]}$	90% Conf. interval
		$(\text{m}^2\,\text{ha}^{-1})$	(%)	$(\text{m}^2\,\text{ha}^{-1})$
ABAM	2	1.4	260.4	−4.7 – 7.4
ALRU	2	0.3	254.8	−1.1 – 1.8
PSME	25	18.7	13.5	14.4 – 23.0
QUGA	1	0.5	519.6	−4.0 – 5.0
THPL	5	1.9	97.7	−1.2 – 5.0
TSHE	13	9.0	39.5	2.9 – 15.1
All species	27	31.8	13.2	24.7 – 38.9

occurrence, because (a) there are fewer trees in each particular species than in all species, and (b) the number of in-trees of each species may vary more, from one sample point to another, than the total number of in-trees of any species. Indeed, the variability is so large for some species shown in Table 8.2 that the lower end of the 90% confidence interval is less than zero.

8.4 Edge effect

All the methods that correct for truncated inclusion probabilities when sampling with fixed-area circular plots work with Bitterlich sampling, too. For example, to apply the mirage method (see §7.5.4, p. 227), we establish a 'mirage point' outside of $\mathcal{A}$ if any tree selected at a sample point, P_s, is so close to the boundary of $\mathcal{A}$ that its inclusion zone may overlap the boundary. The mirage point and P_s are equidistant from, and on a line perpendicular to, the boundary. Up to three mirage points may be needed near a boundary corner. And, of course, only in-trees within $\mathcal{A}$ are tallied from a mirage point. Alternatively, we may apply the walkthrough method (see §7.5.5, p. 229). As in fixed-area plot sampling, a tree is tallied twice if it is closer to the boundary than to P_s.

8.5 Double sampling

Double sampling may be employed to increase the precision of estimation for any attribute of interest that is well and positively correlated with total basal area of trees in $\mathcal{A}$ or basal area per unit land area. Ordinarily, a precise estimate of basal area obtains from a large set of first-phase sample points. The attribute of interest—for example, tree volume—may be measured for all the in-trees at a subset of the first-phase sample points (Bruce 1961), or a subset of the in-trees may be measured at each sample point (e.g., Bell et al. 1983; Iles 1989, 2003).

ASIDE: In Chapter 7 we mentioned the practice of nesting plots of different sizes at the same location. For instance, in Figure 7.1 trees and large woody debris are measured in a 100 m^2 quadrat. Nested within it was a 25 m^2 quadrat on which shrubs and small deadwood are measured. Smaller, more frequently occurring items will be sampled on the smaller plots only, in an attempt to ease the burden of measurement.

For the same reason, Bitterlich sampling often will be combined in an analogous fashion with plot sampling. The sampling design will stipulate a fixed area plot for trees whose diameter is greater than the minimum or threshold diameter, say d_{min}, yet smaller than the minimum diameter, say d_{Bmin}, for which inclusion in the sample will be conducted by Bitterlich sampling. There is no essential difficulty with this practice, provided that the estimator of τ_y that is used computes π_k differently for those trees less than d_{Bmin} from those greater than d_{Bmin}. Indeed for the ONRC inventory (see Table 8.1), trees with diameters less than 20 cm are not shown, because they were only recorded if they were located on a circular plot with radius of 4.75 m.

For a different reason, Bitterlich sampling may be conducted to a stipulated upper diameter limit, d_{Bmax}, only. Trees with diameter greater than d_{Bmax} will be tallied on a fixed area, circular plot with radius, R_{max}, which is fixed in advance of sampling, and which is at least κd_{Bmax} m long (see (8.8)). With this tactic, trees with diameter greater than d_{Bmax} will be selected with inclusion probability $\pi_k = \pi R_{max}^2 / A$, regardless of their exact diameter. This tactic is adopted so as to lessen the risk that a tree with a very large diameter would be mistakenly omitted from the sample at $\mathbb{P}_s$ because its limiting distance from $\mathbb{P}_s$ was so great.

8.5.1 Double sampling with second-phase subset of sample points

Let m_1 be the number of first-phase sample points, and let n_{s_1} be the number of in-trees at the s_1th sample point. Then τ_b is unbiasedly estimated by

$$\hat{\tau}_{b\pi,(1)} = \frac{AF}{m_1} \sum_{s_1=1}^{m_1} n_{s_1} \tag{8.46a}$$

$$= \bar{n}_1 AF, \tag{8.46b}$$

where $\bar{n}_1$ is the average number of in-trees per first-phase sample point.

Let the second phase consist of a subset of the m_1 sample points from the first phase. The number of second-phase sample points is m_2 ($\leq m_1$). If all in-trees are measured both for y and b at each second-phase sample point, then an estimate of τ_y for point s_2 is

$$\hat{\tau}_{y\pi s_2} = AF \sum_{u_k \in \mathbb{P}_{s_2}} \frac{y_k}{b_k}, \tag{8.47}$$

whereas the corresponding estimator of τ_b is

$$\hat{\tau}_{b\pi s_2} = n_{s_2} A F, \tag{8.48}$$

where n_{s_2} is the number of in-trees at sample point s_2.

Therefore, an unbiased estimator of τ_y based solely on the m_2 replicated, second-phase sampling points is

$$\hat{\tau}_{y\pi,(2)} = \frac{1}{m_2} \sum_{s_2=1}^{m_2} \hat{\tau}_{y\pi s_2} \tag{8.49a}$$

$$= \bar{n}_2 A F \bar{v}_{(2)}, \tag{8.49b}$$

where $\bar{n}_2$ is the average number of in-trees at the m_2 second-phase sample points, and $\bar{v}_{(2)}$ is the average y_k/b_k among these trees. The corresponding estimator of τ_b

$$\hat{\tau}_{b\pi,(2)} = \frac{1}{m_2} \sum_{s_2=1}^{m_2} \hat{\tau}_{b\pi s_2} \tag{8.50a}$$

$$= \bar{n}_2 A F. \tag{8.50b}$$

Generally speaking, $\hat{\tau}_{b\pi,(2)} \neq \hat{\tau}_{b\pi,(1)}$.

In analogy to (6.94), the ratio of these two estimators is

$$\hat{R}_{y|b(2)} = \frac{\hat{\tau}_{y\pi,(2)}}{\hat{\tau}_{b\pi,(2)}} \tag{8.51a}$$

$$= \bar{v}_{(2)}, \tag{8.51b}$$

which is the average v_k among all trees selected in the second phase, only.

Provided that the m_2 second-phase points are selected with equal probability from among the m_1 points of the first phase sample, a double-sampling ratio estimator of τ_y which uses information from both phases of sampling is

$$\hat{\tau}_{y,\text{rat,ds}} = \hat{R}_{y|b(2)} \hat{\tau}_{b\pi,(1)} \tag{8.52a}$$

$$= \hat{\tau}_{b\pi,(1)} \bar{v}_{(2)} \tag{8.52b}$$

$$= \bar{n}_1 F A \bar{v}_{(2)}. \tag{8.52c}$$

The similarity of (8.52b) and (8.52c) to (8.35a) and (8.35b), respectively, is obvious.

The variance of $\hat{\tau}_{y,\text{rat,ds}}$ may be estimated by appealing to the variance estimator (6.101). In the notation of this chapter

$$\hat{v}\left[\hat{\tau}_{y,\text{rat,ds}}\right] = \frac{s_y^2}{m_1} + \left(\frac{1}{m_2} - \frac{1}{m_1}\right) \left(\frac{\hat{\tau}_{b\pi,(1)}}{\hat{\tau}_{b\pi,(2)}}\right)^2 s_r^2, \tag{8.53}$$

where s_y^2 is the sample variance of the m_2 point estimates $\hat{\tau}_{y\pi s_2}$, namely

$$s_y^2 = \frac{1}{m_2 - 1} \sum_{s_2=1}^{m_2} (\hat{\tau}_{y\pi s_2} - \hat{\tau}_{y\pi,(2)})^2,$$

and s_r^2 is the sample variance of the m_2 ratio residuals, $\hat{r}_{s_2} = \hat{\tau}_{y\pi s_2} - \hat{R}_{y|b(2)}\hat{t}_{b\pi s_2}$:

$$s_r^2 = \frac{1}{m_2 - 1} \sum_{s_2=1}^{m_2} \hat{r}_k^2. \tag{8.54}$$

Estimates on a per unit area basis follow directly:

$$\hat{\lambda}_{y\pi,\text{rat,ds}} = \frac{\hat{\tau}_{y,\text{rat,ds}}}{A} \tag{8.55a}$$

$$= \bar{n}_1 F \bar{v}_{(2)}; \tag{8.55b}$$

and

$$\hat{v}\left[\hat{\lambda}_{y,\text{rat,ds}}\right] = \frac{s_y^2}{m_1 A^2} + \left(\frac{1}{m_2} - \frac{1}{m_1}\right)\left(\frac{\hat{t}_{b\pi,(1)}}{\hat{t}_{b\pi,(2)}}\right)^2 \frac{s_r^2}{A^2} \tag{8.56a}$$

$$= \frac{s_{y'}^2}{m_1} + \left(\frac{1}{m_2} - \frac{1}{m_1}\right)\left(\frac{\hat{\lambda}_{b\pi,(1)}}{\hat{\lambda}_{b\pi,(2)}}\right)^2 s_{r'}^2, \tag{8.56b}$$

where

$$s_{y'}^2 = \frac{1}{m_2 - 1} \sum_{s_2=1}^{m_2} \left(\hat{\lambda}_{y\pi s_2} - \hat{\lambda}_{y\pi,(2)}\right)^2,$$

and

$$s_{r'}^2 = \frac{1}{m_2 - 1} \sum_{s_2=1}^{m_2} \left(\hat{\lambda}_{y\pi s_2} - \hat{R}_{y|b(2)}\hat{\lambda}_{b\pi s_2}\right)^2. \tag{8.57}$$

In (8.5.1) and (8.57), $\hat{\lambda}_{y\pi s_2} = \hat{\tau}_{y\pi s_2}/A$ and $\hat{\lambda}_{y\pi,(2)} = \hat{\tau}_{y\pi,(2)}/A$.

Iles (2003) advocates a variant of the Bruce formula, (8.38), for estimating the variance of $\hat{\tau}_{y,\text{rat,ds}}$. In the notation of double, point sampling established above, Bruce's formula is

$$\tilde{v}\left[\hat{\tau}_{y,\text{rat,ds}}\right] = \hat{t}_{b\pi,(1)}^2 \hat{v}\left[\bar{v}_{(2)}\right] + \bar{v}_{(2)}^2 \hat{v}\left[\hat{t}_{b\pi,(1)}\right]. \tag{8.58}$$

Substitution of second-phase estimates into (8.40) or (8.42) provides $\hat{v}[\bar{v}_{(2)}]$.

An alternative estimator for the situation where the second-phase sample of points is a subset of the first phase, is the double-sampling regression estimator of τ_y,

$$\hat{\tau}_{y,\text{reg,ds}} = \hat{\tau}_{y\pi,(2)} + \hat{B}_{(2)}\left(\hat{t}_{b\pi,(1)} - \hat{t}_{b\pi,(2)}\right) \tag{8.59a}$$

$$= \hat{A}_{(2)} + \hat{B}_{(2)}\hat{t}_{b\pi,(1)}, \tag{8.59b}$$

where $\hat{B}_{(2)}$ is the estimated slope of the regression line of $\hat{\tau}_{y\pi s_2}$ vs. $\hat{t}_{b\pi s_2}$ for $s_2 = 1, 2, \ldots, m_2$. The variance of $\hat{\tau}_{y,\text{reg,ds}}$ may be estimated by the same expression, (8.56), provided that the regression residuals, $\hat{r}_{s_2} = \hat{\tau}_{y\pi s_2} - A_{(2)} - \hat{B}_{(2)}\hat{t}_{b\pi s_2}$ are used in the calculation of the s_r^2 term.

Example 8.7

The sample data used in Examples 8.5 and 8.6 were part of a double sample. The $m_2 = 27$ second-phase points were a subset of $m_1 = 271$ first-phase sampling points on which the number of trees selected with the $F = 4.59$ m^2 ha^{-1} angle gauge were counted. No other measurements were taken on these sample trees. These data were used to provide

$$\hat{\lambda}_{b\pi,(1)} = 35.0 \text{ m}^2 \text{ ha}^{-1}.$$

These are combined with the results from the second phase (see Example 8.6) to yield

$$\hat{R}_{y|b(2)} = \frac{197,793.3 \text{ kg ha}^{-1}}{31.8 \text{ m}^2 \text{ ha}^{-1}}$$

$$= 6220.0 \text{ kg m}^{-2},$$

which in turn yields

$$\hat{\lambda}_{y\pi,\text{rat,ds}} = 6220.0 \times 35.0 = 217,507.5 \text{ kg ha}^{-1}.$$

Appealing to (8.53),

$$\hat{v}\left[\hat{\lambda}_{y\pi,\text{rat,ds}}\right] = \frac{28,177,925,180}{271} + \left(\frac{1}{27} - \frac{1}{271}\right)\left(\frac{35.0}{31.8}\right)^2 5,156,092,132$$

$$= 311,900,596.6 \left(\text{kg ha}^{-1}\right)^2.$$

As a percentage of $\hat{\lambda}_{y\pi,\text{rat,ds}}$ its estimated standard error is 8.1%. Had λ_y been estimated with the $m_2 = 27$ second-phase points alone, the standard error would have been 16.3% (see Chapter 8, Exercise 2.).

Example 8.8

Johnson (1961, page 5) presented data collected from a double point sample consisting of $m_1 = 62$ first-phase points and $m_2 = 25$ second-phase points. (These data are available in the file Johnson_1961RN_hps.dat.) He reports the number of trees selected at each point with a prism basal area factor of $F_E = 38.55$ ft^2 ac^{-1}. While the diameters of trees on the second-phase sample points must have been measured, Johnson's data only reports on the volume (in board feet measure) to basal area ratio, or vbar, of each of the trees selected in the second phase.

From the $n_1 = 158$ trees selected during the first phase, τ_b is estimated as $\hat{\lambda}_{b\pi,(1)} = 98.2$ ft^2 ac^{-1}. From the second phase data,

$$\hat{R}_{y|b(2)} = \frac{28,226.3 \text{ (bd ft) ac}^{-1}}{94.1 \text{ ft}^2 \text{ ac}^{-1}}$$

$$= 300.1 \text{ (bd ft) ft}^{-2},$$

ASIDE: The type of adjustment suggested by Oderwald (1994) is similar in spirit to a procedure termed *raking* which was introduced, according to Lohr (1999), by Deming & Stephan (1940) to adjust entries in two- or higher-way tables so that they summed to consistent totals in the marginal rows and columns. A brief illustration of raking is given in §8.5.2.2 of Lohr (1999).

---

which, in turn, yields

$$\hat{\lambda}_{y\pi,\text{rat,ds}} = 300.1 \times 98.2 = 29{,}480.1 \text{ (bd ft) ac}^{-1}$$

with an estimated standard error of 9.2%. Had just the 25 second-phase plots been used, the relative standard error would have been 15.0%

The double sampling ratio estimator as presented in (8.52) can be rearranged as a multiplicative adjustment to the second-phase estimator of τ_y:

$$\hat{\tau}_{y,\text{rat,ds}} = \left(\frac{\hat{\tau}_{b\pi,(1)}}{\hat{\tau}_{b\pi,(2)}} \right) \hat{\tau}_{y\pi,(2)}. \tag{8.60}$$

In an obvious fashion, the same multiplicative adjustment may be made to per unit area estimates:

$$\hat{\lambda}_{y\pi,\text{rat,ds}} = \left(\frac{\hat{\lambda}_{b\pi,(1)}}{\hat{\lambda}_{b\pi,(2)}} \right) \hat{\lambda}_{y\pi,(2)}. \tag{8.61}$$

When there is additional interest in obtaining separate estimates by species, as in Example 8.6, or by size classes, Oderwald (1994) suggested that this same multiplicative adjustment be applied to each species or size-class estimate. In this fashion, one may hope to reap the benefits of the double sample rather than rely on estimates by species or size classes solely on the small second-phase sample. With this approach, the additivity of species, or size-class, estimates is retained, as the multiplicatively adjusted individual estimates will sum to the similarly adjusted estimator of τ_y or λ_y. Oderwald (1994) suggested an estimator of variance of these individual estimates that is similar in spirit to $\hat{v}[\hat{\tau}_{y,\text{rat,ds}}]$ in (8.53), but with the expressions for s_y^2 and s_r^2 modified to include only those data belonging to the species or size class of interest. We defer to Oderwald (1994) for details.

As an alternative to applying the same multiplicative factor to each species' (or size-class) estimate from phase two, the $\hat{\tau}_{b\pi,(1)}/\hat{\tau}_{b\pi,(2)}$ (or, $\hat{\lambda}_{b\pi,(1)}/\hat{\lambda}_{b\pi,(2)}$) factor can be based on just those trees sampled in the species (size class) in the two phases. With this approach, additivity is sacrificed, but other advantages may accrue.

Disaggregating τ_y and λ_y into separate estimates by species or size classes following double sampling and regression estimation is more complicated than it is with ratio estimation, as shown by Matney & Parker (1991).

8.5.2 Optimal allocation

With double Bitterlich sampling there is a loss of information when the y characteristic of primary interest is measured only on those trees selected by the second-phase sample points. This loss of information is manifest as an increase in the sampling variance of $\hat{\tau}_{y,\text{rat,ds}}$ compared to the variance of a single-phase estimator, $\hat{\tau}_{y\pi,\text{rep}}$, based on an identical number of sample points.

Conversely, the double sample may cost less, because Bitterlich sampling enables efficient estimation of τ_b simply by counting the number of in-trees at each first-phase point: the comparatively costly measurement of y on the first-phase in-trees is foregone.

In other words, there is a tradeoff between sampling cost and precision of estimation with the point, double sampling design. As with any double sample, it is necessary to allocate the time and effort between the two phases of sampling to achieve an acceptable tradeoff. To this end, Oderwald & Jones (1992) derive optimal sample allocation formulas for double Bitterlich sampling. They depend, inter alia, on being able to stipulate the strength of the linear correlation between y and basal area per land area, as well the average cost of a first-phase sample point and a second-phase sample point.

Typically, when y is the volume or biomass is the bole of a tree, the correlation with basal area on a per unit area basis is quite strong. When this is coupled with the ease with which the first-phase Bitterlich sampling is conducted to estimate basal area per unit land area, double Bitterlich sampling can be very efficient compared to one-phase sampling in which all trees are measured for y.

We defer to Oderwald & Jones (1992) for illustrative examples and for details on deriving the optimal sample sizes in the two phases.

8.5.3 Double sampling with a second-phase subset of trees

Instead of measuring all the trees on a subset of the m_1 first-phase sample points, an alternative tactic is to select a subset of trees for measurement on each first-phase point. By selecting a subset of in-trees at each sample point, the second-phase sample is distributed more evenly over the forested region. Big BAF sampling, as it has become known in the forestry literature, (e.g., Iles 2003; Marshall et al. 2004) provides one way to make the second-phase selections. Choosing the second-phase trees with probability proportional to height provides another way.

With both methods, the first-phase sample trees are selected with a basal area factor, say F_1, and the number of in-trees are counted, possibly by species. As above, no other measurements are taken on the first-phase selections. An estimator of τ_b is available from $\hat{\tau}_{b\pi,(1)}$ as given in (8.46b).

Big BAF sampling

At each point, a subset of the first-phase trees is selected by using a basal area factor, F_2, that is larger than F_1 used in the first phase. This ensures that the trees selected in the second phase are a subset of those already counted in phase one. The size of

F_2 compared to F_1 controls indirectly the number of trees selected at a point in the second phase, n_{s_2}, relative to the number of trees, n_{s_1}, selected at the first phase. If, for example, the objective is to measure, on average, 1/5 of the trees selected in the first phase at a sample point, then $F_2 = 5F_1$. As with one-phase Bitterlich sampling, it is possible that no trees will be measured at some sample points; see, for example, the data used in Example 8.8. The larger F_2 is, the greater is that risk.

The rationale for the 'Big BAF sampling' design is that $\hat{\tau}_{b\pi,(1)}$ is usually much more variable than $\bar{v}_{(2)}$. Therefore a larger sample of trees ought to be selected to estimate $\hat{\tau}_{b\pi,(1)}$ than is required for equally precise estimation of $\bar{v}_{(2)}$. By varying F_1 and F_2 in the manner described, the designer of the forest inventory can account directly for this, while achieving a more evenly distributed selection of 'measured trees' at the same time.

Estimation of τ_y proceeds identically as before by computing $\hat{R}_{y|b(2)}$ in (8.51) from the second-phase sample information, and then multiplying this by $\hat{\tau}_{b\pi,(1)}$ to yield $\hat{\tau}_{y,\text{rat,ds}}$ as given in (8.52). Similarly, the variance of $\hat{\tau}_{y,\text{rat,ds}}$ can be estimated by $\hat{v}\big[\hat{\tau}_{y,\text{rat,ds}}\big]$ in (8.53), although use of (8.58) is an alternative for this purpose.

Subsampling trees with probability proportional to height

Nelson & Gregoire (1994) examined the utility of selecting a subset of the trees selected at a point in phase one with probability proportional to tree height. This method requires that the heights of all trees selected in phase 1 at each point be accumulated into a list so that subsampling may proceed according to the list sampling method presented in § 3.3.1.

To be specific, let $\tau_{h,s}$ denote the total height of all trees selected at the sth point in phase one: Let h_k denotes the height of $\mathcal{U}_k$, then

$$\tau_{h,s} = \sum_{\mathcal{U}_k \in \mathbb{P}_s} h_k \tag{8.62}$$

is the accumulated height of the n_s trees on $\mathbb{P}_s$. From this list, a subsample of $n_{s(2)} < n_s$ trees is selected with replacement using $\pi_{k(2)} = h_k/\tau_{h,s}$ as the selection probability.

Nelson & Gregoire (1994) presented an unbiased estimator of τ_y as

$$\hat{\tau}_{y\pi s} = \frac{1}{n_{s(2)}} \sum_{\mathcal{U}_k \in \mathbb{P}_{s2}} \frac{y_k}{\pi_k \pi_{k(2)}} \tag{8.63a}$$

$$= \frac{FA\tau_{h,s}}{n_{s(2)}} \sum_{\mathcal{U}_k \in \mathbb{P}_{s2}} \frac{y_k}{b_k h_k}. \tag{8.63b}$$

From a replicated sample comprising m sampling points, τ_y is estimated unbiasedly by

$$\hat{\tau}_{y\pi,\text{rep}} = \frac{1}{m} \sum_{s=1}^{m} \hat{\tau}_{y\pi s},$$

as usual. They also provide an exact expression for the variance (see page 252 of that article), which is estimated unbiasedly by (8.31) in the usual fashion. As with the 'big BAF' method, the motivation for subsampling is to reduce the number of trees for which y_k must be measured without incurring an appreciable loss of precision by failing to measure y_k on every in-tree at each sample point. In their simulation experiments conducted with mapped stands of trees, Nelson & Gregoire (1994) found but a modest increase in the standard error of estimation when compared to a comparable one-phase Bitterlich sample. They mention that in stands with little variation in tree heights, the second-phase selection will be almost the same as SRSwR. Also, for the purpose for which the information on tree heights is used, ocular estimation, which can be done quite quickly, should suffice.

Yandle & White (1977) also have investigated the estimator given in (8.63), but for a slightly different sampling design which requires accumulation of trees heights on all m_1 points prior to selecting the subsample.

8.6 Sampling to estimate change in stock

Motivations for sampling to estimate change in τ_y were discussed in 7.7. Ordinarily, we are interested in the change in τ_y between two points in time, t_1 and t_2. For example, τ_y may be the total amount of merchantable volume within the boles of the trees or the amount of carbon sequestered in the organs of the trees. In the former case, τ_y is called the merchantable stock and, in the latter, the carbon stock. Let $\tau_y(t_1)$ denote the stock of interest at time t_1, and let $\tau_y(t_2)$ denote the stock at time t_2. The net change in stock from t_1 to t_2 is thus

$$\Delta_y = \tau_y(t_2) - \tau_y(t_1), \tag{8.64}$$

and a natural estimator of Δ_y is

$$\hat{\Delta}_y = \hat{\tau}_{y\pi,\text{rep}}(t_2) - \hat{\tau}_{y\pi,\text{rep}}(t_1). \tag{8.65}$$

The estimator is very general. Indeed, the sampling may employ fixed-area plots at time t_1 and Bitterlich sample points at time t_2, or vice versa. Or, Bitterlich sample points may be used on both occasions. The sample points used at t_1 may differ in location and number from those used at t_2, or a common set of sample points may be used on both occasions. In the latter case, $\hat{\Delta}_y$ can be expressed as

$$\hat{\Delta}_y = \frac{1}{m} \sum_{s=1}^{m} \left(\hat{\tau}_{y\pi s}(t_2) - \hat{\tau}_{y\pi s}(t_1) \right). \tag{8.66}$$

The sampling variance, $V[\hat{\Delta}_y]$, is expressed as equation (7.30), and an estimator of this variance is expressed as equation (7.31).

Let Δ_b denote the change in basal area from t_1 to t_2, i.e.,

$$\Delta_b = \tau_{b\pi,\text{rep}}(t_2) - \tau_{b\pi,\text{rep}}(t_1). \tag{8.67}$$

If Bitterlich sampling is conducted with the same basal area factor at the same set of

m sample points on both occasions, then Δ_B is estimated by

$$\hat{\Delta}_b = \frac{FA}{m} \sum_{s=1}^{m} [n_s(t_2) - n_s(t_1)] \tag{8.68a}$$

$$= FA\,[\bar{n}(t_2) - \bar{n}(t_1)]\,, \tag{8.68b}$$

where $n_s(t_2)$ is the number of trees selected at P_s and $\bar{n}(t_2)$ is the average number of trees selected per point at the second occasion. Analogous interpretations apply to $n_s(t_1)$ and $\bar{n}(t_1)$. In general, $n_s(t_2) \neq n_s(t_2)$ even when F is the same on both sampling occasions (which need not be the case). The reason for the different numbers of trees selected on the two occasions is due, in part, to the fact that $b_k(t_2) > b_k(t_1)$, unless the tree has stagnated or died. Consequently, $\pi_k(t_2) > \pi_k(t_1)$, if F is held constant.

8.6.1 Conventional components of change

The net change in stock may be decomposed into various *components of change* (e.g., Beers 1962). We shall consider three components that, by convention, are called ingrowth, survivor growth, and loss. As noted in Chapter 7 and §8.2, trees smaller than some minimal diameter (d_{min}) ordinarily are not measured in forest surveys. Any tree, $\mathcal{U}_k$, is 'measurable' if $d_k \geq d_{min}$, and perhaps other restrictions on species of interest and vigor are satisfied. Ingrowth (Δ_{yI}) is the portion of stock at t_2 contributed by those trees which become measurable between t_1 and t_2 and which remain alive at t_2. Survivor growth (Δ_{yS}) is the increase in stock attributable to the growth of trees that are alive and measurable at both t_1 and t_2. Loss (Δ_{yL}) is usually defined as the portion of the stock at t_1 contributed by trees that are alive and measurable at t_1 and either dead or harvested by time t_2. Therefore, the stock at t_2 is

$$\tau_y(t_2) = \Delta_{yI} + \Delta_{yS} + \tau_y(t_1) - \Delta_{yL} \tag{8.69}$$

and the net change in stock from t_1 to t_2 is

$$\Delta_y = \Delta_{yI} + \Delta_{yS} - \Delta_{yL}. \tag{8.70}$$

Let ξ_{kI} indicate whether tree $\mathcal{U}_k$ contributes to ingrowth at t_2, i.e.,

$$\xi_{kI} = \begin{cases} 1, & \text{if } \mathcal{U}_k \text{ contributes to ingrowth;} \\ 0, & \text{otherwise.} \end{cases} \tag{8.71}$$

Then,

$$\Delta_{yI} = \sum_{k=1}^{N} \xi_{kI}\, y_k(t_2). \tag{8.72}$$

Analogously,

$$\xi_{kS} = \begin{cases} 1, & \text{if } \mathcal{U}_k \text{ contributes to survivor growth;} \\ 0, & \text{otherwise} \end{cases} \tag{8.73}$$

and

$$\xi_{kL} = \begin{cases} 1, & \text{if } \mathcal{U}_k \text{ contributes to loss;} \\ 0, & \text{otherwise.} \end{cases} \tag{8.74}$$

Hence,

$$\Delta_{yS} = \sum_{k=1}^{N} \xi_{kS} [y_k(t_2) - y_k(t_1)] \tag{8.75}$$

and

$$\Delta_{yL} = \sum_{k=1}^{N} \xi_{kL} \, y_k(t_1). \tag{8.76}$$

Under a fixed-area plot design, the estimation of the components of change is straightforward (see Example 7.16). Under the Bitterlich design, we presume a common set of m sample points and the same basal area factor at both occasions, and defer to Gregoire (1993) for the treatment of growth estimation when different sets of sample points are used. Any tree, $\mathcal{U}_k$, that is in at the sample point and measurable at t_1 obviously is a 'survivor tree' if that tree is still alive at t_2. And, any tree that is in, but too small for measurement at t_1, is obviously an 'ingrowth tree' if that tree is both alive and measurable at t_2. However, the sets of survivor and ingrowth trees may include more than just the obvious trees at t_2.

Under definitions of Van Deusen et al. (1986), any tree, $\mathcal{U}_k$, which is in and alive at t_2, is an ingrowth tree if $d_k(t_1) < d_{min}$ and $d_k(t_2) \geq d_{min}$, regardless of whether the tree was selected into the sample at t_1. Moreover, any tree, $\mathcal{U}_k$, which is in and alive at t_2, is a survivor if $d_k(t_1) \geq d_{min}$, regardless of whether the tree was in or out at t_1. Any tree, $\mathcal{U}_k$, which was in and alive at t_1 with $d_k(t_1) \geq d_{min}$, is a 'loss tree' if it is dead or harvested at t_2.

Correctly identifying the loss trees is easy, provided the measurable in-trees are numbered or mapped at t_1. It is more difficult to determine whether a tree that is selected at t_2 but not at t_1 belongs to the domain of ingrowth trees or the domain of survivors. This determination is facilitated by measuring the diameters and mapping the locations of any trees at t_1 in the vicinity of $\mathbb{P}_s$ that could conceivably become measurable in-trees at t_2.

The net change in stock from t_1 to t_2 is unbiasedly estimated from Bitterlich samples by

$$\hat{\Delta}_y = \frac{AF}{m} \sum_{s=1}^{m} \left[\sum_{\mathcal{U}_k \in \mathbb{P}_s(t_2)} \frac{y_k(t_2)}{b_k(t_2)} - \sum_{\mathcal{U}_k \in \mathbb{P}_s(t_1)} \frac{y_k(t_1)}{b_k(t_1)} \right]. \tag{8.77}$$

The components of ingrowth and loss, respectively, are unbiasedly estimated by

$$\hat{\Delta}_{yI} = \frac{AF}{m} \sum_{s=1}^{m} \sum_{\mathcal{U}_k \in \mathbb{P}_s(t_2)} \xi_{kI} \left[\frac{y_k(t_2)}{b_k(t_2)} \right] \tag{8.78}$$

and

$$\hat{\Delta}_{yL} = \frac{AF}{m} \sum_{s=1}^{m} \sum_{\mathcal{U}_k \in \mathbb{P}_s(t_1)} \xi_{kL} \left[\frac{y_k(t_1)}{b_k(t_1)} \right]. \tag{8.79}$$

Survivor growth is unbiasedly estimated by

$$\hat{\Delta}_{yS} = \frac{AF}{m} \sum_{s=1}^{m} \left\{ \sum_{\substack{\mathcal{U}_k \in \mathbb{P}_s(t_1) \\ \mathcal{U}_k \in \mathbb{P}_s(t_2)}} \xi_{kS} \left[\frac{y_k(t_2)}{b_k(t_2)} - \frac{y_k(t_1)}{b_k(t_1)} \right] + \sum_{\substack{\mathcal{U}_k \notin \mathbb{P}_s(t_1) \\ \mathcal{U}_k \in \mathbb{P}_s(t_2)}} \xi_{kS} \left[\frac{y_k(t_2)}{b_k(t_2)} \right] \right\}$$

$$\tag{8.80}$$

or, equivalently, by

$$\hat{\Delta}_{yS} = \hat{\Delta}_y - \hat{\Delta}_{yI} + \hat{\Delta}_{yL}. \tag{8.81}$$

Alternative estimators of the components of change—based on alternative definitions of loss-trees, survivors, and ingrowth-trees—were advanced by Grosenbaugh (1958), Beers & Miller (1964), Martin (1982), and Roesch et al. (1989), and all are reviewed by Gregoire (1993). Many of these definitions have been motivated by a desire to preserve additivity, or *compatibility* as it has been called in the forestry literature. Both terms imply that the estimates of the components of change sum identically to the estimate of net change, as provided by (8.77).

However, the conventional components of change generally are not temporally additive (Eriksson 1995a). The results that obtain from adding the component estimates for the time interval $[t_1, t_2]$ to those for the interval $[t_2, t_3]$ differ from the component estimates for $[t_1, t_3]$, which obtain from measurements at t_1 and t_3, omitting measurements at t_2. A tree's contribution to ingrowth, for example, increases from the time the tree becomes measurable until the time that that tree is measured. Thus, a tree that becomes measurable between t_1 and t_2 contributes to the ingrowth component until t_3, if the measurement at t_2 is omitted.

8.6.2 *Eriksson's components of change*

Eriksson (1995a) achieved component and temporal additivity by defining new precise components of change in continuous time. Trees that are measurable during all or part of an interval $[t_1, t_2]$ categorize as eligibility trees, growth trees, and/or depletion trees. An eligibility tree, $\mathcal{U}_k$, is too small for measurement at t_1, but it becomes eligible for measurement when it grows to the minimal diameter ($d_k = d_{\min}$) at time t_{kE}, where $t_1 < t_{kE} \leq t_2$. As $\mathcal{U}_k$ crosses the threshold of eligibility, the stock in $\mathcal{A}$ is incremented by $y_k(t_{kE})$. If an eligibility tree dies or is harvested at t_{kD}, where $t_{kE} < t_{kD} \leq t_2$, then that eligibility tree is also a depletion tree. Moreover, any tree, $\mathcal{U}_k$, that is measurable at t_1 and lost to death or harvest at $t_{kD} \leq t_2$ is also a depletion tree. When any tree, $\mathcal{U}_k$, becomes a depletion tree, the stock in $\mathcal{A}$ is depleted by $y_k(t_{kD})$.

Any tree—while it is alive and measurable during any part of the interval $[t_1, t_2]$—is a growth tree. Thus, growth trees comprise eligibility trees, depletion trees (while alive), and survivors—those trees which are both alive and measurable at both t_1

and t_2. Let w_k denote the earliest point in the time interval $[t_1, t_2]$ that tree $\mathcal{U}_k$ is eligible for measurement, and let v_k denote the last point in the time interval that $\mathcal{U}_k$ is measurable. If $\mathcal{U}_k$ is an eligibility tree, then $w_k = t_{kE}$; otherwise, $w_k = t_1$. If tree $\mathcal{U}_k$ is a depletion tree, then $v_k = t_{kD}$; otherwise $v_k = t_2$. More formally,

$$w_k = \max(t_1, t_{kE}) \quad \text{and} \quad v_k = \min(t_2, t_{kD}).$$

Hence, the growth of any measurable tree $\mathcal{U}_k$ in the interval $[t_1, t_2]$ is

$$\delta_{yk} = y_k(v_k) - y_k(w_k). \tag{8.82}$$

Let $\mathbf{T}$ denote the set of growth trees, i.e., all trees that are measurable at any point in $[t_1, t_2]$. Then, the increase in the stock, Δ_{yT}, from the total growth of all measurable trees in $\mathcal{A}$ is

$$\Delta_{yT} = \sum_{\mathcal{U}_k \in \mathbf{T}} y_k(v_k) - y_k(w_k) \tag{8.83}$$

$$= \sum_{\mathcal{U}_k \in \mathbf{T}} \delta_{yk}. \tag{8.84}$$

Let $\mathbf{E}$ denote the set of trees that become eligible for measurement in $[t_1, t_2]$. The total increment in the stock, Δ_{yE}, attributable to trees crossing the threshold of eligibility is

$$\Delta_{yE} = \sum_{\mathcal{U}_k \in \mathbf{E}} y_k(t_{kE}). \tag{8.85}$$

Finally, let $\mathbf{D}$ denote the set of trees that become depletion trees in $[t_1, t_2]$. The total decrement in the stock, $\Delta_y(D)$, attributable to death or harvest is

$$\Delta_{yD} = \sum_{\mathcal{U}_k \in \mathbf{D}} y_k(t_{kD}). \tag{8.86}$$

Hence,

$$\tau_y(t_2) = \Delta_{yT} + \Delta_{yE} + \tau_y(t_1) - \Delta_{yD} \tag{8.87}$$

and the net change in the stock from t_1 to t_2 is

$$\Delta_y = \Delta_{yT} + \Delta_{yE} - \Delta_{yD}. \tag{8.88}$$

To estimate these components from Bitterlich samples, we must be able to distinguish the eligibility trees and the depletion trees from the other measurable trees that are in at a sample point at time t_2. Depletion trees are obvious and, as it turns out, identification the eligibility trees is easy, because an eligibility tree must be in at the sample point before it is measurable (Eriksson 1995a). If $\mathcal{U}_k$ is in, but too small for measurement at t_1, then it is an eligibility tree at t_2 if $d_k(t_2) \geq d_{\min}$. On the other hand, if $\mathcal{U}_k$ is out at t_1, but in and measurable at t_2, then we determine whether $\mathcal{U}_k$ is an eligibility tree with the plot radius factor, κ, and the distance, ℓ_k, from $\mathcal{U}_k$ to the sample point: to wit, $\mathcal{U}_k$ is an eligibility tree if $\ell_k \leq \kappa d_{\min}$. Conversely, if $\ell_k > \kappa d_{\min}$, then $\mathcal{U}_k$ is not an eligibility because it was measurable before it was in at the sample point.

We do not need to know the point in time, t_{kE}, when $\mathcal{U}_k$ became measurable, but we should be able to calculate or estimate the attribute of interest, $y_k(t_{kE})$, from d_{min}. Similarly, we do not need to know the point in time, t_{kD}, when a depletion tree, $\mathcal{U}_k$, was lost to death or harvest, but we should be able to measure, calculate, or estimate the loss, $y_k(t_{kD})$.

The net change in stock in $[t_1, t_2]$ is estimated with $\hat{\Delta}_y$ as formulated in (8.77). The ingrowth or increment in the stock attributable to trees crossing the threshold of eligibility in $[t_1, t_2]$ is estimated by

$$\hat{\Delta}_{yE} = \frac{AF}{m} \sum_{s=1}^{m} \sum_{\substack{\mathcal{U}_k \in E \\ \mathbb{P}_s(t_{kE}) \ni \mathcal{U}_k}} \frac{y_k(t_{kE})}{b_k(t_{kE})}. \tag{8.89}$$

The decrement in the stock attributable to death or harvest in $[t_1, t_2]$ is estimated by

$$\hat{\Delta}_{yD} = \frac{AF}{m} \sum_{s=1}^{m} \sum_{\substack{\mathcal{U}_k \in D \\ \mathbb{P}_s(t_{kD}) \ni \mathcal{U}_k}} \frac{y_k(t_{kD})}{b_k(t_{kD})}. \tag{8.90}$$

Hence, an estimator of aggregate tree growth in $[t_1, t_2]$ is

$$\hat{\Delta}_{yT} = \hat{\Delta}_y + \hat{\Delta}_{yD} - \hat{\Delta}_{yE}. \tag{8.91}$$

Eriksson's components of change evidently differ from the conventional components. However, if Bitterlich sampling were conducted yearly, and the estimators of Van Deusen et al. (1986) were used to estimate the conventional components, then the estimate of Eriksson's growth component could be quite similar in value to the estimate of conventional survivor growth, and the estimate of Eriksson's depletion component could be similar to the estimate of the conventional loss component. And, naturally, the estimates of eligibility and conventional ingrowth also could be similar. Most definitions for the conventional components ignore trees that (a) become measurable after t_1 and (b) are lost to death or harvest before t_2. Under Eriksson's definitions, these ephemerally measurable trees are eligibility trees, growth trees, and depletion trees. That the history and growth of these tree are accounted for renders Eriksson's components useful in ecological studies. Indeed, if τ_y is aboveground carbon (Mg) in trees, then $(\Delta_{yT}/A)/(t_2 - t_1)$ is the tree component of aboveground net ecosystem productivity (Mg C ha^{-1} yr^{-1}).

8.7 Terms to remember

Angle count sampling	Limiting distance
Angle gauges	Plot radius factor
Basal area factor	Point double sample
Big BAF sampling	Variable radius plot sampling
Components of change	VBAR
Horizontal point sampling	Winkelsählprobe

Table 8.3 *Bitterlich sample data at point s from a forest inventory consisting of* $m = 5$ *sample points using the basal area factor* $F = 3\ m^2\ ha^{-1}$ *(data from de Vries 1986).*

s	Dia.	Height	Volume	s	Dia.	Height	Volume
	(cm)	(m)	$(m^3 \times 10^3)$		(cm)	(m)	$(m^3 \times 10^3)$
1	20	17.3	270	3	26	19.4	497
1	25	19.2	456	3	28	20.0	589
1	25	19.2	456	3	30	20.5	688
1	28	20.0	589	3	31	20.7	739
1	29	20.2	636	3	32	20.9	793
1	33	21.0	845	3	33	21.0	845
1	35	21.3	960	3	40	21.9	1275
1	36	21.5	1020	4	21	17.7	303
1	37	21.6	1080	4	23	18.5	376
1	38	21.7	1145	4	31	20.7	739
2	24	18.8	414	4	32	20.9	793
2	24	18.8	414	4	33	21.0	845
2	28	20.0	589	4	37	21.6	1080
2	30	20.5	688	4	37	21.6	1080
2	30	20.5	688	5	25	19.2	456
2	35	21.3	960	5	25	19.2	456
2	39	21.8	1210	5	28	20.0	589
3	21	17.7	303	5	28	20.0	589
3	25	19.2	456	5	31	20.7	739
3	26	19.4	497	5	34	21.2	904

8.8 Exercises

1. Explain the consequences of $\alpha = 1$.

2. Extend Example 8.5 to include 90% confidence intervals for trees per hectare and biomass per hectare.

3. Repeat Exercise 2. to compute estimates separately for each species that was sampled.

4. Use the data in Table 8.3, previously published in de Vries (1986, §12.6, p. 237), to estimate number of trees, basal area, and volume per hectare.

5. Use the data in Table 8.3, to estimate the average height per tree in the sampled population.

6. Use the data stored in file Ohio_HPS_subset.dat, a portion of which is displayed in Table 8.4, to compute 90% confidence interval estimates for number of trees per acre, basal area per acre, and volume in board feet per acre.

7. Repeat Exercise 6. to compute interval estimates separately for each species tallied in the sample.

8. Repeat Exercise 6. to compute interval estimates separately for each one-inch diameter class.

Table 8.4 *Bitterlich sample data from a forest inventory conducted on m = 99 sample points with a basal area factor* F_E = 10 ft^2 ac^{-1}. *These data comprise the trees tallied at the first* (s = 1) *of the 99 sample points. The complete tally is filed in Ohio_HPS_subset.dat.*

s	Species code	Diameter	Volume	Species code	Diameter	Volume
		(in)	(bd ft)		(in)	(bd ft)
1	BO	14	130	YP	16	284
1	YP	16	236	YP	25	806
1	SM	21	539	WO	18	272
1	WO	16	236	RO	19	307
1	RO	21	492			

Table 8.5 *Tree counts and merchantable volumes per acre at first-phase point s_1 from a double sample consisting of m_1* = 49 *first-phase sample points and m_2* = 17 *second-phase points. The second-phase sample included every third first-phase point. Sampling was conducted with the basal area factor F_E* = 20 ft^2 ac^{-1}.

s_1	No. trees	Vol.	s_1	No. trees	Vol.	s_1	No. trees	Vol.
		(bd ft)			(bd ft)			(bd ft)
1	3	7640	18	4		34	7	7600
2	2		19	2	5160	35	2	
3	3		20	5		36	3	
4	2	2890	21	0		37	5	7040
5	5		22	0	0	38	0	
6	4		23	1		39	0	
7	3	7220	24	4		40	0	0
8	3		25	4	9500	41	0	
9	3		26	1		42	1	
10	5	9560	27	2		43	0	0
11	3		28	2	3880	44	2	
12	3		29	3		45	3	
13	0	0	30	0		46	5	10260
14	0		31	0	0	47	5	
15	4		32	4		48	4	
16	3	5180	33	3		49	6	14340
17	3							

9. Use the double-point-sampling data in Johnson_1961RN_hps.dat to verify the results reported in Example 8.8, and to compute a 90% confidence interval for volume per acre.

10. Use the double-point-sampling data in RGO_cruise.dat and displayed in Table 8.5 to compute 90% confidence interval estimates of basal area per acre and volume in board feet per acre. Note that the volumes shown in this table are the $\hat{\tau}_{y\pi s}$ estimates for each point.

Table 8.6 *Tree counts and sums of merchantable volume:basal area at point s from a double sample consisting if $m_1 = 44$ first-phase sampling points and $m_2 = 44$ second-phase points. Sampling was conducted with the basal area factor $F_E = 9.372\,ft^2 ac^{-1}$ (data from Palley & Horwitz 1961).*

s	No. trees	$\sum y_k/b_k$	s	No. trees	$\sum y_k/b_k$
		$\left[(\text{bd ft})\,\text{ft}^{-2}\right]$			$\left[(\text{bd ft})\,\text{ft}^{-2}\right]$
1	9	2228	23	10	958
2	9	1896	24	10	1789
3	7	991	25	12	2144
4	14	2737	26	6	1021
5	19	2899	27	18	3710
6	17	2560	28	18	3134
7	16	3469	29	11	1299
8	21	4746	30	1	284
9	10	2048	31	8	1329
10	21	5762	32	5	688
11	11	1649	33	19	3000
12	16	2798	34	15	2646
13	25	4466	35	11	1835
14	17	3343	36	19	2661
15	12	3230	37	24	4258
16	22	3067	38	23	3225
17	18	3658	39	11	1818
18	9	1685	40	8	1192
19	18	4268	41	19	2486
20	11	3344	42	25	4860
21	3	609	43	17	3222
22	5	793	44	17	3027

11. In Table 8.6 are displayed the $m_2 = 44$ second-phase point tallies presented in Palley & Horwitz (1961). These and the remaining data from the $m_1 = 44$ first-phase points are filed in Palley_and_Horwitz_1961.dat. The Bitterlich sampling was conducted with a $F_E = 9.372\,\text{ft}^2\text{ac}^{-1}$ prism. Use these data to verify the estimates of Palley and Horwitz of $\hat{\tau}_{y,\text{rat,ds}} = 24{,}033$ bd ft, with a relative standard error of 5.8%.

12. If one treats just the $m_2 = 44$ points from Palley and Horwitz's data (see Exercise 11.) as a single-phase sample, one gets $\hat{\tau}_{y\pi,\text{rep}} = 24{,}033$ bd ft but with a greater standard error than that of $\hat{\tau}_{y,\text{rat,ds}}$. Explain the features of these particular data that results in $\hat{\tau}_{y,\text{rat,ds}} = \hat{\tau}_{y\pi,\text{rep}}$. Also, explain why their standard errors differ, despite this.

CHAPTER 9

Line Intersect Sampling

9.1 Introduction

Line intersect sampling (LIS) is a form of pps sampling in which the sampling unit consists of a line or transect. Also widely known as line intercept sampling, LIS was popularized in forestry and ecology literature by Canfield (1941) in a *Journal of Forestry* article concerned with estimating the density of range vegetation. It is clear, however, that the use of a line as the sampling unit was not novel with this exposition; see, for example, Rosiwal (1898), Clements (1905), Schumacher & Bull (1932), and Bauer (1936). Canfield (1941, p. 388) provided the statistical basis for estimating density "through randomization in the locations of the sampling units." Within the natural resources literature, alone, the range of application of LIS has been very broad, having been used, for example, for the estimation of boundary length, diversity indices, fuelwood loading, plant canopy and area coverage, forest gap size, root length, logging residue, and standing basal area. Presently, LIS is used widely to estimate the abundance of coarse woody debris (CWD) and woody detritus on the forest floor.

Despite the similarity in name and overlap in vocabulary, LIS is not the same as line transect sampling, which is widely used to sample wildlife, fisheries and other mobile populations. For a description of line transect sampling, consult Thompson (2002, Chapter 17) or Seber (1982, Chapters 2, 12).

The large literature on LIS in the applied natural sciences is a bit inconsistent because some authors, e.g., van Wagner (1968), have assumed a Poisson probability model for the distribution of elements in the population of interest. This differs from the design-based approach taken in this book, wherein randomness is insinuated only by the selection of units into the sample, not the realization of the population itself (see Gregoire (1998) for further elaboration of this point). The inconsistencies appear in works which incorrectly claim that the Poisson model must hold in order to permit valid inference. In the design-based framework for statistical inference, this claim is patently incorrect. None of the results which we report for LIS in this chapter rely on any assumption about the location or orientation of the population elements within the region of interest.

While most of the literature on LIS considers transects consisting of a single straight line with fixed length, there has been interest also in transects consisting of more than one segment. Howard & Ward (1972) used two transects joined perpendicularly together in an ell shape as a way to sample elements more robustly. We present LIS in detail for singly segmented transects first, and then we turn our attention to multiply-segmented transects later in the chapter. We also consider the

ASIDE: In some respects LIS is related to a problem posed in 1733 by George Louis Leclerc, a respected French naturalist of the eighteenth century, better known as the Comte de Buffon. The Buffon Needle Problem is widely regarded as the first problem dealing with geometrical probability, according to Uspensky (1937, p. 251). In it, a needle of length l is dropped onto a surface marked with parallel lines spaced a distance d apart. The problem posed by Buffon was to determine the probability that the randomly dropped needle will cross one of the marked lines. Its solution is $p = 2l/\pi d$.

If the needle is thrown m times and $t_{ks} = 1$ if the sth thrown needle intersects a line, $t_{ks} = 0$ otherwise, then t_{ks} is a Bernoulli random variable with probability of success p. A count of m_0 intersections from $m \geq m_0$ throws is a binomial random variable: $m_0 \sim \text{Bin}(m, p)$. Because $\hat{p} = m_0/m$ unbiasedly estimates p, an nearly unbiased estimator of π is $\hat{\pi} = 2l/\hat{p}d$. Lazzerini (1901) conducted an experiment in which the needle was tossed $m = 3408$ times, and he estimated the value of π to within 3×10^{-7}. For further reading about the Buffon needle problem and related matters, consult Perlman & Wichura (1975), Watson (1978), and references cited therein.

case of random length, parallel transects emanating from a baseline and extending to the opposite boundary of the survey region. A Monte Carlo integration approach to LIS is described in Chapter 10.

9.2 LIS with straight-line transects

9.2.1 Sampling protocol

We suppose that within a region $\mathcal{A}$ there are N disjoint elements that constitute the population of interest. As in the previous chapters, the horizontal area of $\mathcal{A}$ is A, and the number of elements comprising the population, N, generally is unknown even at the conclusion of sampling.

Sampling locations, $(x_s, z_s), s = 1, \ldots, m$ are established on $\mathcal{A}$, each of which will serve as the mid-point or the end-point of a transect of length L. To avoid the use of conversion factors in the formulas which follow, let L and A have compatible units of measure in the sense that if L is expressed in m then all other measures of length are expressed also in m, and that A and all other expressions of area have units of m^2. In like manner, volume is in m^3, and so on.

Let the orientation of the sth transect with respect to some reference direction $\theta = 0$ be indicated by $\theta_s, s = 1, \ldots, m$. The sample selected by the sth transect consists of all population elements, $\mathcal{U}_k$, that are completely intersected by the transect, in the sense that the transect crosses the horizontal projection of $\mathcal{U}_k$ onto the floor of $\mathcal{A}$. For example, in the motivating application discussed by Canfield (1941), a plant was selected into the sample if the projection of its canopy onto $\mathcal{A}$ was crossed by the transect. Some elements that are partially intersected are also admitted into the sample according to the protocol established below.

a)

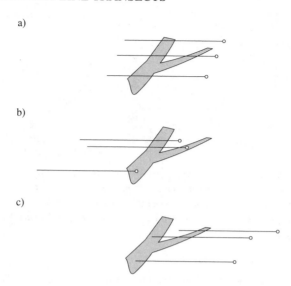

b)

c)

Figure 9.1 *Intersection of a non-convex population element, $\mathcal{U}_k$. In each transect, ○ symbolizes (x_s, z_s), the starting point. a) Examples of complete intersection; b) examples of partial intersection with the back end of the transect; c) examples of partial intersection by the front end of the transect.*

Kaiser (1983) distinguished between full and partial intersections in a manner that is suitable when all $\mathcal{U}_k$ have rather simple shapes. To prescribe a LIS sampling protocol suitable for elements that may have non-convex shapes, multiple lobes, and perhaps interior voids or cavities (e.g., as shown in Figure 9.7 on page 302), it is necessary to distinguish complete from partial intersections in a more exacting fashion, which then may be carried over without alteration to transects comprising multiple segments.

We first consider the case where the transect emanates from (x_s, z_s) and is oriented in direction θ_s, as in Figure 9.1. $\mathcal{U}_k$ is completely intersected only if the transect intersects its boundary as many times as a coincident line of infinite length. This follows the definition established by Affleck et al. (2005). In Figure 9.1, the boundary of $\mathcal{U}_k$ may be crossed by the transect as many as four times. Yet only in Figure 9.1a do the transects completely intersect $\mathcal{U}_k$. The intersections depicted in Figure 9.1b and Figure 9.1c are partial intersections. By convention, (x_s, z_s) defines the starting point or 'back end' of the transect. Naturally, the other end of the transect is called the 'front end.' For partial intersection of $\mathcal{U}_k$ by either end, the transect necessarily must cross the boundary of its projection onto $\mathcal{A}$ at least once, yet not so many times as a complete intersection. Examples of partial intersections by the back end of the transect are displayed in Figure 9.1b; partial intersections by the front end are shown Figure 9.1c.

In the topmost example in Figure 9.1b, where (x_s, z_s) is situated between two segments of a forked element, the intersection is partial, even though the starting

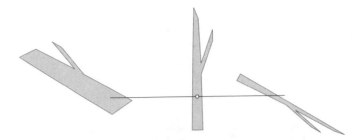

Figure 9.2 *Complete and partial intersections of elements, where* (x_s, z_s) *is the midpoint* (○) *of the straight-line transect. The left element is partially intersected by the front end of a transect segment, and the right element is completely intersected by the other segment. The center element is partially intersected by the back ends of both segments, which meet at the midpoint,* (x_s, z_s).

point, (x_s, z_s), is not located within the projection of $\mathcal{U}_k$ onto $\mathcal{A}$. The intersection meets our definition of a partial intersection because an extension of the transect in the direction $\theta_s + \pi$ (i.e., in the opposite direction of θ_s) would intersect two more boundaries of $\mathcal{U}_k$.

For straight-line transects, where (x_s, z_s) serves as the midpoint, as in Figure 9.2, the notion of a complete intersection by the transect remains unchanged. For partial intersections, it is useful to think of the transect as consisting of two segments joined at (x_s, z_s). Each segment has a front end and a back end, and the back ends of the two segments meet at the starting point, (x_s, z_s). Thus, we may have a partial intersection of an element by the front end of either segment or by both back ends simultaneously. The left element in Figure 9.2 is partially intersected by the front end of a segment; the right element is completely intersected by the other segment; and the center element is partially intersected by both back ends simultaneously.

Our sampling protocol is the same as that proposed by Affleck et al. (2005), namely $\mathcal{U}_k$ is selected into the sample by the sth transect if $\mathcal{U}_k$ is intersected completely or if it is intersected partially by the front end of the transect or by the front end of a segment of the transect. By contrast, an element, $\mathcal{U}_k$, is not selected if it is partially intersected by the back end of a transect or by the back end of a transect

> ASIDE: In some of the literature on LIS, the elements of the population are termed 'particles,' a terminology that has been in use at least since Lucas & Seber (1977). Adoption of such a neutral moniker is tacit recognition that "particles may represent plants, shrubs, tree crowns, nearly stationary animals, animal dens or signs, roads, logs, forest debris, particles on a microscope slide" (Kaiser 1983, p. 966). For sake of consistency in this book, we shall continue to refer to the individuals comprising a discrete population as elements.

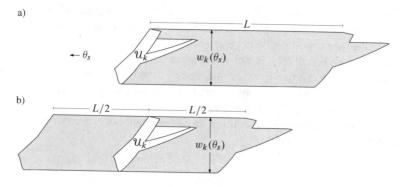

Figure 9.3 *a) Inclusion zone of* $\mathcal{U}_k$ *(gray), where a straight-line transect starts at* (x_s, z_s) *and is oriented in the direction* θ_s *(see Figure 9.1). b) Inclusion zone for the same element where* (x_s, z_s) *is the midpoint of the transect (see Figure 9.2). In both (a) and (b), the element itself, and the white area between the fork is not part of inclusion zone, i.e., if* (x_s, z_s) *occurs here, the element is not selected, since intersections would be partial intersections by the back end(s) of the transect or transect segments.*

segment that emanates from the starting point, (x_s, z_s). Thus, in Figure 9.2, the left and right elements are selected by the transect, but the center element is not.

For a nonconvex element, $\mathcal{U}_k$, such as those shown in Figures 9.1 and 9.2, it is possible for a transect to intersect two or more branches, lobes, or other types of connected sections. Regardless of the number of sections of a single, connected element that are intersected, the element is considered to be intersected but once.

With the proposed sampling protocol, there is nothing to prevent $\mathcal{U}_k$ from being selected from two or more independently placed transects.

9.2.2 Estimation conditioned on transect orientation

When following the first of the two sampling protocols described above, the inclusion zone for a forked element is shown in Figure 9.3a, whereas the inclusion zone for the same element corresponding to the second protocol is shown in Figure 9.3b. Although the inclusion zones differ, they are identical in some important aspects. Each has the same orientation, θ_s, as that of the transect itself. Except near the ends, both inclusion zones have a constant width, $w_k(\theta_s)$, identical to the width of $\mathcal{U}_k$ (expressed in units identical to those used to express L) in the direction perpendicular to θ_s. The area of both inclusion zones is $a_k = w_k(\theta_s)L$.

Regardless of which of the two sampling protocols is used, the (conditional) inclusion probability of $\mathcal{U}_k$ is

$$\pi_k(\theta_s) = \frac{a_k}{A} \qquad (9.1a)$$

$$= \frac{w_k(\theta_s)L}{A} \qquad (9.1b)$$

The notation $w_k(\theta_s)$ and $\pi_k(\theta_s)$ makes explicit the dependence of the element's width and inclusion probability on the transect orientation. In statistical parlance, both measures are conditionally dependent on θ_s, which implies that the conditional inclusion probability of $\mathcal{U}_k$ is not constant unless all transects have a common orientation. That is, for two transects indexed by s and s', respectively, if $\theta_s \neq \theta_{s'}$, then $w_k(\theta_s) \neq w_k(\theta_{s'})$ unless by coincidence or unless $\mathcal{U}_k$ is circular. For that reason, $\pi_k(\theta_s) \neq \pi_k(\theta_{s'})$, generally speaking.

With the above inclusion probabilities, the HT estimator of τ_y based on the sample selected on the sth transect is

$$\hat{\tau}_{y\pi s}^c = \sum_{\mathcal{U}_k \in \mathbb{L}_s} \frac{y_k}{\pi_k(\theta_s)} \tag{9.2a}$$

$$= \frac{A}{L} \sum_{\mathcal{U}_k \in \mathbb{L}_s} \frac{y_k}{w_k(\theta_s)}, \tag{9.2b}$$

where the superscripted 'c' indicates that the inclusion probability of $\mathcal{U}_k$ and the resultant estimate are conditional on the orientation, θ_s, of the sth transect. The estimator based on m replicated transects is

$$\hat{\tau}_{y\pi,\mathrm{rep}}^c = \frac{1}{m} \sum_{s=1}^{m} \hat{\tau}_{y\pi s}^c. \tag{9.3}$$

Being able to estimate τ_y conditionally on the set of transect orientations used in the sampling can be quite convenient in situations where the elements of the population are arrayed on $\mathcal{A}$ with a predominant orientation. This situation occurs sometimes when sampling tree tips and other logging residue, as in Warren & Olsen (1964). In such situations, a feasible LIS design is one which stipulates transect orientations roughly perpendicular to the general orientation of the residue.

Example 9.1

LIS has been used to estimate the number and the aggregate size of gaps in a forest canopy. In Figure 9.4 we show a small section of a 64 ha forest with several gaps, one of which is intersected by the fifth transect of a survey with $m = 40$ transects, each of which is 20 m long. In the direction perpendicular to the transect, the width of the intersected gap is 2.35. The estimated number of gaps in the 64 ha forested region from this sample transect is

$$\hat{\tau}_{y\pi s}^c = \frac{640,000}{20}\left(\frac{1}{2.35}\right) = 13,617$$

or approximately 213 gaps ha^{-1}.

Example 9.2

To measure the area of each gap in Example 9.1 is a time-consuming endeavor in practice. Without going into details of field procedures to take such measure-

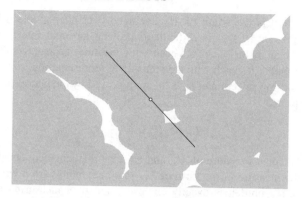

Figure 9.4 *A straight line transect intersects a gap in the forest canopy.*

ments, the area of the gap intersected in the preceding example was determined to be 8.3 m^2. The estimated area of the forest with gaps in the canopy is

$$\hat{\tau}^c_{y\pi s} = \frac{640,000}{20} \left(\frac{8.3}{2.35} \right) = 113,021$$

or approximately 1766 m^2 ha^{-1}.

Example 9.3

As part of a project to estimate the aboveground live, woody biomass of the state of Delaware, U.S.A., Nelson et al. (2004) established $m = 142$ transects, 40 m long, throughout the state. A tree was selected into the sample if the projection of its canopy was intersected by the transect. The diameter of this projected area perpendicular to the transect, $w_k(\theta_s)$, was measured, as was the diameter of the tree bole at breast height. The latter served as input to an allometric regression equation which provided a model-based prediction of the tree's aboveground biomass, y_k.

9.2.3 Unconditional estimation

Suppose that the orientation of each transect had been chosen uniformly at random from the interval 0 to π. That is, $\theta_s \sim U[0, \pi]$, $s = 1, \dots, m$. The probability of intersecting $\mathcal{U}_k$ by a single transect that may be located anywhere within $\mathcal{A}$ and have any orientation is

$$\pi_k = E_\theta \left[\pi_k(\theta_s) \right] \qquad (9.4a)$$

$$= \frac{L}{A} E_\theta \left[w_k(\theta_s) \right], \qquad (9.4b)$$

where $E_\theta [w_k(\theta_s)]$ is the width of $\mathcal{U}_k$ averaged over all possible orientations of a transect. Unlike $\pi_k(\theta_s)$, it is difficult to provide an areal interpretation to π_k corresponding to an inclusion zone, because it constitutes an average of a continuum of inclusion probabilities each with a distinct areal representation.

Following uniform selection of θ_s, a theorem proved by the famous mathematician Cauchy ensures that

$$E_\theta [w_k(\theta_s)] = \frac{c_k}{\pi}, \tag{9.5}$$

where c_k is the convex closure of $\mathcal{U}_k$ and π is the mathematical constant. In other words, c_k is the measure of the girth of $\mathcal{U}_k$ that one obtains from wrapping a cord tightly around $\mathcal{U}_k$ until it is exactly enclosed, and then measuring the length of the cord. If $\mathcal{U}_k$ is not actually on the floor of $\mathcal{A}$, then this measurement of convex closure takes place on its horizontal projection onto the floor. To those familiar with measuring the diameter of a tree bole with a pair of mechanical calipers and with a girth tape, the above result implies that the measurement provided by a tape is the expected value of a random calipering of bole diameter, as discussed by Matérn (1956, p. 6).

Therefore, an alternative to $\hat{\tau}^c_{y\pi s}$ in (9.2) is

$$\hat{\tau}^u_{y\pi s} = \sum_{\mathcal{U}_k \in \mathbb{L}_s} \frac{y_k}{\pi_k} \tag{9.6a}$$

$$= \frac{\pi A}{L} \sum_{\mathcal{U}_k \in \mathbb{L}_s} \frac{y_k}{c_k}. \tag{9.6b}$$

Presuming a replicated sample of m transects, the unconditional estimator of τ_y is

$$\hat{\tau}^u_{y\pi,\text{rep}} = \frac{1}{m} \sum_{s=1}^{m} \hat{\tau}^u_{y\pi s}. \tag{9.7}$$

Example 9.4

Suppose that the population of interest comprises logs lying flat on the forest floor. Each log, $\mathcal{U}_k$, has been trimmed of its branches, and tapers from a diameter of D_k at its larger end to d_k at its smaller end, and its length (on the slant) is l_k. Therefore $c_k \approx 2l_k + d_k + D_k$, so that $E_\theta [w_k(\theta_s)] = (2l_k + d_k + D_k)/\pi$. Figure 9.5 depicts a typical log being intersected by a transect. From the mth transect alone, τ_y is estimated by

$$\hat{\tau}^u_{y\pi s} = \sum_{\mathcal{U}_k \in \mathbb{L}_s} \frac{y_k}{\pi_k} \tag{9.8a}$$

$$= \frac{\pi A}{L} \sum_{\mathcal{U}_k \in \mathbb{L}_s} \frac{y_k}{2l_k + d_k + D_k}. \tag{9.8b}$$

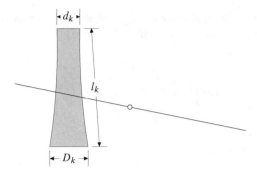

Figure 9.5 *A straight line transect intersects a log with diameters d_k, D_k, and length l_k.*

When log diameters are very much smaller than their lengths, a close approximation to (9.8) is

$$\hat{\tau}^u_{y\pi s} = \frac{\pi A}{2L} \sum_{u_k \in \mathbb{L}_s} \frac{y_k}{l_k},$$ (9.9)

which has appeared in de Vries (1973, p. 7). If the log is tilted at an angle φ to the horizontal, then l_k in (9.9) ought to be replaced by $l_k \cos \varphi_k$.

A further approximation appears when y_k is the volume of the kth log:

$$\hat{\tau}^u_{y\pi s} = \frac{\pi^2 A}{8L} \sum_{u_k \in \mathbb{L}_s} \bar{d}^2_k,$$ (9.10)

in which y_k is approximated by the cylindrical volume $(\pi/4)\bar{d}^2_k l_k$ and $\bar{d}_k$ is the diameter of the cylinder. See, e.g., van Wagner (1968).

Example 9.5

Refer to the preceding example, but suppose that the population parameter of interest was the total length of logs in $\mathcal{A}$. That is $y_k = l_k$, using the notation of Example 9.4. Therefore, using (9.9), it is apparent that

$$\hat{\tau}^u_{y\pi s} = \frac{\pi A n_s}{2L},$$ (9.11)

where n_s is the count of the number of logs intersected by the sth transect. For this purpose, no measurement other than a simple count of the number of intersected logs is required, whereas if the conditional estimator were used, then both l_k and $w_k(\theta_s)$ would need to be measured.

Example 9.6

To unconditionally estimate the number of logs on $\mathcal{A}$, set $y_k = 1$ in (9.9). In this situation, the conditional estimator may be preferable, providing that $w_k(\theta_s)$ is easier to measure than l_k and possibly D_k and d_k.

9.2.4 Estimator variance and variance estimators

The variance of $\hat{\tau}^c_{y\pi s}$ is

$$V\left[\hat{\tau}^c_{y\pi s}\right] = \sum_{k=1}^{N} y_k^2 \left(\frac{1 - \pi_k(\theta_s)}{\pi_k(\theta_s)}\right)$$

$$+ \sum_{k=1}^{N} \sum_{\substack{k'\neq k \\ k'=1}}^{N} y_k y_{k'} \left(\frac{\pi_{kk'}(\theta_s) - \pi_k(\theta_s)\pi_{k'}(\theta_s)}{\pi_k(\theta_s)\pi_{k'}(\theta_s)}\right) \tag{9.12a}$$

$$= V^c, \text{ say,} \tag{9.12b}$$

where $\pi_{kk'}(\theta_s)$ is the joint conditional probability of including both $\mathcal{U}_k$ and $\mathcal{U}_{k'}$ from the sth transect. Consequently,

$$V\left[\hat{\tau}^c_{y\pi,\mathrm{rep}}\right] = \frac{1}{m^2} \sum_{s=1}^{m} V\left[\hat{\tau}^c_{y\pi s}\right] \tag{9.13a}$$

$$= \frac{V^c}{m}, \tag{9.13b}$$

because $\sum_{s=1}^{m} V\left[\hat{\tau}^c_{y\pi s}\right] = mV^c$. $V\left[\hat{\tau}^c_{y\pi,\mathrm{rep}}\right]$ is estimated unbiasedly by

$$\hat{v}\left[\hat{\tau}^c_{y\pi,\mathrm{rep}}\right] = \frac{1}{m(m-1)} \sum_{s=1}^{m} \left(\hat{\tau}^c_{y\pi s} - \hat{\tau}^c_{y\pi,\mathrm{rep}}\right)^2. \tag{9.14}$$

The variance of $\hat{\tau}^u_{y\pi s}$ follows in similar fashion:

$$V\left[\hat{\tau}^u_{y\pi s}\right] = \sum_{k=1}^{N} y_k^2 \left(\frac{1 - \pi_k}{\pi_k}\right) + \sum_{k=1}^{N} \sum_{\substack{k'\neq k \\ k'=1}}^{N} y_k y_{k'} \left(\frac{\pi_{kk'} - \pi_k \pi_{k'}}{\pi_k \pi_{k'}}\right) \tag{9.15a}$$

$$= V^u, \tag{9.15b}$$

where $\pi_{kk'}$ is the joint unconditional probability of including both $\mathcal{U}_k$ and $\mathcal{U}_{k'}$ from the sth transect. This leads to

$$V\left[\hat{\tau}^u_{y\pi,\mathrm{rep}}\right] = \frac{1}{m^2} \sum_{s=1}^{m} V\left[\hat{\tau}^u_{y\pi s}\right] \tag{9.16a}$$

$$= \frac{V^u}{m}, \tag{9.16b}$$

ASIDE: Long before the probabilistic and statistical basis of LIS was articulated, sampling with lines as the sampling unit appeared in various fields of
practical endeavor. For example, both Schumacher & Bull (1932) and Osborne
(1942) investigated the sampling error of parallel line surveys to estimate the
land area of each of several cover types in a region, the overall area of which was
known. The length of the line running through each cover type was measured
and summed. The proportion of the total length of all transect lines combined
that were located in a cover type served as an estimate of the proportional area
of the cover type. Sengupta (1954) investigated a similar use of line transects to
estimate the area of rice paddy in India.

which is estimated unbiasedly by

$$\hat{v}\left[\hat{\tau}^{u}_{y\pi,\mathrm{rep}}\right] = \frac{1}{m(m-1)} \sum_{s=1}^{m} \left(\hat{\tau}^{u}_{y\pi s} - \hat{\tau}^{u}_{y\pi,\mathrm{rep}}\right)^2. \tag{9.17}$$

9.2.5 Interval estimation

An estimated $(1-\alpha)100\%$ confidence interval for τ_y proceeds in the usual fashion,
with a t-value appropriate for $m-1$ degrees of freedom. Therefore an estimated
interval based on the conditional estimator is

$$\hat{\tau}^{c}_{y\pi,\mathrm{rep}} \pm t_{m-1}\sqrt{\hat{v}\left[\hat{\tau}^{c}_{y\pi,\mathrm{rep}}\right]} \tag{9.18a}$$

or

$$\hat{\tau}^{c}_{y\pi,\mathrm{rep}} \pm \frac{t_{m-1}\sqrt{\hat{v}\left[\hat{\tau}^{c}_{y\pi,\mathrm{rep}}\right]}}{\hat{\tau}^{c}_{y\pi,\mathrm{rep}}}100\%. \tag{9.18b}$$

One based on the unconditional estimator is

$$\hat{\tau}^{u}_{y\pi,\mathrm{rep}} \pm t_{m-1}\sqrt{\hat{v}\left[\hat{\tau}^{u}_{y\pi,\mathrm{rep}}\right]} \tag{9.19a}$$

or

$$\hat{\tau}^{u}_{y\pi,\mathrm{rep}} \pm \frac{t_{m-1}\sqrt{\hat{v}\left[\hat{\tau}^{u}_{y\pi,\mathrm{rep}}\right]}}{\hat{\tau}^{u}_{y\pi,\mathrm{rep}}}100\%. \tag{9.19b}$$

9.2.6 Conditional vs. unconditional estimation

Unconditional estimation of τ_y is a viable option in the situation where each transect
is oriented uniformly at random and independently of the orientation of any other

transect, as presented above. It is also appropriate for the situation where a single orientation is selected in this fashion and applied to all m transects. It perhaps is obvious that unconditional estimation of τ_y is not appropriate when transect orientations have not been selected uniformly at random. However, it may be less obvious that one may choose to estimate τ_y conditionally on transect orientation that were assigned even in the situation where these orientations had been selected randomly. In other words, following random orientation of the transects one is free to base estimation and inference on the set of orientations that actually were used, rather than considering the universe of orientations that could have been selected but were not. An important nuance of this result is that one usually must know in advance of sampling whether estimation will proceed conditionally on transect orientation, or not, because this affects whether $w_k(\theta_s)$ must be measured on each intersected element, or whether c_k must be measured.

If V^c in (9.12b) were known to be smaller than V^u in (9.15b), it would be reasonable, perhaps, to favor conditional estimation. In situations where the $\mathcal{U}_k$ do not have an orientation preponderantly in one direction, it would be difficult to discern whether $V^c \leq V^u$, or not. Therefore it is impossible to provide any general recommendation based on the relative precision of conditional versus unconditional estimation. From a practical viewpoint, it might be advisable to choose the method which imposes the lesser burden of field measurement.

9.3 Unit area estimators

Estimators of τ_y prorated to a unit area basis proceeds along the same lines as in the previous two chapters. To establish notation explicitly, the conditional estimator of $\lambda_y = \tau_y/A$ based solely on the sample collected on the sth transect is

$$\hat{\lambda}^c_{y\pi s} = \frac{1}{A}\,\hat{\tau}^c_{y\pi s} \tag{9.20a}$$

$$= \frac{1}{L} \sum_{\mathcal{U}_k \in \mathbb{L}_s} \frac{y_k}{w_k(\theta_s)}. \tag{9.20b}$$

As in the two previous chapters, the area of $\mathcal{A}$ is not needed to estimate the magnitude of y on a unit area basis.

The corresponding estimator of λ_y based on a replicated sample of m transects is

$$\hat{\lambda}^c_{y\pi,\mathrm{rep}} = \frac{1}{A}\,\hat{\tau}^c_{y\pi,\mathrm{rep}} \tag{9.21a}$$

$$= \frac{1}{m} \sum_{s=1}^{m} \hat{\lambda}^c_{y\pi s}. \tag{9.21b}$$

Because there often is special interest in estimating the number of elements per unit area on $\mathcal{A}$, we treat this special case explicitly. The conditional estimator of

$\lambda_N = N/A$ from the sample on the sth transect is

$$\hat{\lambda}^c_{\pi s} = \frac{\hat{N}^c_{\pi s}}{A} \tag{9.22a}$$

$$= \frac{1}{A} \sum_{\mathcal{U}_k \in \mathbb{L}_s} \frac{1}{\pi_k(\theta_s)} \tag{9.22b}$$

$$= \frac{1}{L} \sum_{\mathcal{U}_k \in \mathbb{L}_s} \frac{1}{w_k(\theta_s)}, \tag{9.22c}$$

where $\hat{N}^c_{\pi s}$ is just $\hat{\tau}^c_{y\pi s}$ as in (9.2a) with $y_k = 1$ for all $\mathcal{U}_k$.
The corresponding unconditional estimator is

$$\hat{\lambda}^u_{\pi s} = \frac{\hat{N}^u_{\pi s}}{A} \tag{9.23a}$$

$$= \frac{1}{A} \sum_{\mathcal{U}_k \in \mathbb{L}_s} \frac{1}{\pi_k} \tag{9.23b}$$

$$= \frac{\pi}{L} \sum_{\mathcal{U}_k \in \mathbb{L}_s} \frac{1}{c_k}. \tag{9.23c}$$

Example 9.7

Iles (2003, p. 412–413) shows how one can deduce that a single log of CWD that has length $l_k = 4\,\text{ft}$ implies a density of 142.55 logs per acre when estimating unconditionally over all transect orientations. His solution involved estimating the length per acre of CWD provided by a sample of this 4 ft log on a single 120 ft transect, and then dividing this estimated total length by the length of the log. Our approach yields an identical result by estimating N directly and then prorating to a per acre basis.

Using $\hat{\lambda}^u_{\pi s}$ from (9.23) we calculate $c_k \approx 2l_k = 8\,\text{ft}$ so that $\hat{\lambda}^u_{\pi s} = 0.003272\,\text{logs ft}^{-2}$ or $142.55\,\text{logs ac}^{-1}$. The distinction between Iles' approach to the problem and ours is more apparent than real: it is chiefly in the manner in which the estimators are derived and presented. A comparison of $\hat{\lambda}^u_{\pi s}$ in (9.23c), above, to Iles equation (p. 413) reveals that they are identical algebraically, but for the conversion factor of $43560\,\text{ft}^2\,\text{ac}^{-1}$.

The variance of $\hat{\lambda}^c_{y\pi s}$ is

$$V\left[\hat{\lambda}^c_{y\pi s}\right] = \frac{1}{A^2} V^c, \tag{9.24}$$

and that of $\hat{\lambda}^c_{y\pi,\text{rep}}$ is

$$V\left[\hat{\lambda}^c_{y\pi,\text{rep}}\right] = \frac{1}{A^2} V\left[\hat{\tau}^c_{y\pi,\text{rep}}\right]. \tag{9.25}$$

$V\left[\hat{\lambda}^{c}_{y\pi,\text{rep}}\right]$ is estimated unbiasedly by

$$\hat{v}\left[\hat{\lambda}^{c}_{y\pi,\text{rep}}\right] = \frac{1}{A^2}\,\hat{v}\left[\hat{\tau}^{c}_{y\pi,\text{rep}}\right]. \tag{9.26}$$

An estimated $(1 - \alpha)100\%$ confidence interval for λ_y based on the conditional estimator is

$$\hat{\lambda}^{c}_{y\pi,\text{rep}} \pm t_{m-1}\sqrt{\hat{v}\left[\hat{\lambda}^{c}_{y\pi,\text{rep}}\right]} \tag{9.27a}$$

or

$$\hat{\lambda}^{c}_{y\pi,\text{rep}} \pm \frac{t_{m-1}\sqrt{\hat{v}\left[\hat{\lambda}^{c}_{y\pi,\text{rep}}\right]}}{\hat{\lambda}^{c}_{y\pi,\text{rep}}}100\%. \tag{9.27b}$$

In exactly the same fashion the unconditional estimator of $\lambda_y = \tau_y/A$ based solely on the sample collected on the sth transect is

$$\hat{\lambda}^{u}_{y\pi s} = \frac{1}{A}\,\hat{\tau}^{u}_{y\pi s} \tag{9.28a}$$

$$= \frac{1}{L}\sum_{u_k \in \mathbb{L}_s}\frac{y_k}{w_k(\theta_s)}. \tag{9.28b}$$

The corresponding estimator of λ_y based on a replicated sample of m transects is

$$\hat{\lambda}^{u}_{y\pi,\text{rep}} = \frac{1}{A}\,\hat{\tau}^{u}_{y\pi,\text{rep}} \tag{9.29a}$$

$$= \frac{1}{m}\sum_{s=1}^{m}\hat{\lambda}^{u}_{y\pi s}. \tag{9.29b}$$

The variance of $\hat{\lambda}^{u}_{y\pi s}$ is

$$V\left[\hat{\lambda}^{u}_{y\pi s}\right] = \frac{1}{A^2}\,V^{u}, \tag{9.30}$$

and that of $\hat{\lambda}^{u}_{y\pi,\text{rep}}$ is

$$V\left[\hat{\lambda}^{u}_{y\pi,\text{rep}}\right] = \frac{1}{A^2}\,V\left[\hat{\tau}^{u}_{y\pi,\text{rep}}\right]. \tag{9.31}$$

$V\left[\hat{\lambda}^{u}_{y\pi,\text{rep}}\right]$ is estimated unbiasedly by

$$\hat{v}\left[\hat{\lambda}^{u}_{y\pi,\text{rep}}\right] = \frac{1}{A^2}\,\hat{v}\left[\hat{\tau}^{u}_{y\pi,\text{rep}}\right]. \tag{9.32}$$

An estimated $(1 - \alpha)100\%$ confidence interval for λ_y based on the unconditional estimator is

$$\hat{\lambda}^{u}_{y\pi,\text{rep}} \pm t_{m-1}\sqrt{\hat{v}\left[\hat{\lambda}^{u}_{y\pi,\text{rep}}\right]} \tag{9.33a}$$

or

$$\hat{\lambda}^{\mathrm{u}}_{y\pi,\mathrm{rep}} \pm \frac{t_{m-1}\sqrt{\hat{v}\left[\hat{\lambda}^{\mathrm{u}}_{y\pi,\mathrm{rep}}\right]}}{\hat{\lambda}^{\mathrm{u}}_{y\pi,\mathrm{rep}}} 100\%. \qquad (9.33b)$$

9.4 Estimation with an auxiliary variate

The estimators of τ_y presented so far in this chapter are special cases of a more general estimator that allows for the use of an auxiliary variate, $q_k(\theta_s)$, whose value may depend on the orientation of the transect. Usually we choose an auxiliary variate that allows us to avoid measuring both y_k and $w_k(\theta_s)$. Whenever y_k and $w_k(\theta_s)$ are more difficult or costly to measure than $q_k(\theta_s)$, this has potential to improve estimation of τ_y either by increasing the precision of estimation or lessening the burden of measurement.

The choice of the auxiliary variate will depend on the application. The dimensions of $q_k(\theta_s)$ ordinarily are the dimensions of $y_k/w_k(\theta_s)$. For example, if y_k is the volume of $\mathcal{U}_k$, then we choose $q_k(\theta_s)$ to be the cross-sectional area where $\mathcal{U}_k$ is intersected by the transect; if y_k is the coverage area of $\mathcal{U}_k$, then we choose $q_k(\theta_s)$ to be the length of the intersection through $\mathcal{U}_k$; and, if y_k is the length of $\mathcal{U}_k$, then $q_k(\theta_s)$ is dimensionless, and we may find it expedient to choose $q_k(\theta_s) = 1$.

Defining the binary-valued random indicator variable

$$t_{ks} = \begin{cases} 1, & \text{if } \mathcal{U}_k \text{ is included into the sample for the } s\text{th transect,} \\ 0, & \text{otherwise,} \end{cases}$$

the conditional estimator of τ_y which utilizes the auxiliary variate is

$$\hat{\tau}^{\mathrm{c}}_{yq,\mathrm{rep}} = \frac{1}{m} \sum_{s=1}^{m} \hat{\tau}^{\mathrm{c}}_{yqs} \qquad (9.34)$$

where

$$\hat{\tau}^{\mathrm{c}}_{yqs} = \sum_{\mathcal{U}_k \in \mathbb{L}_s} \frac{q_k(\theta_s) y_k}{E[t_{ks} q_k(\theta_s) \mid \theta_s]}. \qquad (9.35)$$

When $q_k(\theta_s) = 1$ for all $\mathcal{U}_k$, then $\hat{\tau}^{\mathrm{c}}_{yqs} = \hat{\tau}^{\mathrm{c}}_{y\pi s}$, because in this special case $E[t_{ks} q_k(\theta_s) \mid \theta_s] = E[t_{ks} \mid \theta_s] = \pi_k(\theta_s)$.

The corresponding unconditional estimator of τ_y is

$$\hat{\tau}^{\mathrm{u}}_{yq,\mathrm{rep}} = \frac{1}{m} \sum_{s=1}^{m} \hat{\tau}^{\mathrm{u}}_{yqs} \qquad (9.36)$$

where

$$\hat{\tau}^{\mathrm{u}}_{yqs} = \sum_{\mathcal{U}_k \in \mathbb{L}_s} \frac{q_k(\theta_s) y_k}{E[t_{sk} q_k(\theta_s)]}. \qquad (9.37)$$

Similar to the simplification for the conditional estimator when $q_k(\theta_s) = 1$ for all $\mathcal{U}_k$, here $\hat{\tau}^{\mathrm{u}}_{yqs} = \hat{\tau}^{\mathrm{u}}_{y\pi s}$, because $E[t_{ks} q_k(\theta_s)] = E[t_{ks}] = \pi_k$.

Both the conditional and unconditional estimators of τ_y presented in (9.35) and (9.37) were formulated by Kaiser (1983).

Example 9.8

Consider the case where τ_y is the collective area of all population elements projected onto $\mathcal{A}$, such as the aggregate canopy cover of a vegetative population or the area of canopy gaps considered in Example 9.2. Consider further letting $q_k(\theta_s)$ be the length of $\mathcal{U}_k$ along the line that is coincident with the sth transect but unlimited in length, in other words the line containing the transect. For this choice of $q_k(\theta_s)$, $E[t_{ks}q_k(\theta_s) \mid \theta_s]$ is L/A times the projected area of $\mathcal{U}_k$ (see §9.12.2). Therefore,

$$\hat{\tau}_{yqs}^{c} = \sum_{\mathcal{U}_k \in \mathbb{L}_s} \frac{q_k(\theta_s)y_k}{(Ly_k)/A} \tag{9.38a}$$

$$= \frac{A}{L} \sum_{\mathcal{U}_k \in \mathbb{L}_s} q_k(\theta_s). \tag{9.38b}$$

In other words, aggregate projected area, τ_y, can be estimated unbiasedly without having to measure the projected area of a single $\mathcal{U}_k$. By measuring intersection length, $q_k(\theta_s)$, instead, the need to measure both y_k and $w_k(\theta_s)$ is obviated because the conditional expected value of $t_{ks}q_k(\theta_s)$ in the denominator of $\hat{\tau}_{yqs}^{c}$ in (9.35) has a factor that is identically y_k of the numerator. The measurement of length along the transect certainly is simpler than the measurement of the area of an irregularly shaped canopy gap.

For the particular choice of area and intersection length for y_k and $q_k(\theta_s)$, respectively, the result that $E[t_{ks}q_k(\theta_s) \mid \theta_s] = Ly_k/A$ evidently does not depend on the orientation angle, θ_s. Therefore, $E[t_{ks}q_k(\theta_s)]$ in the denominator of $\hat{\tau}_{yqs}^{u}$ in (9.37) is identical to its conditional expected value: $E[t_{ks}q_k(\theta_s)] = E[t_{ks}q_k(\theta_s) \mid \theta_s] = Ly_k/A$. As a result, the two estimators of τ_y coincide for this choice of y_k and $q_k(\theta_s)$:

$$\hat{\tau}_{yqs}^{u} = \hat{\tau}_{yqs}^{c} \tag{9.39a}$$

$$= \frac{A}{L} \sum_{\mathcal{U}_k \in \mathbb{L}_s} q_k(\theta_s). \tag{9.39b}$$

Example 9.9

Extend the previous example by letting y_k be the volume of $\mathcal{U}_k$ and letting $q_k(\theta_s)$ be the area in the vertical plane that contains the sth transect. Again, because $E[t_{ks}q_k(\theta_s) \mid \theta_s] = E[t_{ks}q_k(\theta_s)] = Ly_k/A$, we obtain

$$\hat{\tau}^c_{yqs} = \hat{\tau}^u_{yqs} = \frac{A}{L} \sum_{\mathcal{U}_k \in \mathbb{L}_s} q_k(\theta_s). \tag{9.40}$$

In this example, measurement of the volume of $\mathcal{U}_k$ is obviated by the measurement of the cross-sectional area of $\mathcal{U}_k$ in the vertical plane.

Example 9.10

When dealing with a population of logs, where $\mathcal{U}_k$ tilts at an angle φ_k to the horizontal plane and is intersected at angle γ_k by the transect, the width of $\mathcal{U}_k$ perpendicular to the transect is $w_k(\theta_s) = l_k \cos\varphi_k \sin|\gamma_k|$, where l_k is the length of the central axis of $\mathcal{U}_k$ (i.e., the kth log). Let y_k be log volume, as in the previous example, but now define $q_k(\theta_s)$ as the area in the vertical plane perpendicular to the central axis of the log at the point where the central axis is intersected by the transect, i.e., $q_k(\theta_s) = \pi d_k^2/4$, where d_k is the log diameter. Then from Kaiser (1983),

$$E[t_{ks}q_k(\theta_s) \mid \theta_s] = \frac{Lw_k(\theta_s)y_k}{l_k A}.$$

The conditional estimator of the volume of this population of logs is

$$\hat{\tau}^c_{yqs} = \frac{A}{L} \sum_{\mathcal{U}_k \in \mathbb{L}_s} \frac{q_k(\theta_s)y_k l_k}{y_k w_k(\theta_s)} \tag{9.41a}$$

$$= \frac{A}{L} \sum_{\mathcal{U}_k \in \mathbb{L}_s} \frac{q_k(\theta_s)}{\cos\varphi_k \sin|\gamma_k|} \tag{9.41b}$$

$$= \frac{\pi A}{4L} \sum_{\mathcal{U}_k \in \mathbb{L}_s} \frac{d_k^2}{\cos\varphi_k \sin|\gamma_k|}. \tag{9.41c}$$

The corresponding unconditional estimator of aggregate volume is

$$\hat{\tau}^u_{yqs} = \frac{\pi^2 A}{8L} \sum_{\mathcal{U}_k \in \mathbb{L}_s} \frac{d_k^2}{\cos\varphi_k} \tag{9.42a}$$

or if tilt is, say, less than $10°$,

$$= \frac{\pi^2 A}{8L} \sum_{\mathcal{U}_k \in \mathbb{L}_s} d_k^2, \tag{9.42b}$$

as in van Wagner (1968).

Iles (2003, pp. 392-400) has a very insightful explanation of this estimator, (9.42a), of log volume, which we rewrite as

$$\hat{\tau}_{yqs}^{u} = A \left(\frac{1}{L} \sum_{\mathcal{U}_k \in \mathbb{L}_s} \frac{\pi d_k^2}{4 \cos \varphi_k} \right) \frac{\pi}{2}.$$

In the middle term, $\cos \varphi_k$ corrects for tilt, thereby accounting for the fact that the elliptical cross-sectional area of $\mathcal{U}_k$ is computed as $\pi d_k^2/4$. The sum of these tilt-corrected areas, when divided by L, is the average depth of wood above the transect. The term $\pi/2$ is the correction that applies as the result of averaging over all possible transect orientations in the unconditional estimator. Multiplying by A simply prorates a per-unit-area estimate to one that applies to the entire region $\mathcal{A}$.

Clearly, in this example, the measurement of the angle of intersection, γ_k, is unnecessary with the unconditional estimator, and necessary with the conditional estimator. With either estimator the tilt of the piece from the horizontal is required, even if $\mathcal{U}_k$ lays flat on the surface of sloping ground. For angles of tilt less than $10°$, the correction by $\cos \varphi_k$ is only about 2%, whereas at $20°$, it is about 8%.

Example 9.11

A transect may intersect the boundary of an element, $\mathcal{U}_k$, two or more times, depending on the shape of the element and where the element is intersected (see, e.g., Figure 9.1a). Suppose that the attribute of interest is the total length of the boundaries of all the elements in $\mathcal{A}$. In this application of LIS, it is convenient to use the number of intersections of boundaries by each transect as the auxiliary variate.

This auxiliary variate is particularly useful when we need to estimate the total length of roads, trails, or other linear features in a region, $\mathcal{A}$. The boundary of $\mathcal{A}$, together with, say, the roads within $\mathcal{A}$ may effectively divide or tessellate $\mathcal{A}$ into subregions, in which case an estimate of the total length of the joint boundaries of the elements is also an estimate the total length of roads in $\mathcal{A}$.

Hence, let the attribute of interest, τ_y, be the length of roads in $\mathcal{A}$. An unconditional estimator of the total length of boundary was derived by Kaiser (1983). Applying this estimator to the road-length problem,

$$\hat{\tau}_{yqs}^{u} = \frac{\pi A}{2L} n_s \tag{9.43}$$

where n_s is the total number of intersections of roads with the sth transect. This result may be deduced from the fact that, for this definition of n_s,

$$E\left[n_s\right] = \frac{2L}{\pi A} \tau_y.$$

Generally, the observed number of intersections of a straight-line transect of length L with an open or closed boundary of an element, $\mathcal{U}_k$, unbiasedly estimates the length of the boundary of $\mathcal{U}_k$. In this application, it is important that each intersection of the transect with $\mathcal{U}_k$ be counted, so that the definition of n_s in this example differs from its definition in Example 9.5 on page 287.

The relationship between the number of intersections of straight lines with curves in the plane has long been known: Smith & Guttman (1953) used (9.43) to estimate the total length of crystalline interface in a metallurgical cross-section; Matérn (1964) used it to estimate the aggregate length of forest roads; Newman (1966) used it to estimate the total length of plant root; Skidmore & Turner (1992) used it to estimate the total length of boundary of land-cover polygons delineated on a map; Iles (2003) used it to estimate the total length of coarse woody debris. Further reading about the relation between the number of intersections and the boundary length of the intersected object may be found in Ramaley (1969).

The conditional estimator of the total length of roads or boundaries, using $q_k(\theta_s)$ as the count of the number of intersections, is not useful because it requires an impractical measurement of a special type of projection of each $\mathcal{U}_k$ onto a line perpendicular to the transect.

Example 9.12

As a generalization of the previous example, we consider the estimation of the total quantity of some attribute other than length for pieces of coarse woody debris (i.e., logs) on a forest floor in some region $\mathcal{A}$. Let n_{sk} be the number of times that the sth transect intersects the straight or curved central axis of a log, $\mathcal{U}_k$, and let ℓ_k be the length of the central axis. If the central axis is curved, then ℓ_k is the total length of the curve. Let y_k be an attribute of the log with central axis $\mathcal{U}_k$, $k = 1, 2, \ldots, N$. An estimator of τ_y from de Vries (1986) is

$$\hat{\tau}^u_{yqs} = \frac{\pi A}{2L} \sum_{\mathcal{U}_k \in \mathcal{L}_s} \frac{y_k n_{sk}}{\ell_k}. \tag{9.44}$$

Heuristically, $[\pi A/(2L)] n_{sk}$ is the contribution of $\mathcal{U}_k \in \mathcal{L}_s$ to the estimate of the total length of the N central axes of logs in $\mathcal{A}$, and y_k/ℓ_k is the amount of attribute per unit length of central axis for $\mathcal{U}_k$.

The population comprises the N central axes of logs in $\mathcal{A}$, not the logs themselves, because a log attribute is measured only if the log's central axis is intersected one or more times. The measurement of ℓ_k and the count of intersections, n_{ks}, substitutes for the measurement of c_k, the convex closure of the central axis. However, the estimator reduces to $\hat{\tau}^u_{y\pi s} = \pi A/L \sum_{\mathcal{U}_k \in \mathcal{L}_s} y_k/c_k$ if all central axes are straight, as then $n_{sk} = 1$ for $\mathcal{U}_k \in \mathcal{L}_s$ and $c_k = 2\ell_k$. In the forestry literature, the central axes of logs, whether straight or curved, are often called 'needles.'

When $q_k(\theta_s) = 1$ for all N elements of the population, $E[t_{ks}q_k(\theta_s) \mid \theta_s] = E[t_{ks} \mid \theta_s] = \pi_k(\theta_s)$, and thus $\hat{\tau}^c_{y\pi s}$ in (9.2) can be viewed as just a special case of $\hat{\tau}^u_{yqs}$. Likewise, $\hat{\tau}^u_{y\pi s}$ in (9.6) may be regarded as a special case of $\hat{\tau}^u_{yqs}$.

The variance of $\hat{\tau}^{c}_{yqs}$ in (9.35) is similar to that of $\hat{\tau}^{c}_{y\pi s}$ in (9.2) but with additional factors from the auxiliary variate. It is shown in the Chapter 9 Appendix, along with the variance of $\hat{\tau}^{c}_{yq,\text{rep}}$. An unbiased estimator of the variance of $\hat{\tau}^{c}_{yq,\text{rep}}$ is

$$\hat{v}\left[\hat{\tau}^{c}_{yq,\text{rep}}\right] = \frac{1}{m(m-1)} \sum_{s=1}^{m} \left(\hat{\tau}^{c}_{yqs} - \hat{\tau}^{c}_{yq,\text{rep}}\right)^2, \qquad (9.45)$$

and the usual interval estimator of τ_y, namely

$$\hat{\tau}^{c}_{yq,\text{rep}} \pm t_{m-1} \sqrt{\hat{v}\left[\hat{\tau}^{c}_{yq,\text{rep}}\right]} \quad \text{or} \quad \hat{\tau}^{c}_{yq,\text{rep}} \pm \frac{t_{m-1}\sqrt{\hat{v}\left[\hat{\tau}^{c}_{yq,\text{rep}}\right]}}{\hat{\tau}^{c}_{yq,\text{rep}}}100\%. \qquad (9.46)$$

Analogous results hold for the unconditional estimator, $\hat{\tau}^{u}_{yq,\text{rep}}$, i.e.,

$$\hat{v}\left[\hat{\tau}^{u}_{yq,\text{rep}}\right] = \frac{1}{m(m-1)} \sum_{s=1}^{m} \left(\hat{\tau}^{u}_{yqs} - \hat{\tau}^{u}_{yq,\text{rep}}\right)^2, \qquad (9.47)$$

unbiasedly estimates the variance of $\hat{\tau}^{u}_{yq,\text{rep}}$. The analogous interval estimator of τ_y is

$$\hat{\tau}^{u}_{yq,\text{rep}} \pm t_{m-1} \sqrt{\hat{v}\left[\hat{\tau}^{u}_{yq,\text{rep}}\right]} \quad \text{or} \quad \hat{\tau}^{u}_{yq,\text{rep}} \pm \frac{t_{m-1}\sqrt{\hat{v}\left[\hat{\tau}^{u}_{yq,\text{rep}}\right]}}{\hat{\tau}^{u}_{yq,\text{rep}}}100\%. \qquad (9.48)$$

9.5 Estimating the mean attribute

In order to estimate $\mu_y = \tau_y/N$ from a replicated sample of m line transects, one could use

$$\hat{\mu}^{c}_{yq,\text{rep}} = \frac{\hat{\tau}^{c}_{yq,\text{rep}}}{\hat{N}^{c}_{q,\text{rep}}}, \qquad (9.49)$$

where

$$\hat{N}^{c}_{q,\text{rep}} = \frac{A}{mL} \sum_{s=1}^{m} \sum_{u_k \in \mathbb{L}_s} \frac{q_k(\theta_s)}{w_k(\theta_s)E\left[q_k(\theta_s) \mid \theta_s, t_{ks} = 1\right]};$$

or

$$\hat{\mu}^{u}_{yq,\text{rep}} = \frac{\hat{\tau}^{u}_{yq,\text{rep}}}{\hat{N}^{u}_{q,\text{rep}}}, \qquad (9.50)$$

where

$$\hat{N}^{u}_{q,\text{rep}} = \frac{\pi A}{mL} \sum_{s=1}^{m} \sum_{u_k \in \mathbb{L}_s} \frac{q_k(\theta_s)}{c_k E\left[q_k(\theta_s) \mid t_{ks} = 1\right]}.$$

In (9.49) and (9.50), there is no requirement that $q_k(\theta_s)$ be the same in the numerator and denominator, which allows for the possibility of using $q_k(\theta_s) = 1$

in $\hat{N}^c_{q,\text{rep}}$, which reduces it to

$$\hat{N}^c_{q,\text{rep}} = \frac{A}{mL} \sum_{s=1}^m \sum_{\mathcal{U}_k \in \mathbb{L}_s} \frac{1}{w_k(\theta_s)}. \tag{9.51}$$

In a similar fashion, using $q_k(\theta_s) = 1$ in $\hat{N}^u_{q,\text{rep}}$ simplifies its expression in $\hat{\mu}^u_{yq,\text{rep}}$ to

$$\hat{N}^u_{q,\text{rep}} = \frac{\pi A}{mL} \sum_{s=1}^m \sum_{\mathcal{U}_k \in \mathbb{L}_s} \frac{1}{c_k}. \tag{9.52}$$

Although there is nothing in principle that prevents using the ratio of the unconditional estimator of τ_y in the numerator and the conditional estimator of N in the denominator, or vice versa, such an estimator of μ_y might strike many as perverse.

9.6 Nesting transects of different lengths

In the same way that a single sampling location, (x_s, z_s), can be used to locate two or more plots of different sizes, each of which is used to sample different size classes of elements or different types of populations, so too can line transects of different lengths be nested within each other. See, for example, Brown (1974) and Delisle et al. (1988). Judging from the lack of published literature, there appears to be limited experience using nested transects in LIS. It seems eminently reasonable, however, to consider using transects of appropriate length for smaller, more numerous population elements, and longer transects for larger, less frequently occurring elements.

9.7 Dealing with edge effect in LIS

For $\mathcal{U}_k$ sufficiently close to the edge of $\mathcal{A}$ or the edge of a stratum within it, there is the possibility of boundary overlap of its inclusion zone and the edge of the region where sample transects can be located. Two typical examples are shown in Figure 9.6, the example in frame (a) for transects emanating from the starting point, (x_s, z_s); and the example in frame (b) for transects where (x_s, z_s) is the midpoint. For the first case, a solution is provided by the walkback method of Affleck et al. (2006). For straight-line transects where (x_s, z_s) is the midpoint, a remedy is provided by the reflection method of Gregoire & Monkevich (1994). The reflection method also may be applied to (a) if the orientation of the transect is selected uniformly at random at each sampling location. An alternative tactic is to establish an external peripheral zone around $\mathcal{A}$, as described in §7.5.1, such that no part of any element within $\mathcal{A}$ could have an inclusion zone which extended beyond this zone.

9.7.1 Walkback method

The walkback method is implemented where at sampling location is close to the boundary and the transect is oriented away from the boundary in the direction θ_s. To perform edge correction by the walkback method, we extend the transect in the opposite direction (i.e., in the direction $\theta_s + \pi$) by as much as its length, L. If, and

a)

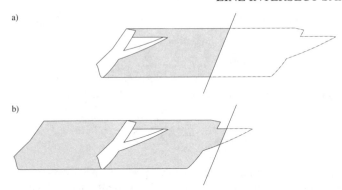

b)

Figure 9.6 *Truncated inclusion zone of* $\mathcal{U}_k$ *when sampling with straight-line transects: a) where* (x_s, z_s) *is the starting point; b) where* (x_s, z_s) *is the midpoint.*

only if, this extension intersects the boundary of $\mathcal{A}$, then we treat the intersection point with the boundary as the starting point (or back end) of the extended transect, and 'walk back' toward and past the original sampling location, measuring all the elements that are intersected by the original transect and its extension. If, on the other hand, the extension of the transect fails to reach the boundary, then the extended portion of the transect and all the elements it intersects are ignored and the sampling is conducted in the usual manner from the original sampling location.

The area of the inclusion zone of any element, $\mathcal{U}_k$, is preserved by the walkback method, though, in expectation, any section of the inclusion zone that falls outside of $\mathcal{A}$ is translated into $\mathcal{A}$. Moreover, no element is intersected more than once by the extended transect, so the estimators that apply in the absence of edge correction continue to apply with edge correction.

9.7.2 Reflection method

To apply the reflection method in LIS, the length of the overlapping portion of the transect that extends outside of $\mathcal{A}$ is reflected into $\mathcal{A}$. Or, to put it another way, the portion that extends outside of $\mathcal{A}$ is folded back at the boundary atop the portion that falls within $\mathcal{A}$. If (x_s, z_s) is at the midpoint of the transect, then no new $\mathcal{U}_k$ will be sampled by the reflected portion of the transect: any $\mathcal{U}_k$ intersected by the reflected portion will already have been sampled and measured, and so it simply needs to be tallied again. A 'multi-tally estimator' of τ_y from the sample at $\mathbb{L}_s$, which includes any elements also tallied from the reflected portion of $\mathbb{L}_s$, is

$$\hat{\tau}^c_{yms} = \sum_{\mathcal{U}_k \in \mathbb{L}_s} \frac{t_{ks} q_k(\theta_s) y_k}{E[t_{ks} q_k(\theta_s) \mid \theta_s]},$$ (9.53a)

which, when $q_k(\theta_s) = 1$ universally, reduces to

$$\hat{\tau}^c_{yms} = \frac{A}{L} \sum_{\mathcal{U}_k \in \mathcal{L}_s} \frac{t_{ks} y_k}{w_k(\theta_s)}, \tag{9.53b}$$

where $t_{ks} = 1$ if $\mathcal{U}_k$ is tallied from portion within $\mathcal{A}$ only, or $t_{ks} = 2$ if $\mathcal{U}_k$ is tallied again from the reflected portion of the transect. The estimator obviously is conditional on θ_s. The unconditional estimator is

$$\hat{\tau}^u_{yms} = \sum_{\mathcal{U}_k \in \mathcal{L}_s} \frac{t_{ks} q_k(\theta_s) y_k}{E[t_{ks} q_k(\theta_s)]}, \tag{9.54a}$$

which reduces to

$$\hat{\tau}^u_{yms} = \frac{\pi A}{L} \sum_{\mathcal{U}_k \in \mathcal{L}_s} \frac{t_{ks} y_k}{c_k} \tag{9.54b}$$

when $q_k(\theta_s) = 1$ universally.

When using the reflection method to counter the truncated inclusion zones of edge elements, the replicated-sampling, conditional estimator of τ_y is

$$\hat{\tau}^c_{ym,\text{rep}} = \frac{1}{m} \sum_{s=1}^{m} \hat{\tau}^c_{yms}, \tag{9.55}$$

the variance of which is estimated unbiasedly by

$$\hat{v}\left[\hat{\tau}^c_{ym,\text{rep}}\right] = \frac{1}{m(m-1)} \sum_{s=1}^{m} \left(\hat{\tau}^c_{yms} - \hat{\tau}^c_{ym,\text{rep}}\right)^2. \tag{9.56}$$

The corresponding unconditional estimator is

$$\hat{\tau}^u_{ym,\text{rep}} = \frac{1}{m} \sum_{s=1}^{m} \hat{\tau}^u_{yms}, \tag{9.57}$$

the variance of which is estimated unbiasedly by

$$\hat{v}\left[\hat{\tau}^u_{ym,\text{rep}}\right] = \frac{1}{m(m-1)} \sum_{s=1}^{m} \left(\hat{\tau}^u_{yms} - \hat{\tau}^u_{ym,\text{rep}}\right)^2. \tag{9.58}$$

9.8 Transects with multiple segments

In some populations, the elements have similar shapes and tend to be oriented in the same direction, for example, a population of trees blown down by a hurricane or felled by a logger. Populations of this sort have given rise to the common perception of an 'orientation bias,' which is thought to occur where transects run more or less parallel to the directional orientation of the elements. To assuage the perceived problem, investigators have proposed and used multi-segmented transects in LIS, where the different segments point in different directions (Gregoire & Valentine

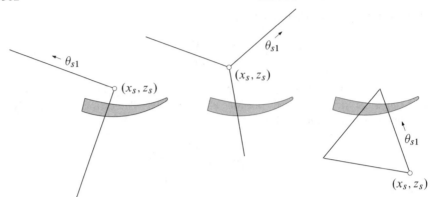

Figure 9.7 *Examples of transects with multiple segments.*

2003). In reality, orientation bias does not exist in the design-based context, so multi-segmented transect designs do not correct for bias, but they may reduce the risk of inaccurate estimation.

In this section, we describe the implementation of LIS with multi-segmented transects, and we provide unbiased estimators that may be used with these designs.

9.8.1 Radial and polygonal transects

Following Affleck et al. (2005), we adopt the term 'radial transect' to refer to a transect consisting of one or more segments directed outwards from a common vertex, as in the case of straight-line, L-shaped, + -shaped, or Y-shaped transects. Polygonal transects are made up of three or more segments forming a closed figure, such as a triangle, square, and more complicated polygonal shapes.

We assume that all m transects in a survey have the same radial or polygonal design, which is almost always the case in practice. In a radial design, the sampling location, (x_s, z_s), is the vertex of the transect, which means that (x_s, z_s) is also the starting point of each the $J \geq 2$ segments.

By contrast, in a polygonal transect, (x_s, z_s) is the starting point of just the first segment. We assume that the second and succeeding segments are established counter-clockwise, so that front end of the first segment connects to the back end of the second segment, and so on, until the front end the Jth segment connects to the back end of the first segment at (x_s, z_s).

Let $\theta_{s1}, \theta_{s2}, \ldots, \theta_{sJ}$, respectively, denote the directional orientations of segments $1, 2, \ldots, J$ of the sth transect. The orientation of the first segment (i.e., $\theta_{s1} \in [0, 2\pi]$) may be selected uniformly at random or fixed in advance of sampling. The predetermined shape of the transect determines all succeeding segment orientations once θ_{s1} has been selected. For example, in both the Y-shaped and triangular transects of Figure 9.7, the orientations of the three segments are $\theta_{s1}, \theta_{s1} + 2\pi/3, \theta_{s1} + 4\pi/3$, respectively. The segments of a transect may vary in length, but we shall assume that

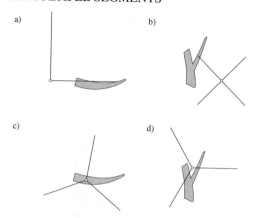

Figure 9.8 *Partial intersections of elements by the front ends of segments of radial transects (a and b); c) partial intersection by an element by the back ends of all the segments and d) partial intersection of a forked element by the back end of the east-oriented segment and complete intersection by the southwest-oriented segment. The elements in (a), (b), and (d) would be included in the sample; the element in (c) would not.*

the length of each of the J segments is identically L/J, so that the total transect length is L.

9.8.2 Sampling protocol for transects having multiple segments

The protocols for radial transects are consistent with those straight transects (§9.2.1). With a radial transect, an element, $\mathcal{U}_k$, may be intersected by more than one segment. An element is included in the sample from (x_s, z_s) only if it is intersected completely by any segment of the transect (Figure 9.7 left, center) or intersected partially by the front end of any segment (Figure 9.8a,b). The element is not included in the sample if all intersections are partial and involve the vertex of the transect. Thus, an element is excluded from the sample if the vertex of the radial transect is inside the boundary of the element's projection onto $\mathcal{A}$ (Figure 9.8c). The east-oriented segment in Figure 9.8d is a partial intersection involving the vertex, yet the element is included in the sample because it is completely intersected by the segment oriented towards the southwest.

In a polygonal transect, the front end of one segment connects to the back end of an adjoining segment. So, with one exception, an element, $\mathcal{U}_k$, is included in the sample from (x_s, z_s) if any portion of any segment of a polygonal transect intersects the element. Hence, an element is included in the sample even if a vertex of the transect falls within the boundary of the element. However, the element is not included in the sample if the boundary of $\mathcal{U}_k$ completely encompasses the transect.

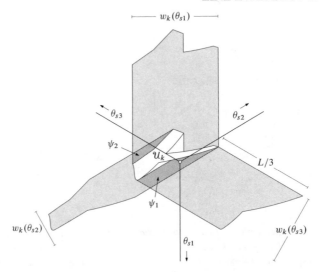

Figure 9.9 *Inclusion zone of an element as defined by a* Y-*shaped radial transect with orientation* θ_{s1}. *If the sampling location* (∘) *is in a light gray region, the element is intersected either completely or partially by the front end of one segment; in a dark gray region, the element is intersected by two segments.*

9.8.3 *Estimation from transects having multiple segments*

With segmented transects there is a possibility that a population element, $\mathcal{U}_k$, will be intersected by two or more segments. For this reason, the probability that an element, $\mathcal{U}_k$, is included in the sample from the sth transect with J segments differs from the element's probability of being intersected by J replicated transects each of length L/J.

Example 9.13

The element depicted in Figure 9.9 is completely intersected by two segments if the vertex of the radial transect is anywhere in either of the two dark grey regions of the inclusion zone, which have areas ψ_1 and ψ_2, respectively. Overall, the area (a_k) of the element's inclusion zone is

$$a_k = [w_k(\theta_{s1}) + w_k(\theta_{s2}) + w_k(\theta_{s3})]\frac{L}{3} - \psi_1 - \psi_2$$

and the probability of $\mathcal{U}_k$ being intersected by the transect is $\pi_k = a_k/A$.

For now, we assume that no auxiliary variate is used, i.e., $q_k(\theta_{sj}) = 1$ for all $\mathcal{U}_k$ and all J segments of the transect. With the sample selected from the transect located at (x_s, z_s), a design-unbiased estimator of τ_y conditional on the orientation, θ_{s1}, of the first segment of the transect is

$$\ddot{\tau}^{c}_{yms} = \sum_{\mathcal{U}_k \in \mathbb{L}_s} \frac{t_{ks\bullet} y_k}{E\left[t_{ks\bullet} \mid \theta_{s1}\right]} \tag{9.59a}$$

$$= \frac{A}{L} \sum_{\mathcal{U}_k \in \mathbb{L}_s} \frac{t_{ks\bullet} y_k}{\bar{w}_k(\theta_s)}, \tag{9.59b}$$

where

$$t_{ks\bullet} = \sum_{j=1}^{J} t_{ksj} \tag{9.59c}$$

and

$$t_{ksj} = \begin{cases} 1, & \text{if } \mathcal{U}_k \in \mathbb{L}_s \text{ owing to intersection with} \\ & \text{the } j\text{th segment of } s\text{th transect;} \\ 0, & \text{otherwise;} \end{cases}$$

and

$$\bar{w}_k(\theta_s) = \frac{1}{J} \sum_{j=1}^{J} w_k(\theta_{sj}), \tag{9.59d}$$

where $w_k(\theta_{sj})$ is the width of $\mathcal{U}_k$ perpendicular to the jth segment of the sth transect. When $J = 1$, $\ddot{\tau}^{c}_{yms}$ is identical to $\hat{\tau}^{c}_{y\pi s}$, which is the HT estimator of τ_y conditional on transect orientation; when $J > 1$, $\ddot{\tau}^{c}_{yms}$ is not the HT estimator of τ_y.

In (9.59a), $t_{ks\bullet}$ is the count of the number of segments that fully or partially intersect $\mathcal{U}_k$. The unbiasedness of $\ddot{\tau}^{c}_{yms}$ was first shown by Gregoire & Valentine (2003) for ell-shaped transects, and thereafter generalized by Affleck et al. (2005) by showing that $E\left[t_{ks\bullet} \mid \theta_s\right] = L\bar{w}_k(\theta_s)/A$. An implication of this result is that the width of $\mathcal{U}_k$ perpendicular to all J segments must be measured, irrespective of which segments intersect $\mathcal{U}_k$, or else the estimator will be biased. The requisite measurements of width are most easily accomplised with L-shaped and $+$-shaped transects.

The variance of $\ddot{\tau}^{c}_{yms}$ is derived in the Chapter 9 Appendix, and with a sample of m replicated sampling locations, the variance of

$$\ddot{\tau}^{c}_{ym,\mathrm{rep}} = \frac{1}{m} \sum_{s=1}^{m} \ddot{\tau}^{c}_{yms} \tag{9.60}$$

is unbiasedly estimated by

$$\hat{v}\left[\ddot{\tau}_{ym,\mathrm{rep}}\right] = \frac{1}{m(m-1)} \sum_{s=1}^{m} \left(\ddot{\tau}^{c}_{y\pi s} - \ddot{\tau}^{c}_{ym,\mathrm{rep}}\right)^2. \tag{9.61}$$

The corresponding unconditional estimator of τ_y when $q_k(\theta_s) = 1$ is

$$\ddot{\tau}^{u}_{yms} = \sum_{\mathcal{U}_k \in \mathbb{L}_s} \frac{t_{ks\bullet} y_k}{E\left[t_{ks\bullet}\right]} \tag{9.62a}$$

$$= \frac{\pi A}{L} \sum_{\mathcal{U}_k \in \mathbb{L}_s} \frac{t_{ks\bullet} y_k}{c_k}. \tag{9.62b}$$

Because $E\left[\bar{w}_k(\theta_s)\right] = c_k/\pi$, $\ddot{\tau}^{u}_{yms}$ multiplies the value y_k/c_k for each intersected $\mathcal{U}_k$ by the number of segments which intersect it. Since c_k need be measured but once, this unconditional estimator of τ_y may require less burdensome field measurements than $\ddot{\tau}^{c}_{yms}$.

The variance of $\ddot{\tau}^{u}_{yms}$ is derived in this chapter's Appendix, also. As with the conditional estimator, a sample of m replicated sampling locations allows one to unbiasedly estimate the variance of

$$\ddot{\tau}^{u}_{ym,\text{rep}} = \frac{1}{m} \sum_{s=1}^{m} \ddot{\tau}^{u}_{yms} \tag{9.63}$$

with

$$\hat{v}\left[\ddot{\tau}_{ym,\text{rep}}\right] = \frac{1}{m(m-1)} \sum_{s=1}^{m} \left(\ddot{\tau}^{u}_{yms} - \ddot{\tau}^{u}_{ym,\text{rep}}\right)^2. \tag{9.64}$$

When $q_k(\theta_{sj}) \neq 1$ because of the potential utility of using an auxiliary variate to obviate measurements of y_k and $w_k(\theta_{sj})$, the conditional estimator is

$$\ddot{\tau}^{c}_{yqs} = \sum_{\mathcal{U}_k \in \mathbb{L}_s} \frac{\sum_{j=1}^{J} t_{ksj} q_k(\theta_{sj})}{\sum_{j=1}^{J} E\left[t_{ksj} q_k(\theta_{sj}) \mid \theta_{s1}\right]} y_k \tag{9.65}$$

and the unconditional estimator is

$$\ddot{\tau}^{u}_{yqs} = \sum_{\mathcal{U}_k \in \mathbb{L}_s} \frac{\sum_{j=1}^{J} t_{ksj} q_k(\theta_{sj})}{\sum_{j=1}^{J} E\left[t_{ksj} q_k(\theta_{sj})\right]} y_k. \tag{9.66}$$

Both (9.65) and (9.66) are design-unbiased estimators of τ_y.

Example 9.14

In an extension of Example 9.8, consider estimating aggregate canopy cover, τ_y, using a transect of J segments, letting $q_k(\theta_{sj})$ be the length of $\mathcal{U}_k$ along the line that is coincident with the jth segment of the sth transect. For this choice of $q_k(\theta_{sj})$, $E[t_{ksj}q_k(\theta_{sj})\,|\,\theta_{sj}]$ is Ly_k/JA. Therefore,

$$\ddot{\tau}^{\,c}_{yqs} = \sum_{\mathcal{U}_k \in \mathbb{L}_s} \frac{\sum_{j=1}^{J} t_{ksj}q_k(\theta_{sj})}{\sum_{j=1}^{J} Ly_k/(JA)}\,y_k \qquad (9.67a)$$

$$= \frac{A}{L} \sum_{\mathcal{U}_k \in \mathbb{L}_s} \sum_{j=1}^{J} t_{ksj}q_k(\theta_{sj}). \qquad (9.67b)$$

In other words, the interception lengths are summed over all the segments which cross $\mathcal{U}_k$, for all elements included in the sample from the sth transect, and then multiplied by A/L.

9.8.4 Averaging estimators for straight transects

Some practitioners interpret segmented transects as clusters of straight transects. Indeed, one may perform LIS with each of the J segments of the sth transect $(s = 1, 2, \ldots, m)$ in isolation, work up an unconditional estimate of τ_y based on each individual segment, and average the J estimates, including any zero estimates, to obtain the final estimate for the sth whole segmented transect (e.g., van Wagner 1968). The average of the J unconditional estimates will equal the estimate of τ_y that obtains from the unconditional estimator for the whole segmented transect. In effect a segmented transect is disassembled into J straight transects for purposes of measurement, then the whole is implicitly reassembled in the estimation phase by the averaging process. Of course, sampling variances should be estimated with the m average estimates, not the estimates for the individual legs.

On the other hand, averaging unbiased conditional estimates for each of the J individual segments also yields an average unbiased estimate of τ_y, but this estimate will differ from, and be less precise than, the estimate that obtains from the unbiased conditional estimator for the segmented transect as a whole. The reduction in precision results from the fact that the inclusion probability of a population unit, $\mathcal{U}_k$, varies with the orientation of each individual segment of the transect.

9.9 Parallel transects of uneven length

There have been some applications of LIS wherein a baseline is established outside the region $\mathcal{A}$ and transects are established perpendicular to this baseline. The orientation of the baseline is not usually chosen in a probabilistic manner, although the location of the transects emanating from the baseline are usually selected randomly, or perhaps systematically with a random start. In order to ensure that every element, $\mathcal{U}_k$, on $\mathcal{A}$ has a non-zero chance of being selected into the sample, it is imperative that the baseline span the breadth of $\mathcal{A}$ (Figure 9.10). Each perpendicular transect would be run until the boundary of $\mathcal{A}$ on the side away from the baseline was encountered. The length of each transect may differ, unless $\mathcal{A}$ is rectangular or has some other regular shape.

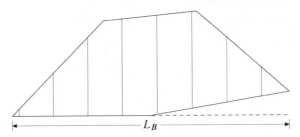

Figure 9.10 *Parallel transects systematically placed from a random start and perpendicular to a defined baseline.*

ASIDE: Kaiser (1983, pp. 973–974) considered the case where each transect traverses $\mathcal{A}$ but with an orientation that is chosen at random. He shows that $\hat{\tau}^c_{yqs}$ and $\hat{\tau}^u_{yqs}$ remain unbiased, despite the randomly varying lengths of the transects.

Seber (1979) apparently was the first to treat this method of LIS statistically. Because of the random length of the transects under this procedure, the estimators proposed above are all biased. He proposed a ratio estimator of $\lambda = N/A$ as

$$\hat{\lambda}_N = \frac{1}{A} \sum_{u_k \in s} \left(\frac{L_s}{\sum_{u_k \in s} L_s} \right) \sum_{u_k \in \mathbb{L}_s} \frac{1}{L_s w_k(\theta_s)/A} \tag{9.68a}$$

$$= \frac{\sum_{u_k \in s} L_s \hat{\lambda}_{Ns}}{\sum_{u_k \in s} L_s}, \tag{9.68b}$$

where $\hat{\lambda}_{Ns} = \sum_{u_k \in \mathbb{L}_s} 1/L_s w_k(\theta_s)$ and L_s is the length of the sth transect. He argues that this ratio estimator is less biased than $\hat{\lambda}^c_{y\pi,\text{rep}}$ when transect lengths are random, and that the jackknife procedure can be used to reduce its bias even more.

However, it is straightforward to estimate τ_y and λ unbiasedly when sampling with transects of random length. With a baseline of fixed orientation and fixed length L_B, the probability that $\mathcal{U}_k$ will be intersected by a transect emanating perpendicularly from it at a randomly chosen location along its length is $w_k(\theta_s)/L_B$. Hence the HT estimator of τ_y with a single transect is just

$$\hat{\tau}_{y\pi\perp} = L_B \sum_{u_k \in \mathbb{L}_s} \frac{y_k}{w_k(\theta_s)}. \tag{9.69}$$

If several transects are spaced systematically, then the estimates from the separate transects are averaged. Muttlak & Sadooghi-Alvandi (1993) and Pontius (1998) discuss this estimator and related estimators for this LIS design.

The utility of this method of sampling large regions is not clear, because the resultant transects would be extremely long.

Table 9.1 *Diameters (cm) of pieces of coarse woody debris (logs) at the point of intersection by transects of length 40 m.*

Transect	Log	Dia.	Transect	Log	Dia.	Transect	Log	Dia.
1	1	13.7	8	1	9.7	13	1	18.8
1	2	9.9	8	2	7.4	13	2	25.9
			8	3	7.4			
2	1	8.9	8	4	4.8	14	1	11.4
2	2	4.3						
			9	1	8.1	15	1	9.6
3	1	15.7	9	2	9.7	15	2	9.4
3	2	7.9				15	3	11.7
			10	1	19.0	15	4	6.9
4	1	7.1						
4	2	9.4	11	1	5.3	16	1	16.8
4	3	20.3	11	2	11.2	16	2	5.8
4	4	18.5	11	3	11.4			
			11	4	21.8	17	-	-
5	1	6.4	11	5	7.6			
			11	6	10.7	18	1	7.6
6	1	8.4	11	7	10.7	18	2	8.4
6	2	8.1				18	3	8.6
			12	1	15.5			
7	1	11.9	12	2	15.0	19	1	37.8
7	2	10.1	12	3	15.2	19	2	11.4
7	3	14.7	12	4	23.9	19	3	7.6
7	4	12.2	12	5	9.4	19	4	18.8
			12	6	12.2	19	5	11.4
						20	1	17.3

9.10 Terms to remember

Conditional estimator	Reflection method
Complete intersection	Radial transect
Multiple segments	Unconditional estimator
Partial intersection	Walkback method
Polygonal transect	

9.11 Exercises

1. Estimate the volume of coarse woody debris per unit land area (m^3 m^{-2}) from the data in Table 9.1, and estimate the sampling error. Hint: see Example 9.10.

$\theta \longrightarrow$

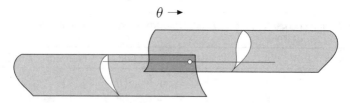

Figure 9.11 *A transect intersects two elements only if the sampling location,* (x_s, z_s), *occurs in their joint inclusion zone (dark grey).*

9.12 Appendix

9.12.1 Joint inclusion zones

A transect intersects more than one element only if (a) the inclusion zones of two more elements overlap to form joint inclusion zones and (b) the sampling location (x_s, z_s) occurs in one of the joint inclusion zones (Figure 9.11).

9.12.2 Derivation of $E[t_{ks}q_k(\theta_s)\,|\,\theta_s]$

Results from statistical theory permit $E[t_{ks}q_k(\theta_s)\,|\,\theta_s]$ to be re-expressed as

$$
\begin{aligned}
E[t_{ks}q_k(\theta_s)\,|\,\theta_s] &= E[t_{ks}q_k(\theta_s)\,|\,\theta_s, t_{ks}=1]\,\mathrm{Prob}(t_{ks}=1) \\
&\quad + E[t_{ks}q_k(\theta_s)\,|\,\theta_s, t_{ks}=1]\,\mathrm{Prob}(t_{ks}=0) \\[2mm]
&= E[t_{ks}q_k(\theta_s)\,|\,\theta_s, t_{ks}=1]\,\pi_k(\theta_s) + 0 \\[2mm]
&= E[q_k(\theta_s)\,|\,\theta_s, t_{ks}=1]\,\frac{Lw_k(\theta_s)}{A}.
\end{aligned}
$$

When y_k is the area of $\mathcal{U}_k$ projected onto $\mathcal{A}$ and $q_k(\theta_s)$ is the length of $\mathcal{U}_k$ in the line containing the sth transect, as discussed on p. 294, results from geometrical probability, as used by Lucas & Seber (1977), show that

$$
E[q_k(\theta_s)\,|\,\theta_s, t_{ks}=1] = \frac{\text{area of } \mathcal{U}_k \text{ projected onto } \mathcal{A}}{w_k(\theta_s)}.
$$

Combining this with the result above provides

$$
\begin{aligned}
E[t_{ks}q_k(\theta_s)\,|\,\theta_s] &= E[q_k(\theta_s)\,|\,\theta_s, t_{ks}=1]\,\frac{Lw_k(\theta_s)}{A} \\[2mm]
&= \left(\frac{\text{area of } \mathcal{U}_k \text{ projected onto } \mathcal{A}}{w_k(\theta_s)}\right)\frac{Lw_k(\theta_s)}{A} \\[2mm]
&= \frac{Ly_k}{A}.
\end{aligned}
\tag{9.70}
$$

9.12.3 Variance of conditional estimator of τ_y with auxiliary variate using transects with a single segment

The variance of $\hat{\tau}^c_{yqs}$, introduced in §9.4, is given by

$$V\left[\hat{\tau}^c_{yqs}\right] = \sum_{k=1}^{N} \frac{E\left[t_{ks}q_k(\theta_s)^2 \mid \theta_s\right] y_k^2}{E\left[t_{ks}q_k(\theta_s) \mid \theta_s\right]^2} \left(\frac{1 - \pi_k(\theta_s)}{\pi_k(\theta_s)}\right)$$

$$+ \sum_{k=1}^{N} \sum_{\substack{k' \neq k \\ k'=1}}^{N} \frac{E\left[t_{ks}q_k(\theta_s)t_{k's}q_{k'}(\theta_s)\right] y_k y_{k'} \left(\pi_{kk'}(\theta_s) - \pi_k(\theta_s)\pi_{k'}(\theta_s)\right)}{E\left[t_{ks}q_k(\theta_s) \mid \theta_s\right] E\left[t_{k's}q_{k'}(\theta_s) \mid \theta_s\right] \pi_k(\theta_s)\pi_{k'}(\theta_s)}$$

$$\tag{9.71a}$$

$$= V^c_q, \text{ say.} \tag{9.71b}$$

Therefore, the variance of $\hat{\tau}^c_{yq,\text{rep}}$ is

$$V\left[\hat{\tau}^c_{yq,\text{rep}}\right] = \frac{1}{m} V^c_q. \tag{9.72}$$

9.12.4 Variance of unconditional estimator of τ_y with auxiliary variate using transects with a single segment

The variance of the unconditional estimator, $\hat{\tau}^u_{yqs}$, is

$$V\left[\hat{\tau}^u_{yqs}\right] = \sum_{k=1}^{N} \frac{E\left[t_{ks}q_k(\theta_s)^2\right] y_k^2}{E\left[t_{ks}q_k(\theta_s)\right]^2} \left(\frac{1 - \pi_k}{\pi_k}\right)$$

$$+ \sum_{k=1}^{N} \sum_{\substack{k' \neq k \\ k'=1}}^{N} \frac{E\left[t_{ks}q_k(\theta_s)t_{k's}q_{k'}(\theta_s)\right] y_k y_{k'} \left(\pi_{kk'} - \pi_k \pi_{k'}\right)}{E\left[t_{ks}q_k(\theta_s)\right] E\left[t_{k's}q_{k'}(\theta_s)\right] \pi_k \pi_{k'}} \tag{9.73a}$$

$$= V^u_q, \text{ say.} \tag{9.73b}$$

The variance of $\hat{\tau}^u_{yq,\text{rep}}$ is

$$V\left[\hat{\tau}^u_{yq,\text{rep}}\right] = \frac{1}{m} V^u_q. \tag{9.74}$$

9.12.5 Conditional expected value and variance of $t_{ks\bullet}$ with transects having multiple segments

Because $t_{ks\bullet} = \sum_{j=1}^{J} t_{ksj}$ and

$$E\left[t_{ksj} \mid \theta_{s1}\right] = \frac{L w_k(\theta_{sj})}{JA} \tag{9.75a}$$

$$= \pi_k(\theta_{sj}), \quad \text{say,} \tag{9.75b}$$

it follows that

$$E\left[t_{ks\bullet} \mid \theta_{s1}\right] = E\left[\sum_{j=1}^{J} t_{ksj} \mid \theta_s\right]$$

$$= \sum_{j=1}^{J} \frac{(L/J)w_k(\theta_{sj})}{A}$$

$$= \frac{L}{A}\bar{w}_k(\theta_s)$$

$$= \sum_{j=1}^{J} \pi_k(\theta_{sj})$$

$$= \tilde{\pi}_k(\theta_s), \qquad (9.76)$$

where $\tilde{\pi}_k(\theta_s)$ is the sum of the segment-by-segment conditional probabilities of including $\mathcal{U}_k$ on a multi-segmented transect with θ_{s1} as the orientation of the first segment.

The conditional variance of $t_{ks\bullet}$ is

$$V\left[t_{ks\bullet} \mid \theta_{s1}\right] = \sum_{j=1}^{J} V\left[t_{ksj} \mid \theta_{s1}\right] + \sum_{j=1}^{J}\sum_{\substack{j' \neq j \\ j'=1}}^{J} C\left[t_{ksj}, t_{ksj'} \mid \theta_{s1}\right], \qquad (9.77)$$

where

$$V\left[t_{ksj} \mid \theta_{s1}\right] = E\left[t_{ksj}^2 \mid \theta_{s1}\right] - \left(E\left[t_{ksj} \mid \theta_{s1}\right]^2\right)$$

$$= \pi_k(\theta_{sj})\left(1 - \pi_k(\theta_{sj})\right),$$

and where $C\left[t_{ksj}, t_{ksj'} \mid \theta_{s1}\right]$ is the conditional covariance between the selection indicator variables for $\mathcal{U}_k$ on distinct segments of the sth transect, namely

$$C\left[t_{ksj}, t_{ksj'} \mid \theta_{s1}\right] = E\left[t_{ksj}t_{ksj'} \mid \theta_{s1}\right] - \pi_k(\theta_{sj})\pi_k(\theta_{sj'}). \qquad (9.78)$$

Therefore,

$$V\left[t_{ks\bullet} \mid \theta_{s1}\right] = \sum_{j=1}^{J} V\left[t_{ksj} \mid \theta_{s1}\right]$$

$$+ \sum_{j=1}^{J} \sum_{\substack{j' \neq j \\ j'=1}}^{J} \left(E\left[t_{ksj} t_{ksj'} \mid \theta_{s1}\right] - \pi_k(\theta_{sj})\pi_k(\theta_{sj'})\right),$$

$$= \sum_{j=1}^{J} \pi_k(\theta_{sj}) \left(1 - \pi_k(\theta_{sj})\right)$$

$$+ \sum_{j=1}^{J} \sum_{\substack{j' \neq j \\ j'=1}}^{J} \left(E\left[t_{ksj} t_{ksj'} \mid \theta_{s1}\right] - \pi_k(\theta_{sj})\pi_k(\theta_{sj'})\right),$$

$$= \sum_{j=1}^{J} \left(\pi_k(\theta_{sj}) \left(1 - \pi_k(\theta_{sj})\right) - \pi_k(\theta_{sj}) \sum_{\substack{j' \neq j \\ j'=1}}^{J} \pi_k(\theta_{sj'}) \right)$$

$$+ \sum_{j=1}^{J} \sum_{\substack{j' \neq j \\ j'=1}}^{J} \frac{\psi_{kjj'}}{A}$$

$$= \sum_{j=1}^{J} \pi_k(\theta_{sj}) \left(1 - \pi_k(\theta_{sj}) - \sum_{\substack{j' \neq j \\ j'=1}}^{J} \pi_k(\theta_{sj'}) \right) + \sum_{j=1}^{J} \sum_{\substack{j' \neq j \\ j'=1}}^{J} \frac{\psi_{kjj'}}{A}$$

$$= \sum_{j=1}^{J} \pi_k(\theta_{sj}) \left(1 - \tilde{\pi}_k(\theta_{s1})\right) + \sum_{j=1}^{J} \sum_{\substack{j' \neq j \\ j'=1}}^{J} \frac{\psi_{kjj'}}{A}$$

$$= \left(1 - \tilde{\pi}_k(\theta_{s1})\right) \sum_{j=1}^{J} \pi_k(\theta_{sj}) + \sum_{j=1}^{J} \sum_{\substack{j' \neq j \\ j'=1}}^{J} \frac{\psi_{kjj'}}{A}$$

$$= \tilde{\pi}_k(\theta_s) \left(1 - \tilde{\pi}_k(\theta_s)\right) + \sum_{j=1}^{J} \sum_{\substack{j' \neq j \\ j'=1}}^{J} \frac{\psi_{kjj'}}{A}, \qquad (9.79)$$

where $\psi_{kjj'}$ is the common area of the overlapping inclusion zones of $\mathcal{U}_k$ along the jth and j'th segments of the sth transect, which implies that $\psi_{kjj'}/A$ is the joint probability of selecting $\mathcal{U}_k$ on both segments.

9.12.6 *Unconditional expected value and variance of $t_{ks\bullet}$ with transects having multiple segments*

Unconditionally, the expected value of $t_{ks\bullet}$ is

$$E_\theta\left[t_{ks\bullet}\right] = \sum_{j=1}^{J} E_\theta\left[t_{ksj}\mid\theta_{s1}\right]$$

$$= \sum_{j=1}^{J} \frac{LE_\theta\left[w_k(\theta_{sj})\right]}{JA}$$

$$= \frac{L}{JA}\sum_{j=1}^{J}\frac{c_k}{\pi}$$

$$= \frac{Lc_k}{\pi A}$$

$$= \pi_k, \tag{9.80}$$

where π_k is the inclusion probability of the kth element, as was derived in §9.2.3.

The variance of $t_{ks\bullet}$ can be derived by expressing it as

$$V\left[t_{ks\bullet}\right] = V_\theta\left[E[t_{ks\bullet}\mid\theta_{s1}]\right] + E_\theta\left[V[t_{ks\bullet}\mid\theta_{s1}]\right] \tag{9.81}$$

where E_θ signifies the expected value over the distribution of θ_s, and V_θ signifies the variance over the distribution of θ_s. Because $E_\theta[\tilde\pi_k(\theta_s)] = \pi_k$, we can use (9.76) and (9.79) to write

$$V\left[t_{ks\bullet}\right] = V_\theta\left[\tilde\pi_k(\theta_s)\right] + E_\theta\left[\tilde\pi_k(\theta_s)\left(1-\tilde\pi_k(\theta_s)\right) + \sum_{j=1}^{J}\sum_{\substack{j'\neq j\\j'=1}}^{J}\frac{\psi_{kjj'}}{A}\right]$$

$$= E_\theta\left[\tilde\pi_k^2(\theta_s)\right] - \left(E_\theta\left[\tilde\pi_k(\theta_s)\right]\right)^2 + E_\theta\left[\tilde\pi_k(\theta_s)\right] - E_\theta\left[\tilde\pi_k^2(\theta_s)\right]$$

$$+ \sum_{j=1}^{J}\sum_{\substack{j'\neq j\\j'=1}}^{J}\frac{E_\theta[\psi_{kjj'}]}{A}$$

$$= \pi_k\left(1-\pi_k\right) + \sum_{j=1}^{J}\sum_{\substack{j'\neq j\\j'=1}}^{J}\frac{c_{kjj'}}{\pi A} \tag{9.82}$$

where $c_{kjj'}/\pi A$ is the unconditional probability of intersecting $\mathcal{U}_k$ along both the jth and j' segments of the sth transect.

9.12.7 Variance of $\ddot{\tau}_{yms}^c$

Because $t_{ks\bullet} = 0$ for all $\mathcal{U}_k$ that are not selected by the sth transect, $\ddot{\tau}_{yms}^c$ in (9.59) can be expressed alternatively as

$$\ddot{\tau}_{yms}^c = \frac{A}{L} \sum_{k=1}^{N} \frac{t_{ks\bullet} y_k}{\bar{w}_k(\theta_s)}. \qquad (9.83)$$

Standard results on the variance of a sum thus provides

$$V\left[\ddot{\tau}_{yms}^c\right] = \frac{A^2}{L^2} \sum_{k=1}^{N} \frac{V\left[t_{ks\bullet} \mid \theta_{s1}\right] y_k^2}{\bar{w}_k^2(\theta_s)} + \frac{A^2}{L^2} \sum_{k=1}^{N} \sum_{\substack{k' \neq k \\ k'=1}}^{N} \frac{C\left[t_{ks\bullet}, t_{k's\bullet} \mid \theta_{s1}\right] y_k y_{k'}}{\bar{w}_k(\theta_s) \bar{w}_{k'}(\theta_s)}. \qquad (9.84)$$

Substituting into the first term of (9.84) from (9.79) provides

$$V\left[\ddot{\tau}_{yms}^c\right] = \frac{A^2}{L^2} \sum_{k=1}^{N} \left(\tilde{\pi}_k(\theta_s)\left(1 - \tilde{\pi}_k(\theta_s)\right) + \sum_{j=1}^{J} \sum_{\substack{j' \neq j \\ j'=1}}^{J} \frac{\psi_{kjj'}}{A} \right) \frac{y_k^2}{\bar{w}_k^2(\theta_s)}$$

$$+ \frac{A^2}{L^2} \sum_{k=1}^{N} \sum_{\substack{k' \neq k \\ k'=1}}^{N} \frac{C\left[t_{ks\bullet}, t_{k's\bullet} \mid \theta_{s1}\right]}{\bar{w}_k(\theta_s) \bar{w}_{k'}(\theta_s)} y_k y_{k'}, \qquad (9.85)$$

where $C\left[t_{ks\bullet}, t_{k's\bullet} \mid \theta_{s1}\right]$ indicates the conditional covariance between the multi-tally count variables of $\mathcal{U}_k$ and $\mathcal{U}_{k'}$ on the sth transect, i.e.,

$$C\left[t_{ks\bullet}, t_{k's\bullet} \mid \theta_{s1}\right] = E\left[t_{ks\bullet} t_{k's\bullet} \mid \theta_{s1}\right] - E\left[t_{ks\bullet} \mid \theta_{s1}\right] E\left[t_{k's\bullet} \mid \theta_{s1}\right]$$

$$= E\left[\sum_{j=1}^{J} t_{ksj} \sum_{j=1}^{J} t_{k'sj} \mid \theta_{s1} \right] - \tilde{\pi}_k(\theta_s) \tilde{\pi}_{k'}(\theta_s)$$

$$= \sum_{j=1}^{J} E\left[t_{ksj} t_{k'sj} \mid \theta_{s1}\right] + \sum_{j=1}^{J} \sum_{\substack{j' \neq j \\ j'=1}}^{J} E\left[t_{ksj} t_{k'sj'} \mid \theta_{s1}\right]$$

$$- \tilde{\pi}_k(\theta_s) \tilde{\pi}_{k'}(\theta_s)$$

$$= \sum_{j=1}^{J} \pi_{kk'j}(\theta_s) + \sum_{j=1}^{J} \sum_{\substack{j' \neq j \\ j'=1}}^{J} \frac{\zeta_{kjk'j's}}{A} - \tilde{\pi}_k(\theta_s) \tilde{\pi}_{k'}(\theta_s), \qquad (9.86)$$

where $\pi_{kk'j}(\theta_s)$ is the analog, when dealing with transects having $J > 1$ segments to $\pi_{kk'}(\theta_s)$ with straight-line transects (see (9.12)). In other words, it is the joint conditional probability of selecting $\mathcal{U}_k$ and $\mathcal{U}_{k'}$ from the jth segment of the sth transect; and $\zeta_{kjk'j's}/A$ is the joint probability of selecting both $\mathcal{U}_k$ and $\mathcal{U}_{k'}$ on the jth and j'th segments of the sth transect.

Upon substitution of (9.86) into (9.85), we get

$$
V\left[\ddot{\tau}^c_{yms}\right] = \frac{A^2}{L^2} \sum_{k=1}^{N} \frac{\tilde{\pi}_k(\theta_s)\,(1 - \tilde{\pi}_k(\theta_s))}{\bar{w}_k^2(\theta_s)} y_k^2 + \frac{A}{L^2} \sum_{k=1}^{N} \sum_{j=1}^{J} \sum_{\substack{j' \neq j \\ j'=1}}^{J} \frac{\psi_{kjj'}}{\bar{w}_k^2(\theta_s)} y_k^2
$$

$$
+ \frac{A^2}{L^2} \sum_{k=1}^{N} \sum_{\substack{k' \neq k \\ k'=1}}^{N} \sum_{j=1}^{J} \frac{\pi_{kk'j}(\theta_s)}{\bar{w}_k(\theta_s)\bar{w}_{k'}(\theta_s)} y_k y_{k'}
$$

$$
+ \frac{A}{L^2} \sum_{k=1}^{N} \sum_{\substack{k' \neq k \\ k'=1}}^{N} \sum_{j=1}^{J} \sum_{\substack{j' \neq j \\ j'=1}}^{J} \frac{\zeta_{kjk'j's}}{\bar{w}_k(\theta_s)\bar{w}_{k'}(\theta_s)} y_k y_{k'}
$$

$$
- \frac{A^2}{L^2} \sum_{k=1}^{N} \sum_{\substack{k' \neq k \\ k'=1}}^{N} \frac{\tilde{\pi}_k(\theta_s)\tilde{\pi}_{k'}(\theta_s)}{\bar{w}_k(\theta_s)\bar{w}_{k'}(\theta_s)} y_k y_{k'}. \tag{9.87}
$$

Replacing $\bar{w}_k(\theta_s)$ and $\bar{w}_{k'}(\theta_s)$ with $A\tilde{\pi}_k(\theta_s)/L$ and $A\tilde{\pi}_{k'}(\theta_s)/L$, respectively, yields

$$
V\left[\ddot{\tau}^c_{yms}\right] = \sum_{k=1}^{N} \frac{y_k^2}{\tilde{\pi}_k(\theta_s)} \left[1 + \left(\sum_{j=1}^{J} \sum_{\substack{j'=1 \\ j' \neq j}}^{J} \frac{\psi_{kjj'}}{A\tilde{\pi}_k(\theta_s)} \right) - \tilde{\pi}_k(\theta_s) \right]
$$

$$
+ \sum_{k=1}^{N} \sum_{\substack{k'=1 \\ k' \neq k}}^{N} \frac{y_k y_{k'}}{\tilde{\pi}_k(\theta_s)\tilde{\pi}_{k'}(\theta_s)}
$$

$$
\times \left[\left(\sum_{j=1}^{J} \pi_{kk'j}(\theta_s) + \sum_{j=1}^{J} \sum_{\substack{j'=1 \\ j' \neq j}}^{J} \frac{\zeta_{kjk'j's}}{A} \right) - \tilde{\pi}_k(\theta_s)\tilde{\pi}_{k'}(\theta_s) \right]. \tag{9.88}
$$

When $J = 1$, the terms in (9.88) involving $\psi_{kjj'}$ and $\zeta_{kjk'j's}$ are nil, and $V\left[\ddot{\tau}^c_{yms}\right]$ coincides with $V\left[\hat{\tau}^c_{y\pi s}\right]$ in (9.12).

9.12.8 Variance of $\ddot{\tau}^{u}_{yms}$

An equivalent expression for $\ddot{\tau}^{u}_{yms}$ to that given in (9.62b) is

$$\ddot{\tau}^{u}_{yms} = \frac{\pi A}{L} \sum_{k=1}^{N} \frac{t_{ks\bullet} y_k}{c_k}. \qquad (9.89)$$

Therefore

$$V\left[\ddot{\tau}^{u}_{yms}\right] = \frac{\pi^2 A^2}{L^2} \sum_{k=1}^{N} \frac{V\left[t_{ks\bullet}\right]}{c_k^2} y_k^2 + \frac{\pi^2 A^2}{L^2} \sum_{k=1}^{N} \sum_{\substack{k' \neq k \\ k'=1}}^{N} \frac{C\left[t_{ks\bullet}, t_{k's\bullet}\right]}{c_k c_{k'}} y_k y_{k'}, \qquad (9.90)$$

where $C[t_{ks\bullet}, t_{k's\bullet}]$ indicates the unconditional covariance between the multi-tally count variables for $\mathcal{U}_k$ and $\mathcal{U}_{k'}$ on the sth transect, namely

$$C[t_{ks\bullet}, t_{k's\bullet}] = E\left[t_{ks\bullet} t_{k's\bullet}\right] - E\left[t_{ks\bullet}\right] E\left[t_{k's\bullet}\right]$$

$$= E\left[\sum_{j=1}^{J} t_{ksj} \sum_{j=1}^{J} t_{k'sj}\right] - \pi_k \pi_{k'}$$

$$= \sum_{j=1}^{J} E\left[t_{ksj} t_{k'sj}\right] + \sum_{j=1}^{J} \sum_{\substack{j' \neq j \\ j'=1}}^{J} E\left[t_{ksj} t_{k'sj'}\right] - \pi_k \pi_{k'}$$

$$= \sum_{j=1}^{J} \pi_{kk'j} + \sum_{j=1}^{J} \sum_{\substack{j' \neq j \\ j'=1}}^{J} \frac{\eta_{kjk'j's}}{A} - \pi_k \pi_{k'}, \qquad (9.91)$$

where $\pi_{kk'j}$ is the analog, when dealing with transects having $J > 1$ segments to $\pi_{kk'}$ with straight-line transects (see (9.15)), namely it is the joint unconditional probability of selecting $\mathcal{U}_k$ and $\mathcal{U}_{k'}$ from the jth segment of the sth transect; and $\eta_{kjk'j's}/A$ is the joint unconditional probability of selecting $\mathcal{U}_k$ and $\mathcal{U}_{k'}$ on the jth and j'th segments of the sth transect.

Substituting from (9.82) for $V[t_{ks\bullet}]$ and from (9.91) for $C[t_{ks\bullet}, t_{k's\bullet}]$ yields

$$V\left[\ddot{\tau}_{yms}^{\mathrm{u}}\right] = \frac{\pi^2 A^2}{L^2} \sum_{k=1}^{N} \frac{\pi_k(1-\pi_k)}{c_k^2} y_k^2 + \frac{\pi^2 A^2}{L^2} \sum_{k=1}^{N} \sum_{j=1}^{J} \sum_{\substack{j'\neq j \\ j'=1}}^{J} \frac{c_{kjj'}/A}{c_k^2} y_k^2$$

$$+ \frac{\pi^2 A^2}{L^2} \sum_{k=1}^{N} \sum_{\substack{k'\neq k \\ k'=1}}^{N} \sum_{j=1}^{J} \frac{\pi_{kk'j}}{c_k c_{k'}} y_k y_{k'}$$

$$+ \frac{\pi^2 A^2}{L^2} \sum_{k=1}^{N} \sum_{\substack{k'\neq k \\ k'=1}}^{N} \sum_{j=1}^{J} \sum_{\substack{j'\neq j \\ j'=1}}^{J} \frac{\eta_{kjk'j's}/A}{c_k c_{k'}} y_k y_{k'}$$

$$- \frac{\pi^2 A^2}{L^2} \sum_{k=1}^{N} \sum_{\substack{k'\neq k \\ k'=1}}^{N} \frac{\pi_k \pi_{k'}}{c_k c_{k'}} y_k y_{k'}. \tag{9.92}$$

Replacing c_k and $c_{k'}$ with $\pi A \pi_k / L$ and $\pi A \pi_{k'} / L$, respectively, in (9.92) yields

$$V\left[\ddot{\tau}_{yms}^{\mathrm{u}}\right] = \sum_{k=1}^{N} \frac{y_k^2}{\pi_k} \left[1 + \left(\sum_{j=1}^{J} \sum_{\substack{j'=1 \\ j'\neq j}}^{J} \frac{c_{kjj'}}{A\pi_k}\right) - \pi_k\right]$$

$$+ \sum_{k=1}^{N} \sum_{\substack{k'=1 \\ k'\neq k}}^{N} \frac{y_k y_{k'}}{\pi_k \pi_{k'}} \left[\left(\sum_{j=1}^{J} \pi_{kk'j} + \sum_{j=1}^{J} \sum_{\substack{j'=1 \\ j'\neq j}}^{J} \frac{\eta_{kjk'j's}}{A}\right) - \pi_k \pi_{k'}\right].$$
$$\tag{9.93}$$

When $J = 1$, the terms in (9.93) involving $c_{kjj'}$ and $\eta_{kjk'j's}$ are nil, and $V\left[\ddot{\tau}_{yms}^{\mathrm{u}}\right]$ coincides with $V\left[\hat{\tau}_{y\pi s}^{\mathrm{u}}\right]$ in (9.15a).

9.12.9 Variance of $\ddot{\tau}_{yqs}^{\mathrm{c}}$

From (9.65)

$$\ddot{\tau}_{yqs}^{\mathrm{c}} = \sum_{u_k \in \mathbb{L}_s} \frac{\sum_{j=1}^{J} t_{ksj} q_k(\theta_{sj})}{\sum_{j=1}^{J} E\left[t_{ksj} q_k(\theta_{sj}) \mid \theta_{s1}\right]} y_k$$

so that

$$V\left[\ddot{\tau}_{yqs}^{c}\right] = \sum_{k=1}^{N} \frac{V\left[\sum_{j=1}^{J} t_{ksj} q_k(\theta_{sj}) \mid \theta_{s1}\right] y_k^2}{\left(\sum_{j=1}^{J} E\left[t_{ksj} q_k(\theta_{sj}) \mid \theta_{s1}\right]\right)^2}$$

$$+ \sum_{k=1}^{N} \sum_{\substack{k' \neq k \\ k'=1}}^{N} \frac{C\left[\sum_{j=1}^{J} t_{ksj} q_k(\theta_{sj}), \sum_{j=1}^{J} t_{k'sj} q_{k'}(\theta_{sj}) \mid \theta_{s1}\right] y_k y_{k'}}{\left(\sum_{j=1}^{J} E\left[t_{ksj} q_k(\theta_{sj}) \mid \theta_{s1}\right]\right)\left(\sum_{j=1}^{J} E\left[t_{k'sj} q_{k'}(\theta_{sj}) \mid \theta_{s1}\right]\right)}.$$

$$(9.94)$$

We examine first the properties of $\sum_{j=1}^{J} t_{ksj} q_k(\theta_{sj})$ by conditioning on both the orientation of the transect and $q_k(\theta_{sj})$. The doubly conditioned expected value of $t_{ksj} q_k(\theta_{sj})$ is

$$E\left[t_{ksj} q_k(\theta_{sj}) \mid \theta_{s1}, q_k(\theta_{sj})\right] = E\left[t_{ksj} \mid \theta_{s1}\right] q_k(\theta_{sj})$$

$$= \pi_k\left(\theta_{sj}\right) q_k(\theta_{sj}), \qquad (9.95)$$

which then leads to

$$E\left[\sum_{j=1}^{J} t_{ksj} q_k(\theta_{sj}) \mid \theta_{s1}, q_k(\theta_{sj})\right] = \sum_{j=1}^{J} E\left[t_{ksj} q_k(\theta_{sj}) \mid \theta_{s1}, q_k(\theta_{sj})\right]$$

$$= \sum_{j=1}^{J} \pi_k\left(\theta_{sj}\right) q_k(\theta_{sj}) \qquad (9.96a)$$

$$= \bar{q}_k(\theta_{s1}). \qquad (9.96b)$$

The corresponding conditional variance is

$$V\left[\left(\sum_{j=1}^{J} t_{ksj} q_k(\theta_{sj})\right) \mid \theta_{s1}, q_k(\theta_{sj})\right]$$

$$= \sum_{j=1}^{J} V\left[t_{ksj} q_k(\theta_{sj}) \mid \theta_{s1}, q_k(\theta_{sj})\right]$$

$$+ \sum_{j=1}^{J} \sum_{\substack{j' \neq j \\ j'=1}}^{J} C\left[t_{ksj} q_k(\theta_{sj}), t_{ksj'} q_k(\theta_{sj'}) \mid \theta_{s1}, q_k(\theta_{sj}), q_k(\theta_{sj'})\right]$$

$$= \sum_{j=1}^{J} V\left[t_{ksj} \mid \theta_{s1}\right] q_k^2(\theta_{sj}) + \sum_{j=1}^{J} \sum_{\substack{j' \neq j \\ j'=1}}^{J} C\left[t_{ksj}, t_{ksj'} \mid \theta_{s1}\right] q_k(\theta_{sj}) q_k(\theta_{sj'})$$

$$= \sum_{j=1}^{J} \pi_k(\theta_{sj}) \left(1 - \pi_k(\theta_{sj})\right) q_k^2(\theta_{sj})$$

$$+ \sum_{j=1}^{J} \sum_{\substack{j' \neq j \\ j'=1}}^{J} \left(E\left[t_{ksj} t_{ksj'} \mid \theta_{s1}\right] - \pi_k(\theta_{sj})\pi_k(\theta_{sj'})\right) q_k(\theta_{sj}) q_k(\theta_{sj'})$$

$$= \sum_{j=1}^{J} \pi_k(\theta_{sj}) \left(1 - \pi_k(\theta_{sj})\right) q_k^2(\theta_{sj})$$

$$\sum_{j=1}^{J} \sum_{\substack{j' \neq j \\ j'=1}}^{J} \pi_k(\theta_{sj})\pi_k(\theta_{sj'}) q_k(\theta_{sj}) q_k(\theta_{sj'}) + \sum_{j=1}^{J} \sum_{\substack{j' \neq j \\ j'=1}}^{J} \frac{\psi_{kjj'}}{A} q_k(\theta_{sj}) q_k(\theta_{sj'}).$$

$$= \sum_{j=1}^{J} \pi_k(\theta_{sj}) q_k(\theta_{sj}) \left[\left(1 - \pi_k(\theta_{sj})\right) q_k(\theta_{sj}) - \sum_{\substack{j' \neq j \\ j'=1}}^{J} \pi_k(\theta_{sj'}) q_k(\theta_{sj'}) \right]$$

$$+ \sum_{j=1}^{J} \sum_{\substack{j' \neq j \\ j'=1}}^{J} \frac{\psi_{kjj'}}{A} q_k(\theta_{sj}) q_k(\theta_{sj'})$$

$$= \sum_{j=1}^{J} \pi_k(\theta_{sj}) q_k(\theta_{sj}) \left[q_k(\theta_{sj}) - \bar{q}_k(\theta_{s1})\right] + \sum_{j=1}^{J} \sum_{\substack{j' \neq j \\ j'=1}}^{J} \frac{\psi_{kjj'}}{A} q_k(\theta_{sj}) q_k(\theta_{sj'}),$$

$$(9.97)$$

where $\psi_{kjj'}/A$ is the probability of intersecting $\mathcal{U}_k$ by the jth and j'th segments of the sth transect.

The variance of $\sum_{j=1}^{J} t_{ksj} q_k(\theta_{sj})$, conditionally on transect orientation only, can be expressed as

$$V\left[\sum_{j=1}^{J} t_{ksj} q_k(\theta_{sj}) \mid \theta_{s1}\right] = V_q\left[E\left[\sum_{j=1}^{J} t_{ksj} q_k(\theta_{sj}) \mid \theta_{s1}, q_k(\theta_{sj})\right]\right] \quad (9.98a)$$

$$+ E_q\left[V\left[\sum_{j=1}^{J} t_{ksj} q_k(\theta_{sj}) \mid \theta_{s1}, q_k(\theta_{sj})\right]\right], \quad (9.98b)$$

where E_q signifies the expected value over the conditional distribution of $q_k(\theta_{sj})$ given transect orientation, θ_{s1}, and V_q signifies the variance of the conditional distribution of $q_k(\theta_{sj})$ given θ_{s1}.

Examining (9.98a) first, we have

$$
V_q \left[E \left[\sum_{j=1}^{J} t_{ksj} q_k(\theta_{sj}) \,\middle|\, \theta_{s1}, q_k(\theta_{sj}) \right] \right]
$$

$$
= V_q \left[\left(\sum_{j=1}^{J} \pi_k(\theta_{sj}) q_k(\theta_{sj}) \right) \,\middle|\, \theta_{s1} \right] \quad \text{from (9.96a)}
$$

$$
= \sum_{j=1}^{J} \pi_k^2(\theta_{sj}) V_q \left[q_k(\theta_{sj}) \,\middle|\, \theta_{s1} \right]
$$

$$
+ \sum_{j=1}^{J} \sum_{\substack{j' \neq j \\ j'=1}}^{J} \pi_k(\theta_{sj}) \pi_k(\theta_{sj'}) C_q \left[q_k(\theta_{sj}), q_k(\theta_{sj'}) \,\middle|\, \theta_{s1} \right], \tag{9.99}
$$

where $C_q \left[q_k(\theta_{sj}), q_k(\theta_{sj'}) \,\middle|\, \theta_{s1} \right]$ is the covariance of $q_k(\theta_{sj})$ and $q_k(\theta_{sj'})$ conditional on transect orientation.

Examining (9.98b) next, we have

$$
E_q \left[V \left[\sum_{j=1}^{J} t_{ksj} q_k(\theta_{sj}) \,\middle|\, \theta_{s1}, q_k(\theta_{sj}) \right] \right]
$$

$$
= E_q \left[\sum_{j=1}^{J} \pi_k(\theta_{sj}) q_k(\theta_{sj}) \left[q_k(\theta_{sj}) - \bar{q}_k(\theta_{s1}) \right] + \sum_{j=1}^{J} \sum_{\substack{j' \neq j \\ j'=1}}^{J} \frac{\psi_{kjj'}}{A} q_k(\theta_{sj}) q_k(\theta_{sj'}) \right]
$$

$$
= \sum_{j=1}^{J} \pi_k(\theta_{sj}) E_q \left[q_k(\theta_{sj}) \left(q_k(\theta_{sj}) - \bar{q}_k(\theta_{s1}) \right) \,\middle|\, \theta_{s1} \right]
$$

$$
+ \sum_{j=1}^{J} \sum_{\substack{j' \neq j \\ j'=1}}^{J} \frac{\psi_{kjj'}}{A} E_q \left[q_k(\theta_{sj}) q_k(\theta_{sj'}) \,\middle|\, \theta_{s1} \right]. \tag{9.100}
$$

Upon combining (9.99) and (9.100), we get

$$
V\left[\sum_{j=1}^{J} t_{ksj} q_k(\theta_{sj}) \,\big|\, \theta_{s1}\right] = \sum_{j=1}^{J} \pi_k^2(\theta_{sj}) \, V_q\left[q_k(\theta_{sj}) \,\big|\, \theta_{s1}\right]
$$

$$
+ \sum_{j=1}^{J} \pi_k(\theta_{sj}) E_q\left[q_k(\theta_{sj})\left(q_k(\theta_{sj}) - \bar{q}_k(\theta_{s1})\right) \,\big|\, \theta_{s1}\right]
$$

$$
+ \sum_{j=1}^{J} \sum_{\substack{j' \neq j \\ j'=1}}^{J} \pi_k(\theta_{sj}) \pi_k(\theta_{sj'}) C_q\left[q_k(\theta_{sj}), q_k(\theta_{sj'}) \,\big|\, \theta_{s1}\right]
$$

$$
+ \sum_{j=1}^{J} \sum_{\substack{j' \neq j \\ j'=1}}^{J} \frac{\psi_{kjj'}}{A} E_q\left[q_k(\theta_{sj}) q_k(\theta_{sj'}) \,\big|\, \theta_{s1}\right]. \quad (9.101)
$$

When $q_k(\theta_{sj}) = 1$ universally, the V_q and C_q terms in (9.101) are nil, whereas $E_q\left[q_k(\theta_{sj})\left(q_k(\theta_{sj}) - \bar{q}_k(\theta_{s1})\right) \big| \theta_{s1}\right] = 1 - \tilde{\pi}_k(\theta_s)$, and $V\left[\sum_{j=1}^{J} t_{ksj} q_k(\theta_{sj}) \big| \theta_{s1}\right]$ in (9.101) coincides with $V[t_{ks\bullet} | \theta_{s1}]$ in (9.79).

The next term from (9.94) that needs to be derived for $V[\ddot{t}_{yqs}^c]$ is that involving the conditional covariance between $\sum_{j=1}^{J} t_{ksj} q_k(\theta_{sj})$ and $\sum_{j=1}^{J} t_{k'sj} q_{k'}(\theta_{sj})$.

$$
C\left[\sum_{j=1}^{J} t_{ksj} q_k(\theta_{sj}), \sum_{j=1}^{J} t_{k'sj} q_{k'}(\theta_{sj}) \,\big|\, \theta_{s1}\right]
$$

$$
= E\left[\sum_{j=1}^{J} t_{ksj} q_k(\theta_{sj}) \sum_{j=1}^{J} t_{k'sj} q_{k'}(\theta_{sj}) \,\big|\, \theta_{s1}\right]
$$

$$
- E\left[\sum_{j=1}^{J} t_{ksj} q_k(\theta_{sj}) \,\big|\, \theta_{s1}\right] E\left[\sum_{j=1}^{J} t_{k'sj} q_{k'}(\theta_{sj}) \,\big|\, \theta_{s1}\right]
$$

$$
= E\left[\sum_{j=1}^{J} t_{ksj} q_k(\theta_{sj}) t_{k'sj} q_{k'}(\theta_{sj})\right]
$$

$$
+ E\left[\sum_{j=1}^{J} \sum_{\substack{j' \neq j \\ j'=1}}^{J} t_{ksj} q_k(\theta_{sj}) t_{k'sj'} q_{k'}(\theta_{sj'}) \,\big|\, \theta_{s1}\right]
$$

$$
- E_q\left[\bar{q}_k(\theta_{s1}) \,\big|\, \theta_{s1}\right] E_q\left[\bar{q}_{k'}(\theta_{s1}) \,\big|\, \theta_{s1}\right]
$$

$$= E_q \left[\sum_{j=1}^{J} \pi_{kk'j} q_k(\theta_{sj}) q_{k'sj}(\theta_{sj}) \,\middle|\, \theta_{s1} \right]$$

$$+ E_q \left[\sum_{j=1}^{J} \sum_{\substack{j' \neq j \\ j'=1}}^{J} \left(\zeta_{kjk'j'} / A \right) q_k(\theta_{sj}) q_{k'} \left(\theta_{k'sj'} \right) \,\middle|\, \theta_{s1} \right]$$

$$- E_q \left[\bar{q}_k(\theta_{s1}) \,\middle|\, \theta_{s1} \right] E_q \left[\bar{q}_{k'}(\theta_{s1}) \,\middle|\, \theta_{s1} \right]. \tag{9.102}$$

Substituting (9.101) and (9.102) into (9.94), the following expression for $V\left[\ddot{\tau}^c_{yqs} \right]$ results:

$$V\left[\ddot{\tau}^c_{yqs} \right] = \sum_{k=1}^{N} \frac{\sum_{j=1}^{J} \pi_k^2(\theta_{sj}) V_q \left[q_k(\theta_{sj}) \,\middle|\, \theta_{s1} \right]}{\left(E_q \left[\bar{q}_k(\theta_{s1}) \,\middle|\, \theta_{s1} \right] \right)^2} y_k^2$$

$$+ \sum_{k=1}^{N} \frac{\sum_{j=1}^{J} \pi_k(\theta_{sj}) E_q \left[q_k(\theta_{sj}) (q_k(\theta_{sj}) - \bar{q}_k(\theta_{s1})) \,\middle|\, \theta_{s1} \right]}{\left(E_q \left[\bar{q}_k(\theta_{s1}) \,\middle|\, \theta_{s1} \right] \right)^2} y_k^2$$

$$+ \sum_{k=1}^{N} \frac{\sum_{j=1}^{J} \sum_{\substack{j' \neq j \\ j'=1}}^{J} \pi_k(\theta_{sj}) \pi_k(\theta_{sj'}) C_q \left[q_k(\theta_{sj}), q_k(\theta_{sj'}) \,\middle|\, \theta_{s1} \right]}{\left(E_q \left[\bar{q}_k(\theta_{s1}) \,\middle|\, \theta_{s1} \right] \right)^2} y_k^2$$

$$+ \sum_{k=1}^{N} \frac{\sum_{j=1}^{J} \sum_{\substack{j' \neq j \\ j'=1}}^{J} \left(\psi_{kjj'} / A \right) E_q \left[q_k(\theta_{sj}) q_k(\theta_{sj'}) \,\middle|\, \theta_{s1} \right]}{\left(E_q \left[\bar{q}_k(\theta_{s1}) \,\middle|\, \theta_{s1} \right] \right)^2} y_k^2$$

$$+ \sum_{k=1}^{N} \sum_{\substack{k' \neq k \\ k'=1}}^{N} \frac{E_q \left[\sum_{j=1}^{J} \pi_{kk'j} q_k(\theta_{sj}) q_{k'sj}(\theta_{sj}) \,\middle|\, \theta_{s1} \right]}{\left(E_q \left[\bar{q}_k(\theta_{s1}) \,\middle|\, \theta_{s1} \right] \right) \left(E_q \left[\bar{q}_{k'}(\theta_{s1}) \,\middle|\, \theta_{s1} \right] \right)} y_k y_{k'}$$

$$- \sum_{k=1}^{N} \sum_{\substack{k' \neq k \\ k'=1}}^{N} \frac{E_q \left[\sum_{j=1}^{J} \sum_{\substack{j' \neq j \\ j'=1}}^{J} \left(\zeta_{kjk'j'} / A \right) q_k(\theta_{sj}) q_{k'}(\theta_{k'sj'}) \,\middle|\, \theta_{s1} \right]}{\left(E_q \left[\bar{q}_k(\theta_{s1}) \,\middle|\, \theta_{s1} \right] \right) \left(E_q \left[\bar{q}_{k'}(\theta_{s1}) \,\middle|\, \theta_{s1} \right] \right)} y_k y_{k'}$$

$$+ \sum_{k=1}^{N} \sum_{\substack{k' \neq k \\ k'=1}}^{N} y_k y_{k'}. \tag{9.103}$$

When $q_k(\theta_s) = 1$ universally, all V_q and C_q terms in (9.103) are nil, and so $E_q\left[\bar{q}_k(\theta_{s1}) \middle| \theta_{s1} \right] = \tilde{\pi}_k(\theta_{s1})$, $E_q\left[\bar{q}_{k'}(\theta_{s1}) \middle| \theta_{s1} \right] = \tilde{\pi}_{k'}(\theta_{s1})$ and thus $V\left[\ddot{\tau}^c_{yqs} \right]$ coincides with $V\left[\ddot{\tau}^c_{yms} \right]$ in (9.88).

9.12.10 Variance of $\ddot{\tau}^u_{yqs}$

From (9.66)

$$\ddot{\tau}^u_{yqs} = \sum_{u_k \in \mathbb{L}_s} \frac{\sum_{j=1}^{J} t_{ksj} q_k(\theta_{sj})}{\sum_{j=1}^{J} E\left[t_{ksj} q_k(\theta_{sj})\right]} y_k$$

so that

$$V\left[\ddot{\tau}^u_{yqs}\right] = E\left[\left(\sum_{k=1}^{N} \frac{\sum_{j=1}^{J} t_{ksj} q_k(\theta_{sj})}{\sum_{j=1}^{J} E\left[t_{ksj} q_k(\theta_{sj})\right]} y_k\right)^2\right] - \tau_y^2$$

$$= \sum_{k=1}^{N} E\left[\left(\frac{\sum_{j=1}^{J} t_{ksj} q_k(\theta_{sj})}{\sum_{j=1}^{J} E\left[t_{ksj} q_k(\theta_{sj})\right]} y_k\right)^2\right]$$

$$+ \sum_{k=1}^{N} \sum_{\substack{k' \neq k \\ k'=1}}^{N} E\left[\frac{\left[\sum_{j=1}^{J} t_{ksj} q_k(\theta_{sj})\right]\left[\sum_{j=1}^{J} t_{k'sj} q_{k'}(\theta_{sj})\right]}{\sum_{j=1}^{J} E\left[t_{ksj} q_k(\theta_{sj})\right] \sum_{j=1}^{J} E\left[t_{k'sj} q_{k'}(\theta_{sj})\right]} y_k y_{k'}\right] - \tau_y^2$$

$$= \sum_{k=1}^{N} E\left[\frac{\sum_{j=1}^{J} t_{ksj}^2 q_k^2(\theta_{sj}) + \sum_{j=1}^{J} \sum_{\substack{j' \neq j \\ j'=1}}^{J} t_{ksj} t_{ksj'} q_k(\theta_{sj}) q_k(\theta_{sj'})}{\left(\sum_{j=1}^{J} E\left[t_{ksj} q_k(\theta_{sj})\right]\right)^2} - 1\right] y_k^2$$

$$+ \sum_{k=1}^{N} \sum_{\substack{k' \neq k \\ k'=1}}^{N} E\left[\frac{\sum_{j=1}^{J} t_{ksj} q_k(\theta_{sj}) t_{k'sj} q_{k'}(\theta_{sj}) + \sum_{j=1}^{J} \sum_{\substack{j' \neq j \\ j'=1}}^{J} t_{ksj} t_{k'sj'} q_k(\theta_{sj}) q_{k'}(\theta_{sj'})}{\sum_{j=1}^{J} E\left[t_{ksj} q_k(\theta_{sj})\right] \sum_{j=1}^{J} E\left[t_{k'sj} q_{k'}(\theta_{sj})\right]} - 1\right] y_k y_{k'}$$

$$= \sum_{k=1}^{N} \left(\frac{\sum_{j=1}^{J} E\left[t_{ksj}^2 q_k^2(\theta_{sj})\right] + \sum_{j=1}^{J} \sum_{\substack{j' \neq j \\ j'=1}}^{J} E\left[t_{ksj} t_{ksj'} q_k(\theta_{sj}) q_k(\theta_{sj'})\right]}{\left(\sum_{j=1}^{J} E\left[t_{ksj} q_k(\theta_{sj})\right]\right)^2} - 1\right) y_k^2$$

$$+ \sum_{k=1}^{N} \sum_{\substack{k' \neq k \\ k'=1}}^{N} \left(\frac{\sum_{j=1}^{J} E\left[t_{ksj} q_k(\theta_{sj}) t_{k'sj} q_{k'}(\theta_{sj})\right] + \sum_{j=1}^{J} \sum_{\substack{j' \neq j \\ j'=1}}^{J} E\left[t_{ksj} t_{k'sj'} q_k(\theta_{sj}) q_{k'}(\theta_{sj'})\right]}{\sum_{j=1}^{J} E\left[t_{ksj} q_k(\theta_{sj})\right] \sum_{j=1}^{J} E\left[t_{k'sj} q_{k'}(\theta_{sj})\right]} - 1\right) y_k y_{k'}$$

$$= \sum_{k=1}^{N} \left(\frac{\sum_{j=1}^{J} \sum_{\substack{j'\neq j \\ j'=1}}^{J} \left(E\left[t_{ksj}t_{ksj'}q_k(\theta_{sj})q_k(\theta_{sj'})\right] - E\left[t_{ksj}q_k(\theta_{sj})\right] E\left[t_{ksj'}q_k(\theta_{sj'})\right] \right)}{\left(\sum_{j=1}^{J} E\left[t_{ksj}q_k(\theta_{sj})\right] \right)^2} \right) y_k^2$$

$$+ \sum_{k=1}^{N} \sum_{\substack{k'\neq k \\ k'=1}}^{N} \left(\frac{\sum_{j=1}^{J} \sum_{\substack{j'\neq j \\ j'=1}}^{J} \left(E\left[t_{ksj}t_{k'sj'}q_k(\theta_{sj})q_{k'}(\theta_{sj'})\right] - E\left[t_{ksj}q_k(\theta_{sj})\right] E\left[t_{k'sj'}q_{k'}(\theta_{sj'})\right] \right)}{\sum_{j=1}^{J} E\left[t_{ksj}q_k(\theta_{sj})\right] \sum_{j=1}^{J} E\left[t_{k'sj'}q_{k'}(\theta_{sj})\right]} \right) y_k y_{k'}$$

$$= \sum_{k=1}^{N} \sum_{j=1}^{J} \sum_{\substack{j'\neq j \\ j'=1}}^{J} \left(\frac{C\left(t_{ksj}q_k(\theta_{sj}), t_{ksj'}q_k(\theta_{sj'})\right)}{\left(\sum_{j=1}^{J} E\left[t_{ksj}q_k(\theta_{sj})\right] \right)^2} \right) y_k^2$$

$$+ \sum_{k=1}^{N} \sum_{\substack{k'\neq k \\ k'=1}}^{N} \sum_{j=1}^{J} \sum_{\substack{j'\neq j \\ j'=1}}^{J} \left(\frac{C\left(t_{ksj}q_k(\theta_{sj}), t_{k'sj'}q_{k'}(\theta_{sj'})\right)}{\sum_{j=1}^{J} E\left[t_{ksj}q_k(\theta_{sj})\right] \sum_{j=1}^{J} E\left[t_{k'sj'}q_{k'}(\theta_{sj})\right]} \right) y_k y_{k'}. \tag{9.104}$$

CHAPTER 10

A Monte Carlo Integration Approach to Areal Sampling

In this chapter we explore a Monte Carlo approach to special designs that have been developed by forest mensurationists and ecologists for the purpose of estimating attributes of forested tracts. Some of these designs can be applied outside the forest, but we describe each of them, for the most part, in the context of its original application.

We have already devoted many pages to descriptions of three special designs for discrete populations, which are widely known by the names plot sampling (Chapter 7), Bitterlich sampling (Chapter 8), and line intersect sampling (Chapter 9). Plot sampling and line intersect sampling are designs with quite general applicability. Bitterlich sampling, by contrast, is primarily a forest sampling design. Within a forested tract, a population of interest often comprises N discrete elements, which are scattered over the landscape. For example, plot sampling or Bitterlich sampling may be used where the discrete units of interest are standing trees, and line intersect sampling where the discrete units of interest may be fallen trees or pieces of coarse woody debris. In any case, attributes of the discrete sample units are measured, and an estimate of the total quantity of each attribute may be calculated for the entire population of N units. This estimate for the N units may also be considered an estimate for a tract, insofar as the N units occur within the closed boundary of the tract.

Designs that use plots, lines, or points, and several related designs may also be formulated as special designs for sampling the areal continuum (e.g., Mandallaz 1991; Eriksson 1995b; Valentine et al. 2001, 2006; Barabesi 2003). In each case, the continuous population comprises the infinitely many location points on the horizontal projection of the land surface of a tract. Sample points are selected uniformly at random within a tract and attribute densities are measured at these sample points. In effect, the special designs for areal sampling reduce to two-dimensional Monte Carlo integration (§4.5).

Whether one chooses to view the special designs for areal sampling from a discrete- or continuous-population perspective is a matter of personal choice. In this chapter, we present the continuous view. If we get down to brass tacks, however, we find that the sampling protocols for the special designs and the measurements taken on elements are unaffected by the choice of perspective.

For ease of presentation, we specify, for this chapter, that all lengths, widths, diameters, and radii are measured in m, and all areas, including the tract area, are measured in m^2.

10.1 Areal sampling

Let A be the horizontal area of tract $\mathcal{A}$ and let τ_ρ be the total amount of some attribute of interest within the boundary of $\mathcal{A}$. By definition, $\tau_\rho = \iint_{\mathcal{A}} \rho(x, z)\, dz\, dx$, where $\rho(x, z)$ is the attribute density at the location point with coordinates (x, z) (see §4.5). In reality, many forest attributes are summations of attributes of an unknown number of trees or other discrete elements. However, the protocols of the areal designs that we consider in this chapter define inclusion zones for discrete elements. Hence any attribute of any element may be converted into a continuous attribute density—the amount of attribute per unit area—simply by prorating the value of the attribute over the horizontal area of the inclusion zone.

Let $\mathcal{I}_k$ denote the inclusion zone of the kth of the N elements in $\mathcal{A}$, and let a_k be the horizontal land area of $\mathcal{I}_k$. In the absence of edge effect, the attribute density at any location point within the inclusion zone is

$$\rho_k(x, z) = \frac{y_k}{a_k}, \qquad (x, z) \in \mathcal{I}_k, \tag{10.1}$$

where y_k is the value of the attribute of the kth element. The attribute density for the kth element is zero everywhere outside of $\mathcal{I}_k$, i.e.,

$$\rho_k(x, z) = 0, \qquad (x, z) \notin \mathcal{I}_k. \tag{10.2}$$

Example 10.1

In plot sampling with round plots, each of the N elements in $\mathcal{A}$ is centered in a round inclusion zone, which is the same size as a plot. If we divide a measurement of an attribute of an element by the horizontal land area of element's inclusion zone, we obtain an attribute density. For example, if the elements are trees and the attribute of interest is basal area, then the attribute density is basal area per unit land area.

As noted in previous chapters, edge effect occurs where an element is located so close to the boundary of $\mathcal{A}$ that part of the element's inclusion zone extends outside of $\mathcal{A}$. While there are different approaches for dealing with edge effect (see §10.7), we presume that the sampling protocol incorporates an edge-correction method, which is designed to 'remap' the inclusion zones of edge elements, reducing the attribute density to zero in any section of the inclusion zone that occurs outside of $\mathcal{A}$ and doubling the density in a section of equal area that occurs inside of $\mathcal{A}$ (see §10.7.3–§10.7.5). In some cases, remapped sections with single or double densities may overlap, in which case the density is tripled or quadrupled at each location point in the overlapping section. Let $t_k(x, z) = 0, 1, 2, 3, 4, \ldots$ be the factor by which the attribute density at a location point (x, z) is multiplied to correct for edge effect. Of course, $t_k(x, z) = 1$ implies no edge correction.

In general, after correction for edge effect, the attribute density for the kth element at any location point (x, z) in $\mathcal{I}_k$ is

$$\rho_k(x, z) = \frac{y_k t_k(x, z)}{a_k} \tag{10.3}$$

and, therefore, integration of the attribute density across the area of $\mathcal{I}_k \cap \mathcal{A}$—the portion of $\mathcal{I}_k$ that occurs in $\mathcal{A}$—yields the attribute, i.e.,

$$\frac{y_k}{a_k} \iint_{\mathcal{I}_k \cap \mathcal{A}} t_k(x, z) \, dz \, dx = \frac{y_k}{a_k} a_k = y_k. \tag{10.4}$$

Inclusion zones of two or more different elements may overlap a given location point, in which case the total attribute density at the location point equals the sum of the attribute densities in the overlapping inclusion zones, i.e.,

$$\rho(x, z) = \sum_{k=1}^{N} \rho_k(x, z). \tag{10.5}$$

If we let $t_k(x, z) = 0$ if $(x, z) \notin \mathcal{I}_k$, then the previous equation can be rewritten as

$$\rho(x, z) = \sum_{k=1}^{N} \frac{y_k t_k(x, z)}{a_k}. \tag{10.6}$$

Hence, the total attribute in $\mathcal{A}$ is equivalent to the integral of the attribute density over the area of $\mathcal{A}$, i.e.,

$$\tau_\rho = \iint_{\mathcal{A}} \rho(x, z) \, dz \, dx$$

$$= \iint_{\mathcal{A}} \sum_{k=1}^{N} \frac{y_k t_k(x, z)}{a_k} \, dz \, dx \tag{10.7}$$

$$= \sum_{k=1}^{N} y_k.$$

This allows us to estimate τ_ρ by Monte Carlo integration, selecting all our sample points from the population of location points in $\mathcal{A}$.

10.1.1 Selection

Recall from §4.5 that a sample point, $\mathbb{P}_s$, at (x_s, z_s), $s = 1, 2, \ldots, m$, is selected independently with probability density $f(x_s, z_s)$, where $f(x, y) > 0$ for all $(x, y) \in \mathcal{A}$, and $\iint_{\mathcal{A}} f(x, z) \, dx \, dz = 1$. Ordinarily, we use a uniform density, i.e., $f(x, z) = 1/A$ for all $(x, y) \in \mathcal{A}$. Thus, the sample point at (x_s, z_s) may be selected with density $f(x_s, z_s) = 1/A$ by the acceptance-rejection method (§4.5.1).

If the sampling protocol indicates that a sample point falls in the inclusion zone of the kth element, we measure y_k and, if need be, a_k. If the kth element is near the edge of $\mathcal{A}$, we use an edge-correction method to determine the value of $t_k(x_s, z_s)$; otherwise, $t_k(x_s, z_s) = 1$.

10.1.2 Estimation

The attribute, τ_ρ, is unbiasedly estimated by

$$\hat{\tau}_{\rho s} = \frac{\rho(x_s, z_s)}{f(x_s, z_s)}. \tag{10.8}$$

If $f(x_s, z_s) = 1/A$, then estimator of τ_ρ simplifies to:

$$\hat{\tau}_{\rho s} = A\rho(x_s, z_s)$$
$$= A \sum_{\mathbb{P}_s \in \mathcal{I}_k} \frac{y_k t_k}{a_k}, \tag{10.9}$$

where $t_k \equiv t_k(x_s, z_s)$ is the 'edge-correction factor' or 'tally' for the kth element at the sample point. The notation $(x_s, z_s) \in \mathcal{I}_k$ indicates that the summation is over all elements in $\mathcal{A}$ whose inclusion zones include the sth sample point.

The mean attribute density in $\mathcal{A}$, i.e., $\mu_\rho = \tau_\rho/A$, is unbiasedly estimated by

$$\hat{\mu}_{\rho s} = \rho(x_s, z_s). \tag{10.10}$$

Combined estimates obtain by averaging across $m \geq 2$ sample points:

$$\hat{\tau}_{\rho,\mathrm{rep}} = \frac{1}{m} \sum_{s=1}^{m} \hat{\tau}_{\rho s} \tag{10.11}$$

and

$$\hat{\mu}_{\rho,\mathrm{rep}} = \frac{1}{m} \sum_{s=1}^{m} \hat{\mu}_{\rho s}. \tag{10.12}$$

The variance of $\hat{\tau}_{\rho,\mathrm{rep}}$ is

$$V\left[\hat{\tau}_{\rho,\mathrm{rep}}\right] = \frac{1}{m} \left(A \iint_{\mathcal{A}} \rho^2(x, z) \, dz \, dx - \tau_\rho^2 \right). \tag{10.13}$$

The calculation of this variance is discussed in the appendix (§10.11.1). An estimator of the variance of $\hat{\tau}_{\rho,\mathrm{rep}}$ is:

$$\hat{v}\left[\hat{\tau}_{\rho,\mathrm{rep}}\right] = \frac{1}{m(m-1)} \sum_{s=1}^{m} \left(\hat{\tau}_{\rho s} - \hat{\tau}_{\rho,\mathrm{rep}}\right)^2 \qquad m \geq 2. \tag{10.14}$$

The variance of $\hat{\mu}_{\rho,\mathrm{rep}}$ is $V[\hat{\mu}_{\rho,\mathrm{rep}}] = V[\hat{\tau}_{\rho,\mathrm{rep}}]/A^2$, and an estimator of this variance is

$$\hat{v}\left[\hat{\mu}_{\rho,\mathrm{rep}}\right] = \frac{1}{m(m-1)} \sum_{s=1}^{m} \left(\hat{\mu}_{\rho s} - \hat{\mu}_{\rho,\mathrm{rep}}\right)^2 \qquad m \geq 2. \tag{10.15}$$

Areal sampling designs differ principally with regard to how inclusion zones are defined and their areas are measured. In the following sections, we describe how to

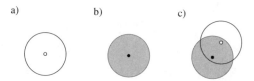

Figure 10.1 a) *A fixed-radius plot fixed about a sample point* (∘) *defines a fixed-radius inclusion zone about the center point* (•) *of any element of interest* (b). *The element occupies a fixed-radius plot if the sample point falls anywhere in the element's inclusion zone* (c).

measure the attribute density at a sample point under five different designs, including plot sampling, Bitterlich sampling, and line intersect sampling. We also introduce point relascope sampling, and perpendicular distance sampling.

10.2 Plot sampling

The Monte Carlo approach to fixed-area plot sampling allows plots of any shape, so it suffices to consider circular plots for illustrative purposes. The target parameter, τ_ρ, is the aggregate quantity of some attribute that is divided among an unknown number of discrete elements in $\mathcal{A}$. Each of these elements is centered—in an unambiguous way—in a circular inclusion zone with radius R and area $a = \pi R^2$. An element's inclusion zone includes a sample point if the distance from the center of the element to the sample point is less than R (Figure 10.1). Obviously, those elements whose inclusion zones overlap the sample point are the elements that would occupy a circular plot with radius R, centered at the sample point.

Ordinarily, the distance from a sample point to an element is determined with a tape. If attributes of standing trees are of interest, then it may be convenient to use a rangefinder to determine which trees' inclusion zones overlap a sample point. If attributes of elongated elements (e.g., fallen trees) are of interest, then a protocol to define the 'center' of such elements is also needed.

Of prime interest is the attribute density the sample point, $\mathbb{P}_s$, which is

$$\rho(x_s, z_s) = \sum_{\mathbb{P}_s \in \mathcal{I}_k} \frac{y_k t_k}{a_k}$$

$$= \frac{1}{a} \sum_{\mathbb{P}_s \in \mathcal{I}_k} y_k t_k, \qquad (10.16)$$

since $a_k = a$ for all k. Hence, substituting into (10.9) and (10.10),

$$\hat{\tau}_{\rho s} = \frac{A}{a} \sum_{\mathbb{P}_s \in \mathcal{I}_k} y_k t_k$$

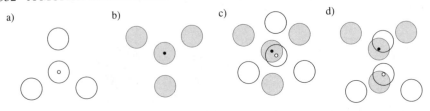

Figure 10.2 a) *A cluster of fixed-radius plots fixed about a sample point* (○) *defines a spatially disjoint inclusion zone about the location* (●) *of any element of interest* (b). *The element occupies a cluster plot if the sample point falls anywhere in the element's inclusion zone* (c) *and* (d).

and

$$\hat{\mu}_{\rho s} = \frac{1}{a} \sum_{\mathbb{P}_s \in \mathcal{J}_k} y_k t_k.$$

The former estimator is equivalent to the multi-tally estimator, (7.23), which was introduced in Chapter 7.

If the attribute of interest is the number of elements in $\mathcal{A}$, then $y_k = 1$ for all k and, therefore, the attribute density at the sample point is

$$\rho(x_s, z_s) = \frac{1}{a} \sum_{\mathbb{P}_s \in \mathcal{J}_k} t_k. \tag{10.17}$$

If (x_s, z_s) is farther than $2R$ from the boundary of $\mathcal{A}$, then $\sum_{\mathbb{P}_s \in \mathcal{J}_k} t_k$ is just the number of elements in the fixed-radius plot, since all the elements are tallied once (i.e., $t_k = 1$). More generally, this sum may include single and multiple tallies, the latter resulting from edge correction. Thus, the number of elements per unit land area at a sample point is 'measured' by summing the tallies of those elements whose inclusion zones include the sample point, i.e., those elements which occur in the plot.

10.2.1 Cluster plots

A cluster plot usually comprises a fixed-radius plot centered at a sample point and one or more satellite plots arranged in some standard configuration. Figure 10.2a, for example, depicts the configuration used by the USDA Forest Service to sample forested lands across the United States. The configuration of a cluster plot defines a spatially disjoint inclusion zone about the center point of any element of interest (Figure 10.2b). A cluster plot includes an element of interest if a sample point at (x_s, z_s) falls anywhere in the element's inclusion zone (Figure 10.2c,d).

If a cluster plot comprises c fixed-radius plots, each with area a, then the total area of an inclusion zone is ca and the total attribute density at the sth sample point is

$$\rho(x_s, z_s) = \sum_{\mathbb{P}_s \in \mathcal{I}_k} \frac{y_k t_k}{a_k}$$

$$= \frac{1}{ac} \sum_{\mathbb{P}_s \in \mathcal{I}_k} y_k t_k. \tag{10.18}$$

Although a cluster plot comprises two or more disjoint plots, it is treated as a single plot because the cluster is tied to a single sample point. Indeed, in the final analysis, the cluster plot merely serves to identify those elements whose spatially disjoint inclusion zones include the sample point.

10.3 Bitterlich sampling

Bitterlich sampling is used in forests around the world. The forest attributes of interest are summations of attributes of trees whose clear boles extend from the ground to breast height or higher. Interest may be restricted to attributes of trees that meet one or more criteria, for example, trees whose diameters at breast height equal or exceed some minimum, trees of certain species, or trees with health issues. In American forestry, breast height is standardized at 4.5 ft or 1.37 m. Elsewhere or in other disciplines, breast height may be standardized at 1.3, 1.37, or 1.4 m.

Trees generally are not circular in cross-section, but they are assumed to be circular for Bitterlich sampling, and so tree radius, r (m), is defined as $r \equiv c/(2\pi)$, where c (m) is circumference (or convex closure) at breast height. A tree with radius r is centered in a circular inclusion zone with radius αr, where α is a constant (Figure 10.3). Therefore, a tree's inclusion zone includes a sample point if the sample point is within a distance of αr (m) from the center of the tree. The cross-sectional area of a tree at breast height is called basal area (b). The area of kth tree's inclusion zone is $a_k = \pi(\alpha r_k)^2 = \alpha^2 b_k$ (m^2). As noted in Chapter 8, the basal area factor, F, is defined as $1/\alpha^2$ (m^{2}m^{-2}) $= 10^{-4}/\alpha^2$ (m^{2}ha^{-1}), so the area of the kth tree's inclusion zone can also be expressed as $a_k = b_k/F$.

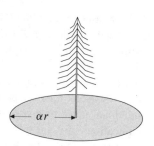

Figure 10.3 *A tree with cross-sectional radius r at breast height has a circular inclusion zone with radius αr.*

The ratio of tree radius to inclusion zone radius, $r/(\alpha r) = 1/\alpha$, can be expressed in terms of an angle v, i.e.,

$$\sin \frac{v}{2} = \frac{1}{\alpha}. \tag{10.19}$$

Consequently, an angle gauge, with angle v, can be used to determine which inclusion zones include a sample point. A sample point occurs in a tree's inclusion zone if the horizontal width of the tree's bole at breast height, when viewed from the sample point, fills the field of view of the angle gauge (see Figure 8.1, p. 247).

The attribute density at a sample point, P_s, is

$$\rho(x_s, z_s) = \sum_{P_s \in \mathfrak{J}_k} \frac{y_k t_k}{a_k}$$

$$= F \sum_{P_s \in \mathfrak{J}_k} \frac{y_k t_k}{b_k}. \tag{10.20}$$

Where the attribute of interest is the total basal area of the trees in the forest, the attribute density at a sample point is basal area per unit land area. In the absence of edge effect, $\rho_k(x, z) = F$ for all $(x, z) \in \mathfrak{J}_k$, independent of k. More generally, $\rho_k(x, z) = F t_k(x, z)$, so the attribute density at P_s is simply

$$\rho(x_s, z_s) = F \sum_{P_s \in \mathfrak{J}_k} t_k, \tag{10.21}$$

so

$$\hat{\tau}_{\rho s} = A F \sum_{P_s \in \mathfrak{J}_k} t_k. \tag{10.22}$$

10.4 Point relascope sampling

Point relascope sampling (Gove et al. 1999b) evolved from a design called transect relascope sampling (Ståhl 1998). Either design may be used in connection with the estimation of aggregate quantities of attributes of fallen trees and large branches. A discrete population approach to point relascope sampling is described in §11.2 and transect relascope sampling is described in §11.4.

Collectively, fallen trees and large branches are called coarse woody debris. An individual piece of coarse woody debris, regardless of origin, is called a 'log.' In applications of point relascope sampling, range poles are erected at each end of a log, which renders the length of the log (ℓ, m) viewable in the same horizon as the sampler's eye. This length is viewed from a sample point with an angle gauge (or relascope). If the log length fills the field of view of the angle gauge, then the inclusion zone of the log includes the sample point.

The angle of an angle gauge may take values in the range $0° < \nu \le 90°$. If $\nu < 90°$, the inclusion zone of a log takes the shape of two identical overlapping circles with the log length ℓ serving as a common chord (Figure 10.4). The horizontal land area (m^2) of the inclusion zone for the kth log is $\varphi \ell_k^2$, where

$$\varphi = \frac{\pi - \tilde{\nu} + \sin \nu \cos \nu}{2 \sin^2 \nu} \tag{10.23}$$

and $\tilde{\nu}$ is the angle of the angle gauge in radians. For a given ℓ_k, increasing values of ν cause the two circles of an inclusion zone to shrink in size and seemingly move closer together. At $\nu = 90°$, the two circles coalesce and so the inclusion zone of the kth log is a circle with diameter ℓ_k and area $a_k = \varphi \ell_k^2 = (\pi/4) \ell^2$ (Figure 10.4).

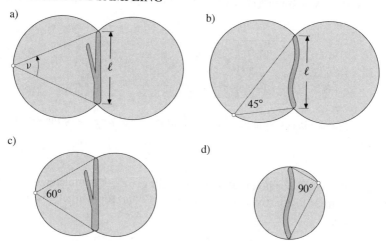

Figure 10.4 *In point relascope sampling with $\nu < 90°$, e.g., $\nu = 45°$, the inclusion zone of a log has the shape of two overlapping circles (a,b). A larger angle, e.g., $\nu = 60°$, yields an inclusion zone with a smaller area (c). When $\nu = 90°$, the two circles of the inclusion zone coalesce to a single circle with a diameter equal to the log's length (d).*

The attribute density of the sample logs at the sample point is:

$$\rho(x_s, z_s) = \sum_{\mathbb{P}_s \in \mathcal{I}_k} \frac{y_k t_k}{a_k}$$
$$= \frac{1}{\varphi} \sum_{\mathbb{P}_s \in \mathcal{I}_k} \frac{y_k t_k}{\ell_k^2}. \tag{10.24}$$

The aggregate log length-square (sum of length-squares) per unit land area ($m^2\ m^{-2}$) at the sample point is

$$\rho(x_s, z_s) = \frac{1}{\varphi} \sum_{\mathbb{P}_s \in \mathcal{I}_k} \frac{\ell_k^2 t_k}{\ell_k^2}$$
$$= \frac{1}{\varphi} \sum_{\mathbb{P}_s \in \mathcal{I}_k} t_k. \tag{10.25}$$

Borderline logs should be carefully checked to see if their inclusion zones include a sample point. The distances from the sample point to the two range poles, δ_1 and δ_2, are measured and a limiting log length (ℓ^*) is calculated (Gove et al. 1999b), i.e.,

$$\ell^* = \sqrt{\delta_1^2 + \delta_2^2 - 2\,\delta_1 \delta_2 \cos \nu}. \tag{10.26}$$

If $\ell \geq \ell^*$, the inclusion zone of the log in question includes the sample point.

The angles of angle gauges used for point relascope sampling (commonly 30° to 90°) are much larger than those used for Bitterlich sampling. At the time of this

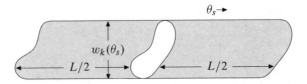

Figure 10.5 *In line intersect sampling, the area of the inclusion zone of the kth element is $a_k = w_k(\theta_s)L$.*

writing, angle gauges for point relascope sampling are not commercially available, but they are simple to construct (see Gove et al. (1999a) for instructions).

10.5 Line intersect sampling

This sampling design is so general that different names (e.g., line intercept sampling, line interception sampling, and planar intersect sampling) have been used in connection with different applications. As a continuous-population design, line intersect sampling may be implemented as an application of either one-dimensional (e.g., Example 4.14) or two-dimensional importance sampling. In the latter case, the size and shape of the inclusion zone of an element generally depends on the length and direction of a transect line and a protocol for handling partial intersections of the element by either end of the transect line. Further complexity results from the use of segmented transect lines (e.g., Gregoire & Valentine 2003; Affleck et al. 2005).

We consider an implementation where a straight transect line with length L and azimuth θ_s is centered at a sample point, $\mathbb{P}_s$. Hence, in effect, the transect has two segments, whose back ends meet at the sample point, and whose front ends occur at distance L from each other. The area of the inclusion zone of the kth element is $a_k = w_k(\theta_s)L$, where $w_k(\theta_s)$ is the 'width' of the inclusion zone. This width is equivalent to the 'projected length' of the kth element, which is the length of the element measured perpendicular to the transect line (Figure 10.5). The inclusion zone of an element includes a sample point, $\mathbb{P}_s$, if the element is intersected completely by either segment of the transect line or if the element is partially intersected by the front end of either segment (Figure 10.6).

The attribute density at a sample point, $\mathbb{P}_s$, obtains from measurements of the intersected elements, i.e.,

$$\rho(x_s, z_s) = \sum_{\mathbb{P}_s \in \mathcal{I}_k} \frac{y_k t_k}{a_k}$$

$$= \frac{1}{L} \sum_{\mathbb{P}_s \in \mathcal{I}_k} \frac{y_k t_k}{w_k(\theta_s)}.$$

Aggregate projected length per unit area at the sample point is measured by simply tallying the intersected elements, i.e.,

$$\rho(x_s, z_s) = \frac{1}{L} \sum_{P_s \in \mathcal{I}_k} \frac{w_k(\theta_s) t_k}{w_k(\theta_s)}$$

$$= \frac{1}{L} \sum_{P_s \in \mathcal{I}_k} t_k.$$

Actual measurement of the projected lengths of the intersected elements is unnecessary.

The measurement of projected length may also be rendered unnecessary in connection with the measurement of some other attribute densities. We simply assume that a transect line intersects a vanishingly thin sliver of each element rather than the whole element. Let Δw be the vanishingly thin width of the sliver perpendicular to the transect, and let l be the length of the intersection across the element (Figure 10.7).

The area of the inclusion zone of a sliver is (Figure 10.7b):

$$\lim_{\Delta w \to 0} \Delta w L.$$

Suppose that the transect line intersects a sliver at a point (·) along the axis of the projected length of the kth element. Let $l_k(\cdot)$ be the length of the intersection across the kth element. The coverage area of the intersected sliver is

$$\lim_{\Delta w \to 0} \Delta w l_k(\cdot),$$

so the aggregate coverage area per unit land area at the sample point is

$$\rho(x_s, z_s) = \sum_{P_s \in \mathcal{I}_k} \frac{\Delta w l_k(\cdot) t_k}{\Delta w L} = \frac{1}{L} \sum_{P_s \in \mathcal{I}_k} l_k(\cdot) t_k.$$

Now let $g_k(\cdot)$ be the vertical cross-sectional area of the element at the point of intersection, then the volume of the sliver is

$$\lim_{\Delta w \to 0} \Delta w g_k(\cdot),$$

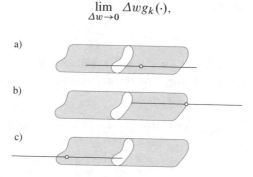

Figure 10.6 a) *The sample point falls in an element's inclusion zone if the element is completely intersected by either segment of the transect line, or if the element is partially intersected by the front end of either segment* (b,c).

a)   b)

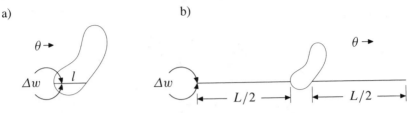

Figure 10.7 a) *Sliver of an element with length l and width* $\Delta w \to 0$. b) *The inclusion zone of a sliver of an element appears as two line segments, each with length* $L/2$ *and width* $\Delta w \to 0$.

and the aggregate volume per unit land area at the sample point is

$$\rho(x_s, z_s) = \sum_{\mathbb{P}_s \in \mathcal{J}_k} \frac{\Delta w g_k(\cdot)\, t_k}{\Delta w L} = \frac{1}{L} \sum_{\mathbb{P}_s \in \mathcal{J}_k} g_k(\cdot)\, t_k.$$

We note that the convenience of measuring slivers of elements rather than whole elements will be countered by an increase in the sampling error of τ_ρ, since both the cross-sectional area and the length of intersection ordinarily will vary along an element's axis of projected length.

10.5.1 Unconditional estimation

Unconditional estimation in LIS also can be formulated under the framework of Monte Carlo integration. If θ_s is selected uniformly at random from $U[0, \pi]$, then the expected width of the kth unit's inclusion zone is

$$E\left[w_k(\theta)\right] = \frac{1}{\pi} \int_0^\pi w_k(\theta)\, d\theta = \bar{w}_k$$

and, therefore, the expected area of $\mathcal{J}_k$ is $\bar{a}_k = \bar{w}_k L$. By design, the attribute density at a location point $(x, z) \in \mathcal{J}_k$ is based on the expected inclusion area for unconditional estimation, so

$$\rho_k(x, z) = \frac{y_k t_k(x, z)}{\bar{w}_k L}.$$

However, when θ_s is selected for $\mathbb{P}_s$, the actual area of $\mathcal{J}_k$ is $a_k = w_k(\theta_s) L$. Integration of the attribute density over the actual area of the inclusion zone, given θ_s, yields

$$\iint_{\mathcal{J}_k} \rho_k(x, z)\, dx\, dz = \frac{y_k}{\bar{w}_k L} \iint_{\mathcal{J}_k} t_k(x, z)\, dx\, dz$$

$$= \frac{y_k}{\bar{w}_k L}\, a_k$$

$$= y_k \frac{w_k(\theta_s)}{\bar{w}_k}.$$

But, since θ is selected uniformly at random,

$$E\left[y_k \frac{w_k(\theta)}{\bar{w}_k}\right] = \frac{y_k}{\bar{w}_k} E\left[w_k(\theta)\right] = y_k.$$

Consequently, τ_ρ is unbiasedly estimated by

$$\hat{\tau}_{\rho_s}^{\mathrm{u}} = \frac{A}{L} \sum_{P_s \in \mathcal{J}_k} \frac{y_k t_k(x_s, z_s)}{\bar{w}_k}$$

$$= \frac{\pi A}{L} \sum_{P_s \in \mathcal{J}_k} \frac{y_k t_k(x_s, z_s)}{c_k},$$

where $c_k = \pi \bar{w}_k$ is the convex closure, which is measured by wrapping a tape tightly about the kth element.

10.6 Perpendicular distance sampling

Perpendicular distance sampling (PDS) is yet another design that is intended to be used where the attributes of interest are summations of attributes of logs on the ground. Under the original prescriptions for PDS (Williams & Gove 2003; Williams et al. 2005), the area of a log's inclusion zone is proportional to log volume. Consequently, aggregate log volume per unit land area can be estimated from tallies of logs at a sample point.

Let H_k denote the horizontal length of the central axis of a log or log-shaped element. If the log curves, a straight central axis may be unambiguously established by erecting range poles at each end of the log. Let h denote an ordinate on the central axis ($0 \le h \le H_k$), and let $g_k(h)$ be the cross-sectional area of the log perpendicular to the central axis at h. If the log is branched, then $g_k(h)$ is the sum of the cross-sectional areas of all the branches intersected by a vertical plane perpendicular to the central axis at h.

The inclusion zone the kth log, $\mathcal{J}_k$, is bilaterally symmetric about the central axis (Figure 10.8). By design, the total width on the inclusion zone on a line perpendicular to the central axis at h is $2\kappa_v g_k(h)$, where the κ_v (m^2 m^{-3}) is a design parameter. Integrating $g_k(h)$ over the length of the central axis gives log volume (v_k), i.e.,

$$v_k = \int_0^{H_k} g_k(h)\, \mathrm{d}h.$$

Thus, the horizontal land area of $\mathcal{J}_k$ is proportional to the log's volume, i.e., $a_k = 2\kappa_v v_k$.

Whether $\mathcal{J}_k$ includes a sample point, P_s, depends on the 'perpendicular distance' from the sample point to the log. The distance from the sample point at (x_s, z_s) to an ordinate h_s on the central axis of the kth log is a perpendicular distance only if the line from (x_s, z_s) to h_s is perpendicular to the central axis. It is possible that no such line exists, in which case the sample point does not fall in $\mathcal{J}_k$. If such a line does exist, we measure L (m), the perpendicular distance from a sample point at (x_s, z_s) to the ordinate h_s. If $L \le \kappa_v g_k(h_s)$, then $\mathcal{J}_k$ includes the sample point. A log is

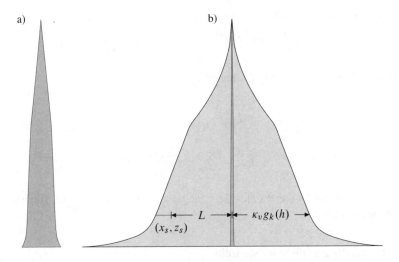

Figure 10.8 a) *A log, scaled with* 10× *diameter relative to length to depict the shape and taper.* b) *The same log, scaled with true diameter relative to length, and its inclusion zone for perpendicular distance sampling* ($\kappa_v = 100$); L *is the perpendicular distance from a sample point to the central axis of the log.*

borderline if $L \approx \kappa_v g_k(h_s)$, in which case care should be taken to locate h_s and measure $g_k(h_s)$ precisely.

The attribute density at a sample point, P_s, is

$$\rho(x_s, z_s) = \sum_{P_s \in \mathcal{I}_k} \frac{y_k t_k}{a_k}$$

$$= \frac{1}{2\kappa_v} \sum_{P_s \in \mathcal{I}_k} \frac{y_k t_k}{v_k}$$

$$= F_v \sum_{P_s \in \mathcal{I}_k} \frac{y_k t_k}{v_k},$$

where $F_v = 1/2\kappa_v$ (m^3 m^{-2}) is called the 'volume factor.' If log volume is the attribute of interest, then log volume per unit land area at P_s is simply

$$\rho(x_s, z_s) = F_v \sum_{P_s \in \mathcal{I}_k} \frac{v_k t_k}{v_k}$$

$$= F_v \sum_{P_s \in \mathcal{I}_k} t_k. \tag{10.27}$$

Hence, the aggregate log volume in $\mathcal{A}$ is unbiasedly estimated by

$$\hat{\tau}_{\rho s} = A F_v \sum_{P_s \in \mathcal{I}_k} t_k. \tag{10.28}$$

For convenience, we may choose to use a volume factor that converts the dimensions of the volume density in an inclusion zone from $m^3\,m^{-2}$ to $m^3\,ha^{-1}$. For example, for $\kappa_v = 100\,m^2\,m^{-3}$, we obtain $1/(2\kappa_v) = 0.005\,m^3\,m^{-2}$, so $F_v = 10000\,m^2\,ha^{-1} \times 0.005\,m^3\,m^{-2} = 50\,m^3\,ha^{-1}$.

10.6.1 Omnibus PDS

PDS would seem to be inconvenient for estimating log attributes other than volume, because the calculation of the attribute density at a sample point requires knowledge of the volumes of the tallied logs. However, a generalization of PDS, called 'omnibus PDS,' renders the measurements of volume unnecessary (Ducey et al. 2007): Measurements of cross-sectional area substitute for the volume measurements.

In omnibus PDS, we divide the inclusion zone of the kth log into vanishingly thin slivers of width ΔH, each sliver perpendicular to the log's central axis. Let $\mathcal{J}_k(h)$ identify the sliver of inclusion zone $\mathcal{J}_k$, whose midpoint is coincident with ordinate h on the central axis. The length of this sliver is $2\kappa_v g_k(h)$, half the length on each side of the central axis.

The attribute of interest may be any log attribute, y, that can be expressed as an integral

$$y_k = \int_0^{H_k} x_k(h)\,dh,$$

provided $x_k(h)$ can be measured at any ordinate h on the central axis. For example, if y_k is coverage area of the kth log, then $x_k(h)$ is the horizontal diameter, which is easily measured with a caliper. If an estimate of the number of logs in $\mathcal{A}$ is desired, then $y_k = 1$ and $x_k(h) = 1/H_k$.

The portion of the attribute y_k in the sliver of log centered at h is $x_k(h)\Delta H$ and the area of the sliver of inclusion zone is $2\kappa_v g_k(h)\Delta H \equiv g_k(h)\Delta H/F_v$. Dividing the attribute for the sliver by the area of the sliver defines the uniform attribute density

$$\lim_{\Delta H \to 0} F_v\left[\frac{x_k(h)\Delta H}{g_k(h)\Delta H}\right] = F_v\left[\frac{x_k(h)}{g_k(h)}\right] \tag{10.29}$$

for each and every point on the line $\mathcal{J}_k(h)$. Thus, allowing for edge correction, the kth log's attribute density at any location point $(x, z) \in \mathcal{A}$ is

$$\rho_k(x, z) \equiv \begin{cases} F_v\,t_k(x, z)\left[\dfrac{x_k(h)}{g_k(h)}\right], & (x, z) \in \mathcal{J}_k(h) \subset \mathcal{J}_k \cap \mathcal{A}; \\ 0, & (x, z) \notin \mathcal{J}_k. \end{cases} \tag{10.30}$$

Consequently,

$$y_k = \iint_{\mathcal{J}_k \cap \mathcal{A}} \rho_k(x, z)\,dz\,dx$$

and, since $\rho(x, z) = \sum_{k=1}^N \rho_k(x, z)$,

$$\tau_\rho = \iint_{\mathcal{A}} \rho(x, z)\,dz\,dx.$$

Therefore, τ_ρ can be estimated by Monte Carlo integration.

The kth log contributes to the attribute density at sample point $\mathbb{P}_s$ if $L \le \kappa_v g_k(h_s)$, i.e., if the perpendicular distance, L, from (x_s, z_s) to an ordinate h_s on the central axis of the kth log does not exceed $\kappa_v g_k(h_s)$. Hence, τ_ρ is unbiasedly estimated by

$$\hat{\tau}_{\rho s} = A\rho(x_s, z_s),$$

where

$$\rho(x_s, z_s) = F_v \sum_{\mathbb{P}_s \in \mathcal{I}_k} t_k \left[\frac{x_k(h_s)}{g_k(h_s)} \right]. \tag{10.31}$$

If y is volume, then $x(h_s) \equiv g(h_s)$, so

$$\rho(x_s, z_s) = F_v \sum_{\mathbb{P}_s \in \mathcal{I}_k} t_k,$$

which is equivalent to (10.27).

Coverage area

Besides simplifying measurements when the area of the inclusion zone of a log is proportional to log volume, omibus PDS also affords other options. Regardless of the attribute of interest, we could abandon the use of inclusion zones that are proportional to volume, and instead formulate omibus PDS with inclusion zones that are, say, proportional to log coverage area (see Ducey et al. 2007). Under this prescription, the length of $\mathcal{I}_k(h)$, is $2\kappa_c d_k(h)$, where $d_k(h)$ is the horizontal log diameter, and κ_c (m^2 m^{-2}) is the design parameter. The area of $\mathcal{I}_k$ is $a_k = 2\kappa_c c_k$, where

$$c_k = \int_0^{H_k} d_k(h)\, dh$$

is the coverage area of the kth log. Moreover, $\mathcal{I}_k$ includes a sample point, $\mathbb{P}_s$, if $L \le \kappa_c d_k(h)$. For the purpose of estimation, we can define a 'coverage area factor,' i.e., $F_c = 1/2\kappa_c$. Hence, $d_k(h)$ and F_c, respectively, substitute for $g_k(h)$ and F_v in (10.29) – (10.31). For example, (10.30) converts to

$$\rho_k(x, z) \equiv \begin{cases} F_c\, t_k(x, z) \left[\dfrac{x_k(h)}{d_k(h)} \right], & (x, z) \in \mathcal{I}_k(h) \subset \mathcal{I}_k \cap \mathcal{A}; \\ 0, & (x, z) \notin \mathcal{I}_k. \end{cases}$$

If volume is the attribute of interest, then $x_k(h) \equiv g_k(h)$. Hence, the aggregate volume of logs in $\mathcal{A}$ is unbiasedly estimated by

$$\hat{\tau}_{\rho s} = AF_c \sum_{\mathbb{P}_s \in \mathcal{I}_k} t_k \left[\frac{g_k(h_s)}{d_k(h_s)} \right].$$

The use of inclusion zones proportional to coverage area is particularly appealing if the aggregate coverage area of the logs in $\mathcal{A}$ is of prime interest, as then $x_k(h) \equiv$

$d_k(h)$, so

$$\rho_k(x, z) \equiv \begin{cases} F_c\, t_k(x, z), & (x, z) \in \mathcal{I}_k(h) \subset \mathcal{I}_k \cap \mathcal{A}; \\ 0, & (x, z) \notin \mathcal{I}_k, \end{cases}$$

and

$$\hat{\tau}_{\rho s} = A F_c \sum_{P_s \in \mathcal{I}_k} t_k.$$

10.7 Edge correction

The problem known variously as edge effect, boundary overlap, or slopover occurs where the inclusion zone of any element of interest slops over the tract boundary. Somewhat analogously, slopover is said to occur where the fire jumps the fireline of a forest fire. In our Monte Carlo approach to areal sampling, slopover is problematic because the attributes of an element are prorated over the element's inclusion zone with uniform density. If some fraction of the area of the inclusion zone is outside the tract, then an equal fraction of the attribute is also outside the tract. Hence, if sample points are constrained to fall only within the tract, the 'slopover portion' of the attribute is ignored, so

$$\iint_{\mathcal{A}} \rho(x, z)\, dz\, dx < \tau_\rho.$$

Because $\hat{\tau}_{\rho,\text{rep}}$ provides an unbiased estimate of the integral, this estimator is expected to provide estimates of τ_ρ that are biased downward.

Corrections for slopover bias include: (*i*) allowing sample points to fall outside the tract, (*ii*) redefining a_k to be the area of the horizontal projection of $\mathcal{I}_k \cap \mathcal{A}$, if $\mathcal{I}_k$ slops over the boundary, and (*iii*) affine transform methods, i.e., methods for reflecting, rotating, or translating the attribute slopover back into the tract, effectively remapping the attribute densities in the inclusion zones near the edge of $\mathcal{A}$. We have prescribed, in our presentation to this point, that a protocol incorporates an appropriate affine transform method for edge correction. Affine transform methods give rise to multiple tallies for edge elements.

Unfortunately, there is no generally easy solution for slopover. This is due, in large part, to the fact that tract boundaries do not always consist of simple defined lines, such as property lines. Tract boundaries may also include natural features such as edges of lakes, rivers, and cliffs, which makes sampling or working outside the tract infeasible. Penner & Otukol (1998) listed 11 methods for dealing with slopover in connection with Bitterlich sampling. Here, we discuss the three approaches for dealing with slopover from the continuous-population perspective.

10.7.1 Buffer method

A 'buffer method,' attributed to (Masuyama 1954), entails allowing sample points to fall outside the boundary of tract $\mathcal{A}$, but within some larger region $\mathcal{A}^*$ that includes all of tract $\mathcal{A}$. For our purpose, region $\mathcal{A}^*$ need only be large enough to encompass all the inclusion zones of the elements that occur in $\mathcal{A}$. Elements outside of tract $\mathcal{A}$ are ignored. Under this protocol, $\rho_k(x, z) = y_k/a_k$ for all $(x, z) \in \mathcal{J}_k$, and $\rho_k(x, z) = 0$ for all $(x, z) \notin \mathcal{J}_k$. The total amount of attribute of interest is:

$$\tau_\rho = \iint_{\mathcal{A}^*} \sum_{k=1}^{N} \rho_k(x, z) \, dz \, dx.$$

Let A^* be the horizontal area of region $\mathcal{A}^*$. Ordinarily, we specify the probability density function $f(x, y) = 1/A^*$ for all $(x, y) \in \mathcal{A}^*$, so $\iint_{\mathcal{A}^*} f(x, z) \, dz \, dx = 1$. Hence, τ_ρ, the total amount of attribute in $\mathcal{A}$, is unbiasedly estimated by

$$\hat{\tau}_{\rho s} = A^* \rho(x_s, z_s). \tag{10.32}$$

The buffer method is applicable with any boundary configuration and any inclusion zone shape. However, the method is not useful where inclusion zones extend into lakes or beyond cliffs, or where working outside the boundary of the tract of interest is prohibited. Moreover, the method tends to inflate the variance of $\hat{\tau}_{\rho s}$ because the average attribute density is lower outside the tract than inside.

The 'toss back method' of Iles (2003) prescribes that sample points, both inside and outside the tract, are arranged on a systematic grid. An attribute density measured at a sample point outside the tract is 'tossed back,' i.e., added, to the attribute density of the nearest sample point inside the tract. Hence, the target parameter, τ_ρ, is unbiasedly estimated with equations (10.9) and (10.11).

10.7.2 Direct measurement of inclusion area

A general, but not particularly easy, solution for slopover involves direct measurement of the area, a'_k, of $\mathcal{J}_k \cap \mathcal{A}$, the portion of $\mathcal{J}_k$ that occurs in $\mathcal{A}$. The attribute density is $\rho'_k(x, z) = y_k/a'_k$ for all $(x, z) \in \mathcal{J}_k \cap \mathcal{A}$ and $\rho'_k(x, z) = 0$ for all $(x, z) \notin \mathcal{J}_k \cap \mathcal{A}$. Hence,

$$\iint_{\mathcal{J}_k \cap \mathcal{A}} \rho'_k(x, z) \, dx \, dz = y_k \tag{10.33}$$

and, since $\rho(x, z) = \sum_{k=1}^{N} \rho'_k(x, z)$,

$$\iint_{\mathcal{A}} \sum_{k=1}^{N} \rho'_k(x, z) \, dx \, dz = \iint_{\mathcal{A}} \rho(x, z) \, dx \, dz = \tau_\rho, \tag{10.34}$$

and, therefore, τ_ρ is unbiasedly estimated from the attribute density at $\mathbb{P}_s$, by

$$\hat{\tau}_{\rho s} = A\rho(x_s, z_s). \tag{10.35}$$

10.7.3 Mirage method

The mirage method of Schmid (1969) is a reflection method (i.e., an affine transform method) that was intended for use with fixed-radius plot sampling or Bitterlich sampling, but it can be used with any of the sampling designs discussed in this chapter—if boundaries are straight.

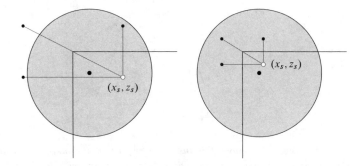

Figure 10.9 *Three mirage points are established at a square corner. In the left diagram, one mirage point falls in a circular inclusion zone, and two mirage points fall outside the inclusion zone, so the element is tallied twice, i.e., $t_k = 2$. In the right diagram, all three mirage points fall inside the inclusion zone, so the element is tallied four times.*

The mechanics of the mirage method were described in §7.5.4. Recall that if a sample point falls near the tract border, we (*i*) measure the distance δ on a perpendicular line from the sample point to the border and (*ii*) establish a mirage point on the same line at a distance δ beyond the border. Near a square corner, a mirage point is established on the perpendicular line across each of the two legs of the boundary, and a third mirage point is established on a line from the sample point through the vertex of the boundary corner. Thus, at a square boundary corner, the sample point and the three mirage points occur at the vertices of a rectangle (Figure 10.9). The tally, t_k, for the kth element's attribute density equals 1 (at the sample point) plus 1 at each the of mirage points that occur in the element's inclusion zone.

In expectation, the mirage method remaps the attribute densities in inclusion zones that slop over a boundary of $\mathcal{A}$, reflecting (or folding) the slopover sections of attribute about the boundary into $\mathcal{A}$. To see how this works, let us suppose that a

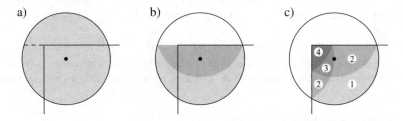

Figure 10.10 *The mirage method effectively folds portions of attribute that fall outside of $\mathcal{A}$ back into $\mathcal{A}$. At square corners (a), two folds are necessary. b) The first fold doubles the attribute density in the darker grey section of the inclusion zone. c) The second fold moves all the attribute into $\mathcal{A}$, creating sections with double, triple, and quadruple attribute densities, as indicated by the numbers. Thus, if the sample point falls, say, in the section with triple density, the mirage method provides a tally for the element of $t_k = 3$.*

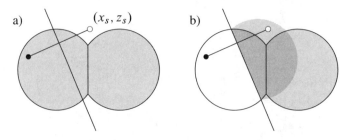

Figure 10.11 *a) A mirage point may fall in the slopover portion of a non-circular inclusion zone, even though the sample point does not fall in the in-tract portion of the inclusion zone. b) The mirage method alters that shape of the inclusion zone and remaps the attribute density. The element is tallied once if the sample point falls in the light gray section, and twice in the darker gray section.*

piece of cloth is tailored to cover the entire inclusion zone of the kth element. The cloth represents the attribute, which is distributed uniformly over the inclusion zone. We fold the slopover section of the cloth that is outside of $\mathcal{A}$ back onto $\mathcal{A}$. The edge of the fold is coincident with the straight boundary line. At a square corner, two folds are necessary to put all the slopover cloth back onto the tract, and a location point may be covered by as many as four layers of cloth (Figure 10.10). The attribute density attributable to the kth element at any given location point (x, z) is $(y_k/a_k) \times t_k(x, z)$, where $t_k(x, z)$ is the number of layers of cloth covering the point. Hence the folded cloth effectively remaps the attribute density over the section of inclusion zone that falls in $\mathcal{A}$. If we paint a spot on each layer of cloth covering a sample point and then unfold cloth to recreate the slopover, we should find that each paint spot in the slopover portion of the inclusion zone covers a mirage point.

Note: for point relascope sampling or any of the line methods, it is possible for a mirage point to fall within different inclusion zones than the sample point. Or, a mirage point may fall within one or more inclusion zones even though the sample point falls in no inclusion zones (Figure 10.11). Therefore, it may be necessary to establish the mirage point(s) even though a zero attribute density was measured at the sample point.

The mirage method preserves the unbiasedness of $\hat{\tau}_{ps}$—where the boundary lines are straight and corners are square. Bias accrues where boundaries are curved. The method requires working outside of $\mathcal{A}$, so lakes, rivers, and cliffs are problematic.

10.7.4 Reflection method for transects

The reflection method for transects is another affine transform method, which originally was devised for line intersect sampling by Gregoire & Monkevich (1994). It may also be used in horizontal line sampling (see §11.3).

Like the mirage method, the reflection method for transects is a folding method, but the folding mechanism is different. Once again, for the sake of illustration, let us assume that a piece of cloth is tailored to cover an inclusion zone perfectly. We

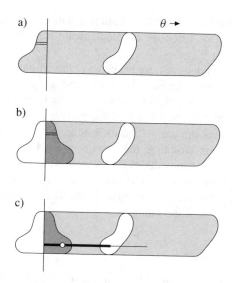

Figure 10.12 *a) The reflection method for transects, in effect, folds vanishingly narrow strips of slopover across the boundary parallel to θ. All the slopover folds neatly and completely back into the inclusion zone doubling the attribute density in the dark grey section (b). If the sample point falls in the double density section (c), the element is intersected by both the straight segment (thin) and the folded segment (thick) of the transect, so* ($t_k = 2$).

cut the portion of cloth that slops over the boundary into narrow strips of width Δw, the strips running parallel to the transect line (Figure 10.12). We fold each of these strips at the boundary back into the tract. This may create a sawtooth edge at the boundary if the strips are not square to the boundary and are cut too wide. However, if we let Δw approach zero, we obtain a smooth edge and the slopover folds neatly and completely back into the tract. Points covered by two layers of cloth have twice the attribute density of points covered by one layer.

In application, we establish a transect line with length L centered at a sample point and oriented with azimuth θ. If a section of the transect line crosses the border, we bend it back upon itself reversing its direction. If this folded section of transect line intersects an element, then the sample point falls in the area where the slopover folds back into the 'in-tract' portion of the element's inclusion zone, in which case the attribute density attributable to the element is doubled and, so, the element is tallied twice.

Implicit in the method is a mirage point along the line defined by the sample point and θ. If the sample point is a distance δ from the border, the mirage point is a distance δ beyond the border. However, since the sample point and the mirage point are equidistant from the border there is really no need to establish the latter. Folding a transect line—which is centered at the sample point—at the border and running the residual back along the same line covers the same ground within the tract as an unfolded transect line centered at the mirage point.

The reflection method for transects is applicable with curved boundaries, including natural feature boundaries, since one can implement the method without leaving the tract.

The reflection method does not, by itself, solve the problem where an element of interest straddles a boundary. In line intersect sampling, we can use a consistent rule that establishes whether a straddler element is in $\mathcal{A}$. And, for calculating the attribute density, we can use the projected width of the portion of the element that is within $\mathcal{A}$. In effect, this shrinks the size of the inclusion zone over which the attribute is spread. In horizontal line sampling, the element is in $\mathcal{A}$ if the center point of the element is in $\mathcal{A}$. The projected width of the element is the tree circle diameter, which is shortened by the length that extends outside the boundary. In either design, the reflection method for transects is applied as usual. The walkthrough method solves the slopover problem in sausage sampling.

10.7.5 Walkthrough method

The walkthrough method is an affine transform method, which was devised by Ducey et al. (2004) for designs that prescribe either bilaterally or radially symmetric inclusion zones. The walkthrough method effectively deals with both defined and natural boundaries. Moreover, the boundaries may be straight or curved, though some small amount of bias may accrue with curved boundaries. The walkthrough method may be characterized as a reflection method and, indeed, it evolved from the 'boundary reflection method' of Gove et al. (1999b).

If an inclusion zone is (i) radially symmetric about the center point of an element or (ii) bilaterally symmetric about the central axis of an element, then any location point in the inclusion zone may be reflected to another location point in the inclusion zone.

For example, in the first case, if we start at any location point in an inclusion zone and walk, say, 3.7 m to the center point of the element, and then continue on this line 'through the element' another 3.7 m, we will end at a location point in the inclusion zone that is 7.4 m from the starting point and 3.7 m from the center point. The starting point is a reflection point of the ending point, and vice versa. The reflection, of course, is about the center point of the element, rather than a boundary.

Now suppose that a portion of an inclusion zone slops over a boundary. For most boundary configurations, each of the location points in the slopover portion of the inclusion zone will have a reflection point in the 'in-tract' portion of the inclusion zone. Thus, we can imagine reflecting the slopover portion of attribute reflects back into the in-tract portion of the inclusion zone, doubling the attribute density where the reflection occurs (Figure 10.13a,b).

In the case where the inclusion zone is bilaterally symmetric about the central axis of an element—for example in perpendicular distance sampling—the reflection point of any location point is on a line perpendicular to the central axis (Figure 10.13c,d).

Operationally, we need only concern ourselves with just one reflection point in each inclusion zone, i.e., the reflection point of a sample point. Suppose that a sample point falls within the inclusion zone of the kth element, which is near a boundary. If

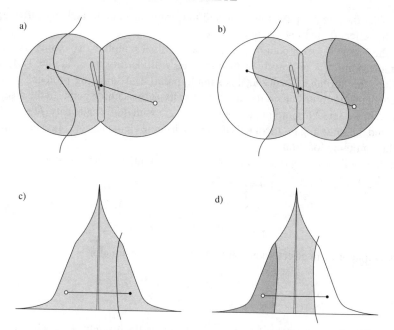

Figure 10.13 *a) A symmetric inclusion zone, where the reflection point of the sample point is outside of A. b) The walkthrough method reflects the slopover section of the inclusion zone, point-by-point, into A, doubling the attribute density in the reflected section. If the sample point falls in the reflection of the slopover, as in (b), the element is tallied twice. In perpendicular distance sampling (c), the reflection is about the central axis of the element instead of the center point. Again, the element is tallied twice, if the sample point falls in the reflection of the slopover (d).*

the reflection point falls in slopover, then the sample point falls within the reflection of the slopover. Hence, the attribute density attributable to the kth element is doubled at the sample point.

The name of the method derives from the fact that we 'walk' from the sample point to the element, recording the distance, and then 'through' and beyond the element an equal distance. If we reach the boundary before we reach the reflection point, we double the element's contribution to the attribute density at the sample point.

The method is unbiased if every location point in the slopover has a reflection point inside the tract. This constraint is not met if more than half of the area of an inclusion zone in outside the tract and it may not be met if the boundary configuration is unusually convoluted (Ducey et al. 2004).

10.8 Redux: Continuous versus discrete

Barabesi (2004) has noted that Monte Carlo integration and Horvitz-Thompson estimation are "two sides of the same coin." We add the proviso that this is the case if (*i*) the sample points are selected uniformly at random and (*ii*) corrections for edge-effect do not involve multi-tallies.

In the Monte Carlo or continuous approach to areal sampling, we ordinarily select a sample point, P_s, uniformly at random, i.e., with density $f(x_s, z_s) = 1/A$, and measure the attribute density at this sample point. The sampling protocol tells us which elements contribute to the attribute density at the sample point. In the discrete approach, we select a sampling location at (x_s, z_s) with uniform density $f(x_s, z_s) = 1/A$, and the sampling protocol tells us which discrete elements constitute the sample for this sampling location.

In the continuous approach, we use a generic Monte Carlo estimator to estimate $\tau_\rho = \iint_{\mathcal{A}} p(x, z) \, dx \, dz = \sum_{k=1}^{N} y_k$, i.e.,

$$\hat{\tau}_{\rho s} = \sum_{P_s \in \mathcal{J}_k} \frac{p(x_s, z_s)}{f(x_s, z_s)} = A \sum_{P_s \in \mathcal{J}_k} \frac{y_k t_k}{a_k}. \tag{10.36}$$

This estimator is, in effect, identical to the multi-tally estimator

$$\hat{\tau}_{yms} = A \sum_{u_k \in P_s} \frac{y_k t_k}{a_k}, \tag{10.37}$$

which provides an estimate of $\tau_y = \sum_{k=1}^{N} y_k$ under the discrete approach. If $t_k = 1$ for all k, then

$$\hat{\tau}_{\rho s} = A \sum_{P_s \in \mathcal{J}_k} \frac{y_k}{a_k}, \tag{10.38}$$

in which case the Monte Carlo estimator provides the same result as the Horvitz-Thompson estimator

$$\hat{\tau}_{y\pi s} = \sum_{u_k \in P_s} \frac{y_k}{\pi_k} = A \sum_{u_k \in P_s} \frac{y_k}{a_k}. \tag{10.39}$$

Thus, under either approach, we measure the same elements and obtain the same unbiased estimates of $\tau_\rho = \tau_y = \sum_{k=1}^{N} y_k$ and $\mu_\rho = \lambda_y = \sum_{k=1}^{N} y_k/A$.

However, besides offering a different perspective on areal designs and edge-correction methods, the Monte Carlo integration approach also offers the possibility for non-uniform selection of sample points, where $f(x, z)$ varies among the location points in $\mathcal{A}$. For example, by design we may prescribe $f(x, z)$ to be a function of mapped elevation. In this case we simply use the more general form of the estimator, i.e.,

$$\hat{\tau}_{\rho,\text{rep}} = \frac{1}{m} \sum_{s=1}^{m} \sum_{P_s \in \mathcal{J}_k} \frac{\rho(x_s, z_s)}{f(x_s, z_s)}$$

$$= \frac{1}{m} \sum_{s=1}^{m} \sum_{P_s \in \mathcal{J}_k} \frac{y_k t_k(x_s, z_s)/a_k}{f(x_s, z_s)},$$

whose variance is

$$V\left[\hat{\tau}_{\rho,\text{rep}}\right] = \frac{1}{m}\left(\iint_{\mathcal{A}} \frac{\rho^2(x,z)}{f(x,z)}\,dz\,dx - \tau_\rho^2\right).$$

10.9 Terms to Remember

Affine transform	Point relascope sampling
Areal sampling	Omnibus PDS
Perpendicular distance sampling	Sliver

10.10 Exercises

1. In Bitterlich sampling, the attribute density, tree basal area per unit land area, is uniform across all inclusion zones. What attribute density is uniform across all inclusion zones in plot sampling? In point relascope sampling?

2. In the original formulation of perpendicular distance sampling, the attribute density, log volume per unit land area, is equivalent to the volume factor. This density is uniform across all inclusion zones. Under what prescription is this statement true for omnibus PDS?

10.11 Appendix

10.11.1 Variance of $\hat{\tau}_{\rho,\text{rep}}$

In the absence of edge effect, the attribute densities for discrete elements are uniform across their respective inclusion zones, i.e., $\rho_k(x,z) = y_k/a_k$ for all $(x,z) \in \mathcal{I}_k$. If sample points are selected uniformly at random, then

$$\hat{\tau}_{\rho,\text{rep}} = \frac{A}{m}\sum_{s=1}^{m}\sum_{\mathbb{P}_s \in \mathcal{I}_k}\frac{y_k}{a_k} \tag{10.40}$$

and the variance of $\hat{\tau}_{\rho,\text{rep}}$ is

$$V\left[\hat{\tau}_{\rho,\text{rep}}\right] = \frac{1}{m}\left(A\iint_{\mathcal{A}}\rho^2(x,z)\,dz\,dx - \tau_\rho^2\right). \tag{10.41}$$

Let $a_{kk'}$ be the area of overlap of the inclusion zones of the kth and k'th elements, then the integral portion of the variance can be calculated as

$$\iint_{\mathcal{A}} \rho^2(x, z)\, dx\, dz = \sum_{k=1}^{N} \sum_{k'=1}^{N} \iint_{\mathcal{I}_k \cap \mathcal{I}_{k'}} \rho_k(x, z)\rho_{k'}(x, z)\, dx\, dz$$

$$= \sum_{k=1}^{N} \sum_{k'=1}^{N} a_{kk'} \rho_k \rho_{k'}$$

$$= \sum_{k=1}^{N} \rho_k^2 a_k + \sum_{k=1}^{N} \sum_{\substack{k'=1 \\ k' \neq k}}^{N} \rho_k \rho_{k'} a_{kk'}$$

$$= \sum_{k=1}^{N} \frac{y_k^2}{a_k} + \sum_{k=1}^{N} \sum_{\substack{k'=1 \\ k' \neq k}}^{N} \frac{y_k}{a_k} \frac{y_{k'}}{a_{k'}} a_{kk'}. \qquad (10.42)$$

Thus, substituting into (10.41),

$$V\left[\hat{\tau}_{\rho,\text{rep}}\right] = \frac{1}{m} \left[A \left(\sum_{k=1}^{N} \frac{y_k^2}{a_k} + \sum_{k=1}^{N} \sum_{\substack{k'=1 \\ k' \neq k}}^{N} \frac{y_k}{a_k} \frac{y_{k'}}{a_{k'}} a_{kk'} \right) - \tau_\rho^2 \right]. \qquad (10.43)$$

Moreover,

$$\tau_\rho^2 = (y_1 + y_2 + \cdots + y_N)^2 = \sum_{k=1}^{N} y_k^2 + \sum_{k=1}^{N} \sum_{\substack{k'=1 \\ k' \neq k}}^{N} y_k y_{k'}. \qquad (10.44)$$

Substitution of (10.44) into (10.43) gives an equivalent formula

$$V\left[\hat{\tau}_{\rho,\text{rep}}\right] = \frac{1}{m} \sum_{k=1}^{N} y_k^2 \left[\frac{1 - (a_k/A)}{a_k/A} \right]$$

$$+ \frac{1}{m} \sum_{k=1}^{N} \sum_{\substack{k'=1 \\ k' \neq k}}^{N} y_k y_{k'} \left[\frac{(a_{kk'}/A) - (a_k/A)(a_{k'}/A)}{(a_k/A)(a_{k'}/A)} \right]. \qquad (10.45)$$

Note that $V[\hat{\tau}_{\rho,\text{rep}}]$ is equivalent to $V[\hat{\tau}_{y\pi,\text{rep}}]$ (see eqns 7.4 and 7.5) and $V[\hat{\tau}_{y\pi,\text{rep}}^{\text{c}}]$ (see eqns 9.12 and 9.13).

If the buffer method is used to correct for edge effect, we substitute A^* for A in (10.40), and in (10.43) or (10.45). Alternatively, if the inclusion zone of the kth element slops over the boundary of $\mathcal{A}$, we can let a_k be the area of the horizontal projection of $\mathcal{A} \cap \mathcal{I}_k$—the portion of $\mathcal{I}_k$ in $\mathcal{A}$—in which case (10.40) is the appropriate estimator and (10.43) is its variance.

10.11.2 Variance with edge correction by walkthrough or mirage

Calculation of the variance of $\hat{\tau}_{\rho s}$ requires more bookkeeping if a reflection method or the walkthrough method is used to correct for edge effect. In either case, the inclusion zone of an edge element divides into different sections, and the attribute density varies among these sections, but is uniform within each one (see, e.g., Figure 10.10c). If a design prescribes non-circular inclusion zones, then correction of edge effect by a reflection method may both section and alter the shape of the inclusion zone (see, e.g., Figure 10.12b).

Let $a_{k_{j'}}$ be the area of the j'th of n_k sections in $\mathcal{I}_k$, so $a_k = \sum_{j'=1}^{n_k} a_{k_{j'}}$. Let $a_{k_{j'}k'_j}$ be the area of overlap of the j'th section of $\mathcal{I}_k$ and jth section of $I_{k'}$, and let $t_{k_{j'}}$ be the tally for the kth element in the j'th section of $\mathcal{I}_k$. Then

$$\iint_{\mathcal{A}} \rho^2(x,z)\,dx\,dz = \sum_{k=1}^{N} \sum_{j'=1}^{n_k} \left(\frac{y_k t_{k_{j'}}}{a_k}\right)^2 a_{k_{j'}}$$

$$+ \sum_{k=1}^{N} \sum_{j'=1}^{n_k} \sum_{\substack{k'=1 \\ k'\neq k}}^{N} \sum_{j=1}^{n_{k'}} \left(\frac{y_k t_{k_{j'}}}{a_k} \frac{y_{k'} t_{k'_j}}{a_{k'}}\right) a_{k_{j'}k'_j}.$$

Substitution of this result into (10.41) gives a formula for the variance where edge effect is corrected by the mirage or walkthrough method, i.e.,

$$V\left[\hat{\tau}_{\rho,\text{rep}}\right] = \frac{1}{m}$$

$$\times \left\{ A \sum_{k=1}^{N} \sum_{j'=1}^{n_k} \left[\left(\frac{y_k t_{k_{j'}}}{a_k}\right)^2 a_{k_{j'}} + \sum_{\substack{k'=1 \\ k'\neq k}}^{N} \sum_{j=1}^{n_{k'}} \left(\frac{y_k t_{k_{j'}}}{a_k} \frac{y_{k'} t_{k'_j}}{a_{k'}}\right) a_{k_{j'}k'_j}\right] - \tau_\rho^2 \right\}.$$

$$(10.46)$$

Alternatively, substituting (10.44) into (10.46), gives an equivalent formula

$$V\left[\hat{\tau}_{\rho,\text{rep}}\right] = \frac{1}{m} \sum_{k=1}^{N} y_k^2 \left[\frac{\left(\sum_{j'=1}^{n_k} t_{k_{j'}}^2 a_{k_{j'}}/a_k\right) - (a_k/A)}{a_k/A}\right]$$

$$+ \frac{1}{m} \sum_{k=1}^{N} \sum_{\substack{k'=1 \\ k'\neq k}}^{N} y_k y_{k'} \left[\frac{\left(\sum_{j'=1}^{n_k} \sum_{j=1}^{n_{k'}} t_{k'_j} t_{k_{j'}} a_{k_{j'}k'_j}/A\right) - (a_k/A)(a_{k'}/A)}{(a_k/A)(a_{k'}/A)}\right].$$

$$(10.47)$$

Note that this formula reduces to (10.45) if $n_k = 1$ for all k, because $t_{k_1} = t_k = 1$ and $a_{k_1 k'_1} = a_{kk'}$.

10.11.3 LIS with segmented transects and correction of edge effect

We consider the case where estimation is conditional upon the orientation (θ_s) of a leg of a segmented transect.

The use of segmented transects creates sets of sections within the inclusion zone of an element. For example, the inclusion zone depicted in Figure 9.9 on page 304 has two sets of sections. Set $\mathcal{I}_{k_1}$, say, comprises the three light gray sections of the inclusion zone, and $\mathcal{I}_{k_2}$ comprises the two dark gray sections. If $\mathbb{P}_s \in \mathcal{I}_{k_1}$, the element is intersected by one leg of the transect and is tallied once. If $\mathbb{P}_s \in \mathcal{I}_{k_2}$, the element is intersected by two legs of the transect and is tallied twice. The total area of this inclusion zone is $a_k = \bar{w}_k(\theta_s)L - (\psi_1 + \psi_2)$. The area of $\mathcal{I}_{k_1}$ is $\bar{w}_k(\theta_s)L - 2(\psi_1 + \psi_2)$ and the area of $\mathcal{I}_{k_2}$ is $(\psi_1 + \psi_2)$. The attribute density at any point (x, z) in $\mathcal{I}_k$ is

$$\rho_k(x, z) = \frac{y_k t_k(x, z)}{\bar{w}_k(\theta_s)L}, \tag{10.48}$$

where $t_k(x, z) = 1$ if $(x, z) \in \mathcal{I}_{k_1}$ and $t_k(x, z) = 2$ if $(x, z) \in \mathcal{I}_{k_2}$. Thus, integration of the attribute density over the area of $\mathcal{I}_k$ yields y_k, i.e.,

$$\iint_{\mathcal{I}_k} \rho_k(x, z) \, dz \, dx = \iint_{\mathcal{I}_{k_1}} \rho_k(x, z) \, dz \, dx + \iint_{\mathcal{I}_{k_2}} \rho_k(x, z) \, dz \, dx$$

$$= \frac{y_k}{\bar{w}_k(\theta_s)L} [\bar{w}_k(\theta)L - 2(\psi_1 + \psi_2)] + \frac{2y_k}{\bar{w}_k(\theta_s)L} (\psi_1 + \psi_2)$$

$$= y_k. \tag{10.49}$$

Equation (10.48) applies generally to transects with one or more legs. However, the number of sections within an inclusion zone depends on the number of legs, their arrangement and orientation, and on the shape of the element. Moreover, corrections for edge effect may increase the number of sections in $\mathcal{I}_k$. Regardless of the number of sections, the attribute density is uniform among the points within each section of $\mathcal{I}_k$, but varies among the sections. Let a_{k_j}, be the area of the j'th of the n_k sections of $\mathcal{I}_k$. The attribute density in the j'th section is $y_k t_{k_j}/(\bar{w}_k(\theta_s)L)$, but t_{k_j}, may take a value greater than 2, owing to the number of legs or to edge correction or both. Hence, an estimator of τ_ρ, given θ, is

$$\hat{\tau}_{\rho s} = A \sum_{\mathbb{P}_s \in \mathcal{I}_k} \frac{y_k t_k(x_s, z_s)}{\bar{w}_k(\theta_s)L} \tag{10.50}$$

and the variance of this estimator is

$$
V[\hat{\tau}_{\rho s}] = A \sum_{k=1}^{N} \sum_{j'=1}^{n_k} \left(\frac{y_k t_{k_{j'}}}{\bar{w}_k(\theta_s)L} \right)^2 a_{k_{j'}}
$$

$$
+ A \sum_{k=1}^{N} \sum_{j'=1}^{n_k} \sum_{\substack{k'=1 \\ k' \neq k}}^{N} \sum_{j=1}^{n_{k'}} \left(\frac{y_k t_{k_{j'}}}{\bar{w}_k(\theta_s)L} \frac{y_{k'} t_{k'_j}}{\bar{w}_{k'}(\theta_s)L} \right) a_{k_{j'} k'_j} - \tau_\rho^2.
$$

$$(10.51)$$

Or, substituting (10.44) and letting $\tilde{\pi}_k = \bar{w}_k(\theta_s)L/A$,

$$
V[\hat{\tau}_{\rho s}] = \sum_{k=1}^{N} y_k^2 \left\{ \frac{\left[\sum_{j'=1}^{n_k} t_{k_{j'}}^2 a_{k_{j'}} / (\tilde{\pi}_k A) \right] - \tilde{\pi}_k}{\tilde{\pi}_k} \right\}
$$

$$
+ \sum_{k=1}^{N} \sum_{\substack{k'=1 \\ k' \neq k}}^{N} y_k y_{k'} \left\{ \frac{\left[\sum_{j'=1}^{n_k} \sum_{j=1}^{n_{k'}} t_{k'_j} t_{k_{j'}} a_{k_{j'} k'_j} / A \right] - \tilde{\pi}_k \tilde{\pi}_{k'}}{\tilde{\pi}_k \tilde{\pi}_{k'}} \right\}.
$$

$$(10.52)$$

CHAPTER 11

Miscellaneous Methods

11.1 Introduction

Some specialized designs for areal sampling have evolved from the widely used designs of Bitterlich sampling and line intersect sampling. In this chapter, we describe four of these specialized designs from a discrete-population perspective. We also provide brief overviews of the ranked set sampling and adaptive cluster sampling designs. More comprehensive information on these latter two designs can be found elsewhere. Finally, we present 3P sampling, a variant of Poisson sampling. 3P sampling is not an areal design, but it is often used in a second-stage of sampling, where the first-stage units are selected with an areal frame.

11.2 Point relascope sampling

Point relascope sampling or PRS (Gove et al. 1999b) can be viewed as Bitterlich sampling of the coarse woody debris (i.e., logs or log-shaped objects with a discernible central axis) on the floor of a forested tract. PRS in other contexts is possible but remains to be explored. Bebber & Thomas (2003) describe an alternative method for sampling coarse woody debris, which also derives from the Bitterlich design.

A Monte Carlo approach to PRS is described in §10.4; in this section, we emphasize the discrete-population approach to the PRS design. As in previous chapters, we denote the tract of interest by $\mathcal{A}$ and we let A be the horizontal land area (ha) of the tract.

In our presentation of Bitterlich sampling (Chapter 8), we defined the circular inclusion zone for a tree, $\mathcal{U}_k$, in terms of a limiting distance, $R_k = \alpha r_k$, where r_k is tree radius at breast height and α is a constant, independent of k. However, we can also define the inclusion zone for a tree in terms of an angle, ν, where $\nu = 2 \arcsin 1/\alpha$ (see §8.2.4). The vertex of the angle is coincident with a point in the inclusion zone of a tree if the rays of the angle can be oriented such that the horizontal width of the tree's stem at breast height subtends the angle (see Figure 8.1, p. 251). The inclusion zone comprises the set of all such location points. This angle-based definition of an inclusion zone is entirely consistent with the definition based on limiting distance, if standing trees with circular stems are the elements of interest. However, the angle-based definition generalizes for use with PRS; the definition based on limiting distance does not.

In point relascope sampling, we use an angle gauge (or relascope) at a sample point, $\mathbb{P}_s \in \mathcal{A}$, to select a sample of logs. Specifically, the practitioner determines

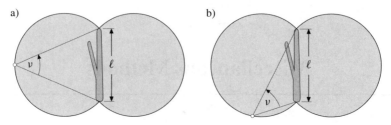

Figure 11.1 *Inclusion zone of a log for point relascope sampling with angle v.*

whether the length of a log—when viewed from the sample point—fills the field of view of the angle gauge. If so, the log is included in the point relascope sample. Range poles erected at each end of a log may facilitate the viewing of the log's length. Whether a borderline log should be included in a sample can be unambiguously determined by a procedure described in §10.4.

The inclusion zone of a given log comprises the set of location points where the length of the log fills the field of view of the angle gauge. The size and shape of the inclusion zone depends on v, the angle of the angle gauge. For $0 < v < \pi/2$, the inclusion zone takes the shape of two overlapping circles of equal size, with the central axis of the log serving as a common chord (Figure 11.1). Small angles provide large inclusion zones and large angles provide small zones. The inclusion zone is circular for $v = \pi/2$ (see Figure 10.4, p. 335). Angles greater than $\pi/2$ are impractical and are not used.

The horizontal land area (m^2) of the inclusion zone for log $\mathcal{U}_k$ is (from Gove et al. 1999b)

$$a_k = \varphi \ell_k^2 \tag{11.1}$$

where ℓ_k is log length (m), and

$$\varphi = \frac{\pi - v + \sin v \cos v}{2 \sin^2 v}. \tag{11.2}$$

The derivation of φ is left as an exercise at the end of the chapter.

In the absence of edge effect, the probability that $\mathcal{U}_k$ is included in a sample at (x_s, z_s) is

$$\pi_k = \frac{10^{-4} a_k}{A} = \left(\frac{10^{-4} \varphi}{A} \right) \ell_k^2. \tag{11.3}$$

Note that the inclusion probability is proportional to the length-square of the log. Similarly, in Bitterlich sampling, the inclusion probability is proportional to the diameter-square (or basal area) of the tree.

The population parameter, τ_y, is estimated in the same fashion that has been elucidated in Chapters 7, 8, and 9, i.e., HT estimation with the data collected at each sampling location, followed by use of the replicated sampling estimator to combine all m individual estimates. With the PRS inclusion probability, the HT estimator is

$$\hat{\tau}_{y\pi s} = \sum_{\mathcal{U}_k \in \mathbb{P}_s} \frac{y_k}{\pi_k}$$

$$= \frac{10^4 A}{\varphi} \sum_{\mathcal{U}_k \in \mathbb{P}_s} \frac{y_k}{\ell_k^2}. \tag{11.4}$$

If $\tau_y = \sum_{k=1}^{N} \ell_k^2$, $\hat{\tau}_{y\pi s}$ requires only a count of the logs selected at each sampling location, i.e.,

$$\hat{\tau}_{y\pi s} = \frac{10^4 A}{\varphi} \sum_{\mathcal{U}_k \in \mathbb{P}_s} \frac{\ell_k^2}{\ell_k^2}$$

$$= \left(\frac{10^4 A}{\varphi} \right) n_s, \tag{11.5}$$

where n_s is the number of logs sampled at $\mathbb{P}_s$.

With information from $m > 1$ sample points, the replicated sampling estimator of τ_y is

$$\hat{\tau}_{y\pi,\text{rep}} = \frac{1}{m} \sum_{s=1}^{m} \hat{\tau}_{y\pi s}. \tag{11.6}$$

The variance of $\hat{\tau}_{y\pi,\text{rep}}$ is estimated unbiasedly with

$$\hat{v} \left[\hat{\tau}_{y\pi,\text{rep}} \right] = \frac{1}{m(m-1)} \sum_{s=1}^{m} \left(\hat{\tau}_{y\pi s} - \hat{\tau}_{y\pi,\text{rep}} \right)^2. \tag{11.7}$$

11.3 Horizontal line sampling

Horizontal line sampling (HLS) was proposed by Strand (1957) for the purpose of estimating attributes of standing trees in a forested tract. The design is used in Taiwan (Yang & Chao 1987), but to our knowledge, it has received little use elsewhere. As usual, we denote the tract of interest by $\mathcal{A}$ and the horizontal projection of its land area by A (ha).

HLS combines features of line intersect sampling (LIS) and Bitterlich sampling. As in LIS, a sampling location, (x_s, z_s), serves as the center point or end point of a transect of length L (m). Unlike LIS, however, a tree may be selected into a horizontal line sample without being directly intersected by the transect. Rather, a tree, $\mathcal{U}_k$, is selected if its perpendicular distance from the transect is within a limiting distance $R_k = \kappa d_k$ (m), where d_k is the tree's diameter (cm), and κ (m cm^{-1}) is a constant limiting-distance factor, independent of k (κ is known as the plot-radius factor in Bitterlich sampling (see § 8.2.3)).

Under the HLS design, the inclusion zone of tree $\mathcal{U}_k$ is rectangular with length L and width $2R_k = 2\kappa d_k$ (m). The tree is centered in its inclusion zone if the transect

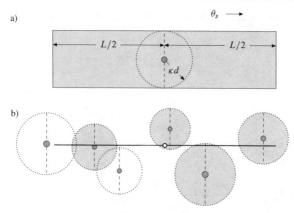

Figure 11.2 *a) Rectangular inclusion zone for a tree prescribed by HLS protocol. The dotted circular region is equivalent to the inclusion zone under the Bitterlich design. b) A tree is selected into a horizontal line sample if the perpendicular distance (along the appropriate dashed line) is $\leq \kappa d_k$. In this example, four trees are selected into the horizontal line sample.*

is centered about the sample location (see Figure 11.2a). The size of the inclusion zone is identical for all orientations of the transect.

Under the Bitterlich design, $R_k = \kappa d_k$ is the radius of the circular inclusion zone of tree $\mathcal{U}_k$. Embedded in the inclusion zone depicted in Figure 11.2a is a dotted circular area, which is equivalent to the inclusion zone under the Bitterlich design. If the sampling location, (x_s, z_s), occurs within the inclusion zone, the transect will intersect, and be perpendicular to, the dashed line that bisects the circular area in the diagram. Any portion of the transect that intersects the circular area is within the limiting distance, κd_k, from tree $\mathcal{U}_k$. However, the only limiting distance that matters in HLS is the perpendicular limiting distance, which is represented by the dashed line on either side of the tree (Figure 11.2b).

In application, the sth transect is established with azimuth θ_s at (x_s, z_s), the latter being selected uniformly at random within $\mathcal{A}$. Trees on either side of the transect are observed with an angle gauge calibrated to produce a predetermined basal area factor F. When sighted perpendicularly from the transect, any tree, $\mathcal{U}_k$, whose stem width at breast height fills the field of view of the angle gauge is within the limiting distance, κd_k, needed to select the tree into the horizontal line sample $\mathbb{L}_s$. As with Bitterlich sampling, an angle gauge provides a convenient means to obviate the actual measurement of the perpendicular distance from the transect to a tree. Care should be taken near the ends of the transect to ensure 'perpendicular sighting' with the angle gauge.

The area (ha) of the inclusion zone for tree $\mathcal{U}_k$ is

$$a_k = 2 \times 10^{-4} \kappa d_k L$$

or, since $\kappa = 1/(2\sqrt{F})$ from (8.13),

$$a_k = \left(\frac{10^{-4}L}{\sqrt{F}}\right)d_k. \tag{11.8}$$

Therefore, the probability that the kth tree is included in a horizontal line sample at (x_s, z_s) is

$$\pi_k = \frac{a_k}{A} = \left(\frac{10^{-4}L}{\sqrt{F}A}\right)d_k. \tag{11.9}$$

Note that the inclusion probability of tree $\mathcal{U}_k$ at each sampling location is proportional to the tree's diameter. From a probabilistic standpoint, this is the chief difference between HLS and Bitterlich sampling, where the inclusion probability of the latter is proportional to the basal area (or diameter-square) of the tree.

The population parameter, τ_y, is estimated in the usual fashion. With the inclusion probabilities that are induced by HLS, the HT estimator at (x_s, z_s) is

$$\hat{\tau}_{y\pi,s} = \sum_{\mathcal{U}_k \in \mathbb{L}_s} \frac{y_k}{\pi_k}$$

$$= \frac{10^4\sqrt{F}A}{L}\sum_{\mathcal{U}_k \in \mathbb{L}_s} \frac{y_k}{d_k}. \tag{11.10}$$

If $y_k = d_k$, then τ_y is the sum of tree diameters, in which case the estimator reduces to

$$\hat{\tau}_{y\pi,s} = \left(\frac{10^4\sqrt{F}A}{L}\right)n_s, \tag{11.11}$$

where n_s is the number of trees selected for the sample at (x_s, z_s). With replicate samples, τ_y is estimated with $\hat{\tau}_{y\pi,\mathrm{rep}}$ and the variance of this estimator may be estimated with $\hat{v}[\hat{\tau}_{y\pi,\mathrm{rep}}]$. Yang & Chao (1987) discussed five alternative methods of estimating conventional components of change (see §8.6). Eriksson's temporally additive components (§8.6.2) could also be estimated.

11.3.1 Sausage sampling

Sausage sampling (Ducey et al. 2002) is a modification of HLS, which was motivated for use in estimating attributes of standing dead trees, though live trees may certainly substitute as the population of interest.

HLS requires careful checking in the field to ensure perpendicular sighting of trees at the ends of the transect. Sausage sampling eliminates the need for such checking. At each end of the sth transect, the sampler performs a 180° sweep with the angle gauge, selecting into the sth sample all trees that fill the angle gauge's field of view. Whereas a horizontal line sample includes only those trees that are within κd_k m, when measured on a perpendicular line from the transect, a sausage sample includes all trees within κd_k m of any point on the transect.

Sausage sampling takes its name from the shape of the inclusion zone of a tree. The sausage includes the rectangular inclusion zone from HLS, but with half circles

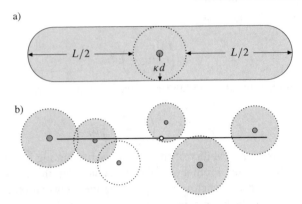

Figure 11.3 *a) Inclusion zone of a tree for sausage sampling. b) Trees are selected for a sausage sample if distance from the tree to the transect $\leq \kappa d_k$.*

appended to each end (see Figure 11.4a). As in HLS, the shape and area of the inclusion zone for a tree, $\mathcal{U}_k$, is unaffected by the orientation of the transect.

Centered about the tree in the inclusion zone depicted in Figure 11.3a is a circular region, which is equivalent in area to an inclusion zone under Bitterlich sampling. If the sampling location, (x_s, z_s), occurs within the inclusion zone, then the tree is within the limiting distance, κd_k, of any point on the portion of the sth transect that intersects this circular 'de facto Bitterlich inclusion zone' (Figure 11.3b).

The area (ha) of the entire sausage-shaped inclusion zone for tree $\mathcal{U}_k$ is

$$a_k = 10^{-4} \left(2\kappa d_k L + \pi \kappa^2 d_k^2 \right)$$

or, expressed in terms of F,

$$a_k = 10^{-4} \left(\frac{L d_k}{\sqrt{F}} + \frac{\pi d_k^2}{4F} \right). \tag{11.12}$$

The probability that tree $\mathcal{U}_k$ is included in the sausage sample at (x_s, z_s) is, of course, $\pi_k = a_k / A$. The population parameter, τ_y, is estimated in the usual manner with $\hat{\tau}_{y\pi,s}$ or $\hat{\tau}_{y\pi,\text{rep}}$.

11.3.2 Modified sausage samping

Sausage sampling can be modified by adopting our intersection protocols for LIS (§9.2.1). That is, a tree is omitted from the modified sausage sample if the sampling location (x_s, z_s) falls within the tree's de facto Bitterlich inclusion zone (see Figure 11.4b). This would entail performing sausage sampling in the usual manner, but omitting any trees which fill the field of view of the angle gauge when sighted from (x_s, z_s).

The inclusion zone, under this protocol, is sausage-shaped, but with a hole in the middle (see Figure 11.4a). The area of the inclusion zone of tree $\mathcal{U}_k$ for modified

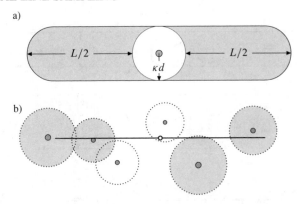

Figure 11.4 *a) Inclusion zone of a tree for modified sausage sampling b) Trees are selected for a sausage sample if distance from the tree to the transect $\leq \kappa d_k$. However, trees selected from the sampling location (○) are omitted. In this example, four trees are selected for the modified sausage sample.*

sausage sampling is

$$a_k = 2 \times 10^{-4} \kappa d_k L, \tag{11.13}$$

so the probability of including tree $\mathcal{U}_k$ in the sth modified sausage sample is proportional to the tree's diameter, i.e.,

$$\pi_k = \left(\frac{2 \times 10^{-4} \kappa L}{A} \right) d_k = \left(\frac{10^{-4} L}{\sqrt{F} A} \right) d_k, \tag{11.14}$$

as in HLS. Consequently, τ_y is estimated with (11.10) and the sum of diameters in $\mathcal{A}$ can be estimated with (11.11).

11.3.3 Redux: LIS

The three line sampling designs—HLS, sausage sampling, and modified sausage sampling—can be viewed as line intersect sampling of de facto Bitterlich inclusion zones (Valentine et al. 2001). Of course this view requires that we define the de facto Bitterlich inclusion zone as the population element for the purpose of selection, even though the tree at the center of the zone is the real element of interest.

When viewed in this context, the three designs differ in regard to how acceptable partial intersections are defined. By an acceptable partial intersection we mean that the tree at the center of the de facto Bitterlich inclusion zone is accepted in the sample. The nuances of partial intersections under our LIS protocols are described in §9.2.1. Recall that for partial intersections it is useful to think of a transect as consisting of two segments joined at (x_s, z_s). Each segment has a front end and a back end, and the back ends of the two segments meet at (x_s, z_s).

In sausage sampling, a tree is included in the sample with any partial intersection (Figure 11.3b). Modified sausage sampling uses our proposed protocols for LIS: Partial intersections by the front end of a transect segment are acceptable, but partial

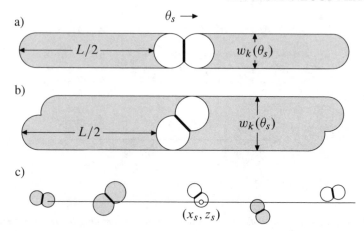

Figure 11.5 *a,b) Inclusion zones of pieces of coarse woody debris (logs) for transect relascope sampling. c) A log is selected for the transect relascope sample if its de facto PRS inclusion zone is intersected by the transect. The log is omitted from the sample if the sample location (○) occurs within the de facto PRS inclusion zone. In this example, three logs are selected by the transect.*

intersections by the back end are not (Figure 11.4b). Hence, if (x_s, z_s) occurs within a tree's de facto Bitterlich inclusion zone, then that tree is omitted from the sample. In HLS, a partial intersection is acceptable only if the transect traverses at least half way across the de facto Bitterlich inclusion zone, which means that the transect must intersect the diameter of the zone that is perpendicular to the transect (Figure 11.2b). In both HLS and modified sausage sampling, a tree is selected in a sample with probability proportional to the diameter of its de facto Bitterlich inclusion zone, which, of course, is proportional to the tree's diameter.

11.4 Transect relascope sampling

Transect relascope sampling (TRS) originated with Ståhl (1998). TRS is similar to HLS, but the elements of interest are pieces of coarse woody debris (logs) on the forest floor, rather than standing trees. And, whereas HLS uses features of LIS and Bitterlich sampling, TRS uses features of LIS and point relascope sampling (PRS).

Departing slightly from the original formulation of Ståhl (1998), we first prescribe a sampling protocol for TRS similar to modified sausage sampling. Stahl's original formulation is described subsequently.

A transect with length L (m) and orientation θ_s is established at a sampling location, $(x_s, z_s) \in \mathcal{A}$, which is selected uniformly at random. The sampler moves along the transect, sighting logs with an angle gauge. As in sausage sampling, the sampler performs 180° sweeps with the angle gauge at both ends of the transect. Any log whose length fills the field of view of the angle gauge is selected for the

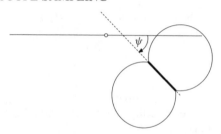

Figure 11.6 *Acute angle (ψ) formed by the central axis of a log and the transect.*

transect relascope sample, unless the field can be filled by the log's length at the sampling location, (x_s, z_s).

Recall that in PRS (see §11.2 and §10.4), the size and shape of a log's inclusion zone depends on log length and the angle, v, of the angle gauge, where $(0 < v \le \pi/2)$. For an angle less than $\pi/2$, the inclusion zone takes the shape of two overlapping circles, with the central axis of the log serving as a common chord (Figure 11.1). With an angle of $\pi/2$, the two circles coalesce (Ståhl 1997), so the inclusion zone is circular, with a diameter equal to the log's length (Figure 10.4, p. 335).

As depicted in Figure 11.5a,b, the shape of a log's TRS inclusion zone depends on the transect's orientation relative to the orientation of the log. The log is centrally located in its inclusion zone if the transect is centered about the sampling location, (x_s, z_s). Embedded within the TRS inclusion zone is a de facto PRS inclusion zone. The log's length fills the field of view of the angle gauge from any point on the portion of the transect that intersects the log's de facto PRS inclusion zone (Figure 11.5c).

The area of TRS inclusion zone for a log, $\mathcal{U}_k$, is $w_k(\theta_s)L$, where $w_k(\theta_s)$ is the width, perpendicular to the transect, of the log's de facto PRS inclusion zone. Inasmuch as the actual measurement of $w_k(\theta_s)$ is impractical, if not impossible, in the field, Ståhl (1998) provided a formula for its calculation. Let ℓ_k be the length of log $\mathcal{U}_k$ measured along its central axis, then

$$w_k(\theta_s) = \ell_k \left(\frac{1}{\sin v} + \cot v \cos \psi \right), \tag{11.15}$$

where ψ is the acute angle formed by the central axis of the log and the transect (Figure 11.6). This angle can be calculated from the azimuth of the log's central axis and θ_s.

The inclusion probability for the $\mathcal{U}_k$, conditional on θ_s, is $\pi_k = 10^{-4} w_k(\theta_s)L/A$, so τ_y may be estimated with the conditional estimator from LIS, i.e., $\hat{\tau}^c_{y\pi,s}$.

However, unconditional estimation is easier, because calculation of the expected width of an inclusion zone does not require the azimuth of the log's central axis. For inference under unconditional estimation, θ_s should be selected at random from $U[0, \pi]$.

The expected width of the inclusion zone of $\log \mathcal{U}_k$ is (from Ståhl 1998)

$$\bar{w}_k = \phi \ell_k,$$

where

$$\phi = \frac{1}{\sin \nu} + \frac{2}{\pi} \cot \nu. \qquad (11.16)$$

Hence, the unconditional probability of including $\log \mathcal{U}_k$ in the sth sample is proportional to log length, i.e.,

$$\pi_k = \left(\frac{10^{-4} \phi L}{A} \right) \ell_k \qquad (11.17)$$

and the unconditional estimator of τ_y is

$$\hat{\tau}^{\mathrm{u}}_{y\pi,s} = \frac{10^4 A}{\phi L} \sum_{\mathcal{U}_k \in \mathbb{L}_s} \frac{y_k}{\ell_k}. \qquad (11.18)$$

If the aggregate log length in $\mathcal{A}$ is of interest, then $y_k = \ell_k$, so

$$\hat{\tau}^{\mathrm{u}}_{y\pi,s} = \frac{A}{\phi L} n_s, \qquad (11.19)$$

where n_s is the number of logs in the sth sample. And, of course, with replicate samples, τ_y is estimated with $\hat{\tau}_{y\pi,\mathrm{rep}}$ and the variance of this estimator may be estimated with $\hat{v}[\hat{\tau}_{y\pi,\mathrm{rep}}]$.

11.4.1 Ståhl's method

Ståhl (1998) envisioned implementing TRS with parallel transects that traverse $\mathcal{A}$, equally spaced along, and perpendicular to, a baseline (see Figure 11.7). As shown on the left side of the figure, it may be necessary to extend transects across the boundary of $\mathcal{A}$ to avoid edge effect.

With a baseline of fixed orientation, $\theta^{\perp}$, and fixed length, L_B, the probability that the de facto PRS inclusion zone of $\log \mathcal{U}_k$ will be intersected by one of t parallel and equally spaced transects, with orientation $\theta_s = \theta^{\perp} \pm \pi/2$, is

$$\pi_k = \frac{w_k(\theta_s)}{L_B/t}. \qquad (11.20)$$

If the orientation of the baseline, $\theta_s^{\perp}$ is chosen at random from $\mathrm{U}[0, \pi]$, then the length of the baseline is denoted by $L_B(\theta_s^{\perp})$, so

$$\pi_k = \frac{\bar{w}}{L_B(\theta_s^{\perp})/t}$$

$$= \frac{\phi \ell_k}{L_B(\theta_s^{\perp})/t}. \qquad (11.21)$$

Collectively, the logs whose de facto PRS inclusion zones are intersected by the parallel transects constitute a single (i.e., the sth) transect relascope sample. Hence,

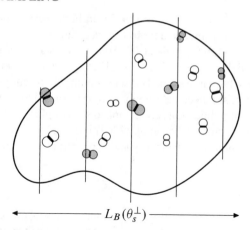

Figure 11.7 *TRS with uniformly spaced transects that traverse* $\mathcal{A}$.

τ_y is unbiasedly estimated by

$$\hat{\tau}_{y\pi,s}^{u} = \frac{L_B(\theta_s^{\perp})}{t\phi} \sum_{u_k \in \mathbb{L}_s} \frac{y_k}{\ell_k}, \tag{11.22}$$

which reduces to

$$\hat{\tau}_{y\pi,s}^{u} = \frac{L_B(\theta_s^{\perp})}{t\phi} n_s, \tag{11.23}$$

if $\ell_k = y_k$. As with the other areal designs, τ_y may be estimated from replicate samples with $\hat{\tau}_{y\pi,\text{rep}}$ and the variance of this estimator is estimated with $\hat{v}[\hat{\tau}_{y\pi,\text{rep}}]$.

11.5 Ranked set sampling

Ranked set sampling (RSS) originally was devised by McIntyre (1952) as an alternative to simple random sampling for the estimation of forage yields in pastures. A ranking procedure reduces a large random sample to a smaller ranked set sample, which is expected to contain units whose attribute values are well representative of the distribution of attribute values found in the population.

Ranked set sampling is an appealing strategy where it is expensive to obtain accurate measurements of the attributes of selected units, but where it is cheap and easy to discern differences among the attributes and rank them by size. Suppose, for example, that a woodland is sampled with small plots for the purpose of estimating the dry weight of browse available to wildlife. It should be easy to rank the relative amounts of browse on different plots by observing the densities and heights of the vegetation. Measurement of the dry weight, however, involves clipping the vegetation off at the ground line and then drying it to a constant weight, a process that may take several days.

The theoretical foundation of ranked set sampling was adduced by Takahasi &

Wakimoto (1968) and Dell & Clutter (1972). In both papers, the authors couched their descriptions and arguments in terms of order statistics, a practice that continues to this day. Patil et al. (1994) provided a thorough and clear description of ranked set sampling for potential users, and Kaur et al. (1995) provided a useful annotated bibliography of the early literature on ranked set sampling. In recent years, mathematical statisticians have found ranked set sampling to be a fruitful topic for study, so much so that the number of publications about ranked set sampling has outpaced the number of reported applications. Chen et al. (2003) provided a monograph that contains a comprehensive summary of the theory and applications of ranked set sampling.

11.5.1 Overview

Ranked set samples usually are described in terms of sets and cycles. A 'set' is a random sample of k units; the selection of k sets in sequence completes a 'cycle.' Balanced rank set sampling entails the selection of k sets in each of n cycles of sampling. Thus, a total of $nk^2 = nk + nk(k - 1)$ units are selected from the population. From these nk^2 units, we extract a ranked set sample of nk units for measurement; the other $nk(k - 1)$ units are discarded.

A ranking procedure is key to the extraction of the nk units that form the ranked set sample. Initially, we select k units from the population, which form the first set of the first cycle. We rank the k units by their attributes, from smallest to largest, either by visual examination, measurement of an auxiliary variate, or by some other quick and easy process. Of the k units in the first set, only the smallest unit—the unit with rank 1—is accepted into the ranked set sample. Next, we select a second set of k units from the population. In this second set of the first cycle, we repeat the ranking procedure and accept only the second smallest unit—the unit with rank 2—into the ranked set sample. We continue in this fashion until we accept the kth smallest unit, i.e., the largest unit, from the kth set. Thus, a total of k units, one unit of each rank, is accepted into the ranked set sample from the first cycle of k sets. This entire procedure is repeated in each of the n cycles, yielding a ranked set sample of nk units, i.e., n units for each of the k ranks. The set size, k, typically is fixed at 3 or 4 to facilitate easy and accurate ranking.

Ranking often involves judgment, and errors do occur. Dell & Clutter (1972) proved that erroneous ranks ordinarily do not lead to biased estimates. However, ranked set sampling is no more efficient that simple random sampling if the ranking is no better than random assignment. Bias may result, however, where the individual doing the ranking gives rank 1 to the unit he or she would most prefer to extract from first sample; rank 2 to a preferred unit in the second sample; and so forth. One way to avoid bias due to such preferential or purposive ranking is to randomize the ranks of the units extracted from the different sets. For example, with $k = 4$, randomization might extract the unit with rank 4 from set 1, rank 2 from set 2, rank 1 from set 3, and rank 3 from set 4. Such randomization could be achieved, for example, by moving otherwise identical coins minted in years ending in 1, 2, 3, and 4 from one pocket to another. With n copies of the four coins, it should be possible to randomize the selection of the ranked set sample across the nk sets.

11.5.2 Traditional applications

Many of the results in the burgeoning literature on ranked set sampling derive from the assumption that sets are random samples of k units from infinite populations. Examples of infinite populations include continuums and discrete populations sampled with replacement.

The original applications of ranked set sampling were concerned with the estimation of forage or browse on tracts of land (McIntyre 1952; Halls & Dell 1966; Martin et al. 1980). In a typical application of this kind, the landscape is sampled with plots, since the plants involved are too small and too numerous to be treated as discrete individuals. Plots also may be used in studies of environmental contaminants, where an attribute of interest is spread more or less continuously, but with varying concentration, over the landscape (e.g., Patil et al. 1994).

Other applications of ranked set sampling have been proposed; indeed, ranked set sampling may be applied to any population from which random samples can be drawn, and the sample units ranked by size. The traditional applications, where a continuous landscape is sampled with plots or sample points, match best with the general theme of this book, and so we describe that usage first.

Let A be the horizontal projection of the area of a tract or region, which we denote by $\mathcal{A}$, and let (x, z) denote the location coordinates of any point in $\mathcal{A}$. Moreover, let $\rho(x, z)$ be the attribute density of interest at any location point $(x, z) \in \mathcal{A}$. If we are dealing with an attribute whose density cannot be ranked and directly measured at a point, then

$$\rho(x, z) = \frac{y(x, z)}{a}, \tag{11.24}$$

where $y(x, z)$ denotes the amount of attribute contained in a plot with area a, the center (or corner) of the plot located at (x, z).

Of prime interest is τ_ρ, the total amount of attribute in $\mathcal{A}$, i.e.,

$$\tau_\rho = \int_{\mathcal{A}} \rho(x, z) \, dx \, dz. \tag{11.25}$$

The mean attribute density (attribute per unit land area),

$$\mu_\rho = \frac{\tau_\rho}{A}, \tag{11.26}$$

and the variance of $\rho(x, z)$,

$$\sigma_\rho^2 = \frac{1}{A} \int_{\mathcal{A}} \left[\rho(x, z) - \mu_\rho \right]^2 \, dx \, dz. \tag{11.27}$$

In balanced sampling, k sample points are selected to form set r of cycle j, where $r = 1, 2, \ldots, k$ and $j = 1, 2, \ldots, n$. Each sample point is selected independently with probability density $1/A$ (see § 4.5, p. 116 and § 10.1, p. 328). The k sample points within set r of cycle j are ranked by their relative attribute densities, possibly with error. The attribute density of the sample point with rank r is denoted by $\rho_{[r]j}$. If plots are established at the sample points, then the amount of attribute on the plot

ASIDE: The notation of ranked set sampling is rooted in the notation of order statistics. Suppose that a sample comprises a set of attributes values y_1, y_2, and y_3. The order statistics are the attribute values ranked by size, smallest to largest. If, for example, $y_2 < y_3 < y_1$, then the order statistics $(y_{(r)}, r = 1, \ldots, k)$ for this set of values are: $y_{(1)} = y_2$, $y_{(2)} = y_3$, and $y_{(3)} = y_1$. Judgmental order statistics may differ from true order statistics, since judgmental ranking is subject to error. Suppose that through a judgmental process we rank $y_2 < y_1 < y_3$. The judgmental order statistics, in this case, are $y_{[1]} = y_2$, $y_{[2]} = y_1$, and $y_{[3]} = y_3$. It is customary in ranked set sampling to denote true order statistics with the rank subscripted in parentheses and judgmental order statistics with the rank subscripted in brackets.

with rank r is denoted by $y_{[r]j}$, in which case the attribute density is

$$\rho_{[r]j} = \frac{y_{[r]j}}{a}. \tag{11.28}$$

Estimation

The mean density, μ_ρ, of the attribute of interest is unbiasedly estimated by

$$\hat{\mu}_{\rho,\text{RSS}} = \frac{1}{nk} \sum_{r=1}^{k} \sum_{j=1}^{n} \rho_{[r]j}. \tag{11.29}$$

The variance of $\hat{\mu}_{\rho,\text{RSS}}$ is $V[\hat{\mu}_{\rho,\text{RSS}}] \leq \sigma_\rho / nk$, depending upon how well assigned ranks agree with true ranks. A consistent estimator of the $V[\hat{\mu}_{\rho,\text{RSS}}]$ is (from Chen et al. 2003, p. 26)

$$\hat{v}[\hat{\mu}_{\rho,\text{RSS}}] = \frac{1}{k^2 n(n-1)} \sum_{r=1}^{k} \sum_{j=1}^{n} \left(\rho_{[r]j} - \bar{\rho}_{[r]}\right)^2 \tag{11.30}$$

where

$$\bar{\rho}_{[r]} = \frac{1}{n} \sum_{j=1}^{n} \rho_{[r]j}. \tag{11.31}$$

The population parameter, τ_ρ, is unbiasedly estimated by

$$\hat{\tau}_{\rho,\text{RSS}} = A\,\hat{\mu}_{\rho,\text{RSS}} \tag{11.32}$$

and the variance of $\hat{\tau}_{\rho,\text{RSS}}$ is estimated by

$$\hat{v}[\hat{\tau}_{\rho,\text{RSS}}] = A^2\,\hat{v}[\hat{\mu}_{\rho,\text{RSS}}]. \tag{11.33}$$

MacEachern et al. (2002) provided an unbiased estimator of σ_ρ^2, the variance of $\rho(x, y)$ in $\mathcal{A}$, i.e.,

$$\hat{\sigma}_{\rho,\text{RSS}}^2 = \frac{1}{k^2 n(n-1)} \sum_{r=1}^{k} \sum_{j=1}^{n} \left(\rho_{[r]j} - \bar{\rho}_{[r]} \right)^2 + \frac{1}{kn} \sum_{r=1}^{k} \sum_{j=1}^{n} \left(\rho_{[r]j} - \hat{\mu}_{\rho,\text{RSS}} \right)^2$$

$$= \hat{v} \left[\hat{\mu}_{\rho,\text{RSS}} \right] + \frac{1}{kn} \sum_{r=1}^{k} \sum_{j=1}^{n} \left(\rho_{[r]j} - \hat{\mu}_{\rho,\text{RSS}} \right)^2 . \tag{11.34}$$

Chen et al. (2003, p. 22) provided alternative unbiased estimators of the population variance. Stokes (1980) provided an asymptotically unbiased estimator that has been widely used, viz.,

$$\hat{\sigma}_{\rho,\text{STOKES}}^2 = \frac{1}{nk-1} \sum_{r=1}^{k} \sum_{j=1}^{n} \left(\rho_{[r]j} - \hat{\mu}_{\rho,\text{RSS}} \right)^2 . \tag{11.35}$$

This estimator tends to overestimate the population variance, but the bias diminishes with increasing sample size. Note, also, that this estimator does not require that we remember the ranks of the sample units.

11.5.3 Discrete applications

We presume a population comprising N discrete elements, where N may be unknown. The population parameters to be estimated are μ_y, and σ_y^2, where y is any attribute amenable to ranking. In addition τ_y may be estimated if N is known.

Let $y_{[r]j}$ be the attribute of interest for the population element with rank r selected in the rth set of the jth cycle of ranked set sampling. Estimators for discrete populations are analogous to those for continuums. To wit: the average amount of attribute per population element is estimated by

$$\hat{\mu}_{y,\text{RSS}} = \frac{1}{nk} \sum_{r=1}^{k} \sum_{j=1}^{n} y_{[r]j}, \tag{11.36}$$

and a consistent estimator of the variance of $\hat{\mu}_{y,\text{RSS}}$ is

$$\hat{v} \left[\hat{\mu}_{y,\text{RSS}} \right] = \frac{1}{k^2 n(n-1)} \sum_{r=1}^{k} \sum_{j=1}^{n} \left(y_{[r]j} - \bar{y}_{[r]} \right)^2 , \tag{11.37}$$

where

$$\bar{y}_{[r]} = \frac{1}{n} \sum_{j=1}^{n} y_{[r]j}.$$

If N is known, then τ_y is estimated by $\hat{\tau}_{y,\text{RSS}} = N \hat{\mu}_{y,\text{RSS}}$, and $\hat{v}[\hat{\tau}_{y,\text{RSS}}]$ is estimated by $N^2 \hat{v} \left[\hat{\mu}_{y,\text{RSS}} \right]$. The population variance, σ_y^2 is estimated by

$$\hat{\sigma}_{y,\text{RSS}}^2 = \hat{v} \left[\hat{\mu}_{y,\text{RSS}} \right] + \frac{1}{kn} \sum_{r=1}^{k} \sum_{j=1}^{n} \left(y_{[r]j} - \hat{\mu}_{y,\text{RSS}} \right)^2. \tag{11.38}$$

Regression estimation of μ_y is an option after RSS, where elements are ranked by the value of an auxiliary variate (see, e.g., Yu & Lam (1997) or Chen et al. (2003) for details).

Efficiency

McIntyre (1952) examined the precision of the estimator of the mean for ranked set sampling relative to the usual estimator for simple random sampling. By definition, relative precision (RP) is the ratio of the sampling variances, i.e.,

$$\text{RP} = \frac{V[\hat{\mu}_{y\pi}]}{V[\hat{\mu}_{y,\text{RSS}}]}.$$

McIntyre found that RP varied with set size, k, and was slightly less than $(k+1)/2$ for several distributions.

Takahasi & Wakimoto (1968) showed that, in the absence of ranking error, $(k+1)/2$ is an upper bound for RP, and this is achieved only for uniform distributions, i.e., where all attribute values or densities occur with equal frequency. The lower bound for RP is 1, which means that ranked set sampling is always at least as precise as simple random sampling. Since the upper bound increases with set size, it would seem that the larger the set size, the better. As was noted, however, set sizes of 3 or 4 units seem to be the norm. This is because RP is diminished by judgmental ranking errors (Dell & Clutter 1972), which are more likely with larger set sizes.

In traditional applications of ranked set sampling, practical considerations often demand that the k plots within a set be located close together, i.e., clustered, especially where the ranking process requires visual comparison. Effective use of digital photography might obviate the need for visual proximity in some applications. However, cluster plots also help minimize travel time, even where visual proximity is unnecessary, e.g., when rankings are based on measurements of auxiliary variates. Unfortunately, RP may also suffer from the use of sets of clustered plots, where the variability between sets exceeds the variability within sets. Cobby et al. (1985) recommended that sets of plots be as spread out as far as practicable.

MacEachern et al. (2002) provided a test of whether a ranking process, with or without clustered sets, provides effective stratification by ranks. They defined MSE and MST, respectively, as the mean-square error and mean-square treatment from a one-way analysis of variance performed on the ranked set sample data with the rank used as the treatment factor, i.e.,

$$\text{MSE} = \frac{1}{k(n-1)} \sum_{r=1}^{k} \sum_{j=1}^{n} \left(y_{[r]j} - \bar{y}_{[r]} \right)^2, \tag{11.39}$$

and

$$\text{MST} = \frac{1}{k-1} \left[\sum_{r=1}^{k} \sum_{j=1}^{n} \left(y_{[r]j} - \hat{\mu}_{y,\text{RSS}} \right)^2 - \sum_{r=1}^{k} \sum_{j=1}^{n} \left(y_{[r]j} - \bar{y}_{[r]} \right)^2 \right]. \quad (11.40)$$

In expectation,

$$E\left[\text{MST}\right] = E\left[\text{MSE}\right] + \frac{n}{k-1} \sum_{r=1}^{n} \left(\bar{y}_{[r]} - \mu_y \right)^2. \quad (11.41)$$

Hence, a ranked set sample is equivalent to a random sample if $\bar{y}_{[r]} = \mu_y$ for all r, i.e., if MST = MSE. On the other hand,

$$V_n = \frac{\text{MST}}{\text{MSE}} > 1 \quad (11.42)$$

measures the degree of effectiveness of the ranking process, the higher the value of V_n, the better. MacEachern et al. (2002) indicated that, for large n, we may test whether the ranking process is significantly better than random, since V_n tends toward an F distribution with k numerator and $k(n-1)$ denominator degrees of freedom.

11.6 Adaptive cluster sampling

Adaptive cluster sampling (ACS) provides for a concentration the sampling effort where the elements of interest are most concentrated, or where the attributes of the elements meet some specified criterion. In areal applications, adaptive strategies may be most useful where the elements or substances of interest tend to be sparse or dilute over much of the landscape, but abundant or concentrated in a few areas. Elements or substances that manifest this kind of dispersion may, but not necessarily, arise from contagion, invasion, or migration processes. Rather thorough treatments of adaptive sampling strategies and extensive lists of references are provided by Thompson (2002) and Thompson & Seber (1996); see also Christman (2000) and Brown (2003). Recent natural resource applications of ACS were reported, e.g., by Smith et al. (2003), Magnussen et al. (2005), Phillipi (2005), and Talvitie et al. (2006).

In this section, we provide only an adumbration of the ACS strategy. Much of our presentation was gleaned from Thompson & Seber (1996), Thompson (2002), and Roesch (1993).

11.6.1 Sampling tessellated landscapes

We presume that (a) elements of interest occur within some tract $\mathcal{A}$ with horizontal area A (Figure 11.8a), and (b) $\mathcal{A}$ tessellates completely into J plots for adaptive cluster sampling (Figure 11.8b). One way to define a tesselation is to specify coordinates, (x, z), of a location point that anchors a square or rectangular grid of points which spans $\mathcal{A}$. The grid points serve as corners of the plots. Without loss of generality, we specify—for this excursus—grid points that define corners of square plots, whose edges run north-south and east-west. Let a_j denote the horizontal area

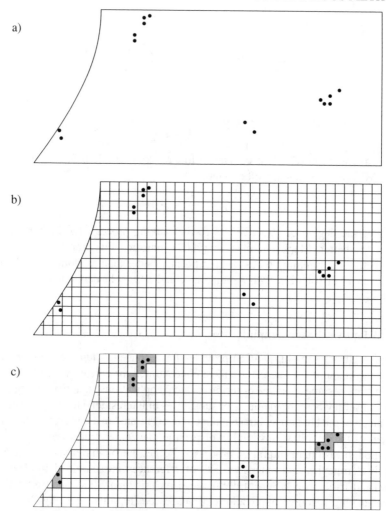

Figure 11.8 *a) An irregular tract containing clustered elements that satisfy an aggregation criterion. b) The tessellated tract, with small irregular plots along the west (left) edge. c) The four multi-plot networks (gray areas) formed by elements that satisfy the the aggregation criterion. Two multi-plot networks share a common corner, but they are separate networks because they do not share a common edge. Two elements that satisfy the aggregation criterion occur in separate single-plot networks, because their plots do not share a common edge.*

of plot $\mathbb{P}_j$, $j = 1, 2, \ldots, J$. Plots in the interior of $\mathcal{A}$ are all the same size. Plots that are truncated by the boundary of $\mathcal{A}$ have smaller areas and possibly irregular shapes (Figure 11.8b).

The plots formed by the tessellation of $\mathcal{A}$ aggregate into T networks according to criteria stipulated by the designer. The networks are mutually exclusive and completely exhaustive (Figure 11.8c). This means that every plot is a member of a network, and no plot may belong to more than one network. Adjacent plots that satisfy the criteria aggregate into a multi-plot network. Hence, a network comprises a single plot if that plot fails to meet the aggregation criteria, or if the plot does meet the aggregation criteria, but its adjacent neighbors do not. All other networks comprise two or more conterminous plots, all plots within each network satisfying the aggregation criteria. Consequently, the size and layout of the network to which a plot belongs is not known prior to sampling. Rather, this critical information is ferreted out with an adaptive sampling process. For identification purposes, we let $N_t, t = 1, 2, \ldots, T$, denote the tth network in $\mathcal{A}$.

Each plot may harbor several elements of interest. Let y_k be amount of attribute y for the element $\mathcal{U}_k$. Then,

$$\tau_y(P_j) = \sum_{\mathcal{U}_k \in P_j} y_k \tag{11.43}$$

is the total amount of y in plot P_j, and

$$\tau_y(N_t) = \sum_{P_j \in N_t} \tau_y(P_j)$$

$$= \sum_{P_j \in N_t} \sum_{\mathcal{U}_k \in P_j} y_k \tag{11.44}$$

is the total amount of y in network N_t. Moreover,

$$\tau_y = \sum_{N_t \in \mathcal{A}} \tau_y(N_t)$$

$$= \sum_{N_t \in \mathcal{A}} \sum_{P_j \in N_t} \sum_{\mathcal{U}_k \in P_j} y_k$$

$$= \sum_{k=1}^{N} y_k \tag{11.45}$$

is the total amount of y in $\mathcal{A}$.

The population comprising the T networks is the population that is actually sampled. Since the networks are unknown at the outset of sampling, we select a plot, and then determine, adaptively, the extent of the network that includes the plot.

Since the usual motivation for adaptive cluster sampling is to concentrate the sampling effort where the elements of interest are concentrated, we typically specify an aggregation criterion such as $\tau_y(P_j) \geq \tau_{\min}$, i.e., the amount of attribute on a plot, $\tau_y(P_j)$, meets or exceeds some stipulated minimum amount or *critical value*, $\tau_{\min}$. Or, we may specify an aggregation criterion such as $\tau_y(P_j)/a_j \geq \lambda_{\min}$, i.e., the attribute density in the plot—the amount of attribute per unit land area—meets or exceeds some stipulated minimum density. Hence, if a sample plot, P_j, meets the aggregation criterion, we measure plots its *neighborhood*, i.e., the adjacent plots to the north, south, east, and west of P_j. Note that these neighboring plots share a common edge boundary with P_j. Any of these neighboring plots that meet the aggregation criterion are members of the same network as plot P_j. Plots to the north, south, east, and west of these additional members are then measured to see which of them, if any, also are members of the network, and so on. Those plots that do not meet the aggregation criterion are omitted from the network. Thompson (2002) calls the omitted plots 'edge units'. A multi-plot network is completely determined when all its edges are bounded by edge units (Figure 11.8c).

11.6.2 Selection and Estimation

Probability proportional to area

The horizontal area of a network comprises the areas of all its plots. We may select a network in $\mathcal{A}$ with probability proportional to its horizontal area by choosing the coordinates, (x_s, z_s), of a location point uniformly at random (see §7.2.1, p. 209). The location point will occur in network N_t with probability

$$p_t = \frac{\sum_{P_j \in N_t} a_j}{A}. \tag{11.46}$$

Since networks must be determined by the adaptive sampling process, we initiate this process in the plot, P_j, that includes the point at (x_s, z_s). The process yields the areal extent of the network to which P_j belongs, and the amount of attribute, $\tau_y(N_t)$, present in the network.

Networks are selected independently with replacement if location points at (x_s, y_s), $s = 1, 2, \ldots, m$, are selected independently. In a sense, we select one network at a time, so the sth sample comprises a single network. Accordingly, the total amount of attribute in $\mathcal{A}$ is unbiasedly estimated from the sth sample with the Hansen-Hurwitz estimator, i.e.,

$$\hat{\tau}_{yp,s} = \frac{\tau_y(N_t)}{p_t}. \tag{11.47}$$

If m networks are selected independently and with replacement, τ_y is unbiasedly estimated by

$$\hat{\tau}_{yp,\text{rep}} = \frac{1}{m} \sum_{s=1}^{m} \hat{\tau}_{yp,s}. \tag{11.48}$$

The variance of $\hat{\tau}_{yp,\text{rep}}$ is

$$V\left[\hat{\tau}_{yp,\text{rep}}\right] = \frac{1}{m} \sum_{t=1}^{T} p_t \left(\frac{\tau_y(N_t)}{p_t} - \tau_y\right)^2. \tag{11.49}$$

This variance is unbiasedly estimated by

$$\hat{v}\left[\hat{\tau}_{yp,\text{rep}}\right] = \frac{1}{m(m-1)} \sum_{s=1}^{m} \left(\hat{\tau}_{yp,s} - \hat{\tau}_{yp,\text{rep}}\right)^2 \qquad m > 1. \tag{11.50}$$

Probability proportional to number of plots

Selection of a network with probability proportional to area may require measurement of the areas of plots that are truncated by the boundary of $\mathcal{A}$. Alternatively, we may consider selecting a network with probability proportional to its number of plots. This would entail selecting a plot, P_j, by SRSwR and then determining the network, N_t, that contains P_j. The selection probability for network N_t, which constitutes the sth sample, is $p_t = n_t / J$, where n_t is the number of plots in N_t.

Alternatively, we can use the HT estimator (Thompson 2002). If the m plots are selected by SRSwR, then the result is a combined sample comprising m (not necessarily distinct) networks. The probability that N_t is included in the resultant sample of m networks is

$$\pi_t = 1 - \left(1 - \frac{n_t}{J}\right)^m . \tag{11.51}$$

The probability that networks N_t and $N_{t'}$ are included in the sample is

$$\pi_{tt'} = 1 - \left[\left(1 - \frac{n_t}{J}\right)^m + \left(1 - \frac{n_{t'}}{J}\right)^m - \left(1 - \frac{n_t + n_{t'}}{J}\right)^m\right]. \tag{11.52}$$

The HT estimator provides an unbiased estimate of τ_y from the m or fewer distinct networks that emerge from the initial sample of m plots,

$$\hat{\tau}_{y\pi,\text{ACS}} = \sum_{N_t \in s} \frac{\tau_y(N_t)}{\pi_t}. \tag{11.53}$$

The variance of $\hat{\tau}_{y\pi,\text{ACS}}$ is

$$V\left[\hat{\tau}_{y,\text{ACS}}\right] = \sum_{t=1}^{T} \sum_{t'=1}^{T} \tau_y(N_t)\tau_y(N_{t'}) \left(\frac{\pi_{tt'} - \pi_t \pi_{t'}}{\pi_t \pi_{t'}}\right), \tag{11.54}$$

which is unbiasedly estimated by

$$\hat{v}\left[\hat{\tau}_{y\pi,\text{ACS}}\right] = \sum_{N_t \in s} \sum_{N_{t'} \in s} \frac{\tau_y(N_t)\tau_y(N_{t'})}{\pi_{tt'}} \left(\frac{\pi_{tt'} - \pi_t \pi_{t'}}{\pi_t \pi_{t'}}\right), \tag{11.55}$$

where $\pi_{tt'} \equiv \pi_t$ for $t = t'$.

Selection of plots without replacement

In this approach, the population comprises the plots in $\mathcal{A}$. We select an initial simple random sample of m plots without replacement, and determine the network that includes each of these plots. The average amount of attribute per plot in N_t is

$$\bar{\tau}_y(P_j) = \frac{\sum_{P_j \in N_t} \tau_y(P_j)}{n_t}, \tag{11.56}$$

so τ_y is unbiasedly estimated by (e.g., Thompson 2002)

$$\hat{\tau}_{y,\text{ACS}} = \frac{J}{m} \sum_{P_j \in s} \bar{\tau}_y(P_j). \tag{11.57}$$

The variance of $\hat{\tau}_{y,\text{ACS}}$,

$$V\left[\hat{\tau}_{y,\text{ACS}}\right] = \frac{(J-m)}{Jm(J-1)} \sum_{j=1}^{J} \left(J\bar{\tau}_y(\mathbb{P}_j) - \tau_y\right)^2 \tag{11.58}$$

is unbiasedly estimated by (Thompson 2002)

$$\hat{v}\left[\hat{\tau}_{y,\text{ACS}}\right] = \frac{(J-m)}{Jm(m-1)} \sum_{\mathbb{P}_j \in s} \left(J\bar{\tau}_y(\mathbb{P}_j) - \hat{\tau}_{y,\text{ACS}}\right)^2. \tag{11.59}$$

Networks may be duplicated even though plots are selected without replacement, as some distinct plots may be members of the same network.

Thompson (2002) also provided the probability that N_t is included a sample of m networks, where m plots are initially selected by SRSwoR, i.e.,

$$\pi_t = 1 - \frac{{}_{J-n_t}C_m}{{}_JC_m}. \tag{11.60}$$

The probability that networks N_t and $N_{t'}$ are included in the sample is (from Thompson 2002)

$$\pi_{tt'} = \pi_t + \pi_{t'} - \left(1 - \frac{{}_{J-n_t-n_{t'}}C_m}{{}_JC_m}\right). \tag{11.61}$$

The HT estimator, (11.53), provides an unbiased estimate of τ_y from the distinct networks in the sample, and its variance is estimated by (11.55) (Thompson 2002).

11.6.3 Sampling landscapes without tessellation

Roesch (1993) devised an adaptive cluster sampling design that avoids tessellation of $\mathcal{A}$. Roesch actually specified an adaptive design that uses Bitterlich sampling, but to generalize the procedure for elements other than trees, we describe a design based on sampling with fixed-area, circular plots.

Two or more elements of interest belong to a common network if (a) all those elements meet the aggregation criteria, and (b) each of those elements is within a distance R of at least one of the other elements. Elements that do not satisfy the aggregation criteria are single-element networks. Moreover, elements that satisfy the aggregation criteria, but have no neighbors within the distance R that satisfy the criteria, are also single-element networks. Hence, each element, $\mathcal{U}_k$, belongs to one and only one network. The total amount of attribute in network N_t is

$$\tau_y(N_t) = \sum_{\mathcal{U}_k \in N_t} y_k. \tag{11.62}$$

In essence, each element, $\mathcal{U}_k$, is centered in a circular 'elemental inclusion zone,' $\mathcal{I}_k$, which has radius R and area $a = \pi R^2$ (Figure 11.9a). Because inclusion zones may overlap the boundary of $\mathcal{A}$, we let a'_k denote area of $\mathcal{I}_k$ that occurs within $\mathcal{A}$, i.e., a'_k is the area of $\mathcal{I}_k \cap \mathcal{A}$.

In the adaptive sampling process, a sample point—chosen uniformly at random at

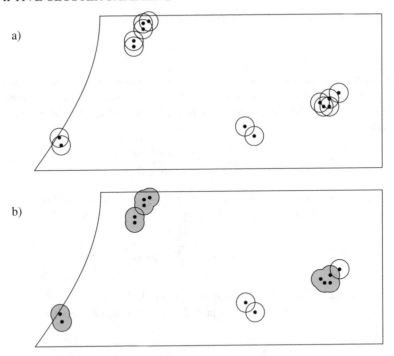

Figure 11.9 *a) Elements of interest and their circular elemental inclusion zones. b) Inclusion zones of single-element (clear) and multi-element (gray) networks. Note that network inclusion zones may overlap each other, and some of them may overlap the boundary of $\mathcal{A}$.*

$(x_s, z_s) \in \mathcal{A}$—serves as the center of a circular plot, $\mathbb{P}_s$, with fixed radius R. The point selects a network of elements into the sample if the location point occurs in the inclusion zone of one of that network's elements. Thus, any network which includes an element in the plot is selected into the sth adaptive cluster sample. The next step, therefore, is to determine the additional members of networks of those elements in the plot which satisfy the aggregation criteria. The networks of such elements may extend well outside the original plot (Figure 11.9b). And, it may turn out that the original plot contains elements of more than one multi-element network, since the distance between two elements in the plot may be greater than R. Elements outside the plot, which do not satisfy the aggregation criteria, are ignored.

Estimation with network inclusion probabilities

For a single-element network, the inclusion zone of the network is equivalent to the inclusion zone of the element. For a multi-element network, the inclusion zone is equivalent to the union of the inclusion zones of all its elements. Let $\mathbb{I}_t$ denote the inclusion zone of network, $\mathcal{N}_t$, then

$$\mathbb{I}_t = \bigcup_{\mathcal{U}_k \in \mathcal{N}_t} \mathcal{J}_k. \tag{11.63}$$

The probability that network N_t is included in the sth sample is

$$\pi_t = \frac{a_t}{A}, \tag{11.64}$$

where a_t is the horizontal area of $I_t \cap A$. The calculation of a_t ordinarily requires the location coordinates of the elements in the network, and possibly points on the boundary of A. The population parameter, τ_y, is unbiasedly estimated by HT estimator,

$$\hat{\tau}_{y\pi s} = \sum_{N_t \in s} \frac{\tau_y(N_t)}{\pi_t}. \tag{11.65}$$

If the sth sample contains no networks, then $\hat{\tau}_{y\pi,s} = 0$. If m samples are selected independently, τ_y is unbiasedly estimated by

$$\hat{\tau}_{y\pi,\text{rep}} = \frac{1}{m} \sum_{s=1}^{m} \hat{\tau}_{y\pi,s}, \tag{11.66}$$

and the variance of $\hat{\tau}_{y\pi,\text{rep}}$ is estimated by $\hat{v}\left[\hat{\tau}_{y\pi,\text{rep}}\right]$.

Taking a slightly different tack, the probability that network N_t occurs in a pooled sample from m sample points is

$$\pi_t = 1 - \left(1 - \frac{a_t}{A}\right)^m. \tag{11.67}$$

Thus, τ_y can be estimated with the alternative HT estimator, $\hat{\tau}_{y\pi,\text{ACS}}$. Estimation of the variance of this estimator requires $a_{tt'}$, the area of overlap of inclusion zones I_t and $I_{t'}$ for the calcuation of $\pi_{tt'}$, i.e.,

$$\pi_{tt'} = 1 - \left[\left(1 - \frac{a_t}{A}\right)^m + \left(1 - \frac{a_{t'}}{A}\right)^m - \left(1 - \frac{a_{tt'}}{A}\right)^m\right]. \tag{11.68}$$

Estimation with element inclusion probabilities

We could have entitled the last section, 'estimation the hard way,' and this section, 'estimation the easy way,' as that is the reality. The easy way utilizes a 'modified estimator' of τ_y, proposed by Roesch (1993), that obviates the need for the calculation of a_t and $a_{tt'}$; the inclusion areas of the elements in P_s are used instead.

The probability that element U_k occurs in plot P_s is

$$\pi'_k = \frac{a'_k}{A} \tag{11.69}$$

The average amount of attribute per element in network N_t is

$$\bar{y}_k = \frac{\tau_y(N_t)}{n_t}, \qquad U_k \in N_t. \tag{11.70}$$

Substituting $\bar{y}_k$ for y_k for all $U_k \in N_t$, τ_y is estimated from the elements in plot P_s

by

$$\tau_{y\pi's} = \sum_{\mathcal{U}_k \in \mathbb{P}_s} \frac{\bar{y}_k}{\pi'_k}$$

$$= A \sum_{\mathcal{U}_k \in \mathbb{P}_s} \frac{\bar{y}_k}{a'_k}. \tag{11.71}$$

Note that the modified estimator, $\tau_{y\pi's}$, is a modified HT estimator, the only modification being the substitution of $\bar{y}_k$ for y_k. Hence, if m plots are selected independently, τ_y may be unbiasedly estimated by

$$\hat{\tau}_{y\pi',\text{rep}} = \frac{1}{m} \sum_{s=1}^{m} \hat{\tau}_{y\pi's}, \tag{11.72}$$

and the variance of $\hat{\tau}_{y\pi'\text{rep}}$ is estimated by

$$\hat{v}\left[\hat{\tau}_{y\pi',\text{rep}}\right] = \frac{1}{m(m-1)} \sum_{s=1}^{m} \left(\hat{\tau}_{y\pi's} - \hat{\tau}_{y\pi'\text{rep}}\right)^2. \tag{11.73}$$

Edge effect

Edge effect occurs where the inclusion zone, $\mathcal{I}_k$, of any element, $\mathcal{U}_k$, overlaps the boundary of $\mathcal{A}$. If edge effect is ignored, estimates of τ_y are expected to be biased downward. Roesch's method provides unbiased estimates, despite boundary overlap, because the inclusion probability, π'_k, is calculated from a'_k, the area of $\mathcal{I}_k \cap \mathcal{A}$, rather than from a, the area of $\mathcal{I}_k$. Consequently, a measurement of a'_k is needed whenever the inclusion zone of an element, $\mathcal{U}_k \in \mathbb{P}_s$, overlaps the boundary.

Alternatively, we could eliminate the need for measurement of a'_k, even where overlap occurs, by (a) incorporating the mirage method or the walkthrough method into the sampling protocol, and (b) substituting an appropriate multi-tally estimator for the modified HT estimator. The mirage method (§7.5.4, p. 227) is applicable if all boundaries of $\mathcal{A}$ are straight or approximately so; the more generally applicable walkthrough method (§7.5.5, p. 229) allows for curved boundaries. Either method would provide a tally, $t_k = 1$ or $t_k = 2$, for the attribute value, $\bar{y}_k$, of any element, $\mathcal{U}_k \in \mathcal{A}$, which occurs in the sth plot. The population parameter, τ_y, is estimated by

$$\hat{\tau}_{ym,\text{rep}} = \frac{1}{m} \sum_{s=1}^{m} \hat{\tau}_{yms}, \tag{11.74}$$

where

$$\hat{\tau}_{yms} = \frac{A}{a} \sum_{\mathcal{U}_k \in \mathbb{P}_s} \bar{y}_k t_k. \tag{11.75}$$

The variance of $\hat{\tau}_{ym,\text{rep}}$ is estimated by

$$\hat{v}\left[\hat{\tau}_{ym,\text{rep}}\right] = \frac{1}{m(m-1)} \sum_{s=1}^{m} \left(\hat{\tau}_{yms} - \hat{\tau}_{ym,\text{rep}}\right)^2. \tag{11.76}$$

11.7 3P sampling

Poisson sampling was mentioned in §3.3.2. To the scientific sampling literature, its introduction by that name appears in Hájek (1958) "...owing to its apparent connection with Poisson's process." During this same period, L. R. Grosenbaugh, a pioneering research scientist with the U. S. Forest Service, conceived of a method of sampling individual trees that relied neither on an areal frame upon which to locate sampling units nor an a priori list frame of trees. Grosenbaugh's method of sampling with *probability proportional to prediction* relied instead on expert judgement to provide the auxiliary information which is used to quantify the inclusion probability of each tree in the population of interest. He then articulated a sample selection procedure in which each tree is included in the sample with the designated inclusion probability and independently of the inclusion of any other trees.

Effectively, 3P sampling as articulated by Grosenbaugh was identical to Hájek's Poisson sampling. Whereas Hájek (1964, p. 1494) dismissed its utility, noting that "Poisson sampling is not very suitable in practice, because it makes the sample size a random variable, and even an empty set may occur," Grosenbaugh (1964) propounded its advantage, since "[t]hree-pee sampling makes objective use of subjective judgment." More than that, Grosenbaugh showed that a multiplicative adjustment by the ratio of the expected sample size to the achieved sample size greatly improved upon the precision of the HT estimator of τ_y. In effect, Grosenbaugh's (1964) adjusted estimator is an early example of what is now called a *calibration estimator*. More recently, Särndal (1996) supported the use of sampling without replacement with probability proportional to size when $\hat{\tau}_{y\pi}$ can be properly calibrated to counter the added variability incurred by random sample size.

In this section we expand upon our earlier presentation in §3.3.2 in which we presented the adjustment in terms of a generalized ratio estimator, $\hat{\tau}_{y\pi,\text{rat}}$ in (3.60), which itself was treated in more general terms in Chapter 6.

11.7.1 Selection

Poisson/3P sampling is conducted sequentially: each $\mathcal{U}_k$ is considered for inclusion into the sample in some sequence. The sequence is immaterial to the sample of size n that is eventually selected. For a specific set of inclusion probabilities, one algorithm for accomplishing the selection of the sample consists of drawing a uniform random number, $u_k \sim U[0, 1]$, for each $\mathcal{U}_k$. If $u_k \leq \pi_k$, then $\mathcal{U}_k$ is included into the sample. Typically, $\pi_k \propto x_k$, where the latter, as usual, is an auxiliary variate that is well and positively correlated with the variate of interest, y_k. Let the constant of proportionality be L^{-1} as in Gregoire & Valentine (1999), so that $\pi_k = x_k/L$. Evidently L must be large enough to ensure that $\pi_k < 1$. Those $\mathcal{U}_k$ with $x_k > L$ are usually placed in a separate stratum, the elements of which are all measured with certainty. The total of y in this *certainty stratum* is denoted by τ_y'.

For the N elements that are not measured with certainty, let τ_x denote their aggregate x value. From $E[n] = \sum_{k=1}^{N} \pi_k$ we can deduce that $\sum_{k=1}^{N} \pi_k = \tau_x/L$, and hence $L = \tau_x/E[n]$.

When τ_x is known in advance of sampling, the inclusion probability can be expressed as $\pi_x = E[n]x_k/\tau_x$. In this situation, $E[n]$, is a parameter of the sampling design, in much the same way that n is parameter in fixed-sample-size designs. Whereas in the latter case one chooses the size of the sample to meet budgetary and practical constraints or requirements of precision, in the former case one chooses the average sample size, $E[n]$, that is desired, in the hope that the size of the sample actually selected will be close to it. For any expected size that is stipulated, it is possible that not a single element will be selected, as noted by Hájek. The probability of such a null sample is $\prod_{k=1}^{N}(1 - \pi_k)$, so that for large N, an empty sample becomes exceedingly unlikely. If it does occur, presumably the entire sampling effort would be repeated. Whereas Grosenbaugh (1965) dealt with the nonnegligible probability of a null sample in his treatise, in this section we ignore the complications it introduces.

One difference between Poisson, as it is customarily implemented, and 3P sampling is that the latter was conceived for a situation in which the absence of an a priori list of the $\mathcal{U}_k$ necessitated assessing each element in sequence, whereupon the assessor's expert judgment would assign a value x_k (hence, the origin of the term 'probability proportional to prediction'). Moreover, practical circumstances dictated that the decision to include $\mathcal{U}_k$ into the 3P sample, or not, had to be made prior to considering the next element in the population. Because τ_x was not known until the conclusion of sampling, obviously it could not be used to determine the inclusion probabilities in advance of sampling. Grosenbaugh (1964, 1965) proposed to determine the inclusion probability of $\mathcal{U}_k$ by setting $L = x_{\max}^* + Z$, where $x_{\max}^*$ is the threshold size for a certainty stratum (aka, the *sure to be measured* stratum), and Z is a factor to ensure the desired expected size of the sample, $E[n]$. After a far-from-obvious exercise, (Grosenbaugh 1965, pp. 25–28) derived that

$$Z \approx 2x_{\max}^* \sqrt{\frac{1}{E[n]} - \frac{1}{N}}. \tag{11.77}$$

ASIDE: The promulgation of 3P sampling by L. R. Grosenbaugh generated some controversy within the field of forest biometry. In the early 1960s Poisson sampling and the work of Hájek were virtually unknown to forest inventory specialists, who nonetheless were quite familiar and comfortable with the technique of unequal probability sampling as incorporated by Bitterlich sampling. Less comfortable were the notions of sampling without relying on an areal or list sampling frame, and designs where the size of the sample could not be fixed in advance. The opening salvo in this controversy may have been a presentation by K. Ware at the 1967 annual meeting of the Society of American Foresters in Ottawa, but the best known challenge to 3P sampling remains the article, "3-P sampling and some alternatives, I," by Schreuder et al. (1968). Grosenbaugh's response appeared later in Grosenbaugh (1976).

Of course, N is unknown prior to sampling, but for those situations where $N \gg E[n]$, (11.77) differs little from $\mathcal{Z} \approx 2x^*_{max} / \sqrt{E[n]}$.

Grosenbaugh (1965, p. 28) encouraged planning for the creation of the sure-to-be-measured stratum as a tactic to increase the precision of estimation. He reasoned that when $\mathcal{Z}$ is determined in the above fashion, 3P sampling is conducted at approximately the desired sampling rate (i.e., $E[n]$), trees having $x_k < x^*_{max}$ constitute a "... a smaller but less variable population" whereas "... all larger trees are measured" with certainty. "This ordinarily results in measuring somewhat more trees than was originally anticipated, but with considerably less sampling error for τ_y, since the sure-to-be-measured trees have no sampling error" (in the preceding quotation, Grosenbaugh's original notation has been changed to accord with the notation used in this book).

In the following, we assume that the measured total, τ'_y, of the sure-to-be-measured stratum will be added to the estimate of the total of the sampled stratum, and that this sum will constitute the estimate of τ_y for the entire population. Inasmuch as τ'_y has zero variance, it suffices to look at the variance of the total of the sampled stratum.

As mentioned in Chapter 6, the variance of $\hat{\tau}_{y\pi}$ following Poisson sampling is

$$V\left[\hat{\tau}_{y\pi}\right] = \sum_{k=1}^{N} y_k^2 \left(\frac{1 - \pi_k}{\pi_k}\right), \tag{11.78}$$

which tends to be quite large inasmuch as it is utterly insensitive to the size of the sample actually selected. Oddly, the HT estimator of τ_y from a Poisson sample is equally precise whether the sample comprises $n = 1$ element or $n = N$ elements. The variance of $\hat{\tau}_{y\pi}$ following Poisson sampling is influenced much more by the variability of n than by the variability of the y_k/π_k ratios (cf. Brewer & Gregoire (2000)).

For this reason the generalized ratio estimator, $\hat{\tau}_{y\pi,\text{rat}}$, is preferred. That is, let $\hat{R}_{y|x} = \hat{\tau}_{y\pi} / \hat{\tau}_{x\pi}$, so that

$$\hat{\tau}_{y\pi,\text{rat}} = \hat{R}_{y|x}\tau_x \tag{11.79a}$$

$$= \frac{\tau_x}{\hat{\tau}_{x\pi}}\hat{\tau}_{y\pi} \tag{11.79b}$$

$$= \frac{E[n]}{n}\hat{\tau}_{y\pi}. \tag{11.79c}$$

As indicated by (11.79c), the generalized ratio estimator following Poisson sampling entails an adjustment of $\hat{\tau}_{y\pi}$ by the ratio of expected to actual sample size. As Furnival et al. (1987) show, this result is identical to using the HT estimator with adjusted inclusion probabilities given by

$$\pi'_k = \frac{\pi_k n}{E[n]}. \tag{11.80}$$

Following the naming convention of Brewer & Gregoire (2000), the estimator

$$\hat{\tau}_{y,3P7} = \sum_{\mathcal{U}_k \in s} \frac{y_k}{\pi'_k} \tag{11.81}$$

is viewed as a conditional form of HT estimation, where the conditioning factor is the size of the sample actually selected. From (11.79c) and (11.80) it is evident that $\hat{\tau}_{y,3P7}$ is just a special case of $\hat{\tau}_{y\pi,\text{rat}}$.

Another form in which $\hat{\tau}_{y,3P7}$ has appeared frequently in the forestry literature is $\hat{\tau}_{y,3P7} = \tau_x \bar{q}$, where $\bar{q}$ is the arithmetic mean of the n ratios $q_k = y_k/x_k$. This form of the estimator has intuitive appeal but tends to obscure its relationship to the HT estimator of τ_y.

11.7.2 Variance estimation following 3P sampling

There has been considerable effort devoted to devising a suitable estimator of the variance of $\hat{\tau}_{y,3P7}$. Two obvious candidates are (6.24) and (6.25) introduced in Chapter 6. In the current context, the latter is

$$\hat{v}_2[\hat{\tau}_{y,3P7}] = \sum_{\mathcal{U}_k \in s} r_k^2 \left(\frac{1 - \pi_k}{\pi_k^2} \right) \tag{11.82a}$$

$$= \sum_{\mathcal{U}_k \in s} (1 - \pi_k) \left(\frac{y_k}{\pi_k} - \frac{\hat{\tau}_{y\pi}}{n} \right)^2, \tag{11.82b}$$

where $\hat{r}_k = y_k - \hat{R}_{y|x} x_k$. The former becomes

$$\hat{v}_1[\hat{\tau}_{y,3P7}] = \left(\frac{\tau_x}{\hat{\tau}_{x\pi}} \right)^2 \hat{v}_2[\hat{\tau}_{y,3P7}] \tag{11.83a}$$

$$= \left(\frac{E[n]}{n} \right)^2 \sum_{\mathcal{U}_k \in s} (1 - \pi_k) \left(\frac{y_k}{\pi_k} - \frac{\hat{\tau}_{y\pi}}{n} \right)^2. \tag{11.83b}$$

Grosenbaugh (1965) reasoned by analogy to the case of unequal probability sampling with replacement and proposed

$$\hat{v}_{3P9}[\hat{\tau}_{y,3P7}] = \frac{n}{n-1} \left[\frac{1}{n^2} \sum_{\mathcal{U}_k \in s} \left(\frac{y_k}{x_k/\tau_x} \right)^2 - \frac{1}{n^3} \left(\sum_{\mathcal{U}_k \in s} \frac{y_k}{x_k/\tau_x} \right)^2 \right] \tag{11.84a}$$

$$= \frac{E[n]^2}{n(n-1)} \sum_{\mathcal{U}_k \in s} \left(\frac{y_k}{\pi_k} - \frac{\hat{\tau}_{y\pi}}{n} \right)^2, \tag{11.84b}$$

where (11.84a) is expressed in a form that resembles that put forth originally by Grosenbaugh (1965) (see the chapter's Appendix for the derivation of the latter expression from it). This estimator can also be written as $\hat{v}_{3P9}[\hat{\tau}_{y,3P7}] = \tau_x^2 s_{\bar{q}}^2$, where

$$s_{\bar{q}}^2 = \frac{1}{n(n-1)} \sum_{u_k \in s} (q_k - \bar{q})^2 \qquad (11.85)$$

Later Grosenbaugh (1976) proposed an alternative and demonstrably more stable alternative estimator that he called vS. Ignoring a term which accounts for the redraw of the sample in the event that a null sample is selected, this variance estimator is

$$\hat{v}_{vS}[\hat{\tau}_{y,3P7}] = \hat{v}_{3P9}[\hat{\tau}_{y,3P7}] \left(1 - \frac{E[n]}{N}\right) \left(\frac{n}{E[n]}\right) \qquad (11.86a)$$

$$= \frac{E[n]^2}{(n-1)} \left(\frac{1}{E[n]} - \frac{1}{N}\right) \sum_{u_k \in s} \left(\frac{y_k}{\pi_k} - \frac{\hat{\tau}_{y\pi}}{n}\right)^2. \qquad (11.86b)$$

In a simulation study Brewer & Gregoire (2000) compared the performance of $\hat{v}_{vS}$ to another alternative estimator:

$$\hat{v}_{BG}[\hat{\tau}_{y,3P7}] = \sum_{u_k \in s} \left(\frac{n - \phi/n}{n-1} - \pi_k'\right) \left(\frac{y_k}{\pi_k'} - \frac{\hat{\tau}_{y,3P7}}{n}\right)^2, \qquad (11.87)$$

where $\phi = \sum_{k=1}^{N} \pi_k^2$. In their study, Brewer & Gregoire (2000) found that $\hat{v}_{BG}[\hat{\tau}_{y,3P7}]$ was less biased than $\hat{v}_{vS}[\hat{\tau}_{y,3P7}]$ following Poisson sampling. In a different study Gregoire & Valentine (1999) found that $\hat{v}_{vS}[\hat{\tau}_{y,3P7}]$ was less biased than $\hat{v}_1[\hat{\tau}_{y,3P7}]$.

Example 11.1

Probably the most intensive and expensive 3P inventory of sawtimber was conducted as a result of the expansion of the Redwood National Park in the late 1970s and early 1980s. This 2800-acre virgin stand is located in Humboldt County, California, and is comprised mostly of redwood and Douglas-fir. It was of such high value that the Arcata Redwood Company, which owned the land and timber, employed the Natural Resources Management Corporation of Eureka, CA to obtain volume estimates for both old-growth redwood and old-growth Douglas-fir, within plus or minus 1% at the 95% level of confidence. The 3P sampling system was used to attempt to achieve this precision.

The 3P method required an estimate of the volume of each tree and careful measurement of sample trees selected with a probability proportional to predicted volume. Each sawtimber tree with any merchantable volume on the tract, over 144 thousand trees, was visited by trained field crews and volume estimates were made using the best volume tables available.

Over 4500 trees (2833 redwoods, 1703 Douglas-fir) were dendrometered for volume determination using the STX program developed by Grosenbaugh. Dendrometry consisted of measuring tree heights and upper-stem diameters at 20 feet and several points up the bole using precise optical dendrometers and professional-grade transits.

A crew of up to 30 foresters worked two years collecting data, although not all of this work related directly to the 3P inventory. The total cost for this work exceeded a million U. S. dollars, and the value of the standing timber exceeded a quarter billion U. S. dollars. A sampling error of 1.1% was obtained for both old-growth redwood and old-growth Douglas-fir, and those volumes were accepted by a court. The total volume, Scribner rule, was about 250 mbf per acre. (This example was excerpted from the notes of Harry V. Wiant, Jr., and personal correspondence from L. R. Grosenbaugh to TGG, dated 14 November 1997.)

Example 11.2

Because of the extensive conversion of commercial forest land to agricultural use in the delta of the Mississippi River during the late 1960s, the U. S. Forest Service conducted a midcycle forest inventory to estimate the timber resources of the state of Mississippi. The goal of the midcycle update "was to quickly and accurately estimate growing-stock volume," and 3P sampling was the technique used in the second stage of sampling to accomplish this goal. The results, reported in Van Hooser (1973), required remeasurement of only 2300 of the 42,000 trees in the preceding survey and revisitation of only one-third of the original sampling locations.

Example 11.3

There is no reason why the objective use of inclusion probabilities based on subjective judgement, i.e., Grosenbaugh's seminal contribution with 3P sampling, can not be applied to natural and environmental resources other than trees. Indeed Ringvall & Kruys (2005) demonstrate its use to estimate the number of substrates that host calicioid lichen species. In this application, y was binary valued (0 or 1) according to the presence of a specific lichen (*Cyphelium inquinans* (Sm.) Trevis.) and the auxiliary variate was probability, as judged by the field sampler, that the species was present on the substrate.

11.7.3 Regression adjustment of inclusion probabilities

With fixed sample size designs, $\hat{\tau}_{y\pi}$ may be expected to estimate τ_y very precisely when y and x are very strongly correlated and the inclusion probability of $\mathcal{U}_k$ is proportional to x_k. Even when those conditions hold, $\hat{\tau}_{y\pi}$ has poor precision following Poisson/3P sampling because it treats the contribution of y_k to the estimate of τ_y the same, regardless of the total number of elements in the sample. Grosenbaugh's estimator, $\hat{\tau}_{y,3P7}$, adjusts the HT estimator to account for the size of the sample actually observed, and by so doing its precision is often dramatically better. (Exceptional cases can be found, however, as pointed out by Williams et al. (1998).) As is clear from (11.80) and (11.81), the same ratio-type adjustment is made to all inclusion probabilities in the $\hat{\tau}_{y,3P7}$ estimator of τ_y. Acting upon a suggestion by Hájek (1981), Furnival et al. (1987) investigated the utility of a regression adjustment

of the inclusion probabilities to more closely approximate the inclusion probability
of each unit conditionally upon the size sample actually selected. Using the notation
of Furnival et al. (1987), the regression-adjusted inclusion probability of $\mathcal{U}_k$ is

$$\tilde{\pi}_k^c = \pi_k + \tilde{\beta}(n - E[n]) \tag{11.88a}$$

$$= \pi_k \left[1 + (1 - \pi_k)\frac{n - E[n]}{V[n]} \right], \tag{11.88b}$$

where $\tilde{\beta} = \pi_k(1 - \pi_k)/V[n]$. Evidently, when π_k is large, the magnitude of its
adjustment is less than it would be for a smaller value. For sake of comparison, the
Grosenbaugh adjustment is

$$\pi_k' = \pi_k \left[1 + \frac{n - E[n]}{V[n]} \right]. \tag{11.89}$$

The improvement offered by the regression adjustment was noteworthy only for
the situation when y and x were poorly correlated, but even in that situation the
advantage of the regression adjustment over the simpler ratio adjustment diminished
with increasing population size. They conjectured that the ratio-estimator-property
of $\hat{\tau}_{y,3P7}$ dampens its mean square error to such a remarkable extent that any other
adjustments provide little additional benefit.

11.7.4 Composite and calibration estimation

At the conclusion of 3P sampling, both N and τ_x are known, whereas $\hat{\tau}_{y,3P7}$
incorporates knowledge only of the latter. Gregoire & Valentine (1999) examined
a composite estimator of τ_y formed as a weighted combination of the $\hat{\tau}_{y,3P7}$ and
$\tilde{\tau}_{y,3P7} = (N/\hat{N}_\pi)\hat{\tau}_{y\pi}$, where the weights are inversely proportional to the variance
of each component. They also investigated a generalized regression, calibration-
estimator. Both alternative estimators had similarly smaller mean squared error than
$\hat{\tau}_{y,3P7}$.

11.8 Terms to Remember

3P sampling	Horizontal line sampling
Adjusted inclusion probability	Point relascope sampling
Calibration estimator	Ranked set sampling
Certainty stratum	Order statistic
Probability proportional to prediction	Rank
Adaptive cluster sampling	Set and cycle
Network	Sausage sampling
Neighborhood	Transect relascope sampling

Table 11.1 *Decay class, small-end (d) and large-end (D) diameter, and length (ℓ) in 16 point relascope samples.*

s	Class	d	D	ℓ	s	Class	d	D	ℓ
		(cm)	(cm)	(m)			(cm)	(cm)	(m)
1	1	5	15	9.50	9	4	12	15	1.65
1	3	6	16	5.85					
1	3	5	12	11.00	10	-	-	-	-
1	2	5	29	11.70					
1	3	11	11	2.35	11	2	5	12	6.26
1	3	16	25	8.60	11	1	5	19	8.21
1	2	5	13	7.70					
1	1	12	26	13.50	12	-	-	-	-
2	-	-	-	-	13	1	5	12	7.83
					13	1	5	26	11.21
3	-	-	-	-	13	2	8	11	5.48
					13	2	5	16	9.73
4	3	5	12	2.92	13	1	5	23	11.32
4	3	5	14	3.13					
4	4	13	13	6.90	14	3	15	21	5.60
					14	1	5	11	5.47
5	-	-	-	-					
					15	2	5	10	3.29
6	1	5	42	17.91	15	2	5	11	5.29
6	2	9	17	7.60	15	2	5	16	7.15
6	2	6	12	9.25	15	1	5	15	9.13
6	5	6	14	4.71	15	1	5	11	4.30
6	4	7	12	6.82					
					16	4	9	22	10.92
7	5	13	21	4.50	16	3	7	11	6.82
7	2	5	12	6.05	16	2	5	10	3.18
					16	2	5	24	12.80
8	2	5	12	6.20	16	4	7	9	1.57
8	4	14	26	6.68					

11.9 Exercises

1. Estimate aggregate log length-square per hectare from the data in Table 11.1. The angle gauge was $53.13° = 0.9273$ rad.

2. Estimate the aggregate log volume per hectare by decay class from the data in Table 11.1.

3. Show how the factor φ in PRS is derived.

4. Show how the expression for the $w_k(\theta_s)$ in (11.15) is derived.

5. Grosenbaugh (1965) presented a small $N = 9$ element population in order to examine the statistical properties of the unadjusted ($\hat{\tau}_{y\pi}$) estimator and adjusted

$(\hat{\tau}_{y\pi,\text{rat}})$ estimators of τ_y. This artificial population is reproduced in the following table. Enumerate the 42 possible Poisson/3P samples that could be selected from this population. In this exercise and those that follow, assume that x^*_{max} exceeds the largest x_k in the population, so that you do not need to have a sure-to-be-measured stratum.

$\mathcal{U}_k$	x_k	y_k
$\mathcal{U}_1$	2	106
$\mathcal{U}_2$	2	106
$\mathcal{U}_3$	2	106
$\mathcal{U}_4$	2	106
$\mathcal{U}_5$	2	106
$\mathcal{U}_6$	2	106
$\mathcal{U}_7$	5	248
$\mathcal{U}_8$	5	248
$\mathcal{U}_9$	10	526

6. Compute the expected sample size when $L = 20$ and sampling from the population of Exercise 5.

7. Compute the probability of selecting a sample of size $n = 5$ when $L = 20$ and sampling from the population of Exercise 5.

8. Select a Poisson/3P sample from the population of Exercise 5, and from it estimate τ_y with $\hat{\tau}_{y\pi,\text{rat}}$.

11.10 Appendix

11.10.1 Details on 3P sampling

The variance of the random sample size, n, is

$$V[n] = \sum_{k=1}^{N} \pi_k(1 - \pi_k) \tag{11.90a}$$

$$= \sum_{k=1}^{N} \pi_k - \sum_{k=1}^{N} \pi_k^2 \tag{11.90b}$$

$$= E[n] - \sum_{k=1}^{N} \pi_k^2. \tag{11.90c}$$

Substituting $\pi_k = E[n]x_k/\tau_x$ yields

$$V[n] = E[n] - \left(\frac{E[n]}{\tau_x}\right)^2 \sum_{k=1}^{N} x_k^2 \qquad (11.91a)$$

$$= E[n] - \left(\frac{E[n]}{\tau_x}\right)^2 \left(N\sigma_x^2 + N\mu_x^2\right) \qquad (11.91b)$$

$$= E[n] - \left(\frac{E[n]}{\tau_x}\right)^2 N\mu_x^2 \left(\frac{\sigma_x^2}{\mu_x^2} + 1\right) \qquad (11.91c)$$

$$= E[n] - \frac{(E[n])^2}{N}\left(\gamma_x^2 + 1\right) \qquad (11.91d)$$

$$\leq E[n] - \frac{(E[n])^2}{N}, \qquad (11.91e)$$

where γ_x is the coefficient of variation of x (see §1.5). The upper bound on right side in (11.91e) is the variance of the random sample size under Bernoulli sampling, i.e., when n has a binomial sampling distribution.

Grosenbaugh indicated that the constant factor Z was a device by which he could ensure that samples would exceed some critical size, n_{crit}, with probability no greater than 0.025. Appealing to the Gaussian approximation to the binomial distribution, he reasoned that

$$n_{crit} \geq E[n] + 2\sqrt{V[n]} \qquad (11.92a)$$

implies that

$$(n_{crit} - E[n])^2 = 4V[n] \qquad (11.92b)$$

$$= 4E[n]\left(1 - \frac{E[n]}{N}\right). \qquad (11.92c)$$

Asserting that nx_{max}^* should be smaller than τ_x in 3P sampling, Grosenbaugh (1965, p. 27) set τ_x/x_{max}^* as the value of n_{crit}. Substituting this into (11.92c) leads to the result that "n will exceed n_{crit} less than 1 time in 40 if"

$$Z \geq 2x_{max}^* \sqrt{\frac{1}{E[n]} - \frac{1}{N}}. \qquad (11.93)$$

de Vries (1986, pp. 302–307) derives this condition in a different manner. The concern with ensuring against an overly large sample, which is evident in the 3P sampling literature, is not prevalent in the literature on Poisson sampling at large.

The text of the 53-page document we cite as Grosenbaugh (1965) was formatted by Grosenbaugh with his own rudimentary word processor, which he coded in the Fortran-4 language. Inasmuch as the document was printed on a line printer equipped only with upper-case letters, the use of subscripts and symbolic notation, e.g., $\hat{\tau}_{y\pi}$, was not possible. As a consequence, Grosenbaugh spelled out the name of estimators. For example, his adjusted 3P estimator was referred to as 3PSEVENTH and is algebraically identical to the estimator we refer to as $\hat{\tau}_{y,3P7}$. Likewise, the estimator of the variance of 3PSEVENTH he proposed as 3PNINTH, and he put it forth as the relative variance, i.e., relative to $\hat{\tau}_{y,3P7}^2$. Expressed in non-relative terms and in a notation consistent with the present text, it is

$$\hat{v}_{3P9}[\,\hat{\tau}_{y,3P7}\,] = \frac{n}{n-1}\left[\frac{1}{n^2}\sum_{\mathcal{U}_k \in s}\left(\frac{y_k}{x_k/\tau_x}\right)^2 - \frac{1}{n^3}\left(\sum_{\mathcal{U}_k \in s}\frac{y_k}{x_k/\tau_x}\right)^2\right] \tag{11.94a}$$

$$= \frac{n}{n-1}\left[\frac{1}{n^2}\sum_{\mathcal{U}_k \in s}\left(\frac{y_k}{\pi_k/E[n]}\right)^2 - \frac{1}{n^3}\left(\sum_{\mathcal{U}_k \in s}\frac{y_k}{\pi_k/E[n]}\right)^2\right] \tag{11.94b}$$

$$= \frac{n}{n-1}\left[\frac{E[n]^2}{n^2}\sum_{\mathcal{U}_k \in s}\left(\frac{y_k}{\pi_k}\right)^2 - \frac{E[n]^2}{n^3}\left(\sum_{\mathcal{U}_k \in s}\frac{y_k}{\pi_k}\right)^2\right] \tag{11.94c}$$

$$= \frac{E[n]^2}{n(n-1)}\left[\sum_{\mathcal{U}_k \in s}\frac{y_k^2}{\pi_k^2} - \frac{\hat{\tau}_{y\pi}^2}{n}\right] \tag{11.94d}$$

$$= \frac{E[n]^2}{n(n-1)}\sum_{\mathcal{U}_k \in s}\left(\frac{y_k}{\pi_k} - \frac{\hat{\tau}_{y\pi}}{n}\right)^2. \tag{11.94e}$$

CHAPTER 12

Two-stage Sampling

Some populations are arranged hierarchically, and sampling may be conducted in a manner that mirrors that hierarchy. For example, in agricultural yield studies, exemplified often in Sukhatme & Sukhatme (1970), villages comprise a number of farms, and each farm has a number of fields planted to wheat, rice, or other crop of interest. It seems natural that a sample of villages is selected, followed by a sample of farms within each sampled village, followed further by a sample of fields within each sampled farm. It is straightforward to imagine a fourth level of hierarchy in this setting which could consist of a tessellation of each field into nonoverlapping cells or strips, such that the ultimate sampling unit would be one of these areal units. Sampling of a population in this hierarchical fashion is known as multistage sampling. When confined to two stages or levels of sampling it is known, appropriately, as two-stage sampling—the principal topic of this chapter. We defer to Sukhatme & Sukhatme (1970) for a more comprehensive treatment of the subject than is provided here.

Each population cluster that serves as the sampling unit in the first stage of sampling is known as a primary sampling unit or first-stage sampling unit. Each element within a primary sample unit is typically referred to as secondary sampling unit, second-stage sampling unit, or subsampling unit. Evidently, the nomenclature is not firm.

An area-based hierarchy of the sampling frame, as in the illustrative example above, is not a necessary condition for two-stage sampling. In other words, the secondary sample units need not be nested physically within the primary sample units. As a counterexample, imagine commercial fishing fleets as primary sample units, and boats within each fleet constituting the secondary sample units. As a second counterexample to an area-based hierarchical structure consider a Bitterlich sample of trees followed by a subsampling of trees at each sampling location. Because the trees in the primary sample unit sample cluster are not confined to a sampling unit delimited by area, the trees selected in the second stage are not either. Finally, many sampling frames are based on lists of primary sample units composed of secondary sample units which depend on lineage, alphabetical ordering, size, or other criterion which is unrelated to geographic, land, areal, or spatial metrics.

In this chapter we present two-stage sampling quite broadly, in contrast to the more customary presentations which restrict consideration to the case where the population comprises a set of N discrete and nonoverlapping primary sample units and secondary sample units occur in one, but no more than one, primary sample unit. We shall begin, however, by presenting two-stage sampling as is customarily found, e.g., in Sukhatme & Sukhatme (1970), Cochran (1977), and Särndal et al. (1992).

In the interest of clarity, the distinction between two-phase and two-stage sampling is worth noting. In two-phase or double sampling as we have presented the technique in §6.9, y_k is measured only on those $\mathcal{U}_k$ that are selected in the second phase of the sample, whereas x_k is measured on all $\mathcal{U}_k$ selected in the first phase. In the two-stage design, there is no implication that an auxiliary variate will be used either to aid sample selection or to improve precision of estimation. Auxiliary information may be used in any two-stage design, but unlike in two-phase sampling, it is not inherent to the design.

The precision with which τ_y or other population parameters may be estimated following a two-stage sampling strategy is almost always less than it would be following a corresponding one-stage sampling design that results in a sample containing the same number of $\mathcal{U}_k$. A heuristic explanation for this result is that elements in the same cluster are usually more similar to each other than they are to elements drawn at large from the population. Consequently, for a specific size of sample selected from a sample of clusters, there is less information about the population than would be obtained from a sample of the same size drawn from the population without regard to clusters. When both precision of estimation and cost of sampling are considered, two-stage sampling can prove to be quite efficient and cost effective.

Sampling in two stages may be an unavoidable necessity when relying on a list sampling frame in situations where a list of clusters is available but not a list of individual elements of the population.

12.1 Customary two-stage sampling

In the customary framework, the population of N elements is exhaustively partitioned into a set of M clusters. There is no overlap in cluster composition, so that $\mathcal{U}_k$ is a member of one and only one cluster. Using c_i to represent the ith cluster, then $\mathcal{P} = \bigcup_{i=1}^{M} c_i$. Let N_i denote the size of the cluster as measured by the number of population elements in c_i. Because clusters are mutually exclusive, $N = \sum_{i=1}^{M} N_i$.

Suppose that the first stage of sampling results in a sample, s_I, of m clusters. In the second stage of sampling, elements of each $c_i \in s_\mathrm{I}$ are selected. Let the subsample from c_i be denoted by s_i, and suppose that s_i comprises $n_i \leq N_i$ elements.

In this framework the inclusion probability of $\mathcal{U}_k \in c_i$ is the product of the conditional probability of including $\mathcal{U}_k$ in s_i given that $c_i \in s_\mathrm{I}$ and the probability of including c_i in the first-stage sample s_I:

$$\pi_k = P\left(\mathcal{U}_k \in s_i \,\middle|\, c_i \in s_\mathrm{I}\right) P\left(c_i \in s_\mathrm{I}\right)$$
$$= \pi_{k \mid c_i} \pi_{c_i}. \tag{12.1}$$

12.1.1 Estimating cluster totals

Upon defining the cluster total as

$$\tau_{yi} = \sum_{\mathcal{U}_k \in c_i} y_k = N_i \mu_{yi}, \tag{12.2}$$

where μ_{yi} is the cluster's mean y_k value, it is clear that τ_{yi} may be estimated unbiasedly by

$$\hat{\tau}_{y\pi i} = \sum_{\mathcal{U}_k \in s_i} \frac{y_k}{\pi_{k\,|\,c_i}}, \tag{12.3}$$

The variance of $\hat{\tau}_{y\pi i}$ is

$$V\left[\hat{\tau}_{y\pi i}\right] = \sum_{\mathcal{U}_k \in c_i} y_k^2 \left(\frac{1 - \pi_{k\,|\,c_i}}{\pi_{k\,|\,c_i}}\right)$$

$$+ \sum_{\mathcal{U}_k \in c_i} \sum_{\substack{k' \neq k \\ k'=1}}^{N_i} y_k y_{k'} \left(\frac{\pi_{kk'\,|\,c_i} - \pi_{k\,|\,c_i}\pi_{k'\,|\,c_i}}{\pi_{k\,|\,c_i}\pi_{k'\,|\,c_i}}\right), \tag{12.4}$$

where $\pi_{kk'\,|\,c_i}$ is the joint probability of subsampling $\mathcal{U}_k$ and $\mathcal{U}_{k'}$ from c_i. This variance is estimated unbiasedly by

$$\hat{v}\left[\hat{\tau}_{y\pi i}\right] = \sum_{\mathcal{U}_k \in s_i} y_k^2 \left(\frac{1 - \pi_{k\,|\,c_i}}{\pi_{k\,|\,c_i}^2}\right)$$

$$+ \sum_{\mathcal{U}_k \in s_i} \sum_{\substack{k' \neq k \\ k'=1}}^{n_i} y_k y_{k'} \left(\frac{\pi_{kk'\,|\,c_i} - \pi_{k\,|\,c_i}\pi_{k'\,|\,c_i}}{\pi_{k\,|\,c_i}\pi_{k'\,|\,c_i}\pi_{kk'\,|\,c_i}}\right). \tag{12.5}$$

12.1.2 Estimating the population total from cluster totals

An unbiased estimator of τ_y inflates each estimated cluster total by the inclusion probability of the cluster, viz.,

$$\hat{\tau}_{y\pi,\text{ts}} = \sum_{c_i \in s_{\text{I}}} \frac{\hat{\tau}_{y\pi i}}{\pi_{c_i}} \tag{12.6a}$$

$$= \sum_{c_i \in s_{\text{I}}} \frac{1}{\pi_{c_i}} \sum_{\mathcal{U}_k \in c_i} \frac{y_k}{\pi_{k\,|\,c_i}} \tag{12.6b}$$

$$= \sum_{c_i \in s_{\text{I}}} \sum_{\mathcal{U}_k \in s_i} \frac{y_k}{\pi_k}. \tag{12.6c}$$

From (12.6a) it is evident that the more closely the probability of including c_i is proportional to $\hat{\tau}_{yi}$ for that cluster, the more precise will be the estimate of τ_y for the population.

Although the variance of $\hat{\tau}_{y\pi,\text{ts}}$ in (12.6c) can be expressed analogously to (12.4), it is more useful to express it as

$$V\left[\hat{\tau}_{y\pi,\text{ts}}\right] = V_{\text{I}} + V_{\text{II}} \tag{12.7}$$

where V_{I} is the variance due to the first-stage selection of clusters and V_{II} is the variance due to the subsampling of elements in clusters. Specifically,

$$V_{\mathrm{I}} = \sum_{i=1}^{M} \tau_{yi}^2 \left(\frac{1 - \pi_{c_i}}{\pi_{c_i}} \right)$$

$$+ \sum_{i=1}^{M} \sum_{\substack{i' \neq i \\ i'=1}}^{M} \tau_{yi} \tau_{yi'} \left(\frac{\pi_{c_i c_{i'}} - \pi_{c_i} \pi_{c_{i'}}}{\pi_{c_i} \pi_{c_{i'}}} \right), \tag{12.8}$$

and

$$V_{\mathrm{II}} = \sum_{i=1}^{M} \frac{V\left[\hat{\tau}_{y\pi i} \right]}{\pi_{c_i}} \tag{12.9}$$

A number of observations are in order at this point:

1. When each primary sample unit selected in stage I is completely sampled, so that $n_i = N_i$ for each $c_i \in s_{\mathrm{I}}$, then $V\left[\hat{\tau}_{y\pi i} \right]$ in (12.4) as well as V_{II} in (12.9) are identically zero. This design is widely known as cluster sampling, a topic that is covered well by Sukhatme & Sukhatme (1970).

2. When all the N_{I} clusters are included in the first-stage sample, the design is stratified sampling of Chapter 5. In this situation, $\pi_{c_i} = 1$, which implies that $V_{\mathrm{I}} = 0$. Consequently, $V\left[\hat{\tau}_{y\pi,\mathrm{ts}} \right]$ in (12.7) coincides with both (12.9) and (5.5).

3. Implicit to the development of the expressions of variance is that subsampling within a cluster does not depend in any way upon the set of primary sample units that are selected into s_{I}. These notions of *independence* and *invariance* of the second stage sampling design are discussed in Särndal et al. (1992), to whom we refer readers wishing to follow the derivation of $V\left[\hat{\tau}_{y\pi,\mathrm{ts}} \right]$ and an estimator of it.

The variance $V\left[\hat{\tau}_{y\pi,\mathrm{ts}} \right]$ in (12.7) is unbiasedly estimated by

$$\hat{v}\left[\hat{\tau}_{y\pi,\mathrm{ts}} \right] = \hat{v}_{\mathrm{I}} + \hat{v}_{\mathrm{II}} \tag{12.10}$$

where

$$\hat{v}_{\mathrm{I}} = \sum_{c_i \in s_{\mathrm{I}}} \hat{\tau}_{yi}^2 \left(\frac{1 - \pi_{c_i}}{\pi_{c_i}^2} \right) + \sum_{c_i \in s_{\mathrm{I}}} \sum_{\substack{i' \neq i \\ i'=1}}^{N_{\mathrm{I}}} \hat{\tau}_{yi} \hat{\tau}_{yi'} \left(\frac{\pi_{c_i c_{i'}} - \pi_{c_i} \pi_{c_{i'}}}{\pi_{c_i} \pi_{c_{i'}} \pi_{c_i c_{i'}}} \right)$$

$$- \sum_{c_i \in s_{\mathrm{I}}} \hat{v}\left[\hat{\tau}_{y\pi i} \right] \frac{1 - \pi_{c_i}}{\pi_{c_i}^2} \tag{12.11}$$

and

$$\hat{v}_{\mathrm{II}} = \sum_{c_i \in s_{\mathrm{I}}} \frac{\hat{v}\left[\hat{\tau}_{y\pi i} \right]}{\pi_{c_i}^2}. \tag{12.12}$$

Combining the latter two results yields

$$\hat{v}\left[\hat{\tau}_{y\pi,\text{ts}}\right] = \sum_{c_i \in s_1} \hat{\tau}_{yi}^2 \left(\frac{1 - \pi_{c_i}}{\pi_{c_i}^2}\right) + \sum_{c_i \in s_1} \sum_{\substack{i' \neq i \\ i'=1}}^{N_\text{I}} \hat{\tau}_{yi} \hat{\tau}_{yi'} \left(\frac{\pi_{c_i c_{i'}} - \pi_{c_i} \pi_{c_{i'}}}{\pi_{c_i} \pi_{c_{i'}}, \pi_{c_i c_{i'}}}\right)$$

$$+ \sum_{c_i \in s_1} \frac{\hat{v}\left[\hat{\tau}_{y\pi i}\right]}{\pi_{c_i}} \tag{12.13}$$

12.1.3 SRSwoR special case

For the special case of SRSwoR at both stages, $\hat{\tau}_{y\pi,\text{ts}}$ and expressions of variance relating to it simplify into more recognizable expressions. If m of the M available clusters are selected by SRSwoR, then $\pi_{c_i} = m/M$ for all M clusters. If n_i of the N_i elements are selected by SRSwoR from c_i, then $\pi_{k|c_i} = n_i/N_i$ for all elements of c_i. Consequently

$$\pi_k = \frac{n_i m}{N_i M} \tag{12.14}$$

and

$$\hat{\tau}_{y\pi,\text{ts}} = \frac{M}{m} \sum_{c_i \in s_1} \frac{N_i}{n_i} \sum_{u_k \in s_i} y_k \tag{12.15a}$$

$$= \frac{M}{m} \sum_{c_i \in s_1} N_i \bar{y}_i, \tag{12.15b}$$

where $\bar{y}_i$ is the average of y_k in the sample from c_i. Furthermore (12.7) reduces to

$$V\left[\hat{\tau}_{y\pi,\text{ts}}\right] = M^2 \left(\frac{1}{m} - \frac{1}{M}\right) \sigma_\text{I}^2 + \frac{M}{m} \sum_{i=1}^{M} N_i^2 \left(\frac{1}{n_i} - \frac{1}{N_i}\right) \sigma_{\text{II},i}^2 \tag{12.16}$$

where

$$\sigma_\text{I}^2 = \frac{1}{M-1} \sum_{i=1}^{M} \left(\tau_{yi} - \frac{\tau_y}{M}\right)^2, \tag{12.17}$$

which may be called the 'between-cluster variance,' and where

$$\sigma_{\text{II},i}^2 = \frac{1}{N_i - 1} \sum_{u_k \in c_i} \left(y_k - \mu_{yi}\right)^2, \tag{12.18}$$

which is the usual variance among the y values within the cluster. From (12.16) it is evident that both V_I and V_II are reduced by increasing the size of stage I sample, whereas increasing the size of the stage II sample only reduces V_II. This is almost always the case irrespective of the sampling design in either stage.

Using (12.13), an unbiased estimator of (12.16) following two-stage SRSwoR is

$$\hat{v}\left[\hat{\tau}_{y\pi,\text{ts}}\right] = M^2 \left(\frac{1}{m} - \frac{1}{M}\right) s_\text{I}^2 + \frac{M}{m} \sum_{i=1}^{M} N_i^2 \left(\frac{1}{n_i} - \frac{1}{N_i}\right) s_{\text{II},i}^2, \tag{12.19}$$

where

$$s_I^2 = \frac{1}{m-1} \sum_{c_i \in s_I} \left(\hat{\tau}_{y\pi i} - \frac{\hat{\tau}_{y\pi,\text{ts}}}{m} \right)^2,$$ (12.20)

and

$$s_{\text{II},i}^2 = \frac{1}{n_i - 1} \sum_{u_k \in c_i} (y_k - \bar{y}_i)^2,$$ (12.21)

12.1.4 Self-weighting SRSwoR designs

If the same sampling fraction, f, is selected in all clusters then

$$\hat{\tau}_{y\pi,\text{ts}} = \frac{Mf}{m} \sum_{c_i \in s_I} \sum_{u_k \in s_i} y_k.$$ (12.22)

All sample y_k values are multiplied by the same constant value, Mf/m. This is an example of a self-weighting design: irrespective of the cluster of origin of y_k, its contribution to the estimator of the population total is identical to that of any other sampled element.

12.1.5 Ppswr special case

When information on the size of each c_i is known, primary sample units commonly are chosen with probability proportional to size, with replacement. The usual practice when employing this design is to subsample a primary sample unit as many times as it is selected in stage I. V_I in (12.7) in this case is

$$V_I = \frac{1}{m} \sum_{i=1}^{M} p_i \left(\frac{\tau_{yi}}{p_i} - \tau_y \right)^2,$$ (12.23)

where p_i is the selection probability of c_i in the first stage of sampling. We defer to Sukhatme & Sukhatme (1970) for a more detailed presentation.

12.1.6 Examples

Example 12.1

Sukhatme & Sukhatme (1970, p. 270) present results of a yield survey on rice in 1946-1947. The first stage of the sampling design consisted of the selection by SRSwoR of $m = 5$ village from $M = 146$ villages in the seventh stratum. Each village belonged to but one stratum. For this situation each village has a probability of $m/M = 5/146$ to be included into the stage I sample. Fields within villages served as second-stage sampling units, and three fields were selected in each. If N_i represents the number of fields in the ith village, the inclusion probability of including a field from c_i is $\pi_k = (5/146)(3/N_i)$.

Example 12.2

In its summary report (Anonymous 1999) the authors explain that the U.S. Natural Resources Inventory is a two-stage stratified sample of the entire country. The primary sample unit is an area/segment of land, and the secondary sample units are points within each primary sample unit. For the 1997 survey there were 300,000 primary sample units and 800,000 sample points.

Example 12.3

Gregoire et al. (1986) computed the variance of an unbiased estimator of total bole volume of a population of 91 trees. Stage I consisted of ppswr list sampling with selection probability proportional to the presumed volume of each tree, followed by importance subsampling of trees selected in the first stage.

Example 12.4

Stehman et al. (2003) report on a stratified two-stage sampling procedure to estimate land-cover change based on satellite imagery at roughly eight year intervals spanning 1973 to 2000. The primary sampling units consisted of 20 km square blocks of terrain, within which were 60 m square pixels which served as secondary sampling units. The primary sampling units were stratified geographically by ecoregion boundaries overlaid on the landscape. Within each stratum 9 to 11 primary sampling units were selected by SRSwoR. While these authors evaluated land-cover changes for each secondary sampling unit within a selected primary unit, an optional design for the future would select a subsample of 60 m pixels instead.

12.1.7 Alternative estimators

The preceding development notwithstanding, estimation of τ_y following two-stage sampling need not be limited to HT estimation principles as in $\hat{\tau}_{y\pi,\text{ts}}$ in (12.6). The general principle is that an estimate of the total of each cluster selected in the first stage of sampling is estimated from units selected in the second stage of sampling. If auxiliary information on all N_i elements of c_i is available or measured when c_i is selected, then ratio or regression estimation using methods of Chapter 6 is a viable alternative to estimation of τ_{yi}. Likewise ratio and regression estimation of τ_y is an alternative if appropriate covariate information as available for all primary sampling units in the population. Details on some of these alternative estimators are provided in Chapter 7 of Sukhatme & Sukhatme (1970).

12.2 General two-stage sampling

A more general two-stage sampling design arises in situations where primary sample units are not mutually exclusive. These occur in ecological and natural resource inventories in which the first stage of sampling relies on an areal sampling frame and which selects population elements into the first-stage sample using plot/quadrat methods of Chapter 7, Bitterlich sampling of Chapter 8, LIS methods of Chapter 9, or one of the special methods presented in Chapter 11. Surely there are other situations where subsampling from overlapping primary sample units occur. In all such cases, let the total number of elements selected in stage II be

$$n_{\mathrm{II}} = \sum_{i=1}^{m} n_i.$$ (12.24)

There is no restriction that the n_{II} are necessarily distinct units.

With general two-stage sampling the inclusion probabilities of units in stage II may depend upon the composition of the sample from stage I. For this more general setup, let $\pi_{k\mathrm{I}}$ represent the probability of including $\mathcal{U}_k$ in the first stage of sampling, and $\pi_{k\mathrm{II}|\mathrm{I}}$ represent the second stage inclusion probability conditional upon s_{I}. Therefore, $\pi_k = \pi_{k\mathrm{II}|\mathrm{I}}\pi_{k\mathrm{I}}$. In the event that the second stage design selects units with replacement, the selection probability of $\mathcal{U}_k$ is $p_k = p_{k\mathrm{II}|\mathrm{I}}\pi_{k\mathrm{I}}$. With reasonably large populations, π_k and p_k will be so nearly equal in magnitude that distinguishing between the two is hardly necessary.

Example 12.5

Yandle & White (1977) proposed a two-stage design which utilized all the trees selected by Bitterlich sampling in a prior forest inventory as the first stage of sampling in a subsequent inventory conducted later. The subsample of trees at the second inventory were selected from the list of first-occasion trees with selection probability proportional to height with replacement. With this design, $\pi_{k\mathrm{I}} = b_k/(FA)$, using the notation of Chapter 8, and $p_{k\mathrm{II}|\mathrm{I}} = h_k/H_{\mathrm{I}}$, where H_{I} is the accumulated height of all trees selected in the m first-stage sample points:

$$H_{\mathrm{I}} = \sum_{i=1}^{m} \sum_{\mathcal{U}_k \in s_i} h_k.$$ (12.25)

In (12.25), the height of $\mathcal{U}_k$ may contribute more than once to the cumulative total, H_{I}, if the first-stage Bitterlich sample selects it from more than one sample location, (x_s, z_s). This is unlikely to occur, but not impossible. More importantly, a tree will be selected into the two-stage sample with a combined probability that is proportional to the volume of a cylinder that of the same diameter and height, which should allow rather precise estimation of the volume of the bole of the tree and other closely-related quantities, as had been pointed out earlier by Grosenbaugh (1971) with regard to two-stage sampling using a 3P sampling design at stage II.

The estimator of τ_y proposed by these authors was

$$\hat{\tau}_{y\text{YW}} = \frac{1}{n_{\text{II}}m} \sum_{i=1}^{m} \sum_{\mathcal{U}_k \in s_i} \frac{y_k I_{ki\text{II}}}{p_k},$$

(12.26)

where

$$p_k = p_{k\text{II}|\text{I}} \pi_{k\text{I}} = \frac{h_k b_k}{H_{\text{I}} FA},$$

(12.27)

and where

$$I_{ki\text{II}} = \begin{cases} 1, & \text{if } \mathcal{U}_k \text{ selected in stage II from the } i\text{th sample point;} \\ 0, & \text{otherwise.} \end{cases}$$

(12.28)

The variance of $\hat{\tau}_{y\text{YW}}$ is

$$V\left[\hat{\tau}_{y\text{YW}}\right] = \frac{1}{m}\left(\sum_{i=1}^{m}\sum_{k=1}^{N}\frac{y_k^2}{\pi_{k\text{I}}} - \tau_y^2\right) + \frac{(n_{\text{II}}-1)}{n_{\text{II}}m}\sum_{k=1}^{N}\sum_{\substack{k' \neq k \\ k'=1}}^{N}\frac{y_k}{\pi_{k\text{I}}}\frac{y_k'}{\pi_{k'\text{I}}}\pi_{kk'\text{I}}$$

$$+ \frac{m-1}{n_{\text{II}}m}\left[\left(\sum_{k=1}^{N}h_k b_k\right)\left(\sum_{k=1}^{N}\frac{y_k^2}{h_k b_k}\right) - \tau_y^2\right]$$

$$+ \frac{1}{n_{\text{II}}m}\sum_{k=1}^{N}\sum_{\substack{k' \neq k \\ k'=1}}^{N}\frac{h_k}{h_{k'}}\frac{y_{k'}^2}{\pi_{k'\text{I}}^2}\pi_{kk'}$$

(12.29)

Example 12.6

Wood & Wiant (1992) modified the Yandle & White procedure of the preceding example by conducting the second-stage ppswr sampling separately at each first-stage sampling location. They proposed the following estimator of τ_y

$$\hat{\tau}_{y\text{WW}} = \frac{1}{m}\sum_{i=1}^{m}\frac{H_{i\text{I}}}{n_i}\sum_{\mathcal{U}_k \in s_i}\frac{y_k I_{ki\text{II}}}{p_k},$$

(12.30)

where p_k is the same as defined above but with $p_{k\text{II}|\text{I}} = h_k/H_{i\text{I}}$, where $H_{i\text{I}}$ is the accumulated height of all trees selected in stage I at ith Bitterlich sampling point. Gregoire (1992) derived the variance of $\hat{\tau}_y$ww:

$$V\left[\hat{\tau}_y\text{ww}\right] = \frac{1}{m}\left(\sum_{i=1}^{m}\sum_{k=1}^{N}\frac{y_k^2}{\pi_{kI}} - \tau_y^2\right)$$

$$+ \frac{1}{\bar{n}_{II}m}\sum_{k=1}^{N}\sum_{\substack{k'\neq k \\ k'=1}}^{N}\frac{y_k}{\pi_{kI}}\frac{\pi_{kk'I}}{h_k}\left(\frac{y_k h_k}{\pi_{kI}} - \frac{y_{k'}h_{k'}}{\pi_{k'I}}\right)$$

$$+ \frac{1}{m}\sum_{k=1}^{N}\sum_{\substack{k'\neq k \\ k'=1}}^{N}\frac{y_k}{\pi_{kI}}\frac{y_{k'}}{\pi_{k'I}}\pi_{kk'I}, \qquad (12.31)$$

where $\bar{n}_{II}$ is the harmonic mean number of trees per first-stage Bitterlich sampling point: $\bar{n}_{II} = \left(m^{-1}\sum_{i=1}^{m}n_i^{-1}\right)^{-1}$.

Nelson & Gregoire (1994) examined the precision of both $\hat{\tau}_y$YW and $\hat{\tau}_y$ww for estimating the aggregate bole volume of tree in a 5.2 hectare forest stand with 4676 trees. Because the locations of all trees had been mapped, they were able to compute the joint inclusion probabilities of all pairs of trees, and therefore able to compute the variances of $\hat{\tau}_y$YW and $\hat{\tau}_y$ww for this stand. Using a spectrum of different sizes of first and second stage samples, they found that $\hat{\tau}_y$ww was always slightly more precise than $\hat{\tau}_y$YW.

Example 12.7

Subsampling is suggested in Chapter 13 following randomized branch sampling of a tree to obviate the need of measuring the volume of each segment of a 'selected path.' In this situation, randomized branch sampling can be considered the first-stage design, and the selected path comprises a sequence of connected internodes extending from the base of the tree to a terminal shoot. The sampling frame at the second stage is the continuous path, and typically this is not discretized into individual internodes. Rather, the entire path is sampled by crude Monte Carlo or importance sampling, so an estimate of the volume of the entire path obtains from a measurement of cross-sectional area at a sample point between the base of the tree and the tip of the terminal shoot.

12.3 Three-stage sampling

In this section we touch briefly on the extension of two-stage sampling to an additional, third stage. The third stage of sampling comprises subsampling of the subunits selected in the second stage. Customarily, as presented in Särndal et al. (1992) and Sukhatme & Sukhatme (1970), the third-stage subunits comprise an exhaustive partition of the second-stage unit in which they are embedded hierarchically. The third-stage sample of population units is denoted by s_{ij}.

In effect, in the first-stage of sampling, we select a sample of clusters, $c_i \in s$; in the second stage, we select a sample of 'sub-clusters,' $t_j \in s_i$, from $c_i \in s$; and in the third stage, we select population units, $\mathcal{U}_k \in s_{ij}$ from $t_j \in s_i$.

In an extension to the notation introduced in §12.1, the conditional probability of including $\mathcal{U}_k$ in the third stage of sampling is denoted as $\pi_{k \mid c_i, t_j}$. An unbiased estimator of $\tau_{y\pi i}$ is provided by

$$\hat{\tau}_{y\pi i} = \sum_{s_{ij} \in s_i} \sum_{\mathcal{U}_k \in s_{ij}} \frac{y_k}{\pi_{k \mid c_i, t_j}\, \pi_{k \mid c_i}}. \tag{12.32}$$

The estimate provided in (12.32) is then substituted into (12.6a) to estimate τ_y.

In keeping with the abbreviated treatment of this topic, we simply note a few observations for the benefit who may wish to delve deeper.

1. Estimation proceeds recursively: the third-stage subunits are used to estimate the total of each second-stage subunit, and these estimated totals are then used to estimate the total of each primary sampling unit. In principle, this may be extended to as many levels or stages as is practicable to undertake.

2. Although we have exemplified estimation following three-stage sampling with an HT-like estimator, there is no necessary presumption that HT estimation will be followed at any stage. In contrast, there is a presumption that the independence and invariance properties mentioned earlier (see p. 396) apply to the third stage of sampling also.

3. As the notation of this section hints, the notational baggage that accompanies multistage sampling with more than two stages becomes increasingly cumbersome. Bookkeeping also becomes burdensome: as the number of stages increases, so too does the difficulty of keeping track of the sundry conditional and unconditional inclusion probabilities which accompany each sampling unit at each stage of the design.

4. In principle the variance of the estimator of τ_y following sampling in three or more stages can be developed in a recursive manner, also. Details are provided in Särndal et al. (1992, §4.4). These authors also present unbiased estimators of the components of variance for the the three-stage estimator presented above, as well as an informative discussion of the formidable computing burden that attends unbiased variance estimation in this context. For those wishing to pursue the latter, Bellhouse (1985) may be a useful starting point.

Extensions of general two-stage sampling as in §12.2 to three or more stages have appeared in various sources, e.g., Valentine et al. (1987) and Gregoire et al. (1993). We are unaware of any comprehensive treatment of this sampling design in the natural resource or environmental literature at this time.

12.4 Terms to Remember

Between-cluster variance	Primary sampling unit
Cluster	Secondary sampling unit
Independence	Self-weighting design
Invariance	Subsample

CHAPTER 13

Randomized Branch Sampling

Randomized branch sampling (RBS) was developed originally by Jessen (1955) to estimate fruit counts on individual orchard trees. This ingenious method of multistage probability sampling can be used to obtain estimates of many different attributes of orchard, forest, and shade trees, and other branched plants. In principle, RBS could be applied to other branched structures such as corals and river systems, though we have not seen any 'non-botanical' applications. Most practitioners implement RBS with replacement at all stages of sampling, though Cancino & Saborowski (2005) have investigated sampling without replacement at the first stage.

The usual objective of RBS is estimation of a total amount of an attribute contained in, or borne by, a tree or branch; for example, the aggregate volume, dry weight, and chemical contents of the woody components (e.g., Valentine et al. 1984; De Gier 1989; Good et al. 2001); the count, surface area, dry weight, and chemical contents of the leaves (e.g., Valentine et al. 1994; Gaffrey & Saborowski 1999; Raulier et al. 2002); the count of tree insects (Furness 1976), and, of course, the count and aggregate mass of fruits and the seeds within the fruits. In addition, RBS, with hydraulic excavation, has potential use in the estimation of the radius, volume, and mass of root systems. In this chapter, we restrict our interest to attributes of the aboveground portion of a tree. Depending of the objective, the implementation of RBS may require felling or climbing the tree.

13.1 Terminology

Our terminology follows Valentine et al. (1984) and Gregoire et al. (1995). We define a 'branch' to be the entire stem system that develops from a single bud (lateral or terminal) and we define a 'branch segment' or, simply, 'segment' to be a part of a branch between two consecutive nodes (or forks). No distinction is made between the segments of the main stem and side branches. The butt of the main stem of a tree is considered a node and the tree is considered a branch for the purpose of RBS. Terminal shoots are considered to be both branches and branch segments. Thus, any tree or branch can be defined as a population of branch segments. Hence, we shall call the tree or branch that we subject to RBS the 'object branch.'

We define a 'path' to be an acropetal stack of connected branch segments. A path may extend from the butt of the main stem to a terminal shoot (Figure 13.1), in which case the number of possible paths equals the number of terminal shoots. However, the terminus need not be a terminal shoot and the starting point of a path need not be the butt of the main stem. For example, a path may extend from the butt of a main

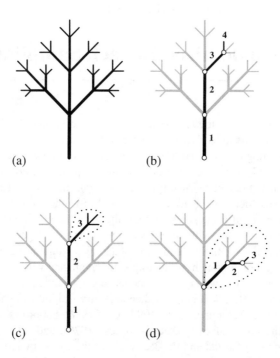

Figure 13.1 *The tree in (a) contains 40 segments and 27 possible paths from the butt of the main stem to a terminal shoot. The four branch segments of one possible path are shown in (b). Randomized branch sampling can be stopped at any node, in which case the selected branch [3 (encircled), in example (c)], is treated as the terminal segment of the path. The sampling can be started from the butt of any branch on the tree [e.g., example (d)], in which case the resultant estimates pertain only to the entire starting branch (encircled), not the entire tree.*

stem to a high-order branch, in which case the entire high-order branch is treated as the terminal segment of the path (Figure 13.1c). A path may extend from the butt of a low-order branch to either a terminal shoot (Figure 13.1d) or a higher-order branch (as in Figure 13.1c), in which case the estimates that derive from RBS are for the entire low-order branch, not the entire tree.

13.2 Path selection

RBS is used to select a path from the butt of an object branch to a terminal segment. The first segment of the path extends from the butt of the object branch, which is defined as the first node, to the second node (Figure 13.2). By convention, the first segment of the path has selection probability q_1. We assign a selection probability to each branch emanating from the second node and choose one at random. The choice of this second branch, with selection probability q_2, fixes the second segment of the path. The second segment is followed to a third node where a branch, and the third

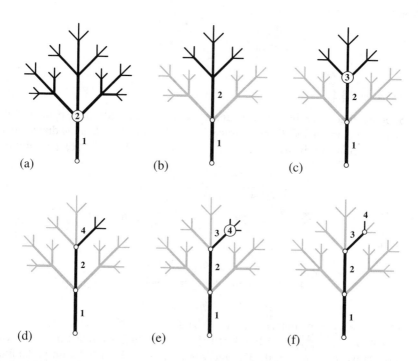

Figure 13.2 *(a) The basal end of the first segment of the object branch is defined as the first node. The distal end of the first segment is the second node (labeled), where one of the three branches is selected with probability q_2, fixing the second segment of the path (b). Proceeding to the third node (c), one of the three branches is selected with probability q_3, which fixes the third segment of the path (d). The sampling continues until a terminal node is reached (e), and the terminal segment of the path is selected (f).*

segment of the path, is chosen with probability q_3. This procedure is repeated until a small branch or terminal shoot is chosen at the final (Rth) node with probability q_R.

The selection probabilities assigned to the branches at each node must sum to one. The first node of a path usually gives rise to one branch, which has selection probability $q_1 = 1$. If there are multiple branches, each is assigned a selection probability, and one is selected with probability $q_1 < 1$. Ordinarily, the selection probabilities are either (a) uniform (i.e., all branches at a node have the same selection probability) or (b) proportional to some measure of the size. For example, letting X_k symbolize the measure of size for the kth of n branches at the rth node, the probability of selection assigned to this branch (q_{r_k}) is:

$$q_{r_k} = \frac{X_k}{\sum_{j=1}^{n} X_j}, \qquad k = 1, 2, \ldots, n.$$

Branch 1 is selected if $u \le q_{r_1}$, where $u \sim U[0, 1]$, or branch k ($k = 2, \ldots, n$) is selected if:

$$\sum_{j=1}^{k-1} q_{r_j} < u \le \sum_{j=1}^{k} q_{r_j}.$$

Example 13.1

Suppose that three branches emanate from the rth node of a path. Let the
measure of size be the diameter-square (d^2) of a branch and let the diameters of
the three branches be $d_1 = 2$, $d_2 = 3$, and $d_3 = 5$. The selection probabilities
are:

$$q_{r_1} = \frac{2^2}{2^2 + 3^2 + 5^2} = \frac{4}{38} = 0.105,$$

$$q_{r_2} = \frac{9}{38} = 0.237, \qquad q_{r_3} = \frac{25}{38} = 0.658.$$

Hence,

$$q_{r_1} = 0.105$$
$$q_{r_1} + q_{r_2} = 0.342$$
$$q_{r_1} + q_{r_2} + q_{r_3} = 1.0$$

We draw a random number u from U[0, 1]. If $u \le 0.105$, we select branch 1 with
probability $q_r \equiv q_{r_1} = 0.105$; if $0.105 < u \le 0.342$, we select branch 2 with
probability $q_r \equiv q_{r_2} = 0.237$; otherwise, we select branch 3 with probability
$q_r \equiv q_{r_3} = 0.658$.

Technically, the selection probability assigned to a branch is the conditional
probability of selecting that branch given that the path has reached the node at which
the branch arises. The unconditional probability of selection for the rth segment of
the path is:

$$Q_r = \prod_{k=1}^{r} q_k, \quad r = 1, 2, \ldots, R, \tag{13.1}$$

i.e.,

$$Q_1 = q_1$$
$$Q_2 = q_1 q_2 = q_2 Q_1$$
$$\vdots$$
$$Q_R = q_1 q_2 \cdots q_R = q_R Q_{R-1}.$$

More than one path may be selected in which case the unconditional probability of
selection of the rth of segment of the ith of $m \ge 2$ paths is denoted by Q_{ir}.

Example 13.2

Table 13.1 contains branch diameters (cm) measured at each of 10 nodes on an
Ocotea guianensis tree in Brazil.

Table 13.1 *Branch diameters and selection probabilities for an Ocatea guianensis tree.*

r	Dia (cm)	q_r	Q_r
1	**9.1**	1.00000	1.00000
2	**4.3**, 2.1	0.87142	$0.87142 \times 1.00000 = 0.87142$
3	1.7, **4.1**	0.91796	$0.91796 \times 0.87142 = 0.79992$
4	1.5, **4.1**	0.93612	$0.93612 \times 0.79992 = 0.74882$
5	1.2, **3.5**, 2.2	0.74247	$0.74247 \times 0.74882 = 0.55598$
6	1.5, **3.2**, 1.8	0.74214	$0.74214 \times 0.55598 = 0.41262$
7	1.0, **2.9**, 2.0	0.69976	$0.69976 \times 0.41262 = 0.28873$
8	1.0, **2.5**, 1.5	0.74501	$0.74501 \times 0.28873 = 0.21511$
9	**2.4**, 1.2	0.86421	$0.86421 \times 0.21511 = 0.18590$
10	**1.6**, 2.0	0.35531	$0.35531 \times 0.18590 = 0.06605$

Branches were selected with probability proportional to $d^{2.67}$. The diameters of the selected branches are in bold face. For example, there were two branches at the third node and the larger one with a diameter of 4.1 cm was selected with conditional selection probability

$$q_3 = \frac{4.1^{2.67}}{1.7^{2.67} + 4.1^{2.67}} = 0.91796.$$

The unconditional probability of selecting this branch was:

$$Q_3 = 1.0 \times 0.87142 \times 0.91796 = 0.79992.$$

13.3 Estimation

Let y_{ir} be the attribute that is measured on the rth segment of the ith path of the object branch, and let τ_y be the target parameter—the total amount of attribute summed over all the segments of the object branch. Then,

$$\hat{\tau}_{y_{Q_i}} = \sum_{r=1}^{R} \frac{y_{ir}}{Q_{ir}} \tag{13.2}$$

is an unbiased estimator for τ_y. The results from $m \geq 2$ paths can be averaged to give a combined estimate:

$$\hat{\tau}_{yQ} = \frac{1}{m} \sum_{i=1}^{m} \hat{\tau}_{y_{Q_i}}. \tag{13.3}$$

The joint probability of selecting all R segments that form the ith path is Q_{iR}. Although the number of segments, R, may vary from path to path, if there are M possible paths with distinct terminal segments, then

$$\sum_{i=1}^{M} Q_{iR} = 1$$

because the unconditional selection probabilities sum to one at every node. There-fore, Q_{iR} is the probability of obtaining the estimate $\hat{\tau}_{y\varrho_i}$, which is one of M possible estimates for the object branch. Accordingly, the variance of $\hat{\tau}_{y\varrho}$ is:

$$V[\hat{\tau}_{y\varrho}] = \frac{1}{m}\left[\sum_{i=1}^{M} Q_{iR}\left(\hat{\tau}_{y\varrho_i} - \tau_y\right)^2\right].$$

The sampled-based estimator of the variance of $\hat{\tau}_{y\varrho}$ is:

$$\hat{v}\left(\hat{\tau}_{y\varrho}\right) = \frac{1}{m(m-1)}\sum_{i=1}^{m}\left(\hat{\tau}_{y\varrho_i} - \hat{\tau}_{y\varrho}\right)^2 \qquad m > 1. \qquad (13.4)$$

Example 13.3

The measurements in Table 13.2 pertain to the *Ocotea guianensis* tree from Example 13.2.

Let τ_y be the fresh weight of the woody tissues and let τ_f be the fresh weight of foliage of the N segments of the tree. Our estimates of the fresh weight of woody tissues from each of the two paths are:

$$\hat{\tau}_{y\varrho_1} = \frac{15100.0}{1} + \frac{75.1}{0.87142} + \cdots + \frac{38.0}{0.18590} + \frac{196.7}{0.06605} = 20789.8$$

$$\hat{\tau}_{y\varrho_2} = \frac{15100.0}{1} + \frac{496.2}{0.12859} = 18958.8$$

The combined estimate is:

$$\hat{\tau}_{y\varrho} = \frac{20789.8 + 18958.8}{2} = 19874.3$$

The standard error of the combined estimate is $|20789.8 - 18958.8|/2 = 915.5$ and the estimated relative standard error is: $100\% \times 915.5/19874.3 = 4.5\%$.

Note that foliage is attached to the last three segments of the first path. Our estimate of the fresh weight of foliage attached to the N segments of the tree is:

$$\hat{\tau}_{f\varrho_1} = \frac{38.1}{0.21511} + \frac{25.9}{0.18590} + \frac{291.0}{0.06605} = 4722.2$$

The last segment of the second path is a subterminal branch. Our second estimate is:

$$\hat{\tau}_{f\varrho_1} = \frac{441.0}{0.12859} = 3429.5$$

The combined estimate is:

$$\hat{\tau}_{f\varrho} = \frac{4722.2 + 3429.5}{2} = 4075.9$$

The standard error of this estimate is: $|4722.2 - 3429.5|/2 = 646.4$ and the relative standard error is: $100\% \times 646.4/4075.9 = 15.9\%$.

Table 13.2 *Fresh weights of woody tissues and foliage by path (i) and segment (r).*

i	r	Q_r	Wood	Foliage
			(g)	(g)
1	1	1.00000	15100.0	0.0
1	2	0.87142	75.1	0.0
1	3	0.79992	412.1	0.0
1	4	0.74882	82.5	0.0
1	5	0.55598	252.6	0.0
1	6	0.41262	209.8	0.0
1	7	0.28873	33.0	0.0
1	8	0.21511	154.6	38.1
1	9	0.18590	38.0	25.9
1	10	0.06605	196.7	291.0
2	1	1.00000	15100.0	0.0
2	2	0.12859	3858.9	441.0

13.3.1 Average stem length

Average stem length may be of interest to investigators of vascular transport, carbon allocation, or meristematic growth. Average stem length is essentially the average length of all M possible paths for which terminal shoots are terminal segments. The radius of a root plate is a similar 'underground parameter.'

Let ℓ_i be total length of the ith path (i.e., the sum of the lengths of the R segments). The sum of the lengths of the M possible paths (τ_ℓ) is unbiasedly estimated by:

$$\hat{\tau}_{\ell Q} = \frac{1}{m} \sum_{i=1}^{m} \frac{\ell_i}{Q_{iR}}.$$

Average stem length, $\mu_\ell = \tau_\ell / M$, is unbiasedly estimated by:

$$\hat{\mu}_\ell = \frac{\hat{\tau}_{\ell Q}}{M}.$$

Ordinarily, M is not known, but it is unbiasedly estimated by:

$$\hat{M} = \frac{1}{m} \sum_{i=1}^{m} \frac{s_{iR}}{Q_{iR}}, \tag{13.5}$$

where s_{iR} is the number of terminal shoots attached to the Rth segment of the ith path (note: $s_{iR} = 1$ if the terminal segment is a terminal shoot). The target parameter, μ_ℓ, is estimated by the ratio (Gregoire and Valentine 1996):

$$\bar{\ell} = \frac{\hat{\tau}_{\ell Q}}{\hat{M}}. \tag{13.6}$$

As with most ratio estimates, $\bar{\ell}$ may be biased. When $m = 1$, we can use $\bar{\ell} = \ell_i$.

When $m \geq 2$, the variance of $\bar{\ell}$ is approximated by (Särndal et al. 1992):

$$\hat{v}(\bar{\ell}) = \frac{1}{\hat{M}^2} \sum_{i=1}^{m} \left(\frac{1 - Q_{iR}}{Q_{iR}^2} \right) (\ell_i - \bar{\ell})^2 \qquad m > 1. \qquad (13.7)$$

13.4 Selection probabilities

As was mentioned, the conditional selection probabilities assigned to the branches at any node of a path must sum to one. Of interest is the method of assignment that yields the most precise estimates. This method is evinced by a multistage estimation process (Valentine et al. 1984).

For illustrative purposes, let us consider the estimation of the fresh weight of an entire tree. The tree is felled and a path to a terminal shoot is selected (Figure 13.3a). Then the segments of the path are separated and weighed. Let y_r be the fresh weight of the rth segment and let w_r be the fresh weight of the branch selected at the rth node. Finally, let τ_{w_r} be the aggregate fresh weight of all the branches at the rth node.

Starting at the last segment (Figure 13.3b) and moving down the path toward the butt we note that

$$\hat{\tau}_{wR} = \frac{y_R}{q_R} = \frac{w_R}{q_R}$$

is an unbiased estimate of τ_{wR}, the fresh weight of all the branches that were attached to the last (Rth) node of our path (Figure 13.3c). Moreover, we note that if $q_R = w_R/\tau_{wR}$, then $\hat{\tau}_{wR} = \tau_{wR}$. In other words, if we had assigned selection probabilities proportional to the actual fresh weights of the respective terminal shoots, then our estimate of the aggregate fresh weight of the terminal shoots would equal the actual weight.

Adding the weight of segment ($R - 1$) gives an unbiased estimate,

$$\hat{w}_{R-1} = \frac{y_R}{q_R} + y_{R-1},$$

of w_{R-1}, the weight of the branch selected at node $R - 1$ (Figure 13.3d). Inflating this estimate of the fresh weight of the branch selected at node $R - 1$ by dividing by q_{R-1}, we obtain an unbiased estimate,

$$\hat{\tau}_{wR-1} = \frac{y_R}{q_R q_{R-1}} + \frac{y_{R-1}}{q_{R-1}} = \frac{\hat{w}_{R-1}}{q_{R-1}},$$

of τ_{wR-1}, the weight of all the branches which were attached at this node (Figure 13.3e). We also note that if $\hat{w}_{R-1} = w_{R-1}$ and $q_{R-1} = w_{R-1}/\tau_{wR-1}$, then $\hat{\tau}_{wR-1} = \tau_{wR-1}$. That is, if we had assigned selection probabilities proportional to the actual fresh weights of the branches at nodes R and $R - 1$, then our estimate of the aggregate fresh weight of the branches at node $R - 1$, would equal the actual fresh weight.

Adding the weight of segment $R - 2$ gives an unbiased estimate,

$$\hat{w}_{R-2} = \frac{y_R}{q_R q_{R-1}} + \frac{y_{R-1}}{q_{R-1} + y_{R-2}},$$

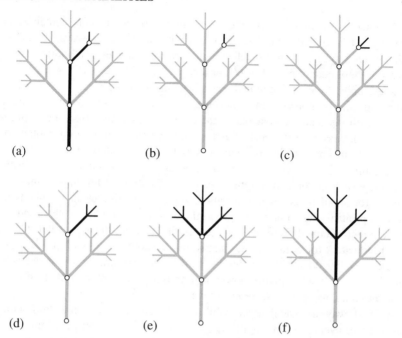

Figure 13.3 *Multistage estimation: (a) A path with 4 segments. (b) The weight of the terminal shoot (fourth segment) divided by q_4 gives an estimate of the aggregate weight of the three terminal shoots at the fourth node (c). Adding the weight of the subordinate (third) segment gives an estimate of the weight of the branch in (d). Dividing the estimate of the weight of the branch in (d) by q_3 gives an estimate an the aggregate weight of the three branches at the third node (e). Adding the weight of the subordinate (second) segment gives an estimate of the weight of the branch in (f).*

of w_{R-2}, the weight of the branch selected at node $R-2$ (Figure 13.3f).

This estimation process can be continued until we arrive at the first node with an unbiased estimate of the fresh weight of the entire tree:

$$\hat{\tau}_{w_1} = \frac{y_R}{q_R q_{R-1} q_{R-2} \cdots q_1} + \frac{y_{R-1}}{q_{R-1} q_{R-2} \cdots q_1} + \cdots + \frac{y_2}{q_2 q_1} + \frac{y_1}{q_1}.$$

This heuristic multistage estimation yields $\hat{\tau}_{w_1} \equiv \hat{\tau}_{y\varrho_i}$, i.e., the estimate of the fresh weight of the whole tree is equivalent to the estimate of the aggregate fresh weight of all the segments in the tree as calculated with (13.2). Thus, this exercise provides an inductive proof of the unbiasedness of $\hat{\tau}_{y\varrho_i}$.

More importantly, however, this exercise reveals the ideal way to assign selection probabilities to branches. If we are estimating the fresh weight of a tree, the selection probability assigned to each branch ideally should equal the fraction of the total fresh weight beyond the node and contained in the branch. More generally, the selection probability assigned to a branch ideally should equal the fraction of the total amount of an attribute that is beyond the node and contained in, or borne by, the branch.

Of course, we cannot discern the exact amount of an attribute contained in, or borne by, a branch; otherwise, we would not be sampling. However, branch-level attributes tend to be strongly correlated with branch diameter (d), branch length (l), powers of these quantities (d^a or l^b), or a product ($d^a l^b$).

For example, if the weight of leaves is of interest, then we may choose to calculate the probabilities proportional to d^2 because weights of leaves borne by branches tend to scale with the diameter-squares of the respective branches (see, e.g., Shinozaki et al. 1964). Jessen (1955) calculated selection probabilities proportional to the diameter-squares of branches in connection with the estimation of fruit counts.

The estimation of woody dry matter or volume is a common objective. Murray (1927) reported that branch weight tends to scale with $d^{2.5}$ for large branches, and with d^3 for small branches like terminal shoots. Valentine et al. (1984) assigned selection probabilities to branches proportional to $d^2 l$. De Gier (1989) used $d^{2.67}$; the exponent obtained from a least-squares fit of an allometric model. Greenhill (1881) indicated, for homogeneous stems, that l should scale with $d^{2/3}$, in which case $d^2 l \propto d^{8/3} \doteq d^{2.67}$. The tendency of branch volume or dry matter to scale with $d^{8/3}$ accords with quarter-power scaling rules (e.g., West et al. 1999), which are said to hold, more or less, for most taxa.

The best way to calculate probabilities of selection in connection with the estimation of average stem length, $\mu_\ell = \tau_\ell / M$, is less obvious. Assignments that are efficient for the estimation of τ_ℓ may not be efficient for M. Branches tend to be area-preserving: the sum of the cross-sectional areas (diameter-squares) of daughter branches tends to equal the cross-sectional area (diameter-square) of the mother branch. Thus, if an area-weighted average stem length is desired, then probability assignments proportional to d^2 are suggested.

Valentine & Hilton (1977) used ocular estimates of the foliage borne by the respective branches to calculate probabilities of selection for the objective of estimating leaf count on standing trees. Ocular estimation is, of course, an option for any RBS objective.

13.5 Tools and tricks of the trade

We assume, for this discussion, that the object branch is a large tree that has been felled to facilitate the RBS. Dead branches may be removed, depending upon the objectives of the sampling. Live branches broken during the felling of the tree are reconstructed and realigned as best as possible. The selection of the path is best performed by two people, one to (wo)man the field computer and the other to measure the branches and mark the path. The calculations for the RBS are easily performed with a calculator, but we prefer a field computer.

During the selection of a path, we usually number the branches emanating from a node with a lumber crayon. We generally have the computer programmed to accept measurements of the diameter and/or length of each branch (or other quantities for the calculation of probabilities of selection) by branch number. The computer applies exponents to the measurements (if appropriate), sets up the probability sampling frame, generates a random number, indicates the number of the branch that is selected, and stores the conditional and unconditional probabilities of selection by segment number. This process is repeated until either a terminal shoot or, more often than not, a sub-terminal branch is selected as the terminal segment of the path.

To avoid large sampling error, small epicormic branches or spur shoots should be ignored during the path selection. The RBS should be confined to those branches that constitute the main architecture of the tree. The ignored shoots or small branches ultimately are treated as parts of the segments to which they are attached.

We find it advisable to number the segments of each path with the crayon and tie a piece of flagging around each terminal segment. All of the material in the tree that is not part of a path may then be cut off and discarded. The bulky mass of the tree is reduced to just the segments of the path(s) and whatever organs are attached to them. Depending upon the quantities of interest, it may be necessary to separate the segments of the path(s) with loppers or a saw before measurements can be taken. On the other hand, quantities of interest whose direct measurement require the separation of the segments (e.g., woody dry matter and, perhaps, volume) could be estimated by subsampling.

Stratified sampling is appealing when the quantity of interest is the weight of leaves. Declining branches that abut the bole in the lower third of a crown often have less foliage per unit cross-sectional area of branch than higher branches. We have had success stratifying crowns of loblolly pines into thirds by length (Valentine et al. 1994). Within each stratum, all the branches abutting the bole were treated as though they emanated from a common node. Two object branches were selected from each of the three strata with probabilitity proportional to d^2. RBS was carried out on each object branch to estimate leaf weight. The resultant estimates of leaf weights and variances, respectively, were summed across strata to give tree-level estimates. We programmed a field computer to direct the entire sampling procedure, including the assignment of identifying codes to foliage samples.

13.6 Subsampling a path

Attributes of interest often include the volume, dry weight, and/or chemical content of the woody components of a tree. Certainly, we can estimate these attributes for the tree (if the entire tree is our object branch) with (13.2) after estimating these attributes for each segment of a path. For example, we could measure the fresh weight of each segment and then select a disk from the segment in some objective fashion for the measurement of the ratio of dry to fresh weight. Obviously, such a procedure could be laborious and time-consuming if it were applied to every path segment. However, there is an alternative. A subsampling strategy, which was introduced by Valentine et al. (1984), eliminates the need to section and measure all the segments of a path.

13.6.1 Subsampling strategy

Let y_{ir} denote the attribute, i.e., volume, weight, or chemical content, of the rth segment of the ith path and let us call y_{ir}/Q_{ir} the *inflated* attribute of the segment.

The inflated attribute for the entire (ith) path is:

$$\hat{\tau}_{y\varrho_i} = \sum_{r=1}^{R} \frac{y_{ir}}{Q_{ir}}.$$

Recall that $\hat{\tau}_{y\varrho_i}$ is an unbiased estimate of τ_y, the total amount of attribute contained in, or borne by, the object branch. However, we do not calculate $\hat{\tau}_{y\varrho_i}$ as that would require measuring the attribute of interest on all the segments of the path, which is what we want to avoid. Instead, we seek an unbiased estimate of this unbiased estimate of τ_y. Valentine et al. (1984) originally used importance sampling as the subsampling method, but other methods can substitute.

The stem length of the path comprises a continuous population of infinitely many points. Let x_r denote the stem length from the basal end of the first segment ($x_0 = 0$) to the distal end of the rth segment of the path. For the moment we assume that the final or Rth segment is a terminal shoot; thus, the total length of the path is x_R. Let $\rho(x)$ be the attribute density at a point x, where $0 \le x \le x_R$ (i.e., $\rho(x)$ is the volume per unit length $\equiv$ cross-sectional area, dry weight per unit length, or chemical content per length at x (see §4.2)). Suppose that x falls on the rth segment of the path. Division of $\rho(x)$ by the unconditional probability of selection of the rth segment yields the inflated attribute density, $\rho^*(x)$, i.e.,

$$\rho^*(x) = \frac{\rho(x)}{Q_r} \qquad x_{r-1} < x \le x_r, \quad r = 1, \dots, R. \tag{13.8}$$

The inflated attribute of the path is equivalent to the inflated attribute density integrated from 0 to x_R, i.e.,

$$\sum_{r=1}^{R} \frac{y_r}{Q_r} \equiv \tau_{\rho^*}(x_R) = \int_{0}^{x_R} \rho^*(x)\,dx.$$

If the terminal (Rth) segment of the path is a sub-terminal branch, we can estimate $\tau_{\rho^*}(x_{R-1})$, the inflated attribute of the path from x_0 to x_{R-1}. To this estimate we add the inflated attribute (y_R/Q_R) of the terminal segment. Thus, $\hat{\tau}_{\rho^*}(x_{R-1}) + y_R/Q_R$ is our estimate of the inflated attribute of the whole path. This estimate substitutes for $\hat{\tau}_{y\varrho_i}$ in equations (13.2) and (13.4).

The integral can be unbiasedly estimated with methods of Monte Carlo integration, e.g., crude Monte Carlo, importance sampling, or sampling with a control variate.

13.6.2 Crude Monte Carlo

Crude Monte Carlo (§4.2) generally is the simplest, but least precise, method. It may suffice, however, if the estimation of volume is the objective. An estimator of the inflated attribute of the path, from $x = 0$ to $x = x_R$, is

$$\hat{\tau}_{\rho^*} = \frac{x_R}{n} \sum_{s}^{n} \rho^*(x_s),$$

where $x_s = u_s x_R$ $(s = 1, 2, \ldots, n)$ and $u_s \sim U[0, 1]$. Efficiency usually is increased by the use of either antithetic (§4.2.4) or systematic selection (§4.2.5) of the sample points. Upon locating a sample point, we must remember that we actually measure the attribute density $\rho(x_s)$. Hence, we must identify the index of the segment, r, where x_s occurs so we can use equation (13.8) to convert $\rho(x_s)$ to the inflated attribute density $\rho^*(x_s)$.

Example 13.4

To estimate the inflated volume of a path with systematic selection (see §4.2.5), we measure or estimate several cross-sectional areas along the path at fixed intervals from a random start. Each cross-sectional area is inflated by dividing by the appropriate Q_r. The inflated volume of the path is estimated by the product of the length of the path and the average inflated cross-sectional area. Cross-sectional area usually is estimated from circumference (or diameter), which is measured with a tape (or diameter tape). Such estimates are slightly biased if the stem is not circular (Matérn 1956). An unbiased estimate obtains from sawing through the stem and averaging cross-sectional areas calculated with radii selected at random.

13.6.3 Importance sampling

Estimation of inflated weight or chemical content of a path is most likely best estimated by either importance sampling or sampling with a control variate. Recall that importance sampling is a continuous analog of sampling with probability proportional to size (§4.3). In order to estimate the inflated attribute of a path, we need to construct a continuous proxy function, $g^*(x)$, that is as nearly proportional to, or coincident with, the inflated attribute density function, $\rho^*(x)$, as is practicable. The inflated attribute is estimated by:

$$\hat{\tau}_{\rho^*} = \frac{G^*(x_R)}{n} \sum_s^n \frac{\rho^*(x_s)}{g^*(x_s)},$$

where x_s $(s = 1, 2, \ldots, n)$ is a sample point, and,

$$G^*(x_R) = \int_0^{x_R} g^*(x)\, dx.$$

We substitute x_{R-1} for x_R if the last segment of the path is a subterminal branch. The sample points can be selected by the inverse-transform method (§4.3.2) or the acceptance-rejection method (§4.3.3).

13.6.4 Constructing a proxy function

Valentine et al. (1984) constructed a segmented linear proxy function from diameters measured at distances $x = L_1, L_2, \ldots, L_t, \ldots, L_T$ from the butt of the path, where

$L_1 = 0$ and $L_T = x_R$ (or x_{R-1}). The placements of L_2 through L_{T-1} are arbitrary, as is the number of measurements, T. Thus, the segments of the proxy function ordinarily are not the same as the original segments of the path. Starting at the butt, we suggest taking measurements every 20 to 25 cm in the region of the butt swell and every 1 to 2 m thereafter depending upon taper and local stem deformities. However, the construction would work if we used just $T = 2$ measurements, viz., $L_1 = 0$ and $L_2 = x_R$ (or x_{R-1}), in which case the segmented linear proxy function would have just one segment.

Denoting the diameter at L_t by $d(L_t)$, we let

$$g^*(L_t) = \frac{d^2(L_t)}{Q_r},$$

where r is the index of the original path segment where $d(L_t)$ is measured. Proxy values between L_{t-1} and L_t are found by linear interpolation, i.e.,

$$g^*(x) = g^*(L_{t-1}) + \left[\frac{g^*(L_t) - g^*(L_{t-1})}{L_t - L_{t-1}}\right][x - L_{t-1}],$$

$$L_{t-1} < x \le L_t.$$

This segmented linear function can be integrated piecewise by the trapezoidal rule, i.e.,

$$G^*(L_t) = G^*(L_{t-1}) + \left[\frac{g^*(L_t) - g^*(L_{t-1})}{2}\right][L_t - L_{t-1}],$$

$$t = 2, 3, \ldots, T.$$

To find a sample point, $x = x_s$, by the acceptance-rejection method (§4.3.3), we require g^*_{max}. Since the proxy function is segmented linear,

$$g^*_{max} = \max\left[g^*(L_1), g^*(L_2), \ldots, g^*(L_T)\right].$$

Drawing u_1 and u_2 from U[0, 1], we let $x_s = u_1 L_T$. If $u_2 \times g^*_{max} \le g^*(x_s)$, then we accept x_s as our sample point; otherwise, we begin again with two new random numbers.

A sample point, x_s, calculated by the inverse-transform method, is between L_{t-1} and L_t if

$$G^*(L_{t-1}) < u_s G^*(L_T) < G^*(L_t).$$

The exact value of x_s is:

$$x_s = L_{t-1} + \left[\frac{-b + \sqrt{b^2 - 4ac}}{2a}\right],$$

where

$$a = \frac{g^*(L_t) - g^*(L_{t-1})}{L_t - L_{t-1}}$$

$$b = 2g^*(L_{t-1})$$

$$c = -2\left[u_s\, G^*(L_t) - G^*(L_{t-1})\right].$$

A field computer can be programmed to accept the measurements of diameters and their locations along the path, formulate the proxy function from the measurements, generate the random number(s), return x_s, prompt the sampler for the index, r, of the original path segment where x_s occurs, and return and/or store $G^*(L_T)/g^*(x_s)$ for the calculation of the estimate.

A computer also affords the option to interpolate the known proxy values with a spline or some other function instead of a segmented-linear function. Of course, the inflated proxy function need not interpolate any measurements; but it seems a good idea to ensure accuracy and precision.

13.6.5 Sampling with a control variate

As noted in §4.4, sampling with a control variate has been shown to have precision equal to, or greater than, importance sampling in connection with the estimation of bole volume. This method should also be efficient for estimating the inflated volume, dry weight, or chemical content of a path. The estimator of the inflated attribute of the path is

$$\hat{\tau}_{\rho^*} = \beta G^*(L_T) + \frac{L_T}{n}\sum_{s=1}^{n}\left[\rho^*(x_s) - \beta g^*(x_s)\right],$$

where $x_s = u_s L_T$ and β is an arbitrary constant. With this method, sample points are selected uniformly along the length of the path. By contrast, importance sampling concentrates sample points where the inflated cross-sectional areas are largest.

The segmented linear interpolation function that serves as the proxy function for importance sampling can also serve as the control variate, $g^*(x)$. Ordinarily, we would let $\beta = \pi/4$ for the estimation of volume and $\beta = \rho_w \pi/4$ for the estimation of dry weight. In the latter case, ρ_w is the dry weight per unit wet volume of a sample disk, which gives $\beta g^*(x)$ dimensions of weight per unit length and $\beta G^*(L_T)$ a dimension of weight.

Example 13.5

If the target parameter is inflated dry weight, then the inflated attribute density at x, $\rho^*(x)$, is dry weight per unit length (g cm^{-1}) divided by the appropriate Q_r. Since we can not measure weight per unit length at a sample point, one option is to cut a disk centered at the sample point and divide the weight of the disk by its thickness. We suggest that disks be cut at least 10 cm thick to diminish the effects of measurement errors. Because the side cuts of the disk may not be parallel, several measurements of length (i.e., the thickness of the disk) should be averaged. Another option (from Van Deusen & Baldwin 1993) is to extract a core with an increment borer at each sample point and multiply the weight per unit volume of the core by the cross-sectional area at the sample point to obtain the requisite weight per unit length. However, heterogeneity in the density of the wood may affect accuracy and unbiasedness because old wood is more intensively sampled than young wood by a core. The thickness of a disk or the volume of a core should be determined before drying. A disk can divided into wedges and one selected with probabilility proportional to actual weight or volume for the measurement the chemical concentration.

It is, of course, possible to select more than one disk per path. However, looking at the big picture, we speculate that multiple paths, each with one disk, may be more efficient than a single path with multiple disks.

13.6.6 Choice of method

Importance sampling or sampling with a control variate is perferred over crude Monte Carlo for estimation of the inflated weight or chemical content of a path because the cutting, handling, drying, and bioassay of disks is kept to a minimum. And, volume can be estimated with essentially no extra cost. Whether it is worthwhile to construct a proxy function to estimate just volume depends on the investigator's preference for unbiased estimators. Volume is estimated from measurements of cross-sectional area with all three methods. If the investigator chooses to estimate cross-sectional area by $\pi d^2(x_s)/4$, then there may be no advantage of importance sampling or sampling with a control variate over crude Monte Carlo with systematic selection. Stems generally are not truly round so $\pi d^2(x_s)/4$ is neither a measurement nor an unbiased estimate of cross-sectional area. Thus, we eschew the unbiasedness of either sampling technique if we substitute $\pi d^2(x_s)/4$ for cross-sectional area. And, we are left with the fact that we need several measurements of diameters at random points to achieve a precise, albeit possibly biased, estimate of volume with crude Monte Carlo. On the other hand, if we use importance sampling or a control variate, we need several measurements of diameter to construct the proxy function to select one or, perhaps, two sample points at random, where we measure diameter yet again. If, however, we make the effort to measure cross-sectional area at the sample point or unbiasedly estimate this area with random radii, then importance sampling or sampling with a control variate makes sense.

13.7 Terms to remember

Conditional selection probability	Object
Inflated attribute	Path
Inflated attribute density	Segment
Node	Unconditional selection probability

13.8 Exercises and projects

1. Suppose a sampler begins selecting branches with probability proportional to $d^{8/3}$ ($d \equiv$ diameter) and then switches to selecting with probability proportional to d^3 at the last three nodes. Would this affect the theoretical unbiasedness of the estimates? Why or why not?

2. In Example 13.3 (also see Example 13.2), the fresh weight of woody tissues is estimated more precisely than the fresh weight of foliage. What is the most likely explanation for this?

13.9 Appendix 13

13.9.1 Proof of the unbiasedness of $\hat{\tau}_{y \varrho_i}$

We claim that the target parameter, τ_y, is unbiasedly estimated by

$$\hat{\tau}_{y \varrho_i} = \sum_{r=1}^{R} \frac{y_{ir}}{\varrho_{ir}}.$$

Ordering all the segments of an object branch from 1 to N, the target parameter is

$$\tau_y = y_1 + y_2 + \cdots + y_N$$

and the estimator can be written thus:

$$\hat{\tau}_{y \varrho_i} = I_1 \frac{y_1}{\varrho_1} + I_2 \frac{y_2}{\varrho_2} + \cdots + I_N \frac{y_N}{\varrho_N}$$

where I_k is an indicator variable, such that $I_k = 1$, if the kth segment is in the ith path; or $I_k = 0$, if otherwise. In expectation,

$$E\left[\hat{\tau}_{y \varrho_i}\right] = E\left[I_1 \frac{y_1}{\varrho_1} + I_2 \frac{y_2}{\varrho_2} + \cdots + I_N \frac{y_N}{\varrho_N}\right]$$

$$= E[I_1] \frac{y_1}{\varrho_1} + E[I_2] \frac{y_2}{\varrho_2} + \cdots + E[I_N] \frac{y_N}{\varrho_N}.$$

In the repeated selection of paths, $E[I_k] = [Q_k \cdot 1] + [(1 - Q_k) \cdot 0] = Q_k$, therefore,

$$E\left[\hat{\tau}_{y \varrho_i}\right] = Q_1 \frac{y_1}{\varrho_1} + Q_2 \frac{y_2}{\varrho_2} + \cdots + Q_N \frac{y_N}{\varrho_N}$$

$$= y_1 + y_2 + \cdots + y_N$$

$$= \tau_y.$$

CHAPTER 14

Sampling with Partial Replacement

14.1 Introduction

Repeated sampling of a population over time raises issues pertinent to both the sampling design and the estimation of population parameters. Regarding the former, one issue is the relationship between the sample at a future occasion to the initial sample. When sampling on just two occasions, three distinct possibilities are suggested: 1) independently selected samples, where overlap of the units selected may occur by chance; 2) the initial sample is retained in its entirety, ensuring complete overlap by design; and 3) partial overlap in sample composition which is ensured by design. When successive sampling on more than two occasions is considered, the number of possible outcomes increases. Analogous to these design issues, alternative estimators of current parameter values and their change through time are possible. In nontechnical language, these matters revolve around two questions: How much of the initial sample should be changed in succeeding samples, and in what way should previously collected data be used to estimate the value of current population parameters?

Jessen (1942) apparently was the first to articulate these matters in the context of successive surveys to obtain 'farm facts,' i.e., information about farming practices and status. Cochran (1953) wrote shortly thereafter, "As confidence in sampling has increased, the practice of relying on samples for the collection of important series of data that are published at regular intervals is becoming more common." Institutions that manage natural and environmental resources periodically sample their resource base (population) with the dual aim of assessing: 1) the current values of population parameters, and 2) the change in values of these parameters since the previous inventory. Indeed, many public or governmental agencies may operate under a legislative mandate to conduct such resource inventories. As a case in point, the McSweeny–McNary Act of 1928 passed by the Congress of the United States resulted in the establishment of a periodic survey of the forest resources in every region of the country. Initiated in 1930, the Forest Survey, as it then was called, has been carried out by the U. S. Forest Service ever since. Its current design has evolved from a nominal 10-year periodic survey to an annual survey conducted separately in each state.

Most of the preceding chapters have presented and explained sampling strategies for the purpose of drawing inference about τ_y, μ_y, λ_y, and so on. Sections of Chapters 7 and 8 were devoted to estimating Δ_y, either from independent samples of plots or points or from a single sample measured at two times. Sampling with partial replacement (SPR) of previously sampled units with newly chosen units elaborates on these techniques.

In this chapter we will consider optimal estimation of μ_y or τ_y on the second occasion; the optimal estimation of Δ_y over two occasions; the simultaneous estimation of both τ_y and Δ_y; and, briefly, the extension to sampling on three of more occasions. Our presentation draws heavily on the pioneering works of Yates (1949), and Patterson (1950) in the statistical literature, and those of Ware (1960) and Ware & Cunia (1962) from the forestry literature. We concentrate on sampling at each occasion with a partial replacement of the sampling units from the preceding occasion in the context of replicated areal designs, such as those discussed in Chapters 7, 8, and 9. A more general treatment, of sample design and estimation from samples that are conducted through time, is contained in Duncan & Kalton (1987).

14.2 Estimation with partially replaced sampling units

Using notation introduced in Chapter 7, let t_1 and t_2 indicate the first and second occasions of sampling on $\mathcal{A}$. Analogously, let m_1 and m_2 indicate the number of sampling locations at t_1 and t_2, respectively. Suppose $\mathcal{S}$ is the set of distinct locations sampled on both occasions. A subset of $\mathcal{S}$, denoted by $\mathcal{S}_{12}$, comprises the m_{12} locations that will be sampled on both occasions. Let $\mathcal{S}_{1\bullet}$ indicate the subset of $\mathcal{S}$ comprising the $m_{1\bullet}$ locations that are sampled at t_1 but not t_2; and let $\mathcal{S}_{\bullet 2}$ indicate the subset of $\mathcal{S}$ comprising the $m_{\bullet 2}$ locations that are sampled at t_2 but not t_1. Therefore, $\mathcal{S} = \mathcal{S}_{1\bullet} \cup \mathcal{S}_{\bullet 2} \cup \mathcal{S}_{12}$, and

$$m_1 = m_{1\bullet} + m_{12} \tag{14.1}$$

and

$$m_2 = m_{\bullet 2} + m_{12}. \tag{14.2}$$

Although the locations in $\mathcal{S}_{\bullet 2}$ 'replace' those in $\mathcal{S}_{1\bullet}$, there is no requirement that $m_{\bullet 2} = m_{1\bullet}$.

We distinguish the following four estimators. Two are based only on the m_{12} locations that are matched on both occasions:

$$\hat{\tau}_{y\pi,\text{rep}}(t_1, m_{12}) = \frac{1}{m_{12}} \sum_{\mathcal{P}_s \in \mathcal{S}_{12}} \hat{\tau}_{y\pi s}(t_1) \tag{14.3}$$

and

$$\hat{\tau}_{y\pi,\text{rep}}(t_2, m_{12}) = \frac{1}{m_{12}} \sum_{\mathcal{P}_s \in \mathcal{S}_{12}} \hat{\tau}_{y\pi s}(t_2). \tag{14.4}$$

The remaining two are based on unmatched locations:

$$\hat{\tau}_{y\pi,\text{rep}}(t_1, m_{1\bullet}) = \frac{1}{m_{1\bullet}} \sum_{\mathcal{P}_s \in \mathcal{S}_{1\bullet}} \hat{\tau}_{y\pi s}(t_1) \tag{14.5}$$

Table 14.1 *Directory of estimators of τ_y and Δ_y that are presented in this chapter.*

Estimator	Description
$\hat{\tau}_{y,\text{SPR}}(t_2:t_1)$	Composite estimator of $\tau_y(t_2)$ based on m_{12} remeasured sample units and $m_{\bullet 2}$ new sample units; see (14.7), (14.8), (14.16), (14.23)
$\hat{\tau}_{y\pi,\text{rep}}(t_2)$	Replicated-sampling estimator of $\tau_y(t_2)$ based on m_2 sample units; see p. 237
$\hat{\tau}_{y\pi,\text{rep}}(t_2, m_{\bullet 2})$	Replicated-sampling estimator of $\tau_y(t_2)$ based on $m_{\bullet 2}$ new sample units; see (14.6)
$\hat{\tau}_{y\pi,\text{rep}}(t_2, m_{12})$	Replicated-sampling estimator of $\tau_y(t_2)$ based on m_{12} remeasured sample units; see (14.4)
$\hat{\tau}_{y,\text{SPR}}(t_1:t_2)$	Composite estimator of $\tau_y(t_1)$ based on m_{12} remeasured sample units and $m_{1\bullet}$ unmatched sample units; see (14.40)
$\hat{\tau}_{y\pi,\text{rep}}(t_1)$	Replicated-sampling estimator of $\tau_y(t_1)$ based on m_1 sample units; see p. 237
$\hat{\tau}_{y\pi,\text{rep}}(t_1, m_{1\bullet})$	Replicated-sampling estimator of $\tau_y(t_1)$ based on $m_{1\bullet}$ unmatched sample units; see (14.5)
$\hat{\tau}_{y\pi,\text{rep}}(t_1, m_{12})$	Replicated-sampling estimator of $\tau_y(t_1)$ based on m_{12} remeasured sample units; see (14.3)
$\hat{\hat{\tau}}_{y,\text{SPR}}(t_2:t_1)$	Composite estimator of $\tau_y(t_2)$ based on m_{12} remeasured sample units and $m_{\bullet 2}$ new sample units using estimated regression coefficient and weights
$\hat{\hat{\tau}}_{y,\text{SPR}}(t_1:t_2)$	Composite estimator of $\tau_y(t_1)$ based on m_{12} remeasured sample units and $m_{1\bullet}$ unmatched sample units using estimated regression coefficient and weights

and

$$\hat{\tau}_{y\pi,\text{rep}}(t_2, m_{\bullet 2}) = \frac{1}{m_{\bullet 2}} \sum_{\mathcal{P}_s \in \mathcal{S}_{\bullet 2}} \hat{\tau}_{y\pi s}(t_2). \qquad (14.6)$$

Treating t_2 as the occasion of principal interest, we seek a linear combination of these four estimators that design-unbiasedly estimates $\tau_y(t_2)$. Namely, we seek values for the coefficients $a, b, c,$ and d to form a *composite estimator* such that

$$\hat{\tau}_{y,\text{SPR}}(t_2:t_1) = a\hat{\tau}_{y\pi,\text{rep}}(t_1, m_{1\bullet}) + b\hat{\tau}_{y\pi,\text{rep}}(t_1, m_{12})$$

$$+ c\hat{\tau}_{y\pi,\text{rep}}(t_2, m_{\bullet 2}) + d\hat{\tau}_{y\pi,\text{rep}}(t_2, m_{12}), \qquad (14.7)$$

while ensuring that $E\left[\hat{\tau}_{y,\text{SPR}}(t_2:t_1)\right] = \tau_y(t_2)$.

There is redundancy in the four coefficients of (14.7), because

$$E\left[\hat{\tau}_{y\pi,\text{rep}}(t_1, m_{1\bullet})\right] = E\left[\hat{\tau}_{y\pi,\text{rep}}(t_1, m_{12})\right] = \tau_y(t_1)$$

implies that $a + b = 0$, and

$$E\left[\hat{\tau}_{y\pi,\text{rep}}(t_2, m_{\bullet 2})\right] = E\left[\hat{\tau}_{y\pi,\text{rep}}(t_2, m_{12})\right] = \tau_y(t_2)$$

implies that $c + d = 1$. Consequently, (14.7) simplifies to

$$\hat{\tau}_{y,\text{SPR}}(t_2 : t_1) = a\hat{\tau}_{y\pi,\text{rep}}(t_1, m_{1\bullet}) - a\hat{\tau}_{y\pi,\text{rep}}(t_1, m_{12})$$

$$+ (1 - d)\hat{\tau}_{y\pi,\text{rep}}(t_2, m_{\bullet 2}) + d\hat{\tau}_{y\pi,\text{rep}}(t_2, m_{12}). \qquad (14.8)$$

To uniquely identify the values of a and d in (14.8) a further constraint is imposed that the values of a and d should be those that minimize the variance of $\hat{\tau}_{y,\text{SPR}}(t_2 : t_1)$, given $m_{1\bullet}$, $m_{\bullet 2}$, and m_{12}. Since the covariance between terms in (14.8) is identically zero for all but the terms containing $\hat{\tau}_{y\pi,\text{rep}}(t_1, m_{12})$ and $\hat{\tau}_{y\pi,\text{rep}}(t_2, m_{12})$, the variance of $\hat{\tau}_{y,\text{SPR}}(t_2 : t_1)$ is

$$V\left[\hat{\tau}_{y,\text{SPR}}(t_2 : t_1)\right] = a^2 \left(V\left[\hat{\tau}_{y\pi,\text{rep}}(t_1, m_{1\bullet})\right] + V\left[\hat{\tau}_{y\pi,\text{rep}}(t_1, m_{12})\right]\right)$$

$$+ (1 - d)^2 V\left[\hat{\tau}_{y\pi,\text{rep}}(t_2, m_{\bullet 2})\right]$$

$$+ d^2 V\left[\hat{\tau}_{y\pi,\text{rep}}(t_2, m_{12})\right] \qquad (14.9)$$

$$- 2ad\, C\left[\hat{\tau}_{y\pi,\text{rep}}(t_1, m_{12}), \hat{\tau}_{y\pi,\text{rep}}(t_2, m_{12})\right].$$

In (14.9), $C\left[\hat{\tau}_{y\pi,\text{rep}}(t_1, m_{12}), \hat{\tau}_{y\pi,\text{rep}}(t_2, m_{12})\right]$ is the covariance between the estimators based on the matched sampling locations, namely

$$C\left[\hat{\tau}_{y\pi,\text{rep}}(t_1, m_{12}), \hat{\tau}_{y\pi,\text{rep}}(t_2, m_{12})\right]$$

$$= \rho_{12}\sqrt{V\left[\hat{\tau}_{y\pi,\text{rep}}(t_1, m_{12})\right]}\sqrt{V\left[\hat{\tau}_{y\pi,\text{rep}}(t_2, m_{12})\right]} \qquad (14.10)$$

where ρ_{12} is the correlation coefficient between $\hat{\tau}_{y\pi,\text{rep}}(t_1, m_{12})$ and $\hat{\tau}_{y\pi,\text{rep}}(t_2, m_{12})$. Imposition of this constraint yields the result (see the Appendix, §14.8) that

$$d = \frac{V\left[\hat{\tau}_{y\pi,\text{rep}}(t_2, m_{\bullet 2})\right]}{V\left[\hat{\tau}_{y\pi,\text{rep}}(t_2, m_{\bullet 2})\right] + V\left[\hat{\tau}_{y\pi,\text{rep}}(t_2, m_{12})\right] - \Upsilon_{12}}, \qquad (14.11)$$

where

$$\Upsilon_{12} = \frac{C^2\left[\hat{\tau}_{y\pi,\text{rep}}(t_1, m_{12}), \hat{\tau}_{y\pi,\text{rep}}(t_2, m_{12})\right]}{V\left[\hat{\tau}_{y\pi,\text{rep}}(t_1, m_{\bullet 2})\right] + V\left[\hat{\tau}_{y\pi,\text{rep}}(t_1, m_{12})\right]}. \qquad (14.12)$$

This reduces to

$$d = \frac{m_{12}}{m_2 - u_1 m_{\bullet 2}\rho_{12}^2}, \qquad (14.13)$$

where $u_1 = m_{1\bullet}/m_1$ is the proportion of sampling locations at t_1 that are not measured at t_2. Also

$$a = \frac{B_{21}m_{12}u_1}{m_2 - u_1 m_{\bullet 2}\rho_{12}^2} \tag{14.14a}$$

$$= B_{21}m_{12}u_1 d, \tag{14.14b}$$

where B_{21} is the linear regression coefficient given by

$$B_{21} = \rho_{12}\sqrt{\frac{\Phi_2}{\Phi_1}}, \tag{14.15}$$

where Φ_1 and Φ_2 are defined in (14.49) and (14.48), respectively, in the Appendix, §14.8.1.

Substituting a and d into (14.8) and using the results presented in the Appendix yields

$$\hat{\tau}_{y,\text{SPR}}(t_2:t_1) = d\,\hat{\tau}_{y\pi,\text{reg}}(t_2:t_1) + (1-d)\hat{\tau}_{y\pi,\text{rep}}(t_2, m_{\bullet 2}), \tag{14.16}$$

where

$$\hat{\tau}_{y\pi,\text{reg}}(t_2:t_1) = \hat{\tau}_{y\pi,\text{rep}}(t_2, m_{12}) + B_{21}\left[\hat{\tau}_{y\pi,\text{rep}}(t_1) - \hat{\tau}_{y\pi,\text{rep}}(t_1, m_{12})\right]. \tag{14.17}$$

This result is somewhat surprising: the optimal estimator of $\tau_y(t_2)$ for specified sample sizes $m_{1\bullet}$, m_{12}, and $m_{\bullet 2}$ is a combination of the regression estimator based on the remeasured data and the replicated sampling estimator based on the $m_{\bullet 2}$ new sample units. The latter is based on current information only, whereas the former is based on previous information which is adjusted through regression to the present occasion. For the purpose of estimating $\tau_y(t_2)$, $\hat{\tau}_{y,\text{SPR}}(t_2:t_1)$ in (14.16) is more precise than the replicated sampling estimator based on all $m_2 = m_{12} + m_{2\bullet}$ samples available at t_2.

Ware (1960) showed that

$$d = \frac{w(t_2:t_1)}{w_{\text{SPR}}}, \tag{14.18}$$

and

$$1 - d = \frac{w(t_2)}{w_{\text{SPR}}}, \tag{14.19}$$

where

$$w(t_2:t_1) = \frac{1}{V\left[\hat{\tau}_{y\pi,\text{reg}}(t_2:t_1)\right]}, \tag{14.20}$$

$$w(t_2) = \frac{1}{V\left[\hat{\tau}_{y\pi,\text{rep}}(t_2, m_{\bullet 2})\right]}, \tag{14.21}$$

and

$$w_{\text{SPR}} = w(t_2:t_1) + w(t_2). \tag{14.22}$$

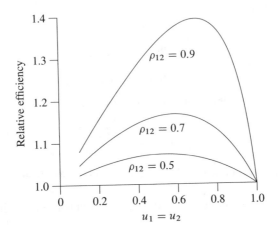

Figure 14.1 *The relative efficiency of $\hat{\tau}_{y,\mathrm{SPR}}(t_2 : t_1)$ compared to $\hat{\tau}_{y\pi,\mathrm{rep}}(t_2)$. As the correlation between the measurements at t_1 and t_2 increases, the relative efficiency of SPR increases.*

Consequently, $\hat{\tau}_{y,\mathrm{SPR}}(t_2 : t_1)$ is a linear combination of two estimators each of which is weighted inversely proportional to its variance:

$$\hat{\tau}_{y,\mathrm{SPR}}(t_2 : t_1) = \frac{w(t_2 : t_1)\hat{\tau}_{y\pi,\mathrm{reg}}(t_2 : t_1) + w(t_2)\hat{\tau}_{y\pi,\mathrm{rep}}(t_2, m_{\bullet 2})}{w_{\mathrm{SPR}}}. \tag{14.23}$$

Combining unbiased estimators by weighting each inversely proportional to its variance is a recurring theme in many statistical applications, as is the issue that these variances are usually unknown (cf. Graybill & Deal 1959).

Concise expressions of the variance of $\hat{\tau}_{y,\mathrm{SPR}}(t_2 : t_1)$ are

$$V\left[\hat{\tau}_{y,\mathrm{SPR}}(t_2 : t_1)\right] = \frac{1}{w_{\mathrm{SPR}}}$$

$$= V\left[\hat{\tau}_{y\pi,\mathrm{rep}}(t_2)\right]\left(\frac{1 - u_1 \rho_{12}^2}{1 - u_1 u_2 \rho_{12}^2}\right), \tag{14.24}$$

where $u_2 = m_{\bullet 2}/m_2$ is the proportion of new sampling units established at t_2 that were not measured at t_1. These results are derived in the Appendix, §14.8.2. Evidently, as $\rho_{12} \to 0$, $V\left[\hat{\tau}_{y,\mathrm{SPR}}(t_2 : t_1)\right]$ approaches $V\left[\hat{\tau}_{y\pi,\mathrm{rep}}(t_2)\right]$ in magnitude. Indeed, inspection of (14.16) reveals that $\hat{\tau}_{y,\mathrm{SPR}}(t_2 : t_1)$ collapses to $\hat{\tau}_{y\pi,\mathrm{rep}}(t_2)$ when $\rho_{12} = 0$.

In Figure 14.1 the relative efficiency of $\hat{\tau}_{y,\mathrm{SPR}}(t_2 : t_1)$ to $\hat{\tau}_{y\pi,\mathrm{rep}}(t_2)$ is shown in the special case where $m_1 = m_2$ and therefore $u_1 = u_2$. Relative efficiency is expressed as the ratio of $V\left[\hat{\tau}_{y\pi,\mathrm{rep}}(t_2)\right]$ to $V\left[\hat{\tau}_{y,\mathrm{SPR}}(t_2 : t_1)\right]$. Relative efficiencies greater than unity indicate the degree to which $\hat{\tau}_{y,\mathrm{SPR}}(t_2 : t_1)$ is more precise than $\hat{\tau}_{y\pi,\mathrm{rep}}(t_2)$. Evidently, as ρ_{12} increases in magnitude, relative efficiency at any given proportion of unmatched sample units increases. The proportion of unmatched sample units at which the maximum relative efficiency is realized also increases directly with ρ_{12}.

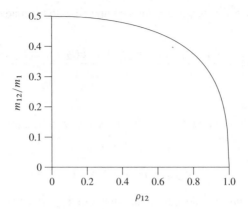

Figure 14.2 *The proportion of the m_1 sample that must be remeasured at t_2 in order to minimize $V\left[\hat{\tau}_{y,\text{SPR}}(t_2:t_1)\right]$. As the correlation between the measurements at t_1 and t_2 increases, the number of remeasured samples decreases at an accelerating rate.*

The relative costs of measuring matched vs. new unmatched plots are not reflected in Figure 14.1.

14.2.1 Optimal number of remeasured units

Ware (1960, p. 24) and references cited therein show how to determine the number of the initial m_1 units that ought to be remeasured to ensure that $V\left[\hat{\tau}_{y,\text{SPR}}(t_2:t_1)\right]$ is a minimum for a sample size, m_2 constrained by the total cost of sampling at occasion t_2. We defer to these authors for details, and present the results only for the special case where the cost of sampling and measuring an unmatched unit is the same as the cost of remeasuring a previously sampled unit. In this situation, the optimal value is

$$m_{12} = m_1 \frac{\sqrt{1-\rho_{12}^2}\left(1-\sqrt{1-\rho_{12}^2}\right)}{\rho_{12}^2}.$$

A graph of m_{12} as a fraction of m_1, shown in Figure 14.2, reveals that the required remeasurement fraction decreases sharply as ρ_{12} increases. Evidently, the more informative the value of $\hat{\tau}_{yns}(t_1)$ is of the value of $\hat{\tau}_{yns}(t_2)$, the fewer the number of units that need be remeasured in order to achieve the best precision for a specified cost of sampling.

14.2.2 Estimation of $\tau_y(t_2)$ with estimated parameter values

Inasmuch as neither B_{12} nor the variance in the weights $w(t_2:t_1)$ and $w(t_2)$ are known, $\hat{\tau}_{y,\text{SPR}}(t_2:t_1)$ is not feasible. In practice estimates of these parameters are inserted into the composite estimator $\hat{\tau}_{y,\text{SPR}}(t_2:t_1)$, thereby yielding an estimator of τ_y with slightly different statistical properties. We introduce $\hat{\hat{\tau}}_{y,\text{SPR}}(t_2:t_1)$ to denote the feasible composite estimator, that is, one computed with estimates of B_{12}, $w(t_2:t_1)$, and $w(t_2)$:

$$\hat{\hat{\tau}}_{y,\text{SPR}}(t_2:t_1) = \frac{\hat{w}(t_2:t_1)\hat{\hat{\tau}}_{y\pi,\text{reg}}(t_2:t_1) + \hat{w}(t_2)\hat{\tau}_{y\pi,\text{rep}}(t_2, m_{\bullet 2})}{\hat{w}_{\text{SPR}}}, \quad (14.25)$$

where

$$\hat{\hat{\tau}}_{y\pi,\text{reg}}(t_2:t_1) = \hat{\tau}_{y\pi,\text{rep}}(t_2, m_{12}) + \hat{B}_{21}\left[\hat{\tau}_{y\pi,\text{rep}}(t_1) - \hat{\tau}_{y\pi,\text{rep}}(t_1, m_{12})\right], \quad (14.26)$$

and

$$\hat{w}_{\text{SPR}} = \hat{w}(t_2:t_1) + \hat{w}(t_2). \quad (14.27)$$

Generally speaking, the design-unbiasedness of $\hat{\tau}_{y,\text{SPR}}(t_2:t_1)$ does not hold for $\hat{\hat{\tau}}_{y,\text{SPR}}(t_2:t_1)$. With reasonably large samples, however, the magnitude of the bias should be negligibly small. Its behavior in this regard is similar to that of the ratio and regression estimators of Chapter 6.

Almost always, the estimator of B_{12} that is used in $\hat{\hat{\tau}}_{y,\text{SPR}}(t_1:t_2)$ is given by (6.83) based on the m_{12} matched units in the sample. When these are selected by simple random sampling, this estimator is identical to $\hat{B}$ of Chapter 6, which we rewrite here as

$$\hat{B}_{12} = \frac{s_y(t_2:t_1)}{s_y^2(t_1, m_{12})}, \quad (14.28)$$

where

$$s_y(t_2:t_1) =$$

$$\frac{1}{m_{12}-1} \sum_{s \in \mathcal{S}_{12}} \left(\hat{\tau}_{y\pi s}(t_1) - \hat{\tau}_{y\pi,\text{rep}}(t_1)\right)\left(\hat{\tau}_{y\pi s}(t_2) - \hat{\tau}_{y\pi,\text{rep}}(t_2)\right) \quad (14.29)$$

and

$$s_y^2(t_1, m_{12}) = \frac{1}{m_{12}-1} \sum_{s \in \mathcal{S}_{12}} \left(\hat{\tau}_{y\pi s}(t_1) - \hat{\tau}_{y\pi,\text{rep}}(t_1)\right)^2. \quad (14.30)$$

Both numerator and denominator in (14.28) are based on estimates from the remeasured plots only.

Because $V\left[\hat{\hat{\tau}}_{y\pi,\text{reg}}(t_2:t_1)\right]$ is unknown, $w(t_2:t_1)$ in (14.20) must be estimated by

$$\hat{w}(t_2:t_1) = \frac{1}{\hat{v}\left[\hat{\hat{\tau}}_{y\pi,\text{reg}}(t_2:t_1)\right]}, \quad (14.31)$$

where $\hat{v}\left[\hat{\hat{\tau}}_{y\pi,\text{reg}}(t_2:t_1)\right]$ is a suitable estimator of $V\left[\hat{\hat{\tau}}_{y\pi,\text{reg}}(t_2:t_1)\right]$. Various estimators have been proposed, one of which is (6.101) adapted to this context:

$$\hat{v}\left[\hat{\hat{\tau}}_{y\pi,\text{reg}}(t_2:t_1)\right] = s_r^2\left(\frac{1}{m_{12}} - \frac{1}{m_2}\right) + \frac{s_y^2}{m_1}, \quad (14.32)$$

where

$$s_r^2 = \frac{1}{m_{12} - 2} \sum_{s \in \mathcal{S}_{12}} r_s^2 \qquad (14.33)$$

and

$$s_y^2 = \frac{1}{m_1 - 1} \sum_{s \in \mathcal{S}_1} \left(\hat{\tau}_{y\pi s}(t_1) - \hat{\tau}_{y\pi,\mathrm{rep}}(t_1) \right)^2, \qquad (14.34)$$

and the regression residual is

$$r_s = \hat{\tau}_{y\pi s}(t_2) - \hat{\tau}_{y\pi,\mathrm{rep}}(t_2, m_{12}) - \hat{B}_{12} \left(\hat{\tau}_{y\pi s}(t_1) - \hat{\tau}_{y\pi,\mathrm{rep}}(t_1, m_{12}) \right). \qquad (14.35)$$

Likewise, because $V\left[\hat{\tau}_{y\pi,\mathrm{rep}}(t_2, m_{\bullet 2}) \right]$ is unknown, $w(t_2)$ in (14.21) must be estimated by

$$\hat{w}(t_2) = \frac{1}{\hat{v}\left[\hat{\tau}_{y\pi,\mathrm{rep}}(t_2, m_{\bullet 2}) \right]}, \qquad (14.36)$$

where $\hat{v}\left[\hat{\tau}_{y\pi,\mathrm{rep}}(t_2, m_{\bullet 2}) \right]$ is a suitable estimator of $V\left[\hat{\tau}_{y\pi,\mathrm{rep}}(t_2, m_{\bullet 2}) \right]$.

Scott (1984) summarizes other estimators that have been proposed of the variances and covariances that are needed in order to compute $\hat{\tau}_{y,\mathrm{SPR}}(t_2 : t_1)$ in (14.25). Citing results from Meier (1953) he proposed

$$\hat{v}\left[\hat{\tau}_{y,\mathrm{SPR}}(t_2 : t_1) \right] = \frac{1 + 4\Psi}{\hat{w}_{\mathrm{SPR}}} \qquad (14.37)$$

as an estimator of $V\left[\hat{\tau}_{y,\mathrm{SPR}}(t_2 : t_1) \right]$ in (14.24). In (14.37),

$$\Psi = \left(\frac{\hat{w}(t_2 : t_1)}{\hat{w}_{\mathrm{SPR}}} \right) \left(\frac{\hat{w}(t_2)}{\hat{w}_{\mathrm{SPR}}} \right) \left(\frac{1}{m_{12} - 1} + \frac{1}{m_{\bullet 2} - 1} \right). \qquad (14.38)$$

14.2.3 Contribution of remeasured sample units

As the time interval $\Delta_t = t_2 - t_1$ increases, the strength of the linear correlation decreases, as does the value of the linear correlation coefficient ρ_{12} in most natural systems. At some point there will be such diminished returns to maintaining a partial replacement of sampling units that SPR should be abandoned. The rate of decay of correlation between successive measurements will vary depending on the type of ecosystem being sampled. Scott (1984) reports experience with forest inventories which suggests that the correlation is sufficiently strong even after 25 years to warrant continued partial replacement of sampling units.

14.3 Estimation of change

The change in τ_y between the two occasions may be estimated in a number of reasonable ways, many of which were presented in Ware & Cunia (1962). We briefly recap a few of these alternative estimators, deferring to these authors for more elaborate details.

14.3.1 Optimal estimator

By following a similar line of development as in the preceding section, the minimum variance estimator of $\Delta_y(t_2, t_1) = \tau_y(t_2) - \tau_y(t_1)$ is

$$\hat{\Delta}_{y,\text{SPR1}}(t_2 : t_1) = \hat{\tau}_{y,\text{SPR}}(t_2 : t_1) - \hat{\tau}_{y,\text{SPR}}(t_1 : t_2) \qquad (14.39)$$

as shown in Ware (1960, pp. 29–40). In (14.39), $\hat{\tau}_{y,\text{SPR}}(t_1 : t_2)$ is the composite estimator of $\tau_y(t_1)$ which uses information from the sample at occasion t_2 to adjust, via regression, the estimate of $\tau_y(t_1)$:

$$\hat{\tau}_{y,\text{SPR}}(t_1 : t_2) = d' \left(\hat{\tau}_{y\pi,\text{rep}}(t_1, m_{12}) + B_{21} \left[\hat{\tau}_{y\pi,\text{rep}}(t_2) - \hat{\tau}_{y\pi,\text{rep}}(t_2, m_{12}) \right] \right)$$

$$+ (1 - d')\hat{\tau}_{y\pi,\text{rep}}(t_1, m_{1\bullet}), \qquad (14.40)$$

where

$$d' = \frac{m_{12}}{m_1 - u_2 m_{1\bullet}\rho_{12}^2}. \qquad (14.41)$$

While $\hat{\Delta}_{y,\text{SPR1}}(t_2 : t_1)$ is optimal in the sense that it the most precise estimator of $\Delta_y(t_2, t_1)$, it is incompatible with $\hat{\tau}_{y\pi,\text{rep}}(t_1, m_{1\bullet})$ in the following sense. The estimator $\hat{\tau}_{y\pi,\text{rep}}(t_1, m_{1\bullet})$ which utilizes all the sample information available at t_1 to estimate $\tau_y(t_1)$, when added to the optimal estimator of change, $\hat{\Delta}_{y,\text{SPR1}}(t_2 : t_1)$, does not yield the optimal estimator of $\tau_y(t_2)$. Inasmuch as $\hat{\tau}_{y\pi,\text{rep}}(t_1, m_{1\bullet})$ may have been published and publicly disseminated, it may be awkward to replace it by $\hat{\tau}_{y,\text{SPR}}(t_1 : t_2)$ at occasion t_2 when computing $\hat{\Delta}_{y,\text{SPR1}}(t_2 : t_1)$. The following estimator of $\Delta_y(t_2, t_1)$ preserves compatibility with $\hat{\tau}_{y\pi,\text{rep}}(t_1, m_{1\bullet})$.

14.3.2 Additive estimator

This estimator ensures additivity in the above sense by construction:

$$\hat{\Delta}_{y,\text{SPR2}}(t_2 : t_1) = \hat{\hat{\tau}}_{y,\text{SPR}}(t_2 : t_1) - \hat{\tau}_{y\pi,\text{rep}}(t_1, m_{1\bullet}) \qquad (14.42)$$

14.3.3 Change estimator based on remeasured sampling locations

Yet another estimator averages the changes observed at the sampling locations measured on both occasions:

$$\hat{\Delta}_{y,\text{SPR3}}(t_2 : t_1) = \hat{\tau}_{y\pi,\text{rep}}(t_2, m_{\bullet 2}) - \hat{\tau}_{y\pi,\text{rep}}(t_1)$$

$$= \frac{1}{m_{12}} \sum_{\mathscr{P}_s \in \mathscr{S}_{12}} \left(\hat{\tau}_{y\pi s}(t_2) - \hat{\tau}_{y\pi s}(t_1) \right) \qquad (14.43)$$

One would generally expect that

$$V\left[\hat{\Delta}_{y,\text{SPR1}}(t_2 : t_1) \right] < V\left[\hat{\Delta}_{y,\text{SPR2}}(t_2 : t_1) \right] < V\left[\hat{\Delta}_{y,\text{SPR3}}(t_2 : t_1) \right].$$

We suspect that $\hat{\Delta}_{y,\text{SPR2}}(t_2 : t_1)$ will always be nearly as precise as $\hat{\Delta}_{y,\text{SPR1}}(t_2 : t_1)$ —a speculation that is supported by the results of a small simulation study by Scott (1984). The degree of precision that is lost by using just the remeasured sample units in $\hat{\Delta}_{y,\mathrm{SPR3}}(t_2 : t_1)$ is highly dependent on the number of remeasured units relative to the number of unmatched units, as well as by the strength of the correlation between measurements on the two occasions. We are skeptical that anything usefully definitive can be said on the matter, other than its reduced sample support will yield less precise estimation than is possible with the optimal or additive estimators of Δ_y.

14.4 SPR with stratification

In principle it is straightforward to apply SPR separately within each stratum of a stratified population. When applied to large-scale, regional inventories of natural and environmental resources, SPR become problematic because the location and extent of areal or land strata defined by cover class change over time. If one opts to preserve the stratification by cover class that was effective at t_1, the correlation between successive measurements may be weakened greatly for those sampling locations whose cover class has changed during Δ_t. Another option is to reclassify sampling locations into strata according to t_2 strata boundaries, a tactic which would introduce bias into the estimators of strata and population totals unless the sampling allocation rule is proportional to strata areas. Scott & Köhl (1994) mention that both tactics have been used in regional forest inventories. There is the further problem that may arise when one or more strata are very small: if the number of remeasured locations within a small stratum are too few, regression estimation of $\tau_y(t_2)$ may have non-negligible bias.

We defer to Scott & Köhl (1994) for a fuller presentation of design considerations and estimation of population parameters with a stratified SPR sampling design.

14.5 SPR for three occasions

The sample at t_2 consists of m_{12} locations that are a subset of the m_1 locations measured at t_1 supplemented by $m_{\bullet 2}$ locations that 'replace' the remaining m_1 locations with new ones, hence the moniker 'partial replacement.' When there is a third sampling of the population on occasion t_3, say, there are more possible combinations of remeasured and newly-established sampling locations. One may opt to:

a) remeasure a portion of the m_{12} locations, i.e., $m_{123} \le m_{12}$;

b) remeasure a portion of the $m_{1\bullet}$ locations that were not remeasured at t_2, i.e., $m_{1\bullet 3} \le m_{1\bullet}$;

c) remeasure a portion of the $m_{\bullet 2}$ locations that were newly established at t_2, i.e., $m_{\bullet 23} \le m_{\bullet 2}$; or

d) establish $m_{\bullet\bullet 3}$ new sampling locations not measured at t_1 or t_2.

Therefore, the number of sampling units measured at t_3 is $m_3 = m_{123} + m_{1\bullet 3} + m_{\bullet 23} + m_{\bullet\bullet 3}$; we indicate this subset of m_{123} sampling locations by $\mathcal{S}_{123}$; the number measured at both t_1 and t_3 is $m_{1\frown 3} = m_{123} + m_{1\bullet 3}$; we indicate this subset of $m_{1\frown 3}$ sampling locations by $\mathcal{S}_{13}$; the number measured at both t_2 and t_3 is $m_{\frown 23} = m_{123} + m_{\bullet 23}$; we indicate this subset of $m_{2\frown 3}$ sampling locations by $\mathcal{S}_{23}$.

One may view $\hat{\tau}_{y,\text{SPR}}(t_2 : t_1)$ as a linear combination of independent estimators, each of which is weighted inversely proportional to its estimated variance. This provides a rationale by which we may extend it to utilize the three successive samples, i.e.,

$$
\begin{aligned}
&\hat{\tau}_{y,\text{SPR}}(t_3 : t_1) \\
&= \frac{\hat{w}(t_3 : t_1)\hat{\tau}_{y\pi,\text{reg}}(t_3 : t_1) + \hat{w}(t_3 : t_2)\hat{\tau}_{y\pi,\text{reg}}(t_3 : t_2) + \hat{w}(t_3)\hat{\tau}_{y\pi,\text{rep}}(t_3, m_{\bullet\bullet 3})}{\hat{w}_{\text{SPR}}},
\end{aligned}
$$

(14.44)

where $\hat{w}_{\text{SPR}}$ is the sum of the estimated weights in the numerator, viz., , $\hat{w}_{\text{SPR}} = \hat{w}(t_3 : t_1) + \hat{w}(t_3 : t_2) + \hat{w}(t_3)$. The terms in the numerator of (14.44) are detailed in the Appendix, §14.8.3.

Scott & Köhl (1994) proposed the following estimator of $V\left[\hat{\tau}_{y,\text{SPR}}(t_3 : t_1)\right]$:

$$
\hat{v}\left[\hat{\tau}_{y,\text{SPR}}(t_3 : t_1)\right] = \frac{1 + 4\Psi'}{\hat{w}_{\text{SPR}}},
$$

(14.45)

where

$$
\begin{aligned}
\Psi' &= \hat{w}(t_3 : t_1)\hat{w}(t_3 : t_2)\left(\frac{1}{m_{1\bullet 3} - 1} + \frac{1}{m_{\bullet 23} - 1}\right) \\
&\quad + \hat{w}(t_3 : t_1)\hat{w}(t_3)\left(\frac{1}{m_{1\bullet 3} - 1} + \frac{1}{m_{\bullet\bullet 3} - 1}\right) \\
&\quad + \hat{w}(t_3 : t_2)\hat{w}(t_3)\left(\frac{1}{m_{\bullet 23} - 1} + \frac{1}{m_{\bullet\bullet 3} - 1}\right).
\end{aligned}
$$

(14.46)

These authors also present estimators of change for three-occasion SPR within a stratified sampling context.

14.6 Concluding remarks

Sampling with partial replacement is undoubtedly a very clever sampling design. When applied on three occasions it begins to get a little unwieldy, and for more than three occasions the practical difficulties make it all but impossible to apply without a herculean bookkeeping effort. It has been suggested that on the fourth occasion, the measurements from the sampling locations of t_1 be dropped, and that the measurements from t_2 be treated as the initial set of measurements for the subsequent replacement panels of t_3 and t_4.

There is a fairly large literature on sampling with partial replacement that has been largely untapped by the brief treatment of the topic in this chapter. Cunia & Chevrou (1969) discuss SPR on three or more occasions; Newton et al. (1974) examined multivariate estimation following SPR sampling; Fuller (1990) has a general treatment of the topic of repeated surveys. Recent bibliographies on the topic have listed more than 80 entries, many of which can be discovered by searching on the internet.

In recent years, Poisson sampling with permanent random numbers (PRN sampling) has been applied widely to maintain a controlled overlap of the sample from one measurement occasion to the next. Treatment of PRN sampling is outside the treatment of this text, however.

14.7 Terms to remember

Composite estimator	Optimal estimation
Matched locations	Successive surveys

14.8 Appendix

14.8.1 Derivation of the composite estimator

$$\frac{\partial V\left[\hat{\tau}_{y,\text{SPR}}(t_2:t_1)\right]}{\partial a} = 2a\left(V\left[\hat{\tau}_{y\pi,\text{rep}}(t_1,m_{1\bullet})\right] + V\left[\hat{\tau}_{y\pi,\text{rep}}(t_1,m_{12})\right]\right)$$

$$- 2dC\left[\hat{\tau}_{y\pi,\text{rep}}(t_1,m_{12}),\hat{\tau}_{y\pi,\text{rep}}(t_2,m_{12})\right]$$

$$\frac{\partial V\left[\hat{\tau}_{y,\text{SPR}}(t_2:t_1)\right]}{\partial d} = -2(1-d)V\left[\hat{\tau}_{y\pi,\text{rep}}(t_2,m_{\bullet 2})\right] + 2dV\left[\hat{\tau}_{y\pi,\text{rep}}(t_2,m_{12})\right]$$

$$- 2aC\left[\hat{\tau}_{y\pi,\text{rep}}(t_1,m_{12}),\hat{\tau}_{y\pi,\text{rep}}(t_2,m_{12})\right].$$

After equating $\partial V\left[\hat{\tau}_{y,\text{SPR}}(t_2:t_1)\right]/\partial a$ and $\partial V\left[\hat{\tau}_{y,\text{SPR}}(t_2:t_1)\right]/\partial d$ each to zero, and then eliminating a, the solution for d is

$$d = \frac{V\left[\hat{\tau}_{y\pi,\text{rep}}(t_2,m_{\bullet 2})\right]}{V\left[\hat{\tau}_{y\pi,\text{rep}}(t_2,m_{\bullet 2})\right] + V\left[\hat{\tau}_{y\pi,\text{rep}}(t_2,m_{12})\right] - \Upsilon_{12}}, \tag{14.47}$$

as in (14.11).

From (7.4) and (7.5) we can express $V\left[\hat{\tau}_{y\pi,\text{rep}}(t_2,m_{\bullet 2})\right]$ as $\Phi_2/m_{\bullet 2}$, where

$$\Phi_2 = \sum_{k(2)=1}^{N_2} y_{k(2)}^2\left(\frac{1-\pi_{k(2)}}{\pi_{k(2)}}\right)$$

$$+ \sum_{k(2)=1}^{N_2}\sum_{\substack{k(2)'\neq k(2) \\ k(2)'=1}}^{N_2} y_{k(2)}y_{k(2)'}\left(\frac{\pi_{k(2)k(2)'} - \pi_{k(2)}\pi_{k(2)'}}{\pi_{k(2)}\pi_{k(2)'}}\right), \tag{14.48}$$

where $y_{k(2)}$ is the value associated with $\mathcal{U}_k$ at t_2, and $\pi_{k(2)}$ is the corresponding inclusion probability.

Likewise, $V\left[\hat{\tau}_{y\pi,\text{rep}}(t_2,m_{12})\right] = \Phi_2/m_{12}$, and $V\left[\hat{\tau}_{y\pi,\text{rep}}(t_1,m_{1\bullet})\right] = \Phi_1/m_{1\bullet}$,

where

$$\Phi_1 = \sum_{k(1)=1}^{N_1} y_{k(1)}^2 \left(\frac{1 - \pi_{k(1)}}{\pi_{k(1)}} \right)$$

$$+ \sum_{k(1)=1}^{N_1} \sum_{\substack{k(1)' \neq k(1) \\ k(1)'=1}}^{N_1} y_{k(1)} y_{k(1)'} \left(\frac{\pi_{k(1)k(2)'} - \pi_{k(1)}\pi_{k(1)'}}{\pi_{k(1)}\pi_{k(1)'}} \right). \qquad (14.49)$$

Substituting these results into the expression for d yields

$$d = \frac{m_{12}}{m_2 - u_1 m_{\bullet 2} \rho_{12}^2},$$

as presented in (14.13).

Substituting this result for d into $\partial V \left[\hat{\tau}_{y,\mathrm{SPR}}(t_2 : t_1) \right] / \partial a = 0$ gives

$$a = \frac{B_{21} m_{12} u_1}{m_2 - u_1 m_{\bullet 2} \rho_{12}^2},$$

as presented in (14.14a); evidently this is identical to $a = B_{21} m_{12} u_1 d$.

As in Chapters 7 and 8, let $\hat{\tau}_{y\pi,\mathrm{rep}}(t_1)$ denote the estimator of $\tau_y(t_1)$ which is based on all m_1 sampling locations. The identity

$$\hat{\tau}_{y\pi,\mathrm{rep}}(t_1) = \frac{m_{1\bullet} \hat{\tau}_{y\pi,\mathrm{rep}}(t_1, m_{1\bullet}) + m_{12} \hat{\tau}_{y\pi,\mathrm{rep}}(t_1, m_{12})}{m_{1\bullet} + m_{12}} \qquad (14.50a)$$

$$= u_1 \hat{\tau}_{y\pi,\mathrm{rep}}(t_1, m_{1\bullet}) + (1 - u_1) \hat{\tau}_{y\pi,\mathrm{rep}}(t_1, m_{12}) \qquad (14.50b)$$

can be arranged to yield

$$\hat{\tau}_{y\pi,\mathrm{rep}}(t_1) - \hat{\tau}_{y\pi,\mathrm{rep}}(t_1, m_{12})$$

$$= u_1 \left(\hat{\tau}_{y\pi,\mathrm{rep}}(t_1, m_{1\bullet}) - \hat{\tau}_{y\pi,\mathrm{rep}}(t_1, m_{12}) \right). \qquad (14.51)$$

Substituting this result into (14.8) yields

$$\hat{\tau}_{y,\mathrm{SPR}}(t_2 : t_1) = d \left(\hat{\tau}_{y\pi,\mathrm{rep}}(t_2, m_{12}) + B_{21} \left[\hat{\tau}_{y\pi,\mathrm{rep}}(t_1) - \hat{\tau}_{y\pi,\mathrm{rep}}(t_1, m_{12}) \right] \right)$$

$$+ (1 - d) \hat{\tau}_{y\pi,\mathrm{rep}}(t_2, m_{\bullet 2}) \qquad (14.52)$$

14.8.2 Variance of the composite estimator

From (14.23),

$$V \left[\hat{\tau}_{y,\mathrm{SPR}}(t_2 : t_1) \right]$$
$$= \frac{w(t_2 : t_1)^2 V \left[\hat{\tau}_{y\pi,\mathrm{reg}}(t_2 : t_1) \right] + w(t_2)^2 V \left[\hat{\tau}_{y\pi,\mathrm{rep}}(t_2, m_{\bullet 2}) \right]}{w_{\mathrm{SPR}}^2} \qquad (14.53)$$

and substituting from (14.2)

$$= \frac{V\left[\hat{\tau}_{y\pi,\text{reg}}(t_2:t_1)\right]}{V\left[\hat{\tau}_{y\pi,\text{reg}}(t_2:t_1)\right]^2 w_{\text{SPR}}^2} + \frac{V\left[\hat{\tau}_{y\pi,\text{rep}}(t_2,m_{\bullet2})\right]}{V\left[\hat{\tau}_{y\pi,\text{rep}}(t_2,m_{\bullet2})\right]^2 w_{\text{SPR}}^2}$$

$$= \left(\frac{1}{V\left[\hat{\tau}_{y\pi,\text{reg}}(t_2:t_1)\right]} + \frac{1}{V\left[\hat{\tau}_{y\pi,\text{rep}}(t_2,m_{\bullet2})\right]}\right) \frac{1}{w_{\text{SPR}}^2}$$

$$= \frac{1}{w_{\text{SPR}}}. \tag{14.54}$$

Using results from Ware (1960, p. 18) and Ware & Cunia (1962, p. 9, eqn. (12)), this result can be expressed equivalently in terms of the variance of the replicated sampling estimator based on m_2 sampling locations at t_2:

$$V\left[\hat{\tau}_{y,\text{SPR}}(t_2:t_1)\right] = \frac{V\left[\hat{\tau}_{y\pi,\text{rep}}(t_2)\right]\left(1 - u_1\rho_{12}^2\right)}{1 - u_1 u_2 \rho_{12}^2}, \tag{14.55}$$

where $u_2 = m_{\bullet2}/m_2$ is the proportion of unmatched, or new, sampling occasions established at t_2.

14.8.3 Components of the composite estimator $\hat{\tau}_{y,\text{SPR}}(t_3:t_1)$

The weights in the numerator of (14.44) are:

$$\hat{w}(t_3:t_1) = \frac{1}{\hat{v}\left[\hat{\tau}_{y\pi,\text{reg}}(t_3:t_1)\right]}, \tag{14.56}$$

$$\hat{w}(t_3:t_2) = \frac{1}{\hat{v}\left[\hat{\tau}_{y\pi,\text{reg}}(t_3:t_2)\right]}, \tag{14.57}$$

$$\hat{w}(t_3) = \frac{1}{\hat{v}\left[\hat{\tau}_{y\pi,\text{rep}}(t_3,m_{\bullet\bullet3})\right]}, \tag{14.58}$$

where $\hat{v}\left[\hat{\tau}_{y\pi,\text{reg}}(t_3:t_1)\right]$, $\hat{v}\left[\hat{\tau}_{y\pi,\text{reg}}(t_3:t_2)\right]$, and $\hat{v}\left[\hat{\tau}_{y\pi,\text{rep}}(t_3,m_{\bullet\bullet3})\right]$, respectively, are suitable estimators of $V\left[\hat{\tau}_{y\pi,\text{reg}}(t_3:t_1)\right]$, $V\left[\hat{\tau}_{y\pi,\text{reg}}(t_3:t_2)\right]$, and $V\left[\hat{\tau}_{y\pi,\text{rep}}(t_3,m_{\bullet\bullet3})\right]$. The estimators in the numerator of (14.44) are:

$$\hat{\tau}_{y\pi,\text{reg}}(t_3:t_1) = \hat{\tau}_{y\pi,\text{rep}}(t_3,m_{1\frown3})$$

$$+ \hat{B}_{13}\left[\hat{\tau}_{y\pi,\text{rep}}(t_1) - \hat{\tau}_{y\pi,\text{rep}}(t_1,m_{1\frown3})\right], \tag{14.59}$$

$$\hat{\tau}_{y\pi,\text{reg}}(t_3:t_2) = \hat{\tau}_{y\pi,\text{rep}}(t_3,m_{\frown23})$$

$$+ \hat{B}_{23}\left[\hat{\tau}_{y\pi,\text{rep}}(t_2) - \hat{\tau}_{y\pi,\text{rep}}(t_2,m_{\frown23})\right], \tag{14.60}$$

and

$$\hat{\tau}_{y\pi,\text{rep}}(t_3,m_{\bullet\bullet3}) = \frac{1}{m_{\bullet\bullet3}} \sum_{\mathscr{P}_s \in \mathscr{S}_{\bullet\bullet3}} \hat{\tau}_{y\pi s}(t_3), \tag{14.61}$$

where

$$\hat{B}_{13} = \frac{s_y(t_3 : t_1)}{s_y^2(t_1, m_{1 \frown 3})}, \tag{14.62}$$

$$\hat{B}_{23} = \frac{s_y(t_3 : t_2)}{s_y^2(t_2, m_{\frown 23})}, \tag{14.63}$$

$s_y(t_3 : t_1)$

$$= \frac{1}{m_{1 \frown 3} - 1} \sum_{\mathcal{P}_s \in \mathcal{S}_{13}} \left(\hat{\tau}_{y\pi s}(t_1) - \hat{\tau}_{y\pi,\text{rep}}(t_1) \right) \left(\hat{\tau}_{y\pi s}(t_3) - \hat{\tau}_{y\pi,\text{rep}}(t_3) \right), \tag{14.64}$$

$s_y(t_3 : t_2)$

$$= \frac{1}{m_{\frown 23} - 1} \sum_{\mathcal{P}_s \in \mathcal{S}_{23}} \left(\hat{\tau}_{y\pi s}(t_2) - \hat{\tau}_{y\pi,\text{rep}}(t_2) \right) \left(\hat{\tau}_{y\pi s}(t_3) - \hat{\tau}_{y\pi,\text{rep}}(t_3) \right), \tag{14.65}$$

and where

$$s_y^2(t_1, m_{1 \frown 3}) = \frac{1}{m_{1 \frown 3} - 1} \sum_{\mathcal{P}_s \in \mathcal{S}_{13}} \left(\hat{\tau}_{y\pi s}(t_1) - \hat{\tau}_{y\pi,\text{rep}}(t_1) \right)^2, \tag{14.66}$$

$$s_y^2(t_2, m_{\frown 23}) = \frac{1}{m_{\frown 23} - 1} \sum_{\mathcal{P}_s \in \mathcal{S}_{23}} \left(\hat{\tau}_{y\pi s}(t_2) - \hat{\tau}_{y\pi,\text{rep}}(t_2) \right)^2. \tag{14.67}$$

Bibliography

Affleck, D. L. R., Gregoire, T. G., & Valentine, H. T. 2005. Design unbiased estimation in line intersect sampling using segmented transects. *Environmental and Ecological Statistics*, **12(2)**, 139–154.

Affleck, D. L. R., Gregoire, T. G., & Valentine, H. T. 2006. Edge effects in line intersect sampling with segmented transects. *Journal of Agricultural, Biological, and Environmental Statistics*, **10**, 460–477.

Anonymous. 1999. *Summary Report 1997 National Resources Inventory*. Tech. rept. United States Department of Agriculture, Natural Resources Conservation Service.

Assmann, E. 1970. *The Principles of Forest Yield Study*. Oxford: Pergamon Press.

Avery, T. E., & Burkhart, H. E. 2002. *Forest Measurements*. 5 ed. Series in Forest Resources. Boston: McGraw-Hill.

Bankier, M. 1988. Power allocations: determining sample sizes for subnational areas. *The American Statistician*, **42**, 174–177.

Barabesi, L. 2003. A Monte Carlo integration approach to Horvitz-Thompson estimation in replicated environmental designs. *Metron*, **LXI**, 355–374.

Barabesi, L. 2004. Replicated environmental sampling designs and Monte Carlo integration methods: two sides of the same coin. In: *Proceedings of the XLII Meeting of the Italian Statistical Society*. Bari: Italian Statistical Society.

Barabesi, L., & Fattorini, L. 1998. The use of replicated plot, line and point sampling for estimating species abundance and ecological diversity. *Environmental and Ecological Statistics*, **5**, 353–370.

Barabesi, L., & Pisani, C. 2004. Steady-state ranked set sampling for replicated environmental sampling designs. *Environmetrics*, **15**, 45–56.

Barrett, J. P., & Nutt, M. E. 1979. *Survey Sampling in the Environmental Sciences: A Computer Approach*. Wentworth, NH: COMPress, Inc.

Bartlett, R. F. 1986. Estimating the total of a continuous population. *Journal of Statistical Planning and Inference*, **13**, 51–66.

Bauer, H. L. 1936. Moisture relations in the chaparral of the Santa Monica Mountains, California. *Ecological Monograph*, **6(3)**, 409–454.

Bebber, D. P., & Thomas, S. C. 2003. Prism sweeps for coarse woody debris. *Canadian Journal of Forest Research*, **33**, 1737–1743.

Bebbington, A. C. 1975. A simple method of drawing a sample without replacement. *Applied Statistics*, **24**, 136.

Beers, T. W. 1962. Components of forest growth. *Journal of Forestry*, **60**, 245–248.

Beers, T. W., & Gingrich, S. F. 1958. Construction of cubic-foot volume tables for red oak in Pennsylvania. *Journal of Forestry*, **56**, 210–214.

Beers, T. W., & Miller, C. I. 1964. *Point Sampling: Research Results, Theory and Application*. Agricultural Experiment Station Bulletin No. 786. Lafayette, Indiana: Purdue University.

Bell, J. F., & Alexander, L. B. 1957. *Application of the variable plot method of sampling forest stands*. Tech. rept. 30. Oregon State Board of Forestry.

Bell, J. F., Iles, K., & Marshall, D. D. 1983. Balancing the ratio of tree count-only sample points and VBAR mesaurements in variable plot sampling. Pages 699–702 of: Bell, J. F., & Atterbury, T. (eds), *Renewable resource inventories for monitoring changes and trends*. Corvallis, Oregon: College of Forestry, Oregon State University.

Bellhouse, D. R. 1985. Computing methods for variance estimation in complex surveys. *Journal of Official Statistics*, **1(3)**, 323–329.

Bellhouse, D. R. 1988. A brief history of random sampling methods. Pages 1–14 of: Patil, G. P., & Rao, C. R. (eds), *Handbook of Statistics, Volume 6*. Amsterdam: North Holland.

Bethel, J. 1989. Sample allocation in multivariate surveys. *Survey Methodology*, **15**, 47–57.

Bethlehem, J. G., & Keller, W. J. 1987. Linear weighting of sample survey data. *Journal of Official Statistics*, **3**, 141–153.

Bissell, A. F. 1986. Ordered random selection without replacement. *Applied Statistics*, **35**, 73–75.

Bitterlich, W. 1949. Die Winkelzählprobe. *Allgemeine Forst- und Holzwirtschaftliche Zeitung*, **59(1/2)**, 4–5.

Bitterlich, W. 1984. *The Relascope Idea: Relative Measurements in Forestry*. Slough, England: Commonwealth Agricultural Bureaux.

Bliss, C. I. 1941. Statistical problems in estimating populations of Japanese beetle larvae. *Journal of Economic Entomology*, **34**, 221–232.

Bormann, F. H. 1953. The statistical efficiency of sample plot size and shape in forest ecology. *Ecology*, **34(3)**, 474–487.

Brewer, K. R. W. 2002. *Combined Survey Sampling Inference: Weighing Basu's Elephants*. London: Arnold.

Brewer, K. R. W., & Gregoire, T. G. 2000. Estimators for use with Poisson sampling and related selection procedures. Pages 279–288 of: *Proceedings of the Second International Conference on Establishment Surveys*. Alexandria, Virginia: American Statistical Association.

Brewer, K. R. W., & Hanif, M. 1982. *Sampling With Unequal Probabilities*. New York: Springer-Verlag.

Brooks, J. R. 2006. An evaluation of big basal area factor sampling in Appalachian hardwoods. *Northern Journal of Applied Forestry*, **23(1)**, 62–65.

Brown, J. A. 2003. Designing an efficient adaptive cluster sample. *Environmental and Ecological Statistics*, **10**, 95–105.

Brown, J. K. 1974. *Handbook for inventorying downed woody material*. Tech. rept. INT-16. USDA Forest Service Intermountain Forest and Range Experiment Station.

Bruce, D. 1961. *Prism cruising in the western United States and volume tables for use therewith*. Portland, Oregon: Mason, Bruce & Girard, Inc.

Cancino, J., & Saborowski, J. 2005. Comparison of randomized branch sampling with and without replacement at the first stage. *Silva Fennica*, **39**, 201–216.

Canfield, R. H. 1941. Application of the line interception method in sampling range vegetation. *Journal of Forestry*, **39**, 34–40.

Cassel, C.-M., Särndal, C.-E., & Wretman, J. H. 1977. *Foundations of Inference in Survey Sampling*. New York: Wiley.

Chaudhuri, A., & Stenger, H. 1992. *Survey Sampling: Theory and Methods*. New York: Marcel Dekker.

Chaudhuri, A., & Vos, J. W. E. 1988. *Unified Theory and Strategies of Survey Sampling*. Amsterdam: North Holland.

Chen, Z., Bai, Z., & Sinha, B. K. 2003. *Ranked Set Sampling: Theory and Applications*. Lecture Notes in Statistics, 176. New York: Springer-Verlag.

Christman, M. C. 2000. A review of quadrat-based sampling of rare, geographically clustered populations. *Journal of Agricultural, Biological, and Environmental Statistics*, **5**, 168–201.

Clapham, A. R. 1932. The form of the observational unit in quantitative ecology. *Journal of Ecology*, **20**, 192–197.

Clark, III, A., Souter, R. A., & Schlaegel, B. E. 1991. *Stem profile equations for southern tree species*. Research Paper SE-282. Washington, D.C.: USDA Forest Service.

Clark, S. J., & Perry, J. N. 1994. Small sample estimation of Taylor's power law. *Environmental and Ecological Statistics*, **1**, 287–302.

Clements, F. E. 1905. *Research Methods in Ecology*. Lincoln, Nebraska (U.S.A): The University Publishing Company.

Cobby, M. L., Ridout, M. S., Bassett, P. J., & Large, R. V. 1985. An investigation into the use of ranked set sampling on grass and grass-clover swards. *Grass and Forage Science*, **40**, 257–263.

Cochran, W. G. 1953. *Sampling Techniques*. New York: Wiley.

Cochran, W. G. 1977. *Sampling Techniques*. New York: Wiley.

Cochran, W. G., Mosteller, F., & Tukey, J. W. 1954. Principles of sampling. *Journal of the American Statistical Association*, **49**, 13–35.

Cordy, C. 1993. An extension to the Horvitz-Thompson theorem to point sampling from a continuous population. *Probability and Statistics Letters*, **18**, 353–362.

Cunia, T. 1979. *Basic Designs for Survey Sampling: Simple, Stratified, Cluster, and Systematic Sampling*. Forest Biometry Monograph Series 3. College of Environmental Science and Forestry, State University of New York, Syracuse, New York, USA.

Cunia, T., & Briggs, R. D. 1984. Forcing additivity of biomass tables: some empirical results. *Canadian Journal of Forest Research*, **14**, 376–384.

Cunia, T., & Chevrou, R. B. 1969. Sampling with partial replacement on three or more occsaions. *Forest Science*, **15**, 204–224.

Curtis, R. O., & Marshall, D. D. 2005. *Permanent-plot procedures for silvicultural research*. General Technical Report, PNW-634. Washington, D.C.: USDA Forest Service.

Dalenius, T. 1957. *Sampling in Sweden: Contributions to the Methods and Theories of Sample Survey Practice*. Stockholm: Almquist and Wiskell.

Dalenius, T., Hájek, J., & Zubrzycki, S. 1961. On plane sampling and related geometrical problems. Pages 125–150 of: *Proceedings of the 4th Berkeley Symposium on Probability and Mathematical Statistics, Vol. 1*. Berkeley: University of California Press.

Daubenmire, R. 1959. A canopy-coverage method of vegetational analysis. *Northwest Science*, **33(1)**, 43–64.

Davison, A. C., & Hinkley, D. V. 1997. *Bootstrap Methods and Their Application*. Cambridge: Cambridge University Press.

De Gier, A. 1989. *Woody Biomass for Fuel: Estimating the Supply in Natural Woodlands and Shrublands*. ITC Publication Number 9. Enschede, The Netherlands: International Institute for Aerospace Survey and Earth Sciences (ITC).

de Gruijter, & ter Braak, C. J. F. 1990. Model-free estimation from spatial samples: a reappraisal of classical sampling theory. *Mathematical Geology*, **22**, 407–415.

de Vries, P. G. 1973. *A general theory on line intersect sampling, with application to logging residue inventory*. Tech. rept. 73-11. Mededelingen Landbouwhogeschool, Wageningen, The Netherlands.

de Vries, P. G. 1986. *Sampling Theory for Forest Inventory*. Berlin: Springer-Verlag.

Delisle, G. P., Woodard, P. M., Titus, S. J., & Johnson, A. F. 1988. Sample size and variability of fuel weight estiamtes in natural stands of lodgepole pine. *Canadian Journal of Forest Research*, **18**, 649 – 652.

Dell, T. R., & Clutter, J. L. 1972. Ranked set sampling theory with order statistics background. *Biometrics*, **28**, 545–555.

Deming, W. E. 1950. *Some Theory of Sampling*. New York: Wiley.

Deming, W. E., & Stephan, F. F. 1940. On a least squares adjustment of a sample frequency tables when the expected marginal totals are known. *Annals of Mathematical Statistics*, **11**, 427–444.

Dixon, P. M. 2002. Bootstrap resampling. Pages 212–220 of: *Encyclopedia of Environmetrics, Volume 1*. Chichester: Wiley.

Ducey, M. J., Jordan, G. J., Gove, J. H., & Valentine, H. T. 2002. A practical modfication of horizontal line sampling for snag and cavity tree inventory. *Canadian Journal of Forest Research*, **32**, 1217–1224.

Ducey, M. J., Gove, J. H., & Valentine, H. T. 2004. A walk-through solution to the boundary overlap problem. *Forest Science*, **50**, 427–435.

Ducey, M. J., Williams, M. S., Gove, J. H., & Valentine, H. T. 2007. Simultaneous unbiased estimates of multiple downed wood attributes in perpendicular distance sampling. *Submitted to: Canadian Journal of Forest Research*.

Duncan, G. J., & Kalton, G. 1987. Issues of design and analysis of surveys across time. *International Statistical Review*, **55**, 97–117.

Efron, B., & Tibshirani, R. 1986. *An Introduction to the Bootstrap*. New York: Chapman and Hall.

Eriksson, M. 1995a. Compatible and time-additive change component estimators for horizontal-point-sampled data. *Forest Science*, **41**, 796–822.

Eriksson, M. 1995b. Design-based approaches to horizontal-point sampling. *Forest Science*, **41**, 890–907.

Evans, M., & Swartz, T. 2000. *Approximating integrals via Monte Carlo and Deterministic Methods*. Oxford: Oxford University Press.

FAO. 1990. *Forest Resources Assessment 1990*. FAO Forestry paper 124. Rome: United Nations Food and Agriculture Organization.

FHM. 1998. *Forest Health Monitoring Methods Guide*. Research Triangle Park, North Carolina: USDA Forest Service, National Forest Health Monitoring Program.

Flewelling, J., & Iles, K. 2004. Area-independent sampling for total basal area. *Forest Science*, **50(4)**, 512–517.

Fowler, G. W., & Hauke, D. 1979. *A distribution-free method for interval estimation and sample size determination*. Resource Inventory Notes BLM-19. Denver, Colorado: Bureau of Land Management, U. S. Department of Interior.

Frayer, W. E. 1978. *Stratification in double sampling – The easy way out may sometimes be the best way*. Resource Inventory Notes BLM-10. Denver, Colorado: Bureau of Land Management, U. S. Department of Interior.

Freese, F. 1961. Relation of plot size to variability. *Journal of Forestry*, **59**, 679.

Fuller, W. A. 1990. Analysis of repeated surveys. *Statistical Methodology*, **16(2)**, 167–180.

Fuller, W. A. 1995. Estimation in the presence of measurement error. *International Statistical Review*, **63**, 121–147.

Fuller, W. A. 1999. Environmental surveys over time. *Journal of Agricultural, Biological, and Environmental Statistics*, **4(4)**, 331–345.

Fuller, W. A. 2002. Regression estimation for survey samples. *Survey Methodology*, **28**, 5–23.

Furness, G. O. 1976. The dispersal, age-structure and natural enemies of the long-tailed mealy bug, *Pseudococcus longispinus* (Targoni-Tozzetti) in relation to sampling and control. *Australian Journal of Zoology*, **24**, 237–247.

Furnival, G. M., Gregoire, T. G., & Grosenbaugh, L. R. 1987. Adjusted inclusion probabilities with 3P sampling. *Forest Science*, **33(3)**, 617–631.

Gaffrey, D., & Saborowski, J. 1999. RBS, ein mehrstufiges Inventurverfahren zur Schätzung von Baummerkmalen I. Schätzung von Nadel- und Asttrockenmassen bei 66-jährigen Douglasien. *Allg. Forst. u.J.-Ztg.*, **170**, 177–183.

Gilbert, R. O., & Eberhardt, L. L. 1976. An evaluation of double sampling for estimating plutonium inventory in surface soil. Pages 157–163 of: Cushing, C. E. et. al. (ed), *Radioecology and energy resources : proceedings of the Fourth National Symposium on Radioecology.* The Ecological Society of America, Special Publication No. 1. Stroudsburg, Pennsylvania: Dowden, Hutchinson and Ross, Inc.

Godambe, V. P. 1955. A unified theoy of sampling from finite populations. *Journal of the Royal Statistical Society, Series B*, **17**, 269–278.

Good, N. M., Paterson, M., Brack, C., & Mengersen, K. 2001. Estimating tree component biomass using variable probability sampling methods. *Journal of Agricultural, Biological and Environmental Statistics*, **6**, 258–267.

Goodman, L. A. 1949. On the estimation of the number of classes in a population. *Annals of Mathematical Statistics*, **20**, 572–579.

Goodman, L. A. 1960. On the exact variance of products. *Journal of the American Statistical Association*, **55**, 708–713.

Gove, J. H., Ducey, M. J., Ståhl, G., & Ringvall, A. 1999a. Point relascope sampling: a new way to assess downed coarse woody debris. *Journal of Forestry*, **99**, 4–11.

Gove, J. H., Ringvall, A., Ståhl, G., & Ducey, M. J. 1999b. Point relascope sampling for downed coarse woody debris. *Canadian Journal of Forest Research*, **29**, 1718–1726.

Gray, H. R. 1943. Volume measurement of forest crops. *Australian Forestry*, **7**, 48–74.

Graybill, F. A., & Deal, R. B. 1959. Combining unbiased estimators. *Biometrics*, **15**, 543–550.

Greenhill, A. H. 1881. Determination of the greatest height consistent with stability that a vertical pole or mast can be made, and the greatest height to which a tree of given proportions can grow. *Proceedings of the Cambridge Philosophical Society*, **4**, 65–73.

Gregoire, T. G. 1982. The unbiasedness of the mirage correction procedure for boundary overlap. *Forest Science*, **28(3)**, 504–508.

Gregoire, T. G. 1984. The jackknife: an introduction with applications in forestry data analysis. *Canadian Journal of Forest Research*, **14**, 493–497.

Gregoire, T. G. 1992. Analytical derivation of the variance of the modified point-list sampling estimator. Pages 404–415 of: Wood, G., & Turner, B. (eds), *Integrating Forest Information Over Space and Time*. Proceedings of IUFROM S4.01 & S4.02 conference. Canberra, Australia: ANUTECH Pty. Ltd.

Gregoire, T. G. 1993. Estimation of growth from successive surveys. *Forest Ecology and Management*, **56**, 267–278.

Gregoire, T. G. 1998. Design-based and model-based inference in survey sampling: appreciating the difference. *Canadian Journal of Forest Research*, **28**, 1429–1447.

Gregoire, T. G., & Monkevich, N. S. 1994. The reflection method of line intercept sampling to eliminate boundary bias. *Environmental and Ecological Statistics*, **1**, 219–226.

Gregoire, T. G., & Schabenberger, O. S. 1999. Sampling skewed biological populaitons: behavior of confidence intervals for the population total. *Ecology*, **80**, 1056–1065.

Gregoire, T. G., & Scott, C. T. 1990. Sampling at the stand boundary: a comparison of the statistical performance among eight methods. Pages 78–85 of: Burkhart, H. E., Bonnor, G. M., & Lowe, J. J. (eds), *Research in Forest Inventory, Monitoring, Growth and Yield*. Publ. FWS-3-90. School of Forestry and Wildlife Resources, Virginia Polytechnic Institute and State University, Blacksburg, Virginia (USA).

Gregoire, T. G., & Scott, C. T. 2003. Altered selection probabilities caused by avoiding the edge in field surveys. *Journal of Agricultural, Biological and Environmental Statistics*, **8**, 36–47.

Gregoire, T. G., & Valentine, H. T. 1999. Composite and calibration estimation following 3P sampling. *Forest Science*, **45(2)**, 179–185.

Gregoire, T. G., & Valentine, H. T. 2003. Line intersect sampling: ell-shaped transects and multiple intersections. *Environmental and Ecological Statistics*, **10**, 263–279.

Gregoire, T. G., Valentine, H. T., & Furnival, G. M. 1986. Estimation of bole volume by importance sampling. *Canadian Journal of Forest Research*, **16**, 554–557.

Gregoire, T. G., Valentine, H. T., & Furnival, G. M. 1993. Two-stage and three-stage sampling strategies to estimate aggregate bole volume in the forest. Pages 201–211 of: Nyyssonen, A., Poso, S., & Rautala, J. (eds), *Ilvessalo Symposium on National Forest Inventories*. Research Paper 444. Helsinki, Finland: The Finnish Forest Research Institute.

Gregoire, T. G., Valentine, H. T., & Furnival, G. M. 1995. Sampling methods to estimate foliage and other characteristics of individual trees. *Ecology*, **76**, 1181–1194.

Grosenbaugh, L. R. 1952. Plotless, timber estimates – new, fast, easy. *Journal of Forestry*, **50(1)**, 32–37.

Grosenbaugh, L. R. 1958. *Point-sampling and line-intercept sampling: probability theory, geometric implications, synthesis*. Southern Forest Experiment Station Occasional Paper SO-160. Washington, D.C.: USDA Forest Service.

Grosenbaugh, L. R. 1964. Some suggestions for better sample-tree measurement. Pages 36–42 of: anon. (ed), *Proceedings, Society of American Foresters*. Boston, Massachusetts: Society of American Foresters.

Grosenbaugh, L. R. 1965. *THREE-PEE SAMPLING THEORY and program 'THRP' for computer generation of selection criteria*. Research Paper, PSW-21. Berkeley, California: USDA Forest Service.

Grosenbaugh, L. R. 1971. *STX 1-11-71 for Dendrometry of Multistage 3P Samples*. USDA Forest Service.

Grosenbaugh, L. R. 1976. Approximate sampling variance of adjusted 3P estimates. *Forest Science*, **22**, 173–176.

Gupta, P. C. 1972. A Monte Carlo integration approach to Horvitz-Thompson estimation in replicated environmental designs. *Australian Journal of Statistics*, **14**, 123–128.

Hájek, J. 1958. Some contributions to the theory of probability sampling. *Bulletin of the International Statistical Institute*, **36**, 127–134.

Hájek, J. 1964. Asymptotic theory of rejective sampling with varying probabilities from a finite population. *Annals of Mathematical Statistics*, **35(4)**, 1491–1523.

Hájek, J. 1981. *Sampling from a finite population*. New York: Marcel Dekker.

Halls, L. S., & Dell, T. R. 1966. Trial of ranked set sampling for forage yields. *Forest Science*, **12**, 22–26.

Hammersley, J. M., & Handscomb, D. C. 1979. *Monte Carlo Methods*. London: Chapman and Hall.

Hammersley, J. M., & Morton, K. W. 1956. A new Monte Carlo technique: antithetic variates. *Proceedings of the Cambridge Philosophical Society*, **52**, 449–475.

Hansen, M. H., & Hurwitz, W. N. 1943. On the theory of sampling from finite populations. *Annals of Mathematical Statistics*, **14**, 333–362.

Hansen, M. H., Hurwitz, W. N., & Madow, W. G. 1953. *Sample Survey Methods and Theory, Volume 1*. New York: Wiley.

Hansen, M. R., Dalenius, T., & Tepping, B. J. 1985. The development of sample surveys of finite populations. Pages 327–354 of: Atkinson, A. C., & Feinberg, S. E. (eds), *A Celebration of Statistics*. New York: Springer-Verlag.

Hanson, H. C. 1934. A comparison of methods of botanical analysis of the native prairie in western North Dakota. *Journal of Agricultural Research*, **49(9)**, 815–842.

Hartley, H. O., & Rao, J. N. K. 1962. Sampling with unequal probabilities and without replacment. *Annals of Mathematical Statistics*, **33**, 350–374.

Holt, D., & Smith, T. M. F. 1979. Post stratification. *Journal of the Royal Statistical Society, Series A*, **142**, 33–46.

Horvitz, D. G., & Thompson, D. J. 1952. A generalization of sampling without replacement from a finite universe. *Journal of the American Statistical Association*, **47**, 663–685.

Howard, J. O., & Ward, F. R. 1972. *Measurement of Logging Residue – Alternative Applications of the Line Intersect Method.* Research Note PNW-183. Washington, D.C.: USDA Forest Service.

Iles, K. 1989. Subsampling for VBAR. *Inventory Cruising News,* **8**, 2–3.

Iles, K. 2003. *A Sampler of Inventory Topics.* Nanaimo, BC, Canada: Kim Iles & Associates.

Jenkins, J. C., Chojnacky, D. C., Heath, L. S., & Birdsey, R. A. 2004. *Comprehensive database of diameter-based biomass regressions for North American tree species.* General Technical Report, NE-319. Washington, D.C.: USDA Forest Service.

Jessen, R. J. 1942. *Statistical Investigation of a sample survey for obtaining farm facts.* Research Bulletin 304. Ames, Iowa: Iowa Agricultural Experiment Station.

Jessen, R. J. 1955. Determining the fruit count on a tree by randomized branch sampling. *Biometrics,* **11**, 99–109.

Jessen, R. J. 1978. *Statistical Survey Techniques.* New York: Wiley.

Johnson, F. A. 1961. *Standard error of estimated average timber volume per acre under point sampling when trees are measured for volume on a subsample of all points.* Research Note, PNW-201. Portland, Oregon: USDA Forest Service.

Johnson, F. A., & Hixon, H. J. 1952. The most efficient size and shape plot to use for cruising in old-growth Douglas-fir timber. *Journal of Forestry,* **50(1)**, 17–20.

Jones, H. L. 1958. Inadmissable samples and confidence limits. *Journal of the American Statistical Association,* **53**, 482–490.

Jönrup, H., & Rennermalm, B. 1976. Regression analysis in samples from finite populations. *Scandinavian Journal of Statistics,* **3**, 33–36.

Kaiser, L. 1983. Unbiased estimation in line-intercept sampling. *Biometrics,* **39**, 965–976.

Kaur, A., Patil, G. P., Sinha, A. K., & Tallie, C. 1995. Ranked set sampling: an annotated bibliography. *Environmental and Ecological Statistics,* **2**, 25–54.

Koop, J. C. 1990. Systematic sampling of two-dimensional surfaces and related problems. *Communications in Statistics–Theory and Methods,* **19**, 1701–1759.

Kraft, K. M., Johnson, D. H., Samuleson, J. M., & Allen, S. H. 1995. Using known populations of pronghorn to evaluate sampling plans and estimators. *Journal of Wildlife Management,* **59**, 129–137.

Kruskal, W. H., & Mosteller, F. 1979a. Representative sampling, I. Nonscientific literature. *International Statistical Review,* **47**, 13–24.

Kruskal, W. H., & Mosteller, F. 1979b. Representative sampling, II. Scientific literature, excluding statistics. *International Statistical Review,* **47**, 111–127.

Kruskal, W. H., & Mosteller, F. 1979c. Representative sampling, III. The current statistical literature. *International Statistical Review,* **47**, 245–265.

Kulow, D. R. 1966. Comparison of forest sampling designs. *Journal of Forestry,* **64**, 469–474.

448

Lahiri, D. B. 1951. A method of sample selection providing unbiased ratio estimates. *Bulletin the International Statistcal Institute*, **33**, 133–140.

Lazzerini, M. 1901. Un applicazione del calculo della probabilita alla ricerca di un valore approsimato di π. *Periodico di Matematiche*, **4**, 140–143.

Leemis, L. M., & Trivedi, K. S. 1996. A comparison of approximate interval estimators for the Bernoulli parameter. *The Amercian Statistician*, **50**, 63–68.

Lohr, S. L. 1999. *Sampling: Design and Analysis*. Pacific Grove, California: Duxbury Press.

Lucas, H. A., & Seber, G. A. F. 1977. Estimating coverage and particle density using the line intercept method. *Biometrika*, **64**, 618–622.

MacEachern, S. N., Öztürk, Ö., & Wolfe, D. A. 2002. A new ranked set sample estimator of variance. *Journal of the Royal Statistical Society, B*, **64, Part 2**, 177–188.

MacLean, C. D. 1972a. Improving inventory volume estimates by double sampling on aerial photographs. *Journal of Forestry*, **70**, 748–749.

MacLean, C. D. 1972b. *Photo stratification improves Northwest timber volume estimates*. Research Paper, PNW-150. Washington, D.C.: USDA Forest Service.

Magnussen, S., Kurz, W., & Leckie, D. P. 2005. Adaptive cluster sampling for estimation of deforestation rates. *European Journal of Forest Research*, **124**, 207–220.

Mahalanobis, P. C. 1946. Recent experiments in statistical sampling in the Indian Statistical Institute. *Journal of the Royal Statistical Society, A*, **109**, 326–378.

Mandallaz, D. 1991. *A Unified Approach to Sampling Theory for Forest Inventory Based on Infinite Population and Superpopulation Models*. Zurich: Chair of Forestry Inventory and Planning, Swiss Federal Institute of Technology (ETH).

Marshall, A. W. 1956. The use of multi-stage sampling schemes in Monte Carlo computations. Pages 123–140 of: Meyer, M. A. (ed), *Symposium on Monte Carlo Methods*. New York: Wiley.

Marshall, D. D., Iles, K., & Bell, J. F. 2004. Using a large-angle gauge to select trees for measurement in variable plot sampling. *Canadian Journal of Forest Research*, **34**, 840–845.

Martin, G. L. 1982. A method for estimating ingrowth on permanent horizontal sample points. *Forest Science*, **28**, 110–114.

Martin, W. L., Sharik, T. L., Oderwald, R. G., & Smith, D. W. 1980. *Evaluation of Ranked Set Sampling for Estimating Shrub Phytomass in Appalachian Oak Forests*. Publication No. FWS-4-80. Blacksburg: School of Forestry and Wildlife Resources, Virginia Polytechnic Institute and State University.

Masuyama, M. 1954. On the error in crop cutting experiment due to the bias on the border of the grid. *Sankhyā*, **14**(3), 181–186.

Matérn, B. 1956. On the geometry of the cross-section of a stem. *Meddelanden Från Statens Skogsforskningsinstitut*, **Band 46 Nr(11)**, 1–28.

Matérn, B. 1964. A method of determining the total length of roads by means of a line survey. *Studia Forestalia Suecica*, **18**, 68–70.

Matney, T. G., & Parker, R. C. 1991. Stand and stock tables from double-point samples. *Forest Science*, **37**, 1605–1613.

McIntyre, G. A. 1952. A method for unbiased selective sampling, using ranked sets. *Australian Journal of Agricultural Research*, **3**, 385–390.

McLeod, A. I., & Bellhouse, D. R. 1983. A convenient algorithm for drawing a simple random sample. *Applied Statistics*, **32**, 182–184.

Meier, P. 1953. Variance of a weighted mean. *Biometrics*, **9**, 59–73.

Mendenhall, W., & Schaeffer, R. L. 1973. *Mathematical Statistics with Applications*. North Scituate, Massachusetts: Duxbury Press.

Mesavage, C., & Grosenbaugh, L. R. 1956. Efficiency of several cruising designs on small tracts in north Arkansas. *Journal of Forestry*, **54(9)**, 569–576.

Meyer, H. A. 1948. Cruising by narrow strips. *Pennsylvania State Forestry School Research Paper 12*.

Meyer, H. A. 1958. The calculation of the sampling error of a cruise from the mean square successive difference. *Journal of Forestry*, **54**, 341.

Monkevich, N. S. 1994. *Assessing petrified wood changes in Petrifed Wood National Park*. M.Phil. thesis, Virginia Polytechnic Institute and State University.

Murray, C. D. 1927. A relationship between circumference and weight in trees and its bearing on branching angles. *Journal of General Physiology*, **10**, 725–739.

Murthy, M. N. 1964. Product method of estimation. *Sankhyā, Series A*, **26**, 69–74.

Murthy, M. N. 1967. *Sampling: Theory and Methods*. Calcutta: Statistical Publishing Association.

Muttlak, H. A., & Sadooghi-Alvandi, S. M. 1993. A note on the line intercept sampling method. *Biometrics*, **49**, 1209–1215.

Nelson, R., & Gregoire, T. G. 1994. Two-stage forest sampling: a comparison of three procedures to estimate aggregate volume. *Forest Science*, **40(2)**, 247–266.

Nelson, R., Short, A., & Valenti, M. 2004. Measuring biomass and carbon in Delaware using an airborne profiling LIDAR. *Scandinavian Journal of Forest Research*, **19(6)**, 500–511.

Newman, E. I. 1966. A method of estimating the total length of root in a sample. *Journal of Applied Ecology*, **3(1)**, 139–145.

Newton, C. M., Cunia, T., & Bickford, C. A. 1974. Multivariate estimators for sampling with partial replacement on two occsaions. *Forest Science*, **20**, 106–116.

Neyman, J. 1934. On the two different aspects of the representative method: the method of stratified sampling and the method of purposive selection. *Journal of the Royal Statistical Society*, **97**, 558–606.

Neyman, J. 1938. Contribution to the theory of sampling human populations. *Journal of the American Statistical Association*, **33**, 101–116.

Nusser, S. M., & Goebel, J. J. 1997. The National Resources Inventory: a long-term multi-resource monitoring programme. *Environmental and Ecological Statistics*, **4**, 181–204.

Nusser, S. M., Breidt, F. J., & Fuller, W. A. 1998. Design and estimation for investigating the dynamics of natural resources. *Ecological Applications*, **8**, 234–245.

Oderwald, R. G. 1993. Stratification gains in cruising: proportional allocation. *Southern Journal of Applied Forestry*, **17**, 96–99.

Oderwald, R. G. 1994. Stock and stand tables for point, double sampling with a ratio of means estimator. *Canadian Journal of Forest Research*, **24**, 2350–2352.

Oderwald, R. G., & Jones, E. 1992. Sample sizes for point, double sampling. *Canadian Journal of Forest Research*, **22**, 980–983.

Olsen, A. R., Sedransk, J., Edwards, D., Gotway, C. A., Liggett, W., Rathbun, S., Reckhow, K. H., & Young, L. J. 1999. Statistical issues for monitoring ecological and natural resources in the United States. *Environmental Monitoring and Assessment*, **54(1)**, 1–45.

Osborne, J. G. 1942. Sampling errors of systematic and random surveys of cover-type areas. *Journal of the American Statistical Association*, **37**, 256–264.

Overton, W. S., & Stehman, S. V. 1993. Properties of designs for sampling continuous spatial resources from a triangular grid. *Communications in Statistics–Theory and Methods*, **22**, 2641–2660.

Palley, M. N., & Horwitz, L. G. 1961. Properties of some random and systematic point sampling estimators. *Forest Science*, **7**, 52–65.

Patil, G. P., Sinha, A. K., & Tallie, C. 1994. Ranked set sampling. Pages 167–200 of: Patil, G. P., & Rao, C. R. (eds), *Handbook of Statistics, Volume 12: Environmental Statistics*. Amsterdam: North Holland.

Patterson, H. D. 1950. Sampling on successive occasions with partial replacement of units. *Journal of the Royal Statistical Society, B*, **12(2)**, 241–255.

Penner, M., & Otukol, S. 1998. Boundary correction for variable radius plots—simulation results. Pages 148–157 of: Hansen, M., & Burk, T. (eds), *Inetgrated Tools for Natural Resources Inventories in the 21st Century*. St. Paul, Minnesota: USDA Forest Service.

Perlman, M. D., & Wichura, M. J.. 1975. Sharpening Buffon's Needle. *The American Statistician*, **29(4)**, 157–163.

Pfeffermann, D., & Krieger, A. M. 1991. Poststratification using regression estimators when information on strata means and sizes is missing. *Biometrika*, **78**, 409–419.

Phillipi, T. 2005. Adaptive cluster sampling for estimation of abundances within local populations of low-abundance plants. *Ecology*, **86**, 1091–1100.

Pinkham, R. S. 1987. An efficient algorithm for drawing a simple random sample. *Applied Statistics*, **36**, 370–372.

Ponce-Hernandez, R. 2004. *Assessing carbon stocks and modelling win-win scenarios of carbon sequestation through land-use changes.* Rome: Food and Agriculture Organization, United Nations.

Pontius, J. S. 1996. Forest spatial surveys using the Rao-Hartley-Cochran sampling design. In: *Spatial Accuracy Assessment in Natural Resources and Environmental Sciences: Second International Symposium.* General Technical Report RM-GTR-277. Washington, D.C.: USDA Forest Service.

Pontius, J. S. 1998. Estimation of the mean in line intercept sampling. *Environmental and ecological statistics,* **5**, 371–379.

Quenouille, M. H. 1949. Problems in plane sampling. *Annals of Mathematical Statistics,* **20**, 355–375.

Quenouille, M. H. 1956. Notes on bias in estimation. *Biometrika,* **43**, 353–360.

Radtke, P. J., & Bolstad, P. V. 2001. Laser point-quadrat sampling for estimating foliage-height profiles in broad-leaved forests. *Canadian Journal of Forest Research,* **31**, 410–418.

Raj, D. 1968. *Sampling Theory.* New York: McGraw-Hill.

Ramaley, J. F. 1969. Buffon's noodle problem. *The American Mathematical Monthly,* **76(8)**, 916–918.

Rao, J. N. K. 2003. *Small Area Estimation.* Hoboken, New Jersey: Wiley-Interscience.

Rao, J. N. K., & Ramachandran, V. 1974. Comparison of the separate and combined ratio estimators. *Sankhyā,* **36**, 151–156.

Rao, J. N. K., Hartley, H. O., & Cochran, W. G. 1962. On a simple procedure of unequal probability sampling without replacement. *Journal of the Royal Statistical Society, Series B,* **24**, 482–491.

Rao, P. S. R. S. 1988. Ratio and regression estimators. Pages 449–468 of: Krishnaiah, P. R., & Rao, C. R. (eds), *Handbook of Statistics 6: Sampling.* Amsterdam: North Holland.

Raulier, F., Bernier, P. Y., Ung, C.-H., & Boutin, R. 2002. Structural differences and functional similarities between two sugar maple (*Acer saccharum*) stands. *Tree Physiology,* **22**, 1147–1156.

Rennolls, K. 1989. *Design of the Census of Woodlands and Trees 1979-82.* Edinburgh: Occasional Paper 18, Forestry Commission.

Ringvall, A., & Kruys, N. 2005. Sampling of sparse species with probability proportional to prediction. *Environmental Monitoring and Assessment,* **104**, 131–146.

Risebrough, R. W. 1972. Effects of pollutants upon animals other than man. In: *Proceedings of the 6th Berkeley Symposium on Mathematics and Statistics, VI.* Berkeley: University of California Press.

Robson, D. S. 1957. Applications of multivariate polykays to the theory of unbiased ratio-type estimation. *Journal of the American Statistical Association,* **52**, 511–522.

Roesch, Jr., F. A., Green, E. J., & Scott, C. T. 1989. New compatible estimators for survivor growth and ingrowth from remeasured horizontal point samples. *Forest Science*, **35**, 281–293.

Roesch, F. A., Jr. 1993. Adaptive cluster sampling for forest inventories. *Forest Science*, **39**, 655–669.

Rosiwal, A. 1898. Ueber geometrische Gesteinsaualysen. Ein einfacher Weg zur ziffremassigen Foxtstellung des Quantitätsverhäitnissos der Mineralbestandtheile gemengter Geneine. *Verhandlungen*, **5(6)**, 143–174.

Rubinstein, R. Y. 1981. *Simulation and the Monte Carlo Method*. New York: Wiley.

Rudis, V. A. 1991. *Wildlife Habitat, Range, Recreation, Hydrology, and Related Research Using Forest Inventory and Analysis Surveys: A 12-Year Compendium*. General Technical Report SO-84. Washington, D.C.: USDA Forest Service.

Särndal, C.-E. 1978. Design-based and model-based inference in survey sampling. *Scandinavian Journal of Statistics*, **5**, 27–52.

Särndal, C.-E. 1982. Implications of survey design for generalized regression estimation of linear functions. *Journal of Statistical Planning and Inference*, **7**, 155–170.

Särndal, C.-E. 1996. Efficient estimators with simple variance in unequal probability sampling. *Journal of the American Statistical Association*, **91**, 1289–1300.

Särndal, C.-E., Swensson, B., & Wretman, J. 1992. *Model Assisted Survey Sampling*. New York: Springer-Verlag.

Schabenberger, O. S., & Gregoire, T. G. 1994. Competitors to genuine πps designs: a comparison. *Survey Methodology*, **20**, 185–192.

Schmid, V. P. 1969. Stichproben am waldtrand (Sample plots at forest stand margins). *Schweizerische Anstalt für das Forstliche Versuchswesen, Mitteilungen*, **45(3)**, 234–303.

Schreuder, H. T., Sedransk, J., & Ware, K. D. 1968. 3-P sampling and some alternatives, I. *Forest Science*, **14**, 429–454.

Schreuder, H. T., Gregoire, T. G., & Wood, G. B. 1993. *Sampling Methods for Multiresource Forest Inventory*. New York: Wiley.

Schumacher, F. X., & Bull, H. 1932. Determination of the errors of estimate of a forest survey, with special reference to the bottom-land hardwood forest region. *Journal of Agricultural Research*, **45(12)**, 741–756.

Scott, C. T. 1984. A new look at sampling with partial replacement. *Forest Science*, **30**, 157–166.

Scott, C. T., & Köhl, M. 1994. Sampling with partial replacement and stratification. *Forest Science*, **40**, 30–46.

Scott, C. T., Köhl, & Schnellbächer, H. J. 1999. A comparison of periodic and annual forest inventories. *Forest Science*, **45(3)**, 433–451.

Seber, G. A. F. 1979. Transects of random length. Pages 183–192 of: McCormack, R. M., Patil, G. P., & Robson, D. S. (eds), *Sampling Biological Populations*. Fairfield, Maryland, USA: International Cooperative Publishing House.

Seber, G. A. F. 1982. *The Estimation of Animal Abundance and Related Parameters.* London: Charles Griffin & Company Ltd.

Sengupta, J. M. 1954. Some experiments with different types of area sampling for winter paddy in Giridh, Bihar: 1945. *Sankhyā,* **13**, 235–240.

Shinozaki, K., Yoda, K., Hozumi, K., & Kira, T. 1964. A quantitative analysis of plant form: the pipe model theory. I. Basic analysis. *Japanese Journal of Ecology,* **14**, 97–105.

Singh, H., & Espejo, M. R. 2003. On linear regression and ratio-product estimation of finite population mean. *The Statistician,* **52**, 59–67.

Singh, S., & Horn, S. 1998. An alternative estimator for multicharacter surveys. *Metrika,* **48**, 99–107.

Sitter, R. R. 1992. A resampling procedure for complex survey data. *Journal of the American Statistical Association,* **87**, 755–765.

Skidmore, A. K., & Turner, B. J. 1992. Map accuracy assessment using line intersect sampling. *Photogrammetric Engineering and Remore Sensing,* **58(10)**, 1453–1457.

Smith, C., & Guttman, L. 1953. Measurement of internal boundaries in three-dimensional structures by random sampling. *Journal of Metals,* **5(1)**, 81–87.

Smith, D. R., Villela, R. F., & Lamarié, D. P. 2003. Application of adaptive cluster sampling to low-density populations of freshwater mussels. *Environmental and Ecological Statistics,* **10**, 7–15.

Smith, H. F. 1938. An empirical law describing heterogeneity in the yields of agricultural crops. *Journal of Agricultural Sciences,* **28**, 1–23.

Smith, S. J., & Gavaris, S. 1993. Improving the precision of abundance estimates of eastern Scotion shelf Atlantic cod from bottom trawl surveys. *North American Journal of Fisheries Management,* **13**, 35–47.

Smith, T. M. F. 1991. Post stratification. *The Statistician,* **40**, 315–323.

Smith, V. G., & Aird, P. L. 1975. *Canadian Forest Inventory Methods.* Toronto: University of Toronto Press.

Spetich, M. A., & Parker, G. R. 1998. Plot size recommendations for biomass estiamtion in midwestern old-growth forest. *Northern Journal of Applied Forestry,* **15(4)**, 165–168.

Stage, A. R., & Rennie, J. C. 1994. Fixed-radius plots or variable-radius plots? *Journal of Forestry,* **92(12)**, 20–24.

Ståhl, G. 1997. Transect relascope sampling for assessing coarse woody debris: the case for a $\pi/2$ relascope angle. *Scandinavian Journal of Forest Research,* **12**, 375–381.

Ståhl, G. 1998. Transect relascope sampling—a method for the quantification of coarse woody debris. *Forest Science,* **44**, 58–63.

Stehman, S. V., & Overton, W. S. 2002. Environmental Sampling. Pages 1914–1937 of: *Encyclopedia of Environmetrics, Volume 4.* Chichester: Wiley.

Stehman, S. V., & Salzer, D. W. 2000. Estimating density from surveys employing unequal-area belt transects. *Wetlands*, **20**, 512–519.

Stehman, S. V., Sohl, T. L., & Loveland, T. R. 2003. Statistical sampling to characterize recent United States land-cover change. *Remore Sensing of Environment*, **20**, 512–519.

Stevens, Jr., D. L. 1997. Variable density grid-based sampling designs for continuous spatial populations. *Environmetrics*, **8**, 167–195.

Stevens, Jr., D. L., & Olsen, A. R. 1991. Statistical issues in environmental monitoring and assessment. Pages 76–85 of: *Proceedings of the Section on Statistics and the Environment*. Alexandria, Virginia: American Statistical Association.

Stokes, S. L. 1980. Estimation of variance using judgment ordered rank set samples. *Biometrics*, **36**, 35–42.

Strand, L. 1957. "Relaskopisk" hoyde- og kubikkmassebestemmelse. *Norsk Skogbruk.*, **3**, 535–538.

Stuart, A. 1962. *Basic Ideas of Scientific Sampling*. New York: Hafner Press.

Sudakar, K. 1978. A note of circular systematic sampling design. *Sankhyā, Series C*, **40**, 72–73.

Sukhatme, P. V. 1947. The problem of plot size in large-scale yield surveys. *Journal of the American Statistical Association*, **42**, 297–310.

Sukhatme, P. V., & Sukhatme, B. V. 1970. *Sampling Theory of Surveys with Applications*. Second ed. Ames: Iowa State University Press.

Sukhatme, P. V., Sukhatme, B. V., Sukhatme, S., & Asok, C. 1984. *Sampling Theory of Surveys with Applications*. Third ed. Ames: Iowa State University Press.

Sunter, A. 1986. Solutions to the problem of unequal probability sampling without replacement. *International Statistical Review*, **54**, 33–50.

Sunter, A. 1989. Updating size measures in a PPSWOR design. *Survey methodology*, **15**, 253–260.

Swindel, B., & Yandle, D. O. 1972. Allocation in stratified sampling as a game. *Journal of the American Statistical Association*, **67**, 684–686.

Takahasi, K., & Wakimoto, K. 1968. On unbiased estimates of the population mean based on the sample stratified by means of ordering. *Annals of the Institute of Statistical Mathematics*, **20**, 1–31.

Talvitie, M., Leino, O., & Holopainen, M. 2006. Inventory of sparse forest populations using adaptive cluster sampling. *Silva Fennica*, **40**, 101–108.

Taylor, L. R. 1971. Aggregation, variation, and the mean. *Nature*, **189**, 2732–735.

Thompson, S. K. 2002. *Sampling*. Second ed. New York: Wiley.

Thompson, S. K., & Seber, G. A. 1996. *Adaptive Sampling*. New York: Wiley.

Tillé, Y. 1998. Estimation in surveys using conditional inclusion probabilities: simple random sampling. *International Statistical Review*, **66(3)**, 303–322.

Tschuprow, A. A. 1922. On the mathematical expectation of the moments of frequency distributions in the case of correlated observations. *Metron*, **2**, 461–493, 646–683.

Uspensky, J. V. 1937. *Introduction to Mathematical Probability*. New York: McGraw-Hill Book Company, Inc.

Valentine, H. T., & Hilton, S. J. 1977. Sampling oak foliage by the randomized-branch method. *Canadian Journal of Forest Research*, **7**, 295–298.

Valentine, H. T., Tritton, L. M., & Furnival, G. M. 1984. Subsampling trees for biomass, volume, or mineral content. *Forest Science*, **30**, 673–681.

Valentine, H. T., Gregoire, T. G., & Furnival, G. M. 1987. Unbiased estimation of total tree weight by three stage sampling with probability proportional to size. Pages 129–132 of: Wharton, E. H., & Cunia, T. (eds), *Tree Biomass Regression Functions and their Contribution to the Error of Forest Inventory Estimates*. General Technical Report NE-GTR-117. Broomall, Pennsylvania: USDA Forest Service.

Valentine, H. T., Bealle, C., & Gregoire, T. G. 1992. Comparing vertical and horizontal modes of importance and control-variate sampling for bole volume. *Forest Science*, **38**, 160–172.

Valentine, H. T., Baldwin, Jr., V. C., Gregoire, T. G., & Burkhart, H. E. 1994. Surrogates for foliar dry matter in loblolly pine. *Forest Science*, **40**, 576–585.

Valentine, H. T., Gove, J. H., & Gregoire, T. G. 2001. Monte Carlo approaches to sampling forested tracts with lines or points. *Canadian Journal of Forest Research*, **31**, 1410–1424.

Valentine, H. T., Ducey, M. J., Gove, J. H., Lanz, A., & Affleck, D. L. R. 2006. Corrections for cluster plot slop. *Forest Science*, **52**, 55–66.

Valiant, R. 1990. Comparions of variance estimators in stratifed random and systematic sampling. *Journal of Official Statistics*, **6**, 115–131.

Van Deusen, P. C. 1990. Critical height versus importance sampling for log volume: does critical height prevail? *Forest Science*, **36**, 930–938.

Van Deusen, P. C., & Baldwin, Jr., V.C. 1993. Sampling and predicting tree dry weight. *Canadian Journal of Forest Research*, **23**, 1826–1829.

Van Deusen, P. C., Dell, T. R., & Thomas, C. E. 1986. Volume growth estimation from permanent horizontal points. *Forest Science*, **32**, 415–422.

Van Hooser, D. D. 1973. *Midcycle evalaution of Mississippi timber resources*. Resource Bulletin, SO-44. New Orleans, Louisiana: USDA Forest Service.

van Wagner, C. E. 1968. The line intersect method in forest fuels sampling. *Forest Science*, **14(1)**, 267–276.

Ware, K. D. 1960. *Optimum regression sampling design for sampling of forest populations on successive occasions*. Tech. rept. Yale University.

Ware, K. D., & Cunia, T. 1962. Continuous forest inventory with partial replacement of samples. *Forest Science Monograph*.

Warren, W. G., & Olsen, P. F. 1964. A line intersect technique for assessing logging waste. *Forest Science*, **10**, 267–276.

Watson, G. S. 1978. Characteristics statistical problems of stochastic geometry. Pages 215–234 of: Miles, R. E., & Serra, J. (eds), *Geometrical Probability and Biological Structures: Buffon's 200th Anniversary*. Lecture Notes in Biomathematics, no. 23. Springer-Verlag.

West, G. B., Brown, J. H., & Enquist, B. J. 1999. A general model for the structure and allometry of plant vascular systems. *Nature*, **400**, 664–667.

Wiegert, R. G. 1962. The selection of an optimum quadrat size for sampling the standing crop of grasses and forbs. *Ecology*, **43(1)**, 125–129.

Wilk, S. J., Morse, W. W., Ralph, D. E., & Azarovitz, T. R. 1977. *Fishes and associated environmental data collected in New York Bight, June 1974–June 1975*. NOAA Technical Report NMFS SSRF-716. Washington, D.C.: U.S. Department of Commerce.

Williams, M. S., & Gove, J. H. 2003. Perpendicular distance sampling: an alternative method for sampling downed coarse woody debris. *Canadian Journal of Forest Research*, **33**, 1564–1579.

Williams, M. S., Schreuder, H. T., & Terrazas, G. H. 1998. *Poisson sampling —the adjusted and unadjusted estimator*. Research Note RMRS-RN-4. Fort Collins, Colorado: USDA Forest Service.

Williams, M. S., Valentine, H. T., Gove, J. H., & Ducey, M. J. 2005. Additional results for perpendicular distance sampling. *Canadian Journal of Forest Research*, **35**, 961–966.

Wolf, A. T., Burk, T. E., & Isebrands, J. G. 1995. Estimation of daily and seasonal whole-tree photosynthesis using Monte Carlo integration techniques. *Canadian Journal of Forest Research*, **25**, 253–260.

Wolter, K. M. 1985. *Introduction to Variance Estimation*. New York: Springer-Verlag.

Wood, G. B., & Wiant, H. V. 1992. Inventorying a forest using point-list sampling during a single visit to the field. Pages 397–403 of: Wood, G., & Turner, B. (eds), *Integrating Forest Information Over Space and Time*. Proceedings of IUFROM S4.01 & S4.02 conference. Canberra, Australia: ANUTECH Pty. Ltd.

Yandle, D. O, & White, F. M. 1977. An application of two-stage forest sampling. *Southern Journal of Applied Forestry*, **1(3)**, 27–32.

Yang, Y.-C., & Chao, S.-L. 1987. Comparison of volume growth calculation mehtods for remeasured horizontal line sampling. *Forest Science*, **33**, 1062–1067.

Yates, F. 1949. *Sampling Methods for Censuses and Surveys*. London: Charles Griffin & Co.

Yu, P. L. H., & Lam, K. 1997. Regression estimator in ranked set sampling. *Biometrics*, **53**, 1070–1080.

Zeide, B. 1980. Plot size optimization. *Forest Science*, **26(2)**, 251–257.

Zhang, L.-C. 2000. Post-stratification and calibration—a synthesis. *Journal of the American Statistical Association*, **54**, 178–184.

Author Index

Subject Index